Control of Primary Metabolism in Plants

Annual Plant Reviews

A series for researchers and postgraduates in the plant sciences. Each volume in this series focuses on a theme of topical importance and emphasis is placed on rapid publication.

Editorial Board:

Professor Jeremy A. Roberts (Editor-in-Chief), Plant Science Division, School of Biosciences, University of Nottingham, Sutton Bonington Campus, Loughborough, Leicestershire, LE12 5RD, UK; **Dr David Evans**, School of Biological and Molecular Sciences, Oxford Brookes University, Headington, Oxford, OX3 0BP; **Professor Hidemasa Imaseki**, Obata-Minami 2419, Moriyama-ku, Nagoya 463, Japan; **Dr Michael T. McManus**, Institute of Molecular BioSciences, Massey University, Palmerston North, New Zealand; **Dr Jocelyn K.C. Rose**, Department of Plant Biology, Cornell University, Ithaca, New York 14853, USA.

Titles in the series:

1. **Arabidopsis**   Edited by M. Anderson and J.A. Roberts
2. **Biochemistry of Plant Secondary Metabolism**   Edited by M. Wink
3. **Functions of Plant Secondary Metabolites and their Exploitation in Biotechnology**   Edited by M. Wink
4. **Molecular Plant Pathology**   Edited by M. Dickinson and J. Beynon
5. **Vacuolar Compartments**   Edited by D.G. Robinson and J.C. Rogers
6. **Plant Reproduction**   Edited by S.D. O'Neill and J.A. Roberts
7. **Protein–Protein Interactions in Plant Biology**   Edited by M.T. McManus, W.A. Laing and A.C. Allan
8. **The Plant Cell Wall**   Edited by J.K.C. Rose
9. **The Golgi Apparatus and the Plant Secretory Pathway**   Edited by D.G. Robinson
10. **The Plant Cytoskeleton in Cell Differentiation and Development**   Edited by P.J. Hussey
11. **Plant–Pathogen Interactions**   Edited by N.J. Talbot
12. **Polarity in Plants**   Edited by K. Lindsey
13. **Plastids**   Edited by S.G. Moller
14. **Plant Pigments and their Manipulation**   Edited by K.M. Davies
15. **Membrane Transport in Plants**   Edited by M.R. Blatt
16. **Intercellular Communication in Plants**   Edited by A.J. Fleming
17. **Plant Architecture and its Manipulation**   Edited by C. Turnbull
18. **Plasmodesmata**   Edited by K.J. Oparka
19. **Plant Epigenetics**   Edited by P. Meyer
20. **Flowering and its Manipulation**   Edited by C. Ainsworth
21. **Endogenous Plant Rhythms**   Edited by A. Hall and H. McWatters
22. **Control of Primary Metabolism in Plants**   Edited by W.C. Plaxton and M.T. McManus
23. **Biology of the Plant Cuticle**   Edited by M. Riederer

Control of Primary Metabolism in Plants

Edited by

WILLIAM C. PLAXTON
Department of Biology
Queen's University
Kingston, Ontario
Canada

and

MICHAEL T. McMANUS
Institute of Molecular BioSciences
Massey University
Palmerston North
New Zealand

Blackwell Publishing

© 2006 by Blackwell Publishing Ltd

Editorial Offices:
Blackwell Publishing Ltd, 9600 Garsington Road, Oxford OX4 2DQ, UK
  Tel: +44 (0)1865 776868
Blackwell Publishing Professional, 2121 State Avenue, Ames, Iowa 50014-8300, USA
  Tel: +1 515 292 0140
Blackwell Publishing Asia, 550 Swanston Street, Carlton, Victoria 3053, Australia
  Tel: +61 (0)3 8359 1011

The right of the Author to be identified as the Author of this Work has been asserted in accordance with the Copyright, Designs and Patents Act 1988.

All rights reserved. No part of this publication may be reproduced, stored in a retrieval system, or transmitted, in any form or by any means, electronic, mechanical, photocopying, recording or otherwise, except as permitted by the UK Copyright, Designs and Patents Act 1988, without the prior permission of the publisher.

First published 2006 by Blackwell Publishing Ltd

ISBN-13: 978-14051-3096-7
ISBN-10: 1-4051-3096-2

Library of Congress Cataloging-in-Publication Data

Control of primary metabolism in plants / edited by William C. Plaxton
  and Michael T. McManus.
    p. ; cm.
  Includes bibliographical references and index.
  ISBN-13: 978-1-4051-3096-7 (hardback : alk. paper)
  ISBN-10: 1-4051-3096-2 (hardback : alk. paper)
  1. Plants–Metabolism. I. Plaxton, William C. II. McManus, Michael T.
  [DNLM: 1. Plants–metabolism. 2. Plants–enzymology.
QK 881 C764 2006]
QK881.C664 2006
572'.42–dc22
                                                           2005021045

A catalogue record for this title is available from the British Library

Set in 10/12 Times
by TechBooks, New Delhi, India
Printed and bound in India
by Replika Press Pvt. Ltd, Kundli

The publisher's policy is to use permanent paper from mills that operate a sustainable forestry policy, and which has been manufactured from pulp processed using acid-free and elementary chlorine-free practices. Furthermore, the publisher ensures that the text paper and cover board used have met acceptable environmental accreditation standards.

For further information on Blackwell Publishing, visit our Web site:
www.blackwellpublishing.com

Contents

Contributors   xv

Preface   xvii

1 Evaluation of the transcriptome and genome to inform the study of metabolic control in plants   1
OLIVER THIMM, OLIVER E. BLÄSING, BJORN USADEL and YVES GIBON
  1.1 Introduction   1
  1.2 Transcript profiling technologies   2
  1.3 Transcript profiling workflow   3
    1.3.1 Data generation   4
    1.3.2 Data management   5
    1.3.3 Data processing   6
      1.3.3.1 Raw data handling   6
      1.3.3.2 Normalisation   7
    1.3.4 Data analysis   9
      1.3.4.1 Differential expression   9
      1.3.4.2 Data mining   10
      1.3.4.3 Functional categorisation   12
    1.3.5 Data visualisation   13
  1.4 What can we learn from transcript profiles performed in a starchless mutant?   15
  1.5 Conclusion/perspectives   17
  Acknowledgements   18
  References   19

2 The use of proteomics in the study of metabolic control   24
LEE J. SWEETLOVE
  2.1 Introduction   24
  2.2 Proteomic methodologies   25
    2.2.1 Extraction of proteins from plant tissue   26
    2.2.2 Separation, display and quantification of proteins   27
    2.2.3 Identification of proteins by mass spectrometry   28
    2.2.4 Gel-free proteomic approaches   30
  2.3 Cataloging protein localization   31
    2.3.1 Localizing proteins to different tissues   31
    2.3.2 Establishing subcellular protein localization: methodologies   33

	2.3.3	Mitochondrial and chloroplast proteomes	35
	2.3.4	Other subcellular proteomes	39
	2.3.5	A stamp of authenticity for the subcellular protein postcode?	40
2.4	Quantitative analyses of the proteome		41
	2.4.1	Examples of quantitative proteomics	42
	2.4.2	The use of high-throughput measurements of enzyme activity as a proxy for quantitative proteomics	44
2.5	The use of proteomics to investigate post-translational modification of proteins		45
	2.5.1	Systematic identification of phosphorylated proteins	46
	2.5.2	Systematic identification of protein redox modifications	47
2.6	The use of proteomics to investigate protein–protein interactions		48
2.7	Future perspectives		50
References			52

3 Study of metabolic control in plants by metabolomics — 60
OLIVER FIEHN

3.1	Introduction			60
	3.1.1	What is metabolomics?		60
	3.1.2	Systemic properties in metabolic networks		61
3.2	Metabolomic methods			62
	3.2.1	Historic perspective of plant metabolite analysis		62
	3.2.2	Modern instrumentation in metabolite analysis		63
	3.2.3	Sample preparation for metabolomics		64
	3.2.4	Metabolome coverage		66
		3.2.4.1	The quest for combining sensitivity and selectivity	66
		3.2.4.2	Cellular and subcellular metabolomics	68
		3.2.4.3	Compound identification	69
	3.2.5	Quality control		70
3.3	Metabolomic databases			71
3.4	Pathways, clusters and networks: applications of plant metabolomics			72
	3.4.1	Bioengineering of metabolism		73
	3.4.2	Plant biochemistry		74
		3.4.2.1	Pathway analysis	74
		3.4.2.2	Flux measurements	75
	3.4.3	Physiological studies		76
	3.4.4	Plant metabolomic methods		77
	3.4.5	Food science		78
3.5	Outlook			80
References				80

4 Metabolite transporters in the control of plant primary metabolism   85
MECHTHILD TEGEDER and ANDREAS P. M. WEBER

- 4.1 Introduction   85
- 4.2 Photoassimilation and assimilate transport in source cells   86
  - 4.2.1 Carbon assimilation by the reductive pentose-phosphate pathway (Calvin cycle)   86
  - 4.2.2 The plastidic triose-phosphate pool – a metabolic crossway   86
    - 4.2.2.1 Communication between the starch and sucrose biosynthetic pathways via TPT   87
  - 4.2.3 Allocation of recently assimilated carbon to other pathways   90
- 4.3 Nitrogen assimilation   90
- 4.4 Amino acid and isoprenoid metabolism   93
  - 4.4.1 Methionine and S-adenosylmethionine metabolism   94
  - 4.4.2 Shikimic acid pathway and aromatic amino acid biosynthesis   94
  - 4.4.3 Isoprenoid synthesis via the deoxy-xylulose 5-phosphate pathway   96
- 4.5 Sucrose and amino acid loading into the phloem for long-distance transport   97
  - 4.5.1 Mobilization of stored carbon and nitrogen   97
  - 4.5.2 Mechanisms of phloem loading and involvement of transporter proteins   98
  - 4.5.3 Sucrose transporters   99
  - 4.5.4 Amino acid transporters   100
  - 4.5.5 Genetic modification of phloem loading with assimilates   100
- 4.6 Phloem unloading in sinks and assimilate transport to developing seeds   101
  - 4.6.1 Assimilate distribution and transport in seed coats   102
  - 4.6.2 Uptake of sucrose and amino acids by the developing embryo   102
  - 4.6.3 Specialized sites of import   103
  - 4.6.4 Sucrose and amino acid import into developing embryos/cotyledons   103
  - 4.6.5 Genetic modification of assimilate transport in seeds   104
- 4.7 Assimilate transport and metabolism in sink cells   104
  - 4.7.1 The role of hexose-phosphate import into nongreen plastids   105
  - 4.7.2 The role of ATP-transport into nongreen plastids   106
  - 4.7.3 Knockout of NTTs in Arabidopsis   106
  - 4.7.4 Antisense repression and overexpression of NTTs in potato   106
  - 4.7.5 A novel role for Rubisco in developing oilseeds   107
- 4.8 Concluding remarks   108
- Acknowledgements   108
- References   109

5 Role of protein kinases, phosphatases and 14-3-3 proteins in the control of primary plant metabolism 121
GREG B. G. MOORHEAD, GEORGE W. TEMPLETON and HUE T. TRAN

5.1	Introduction	121
5.2	Protein kinases	122
	5.2.1 Phosphoinositide 3-kinase-like kinases	123
5.3	Protein phosphatases	124
	5.3.1 Protein phosphatase 1	125
	5.3.2 Protein phosphatase 2A	126
	5.3.3 Protein phosphatase 2C	127
	5.3.4 Novel protein phosphatases	128
	5.3.5 The tyrosine and dual specificity protein phosphatases	130
	5.3.5.1 Class I cysteine-based protein tyrosine phosphatases	130
	5.3.5.2 Class II cysteine-based protein tyrosine phosphatase	130
	5.3.5.3 Class III cysteine-based protein tyrosine phosphatases	130
	5.3.5.4 Class IV protein tyrosine phosphatases	131
	5.3.6 RNA polymerase II phosphatases-FCP1 and SCP	131
	5.3.7 Histidine phosphatases	132
5.4	A Multitude of phosphospecific binding modules	132
	5.4.1 Phosphospecific binding modules	132
	5.4.2 14-3-3 proteins	133
	5.4.2.1 14-3-3 structures and function	134
	5.4.2.2 14-3-3 roles and control	136
5.5	The role of protein phosphorylation in the control of plant primary metabolism	137
	5.5.1 Nutrient sensing and signalling through conserved protein kinases	137
	5.5.2 Nitrate reductase	139
	5.5.3 Sucrose synthase	140
	5.5.4 Sucrose phosphate synthase and trehalose phosphate synthase	140
	5.5.5 6-phosphofructo-2-kinase/fructose2,6-bisphosphatase	141
	5.5.6 Starch synthase and starch branching enzyme	141
	5.5.7 Glutamine synthetase (GS_1 and GS_2)	142
	5.5.8 Nonphosphorylating glyceraldehyde-3-phosphate dehydrogenase	142
	5.5.9 Phosphoenolpyruvate carboxylase and PEPC kinase	143
5.6	Summary	143
	Acknowledgements	143
	References	144

6	**Redox signal transduction in plant metabolism**	**150**
	SANTIAGO MORA-GARCIA, FABIANA G. STOLOWICZ and	
	RICARDO A. WOLOSIUK	
	6.1 Introduction	150
	6.2 The reactivity of the sulfhydryl group	151
	6.3 Protein-disulfide oxido-reductases	155
	6.4 Thioredoxins	155
	6.4.1 Thioredoxin isoforms	156
	6.4.2 Reductants of thioredoxins (sources of reducing power)	158
	6.4.3 Targets of thioredoxins (oxidants of thioredoxin)	161
	6.4.4 Control of chloroplast enzymes by thioredoxin	162
	6.4.5 Translation of chloroplast mRNA	165
	6.4.6 Phosphorylation of chloroplast proteins	166
	6.4.7 Control of mitochondrial proteins	166
	6.4.8 Removal of reactive oxygen species	167
	6.4.9 Seed germination	169
	6.4.10 Modulation of receptor functions	170
	6.5 Glutaredoxins	170
	6.6 Protein-disulfide isomerases	173
	6.7 Concluding remarks	175
	Acknowledgements	175
	References	175
7	**Control of carbon fixation in chloroplasts**	**187**
	BRIGITTE GONTERO, LUISANA AVILAN and	
	SANDRINE LEBRETON	
	7.1 Introduction	187
	7.2 Ribulose-1,5-bisphosphate carboxylase-oxygenase	190
	7.3 Glyceraldehyde-3-phosphate dehydrogenase	194
	7.4 Fructose-1,6-bisphosphatase and sedoheptulose-1,7-bisphosphatase	197
	7.5 Phosphoribulokinase	199
	7.6 Other important enzymes in the Calvin cycle	201
	7.6.1 Transketolase	201
	7.6.2. Aldolase	202
	7.7 Supramolecular complexes of the Calvin cycle	203
	7.8 Conclusions	206
	Acknowledgement	207
	References	207
8	**Control of phosphoenolpyruvate carboxylase in plants**	**219**
	HUGH G. NIMMO	
	8.1 Introduction	219
	8.2 PPCK genes and their roles	220

	8.3	Signalling pathways that control *PPCK* expression in CAM and C$_4$ plants	224
	8.4	The 'bacterial-type' PEPC	228
	8.5	Conclusions	229
	Acknowledgements		230
	References		230
9	**Control of sucrose biosynthesis**		**234**
	ELSPETH MACRAE and JOHN LUNN		
	9.1	Introduction	234
	9.2	Pathways of sucrose biosynthesis in leaves	234
		9.2.1 Sucrose synthesis in leaves during the day	236
		9.2.2 Sucrose synthesis in leaves at night	237
	9.3	Control of sucrose biosynthesis – the precursors	238
		9.3.1 The conversion of triose-phosphate to hexose-phosphate	238
		9.3.2 The hexose-phosphate pool and UDP-glucose pyrophosphorylase	240
	9.4	The committed enzymes of sucrose biosynthesis	241
		9.4.1 Sucrose-phosphate synthase	241
		9.4.2 Sucrose-phosphatase	245
		9.4.3 Evidence for a metabolon in sucrose biosynthesis	246
	9.5	Integrated pathway control	247
	9.6	Future perspectives	248
	References		250
10	**Control of starch biosynthesis in vascular plants and algae**		**258**
	MATTHEW K. MORELL, ZHONGYI LI, AHMED REGINA, SADIQ RAHMAN, CHRISTOPHE D'HULST and STEVEN G. BALL		
	10.1	Introduction	258
	10.2	Synthesis of bacterial glycogen	259
	10.3	Synthesis of starch in vascular plants	260
		10.3.1 Substrate supply and activation	260
		10.3.2 Amylose synthesis	263
		10.3.3 Amylopectin synthesis	264
	10.4	Starch synthesis and breakdown in leaves and tubers	266
		10.4.1 Isoamylases are directly involved in the synthesis of amylopectin but also during starch mobilization	267
		10.4.2 Glucan phosphorylation is the key signal of starch degradation in both leaves and tubers	268
		10.4.3 Starch degradation essentially occurs through β-amylolysis in leaves	268
		10.4.4 Starch metabolism is tightly controlled by several levels of regulation	269
	10.5	Control of starch biosynthesis in monocotyledonous species	269
		10.5.1 Genes in the rice genome	271

10.6	Starch synthesis in green algae		272
	10.6.1	*Chlamydomonas reinhardtii* defines the best microbial system to study plant starch metabolism	272
	10.6.2	The Chlamydomonas single cell can account for both transitory or storage starch synthesis	273
	10.6.3	What have we learned from Chlamydomonas?	274
	10.6.4	Similarity and differences between starch metabolism in plants and algae	275
	10.6.5	The future of starch research in green algae	276
10.7	Starch synthesis in other systems		276
	10.7.1	Bacterial cells may have a primitive starch synthesizing machinery	276
	10.7.2	UDPglucose-based systems that produce starch	278
10.8	Control of starch biosynthesis		279
10.9	Opportunities for the manipulation of starch synthesis and structure		280
10.10	Conclusions		281
References			282

11 The organization and control of plant mitochondrial metabolism — **290**
ALLISON E. MCDONALD and GREG C. VANLERBERGHE

11.1	Introduction		290
11.2	Organization of the tricarboxylic acid cycle and mitochondrial electron transport chain		290
	11.2.1	Tricarboxylic acid cycle and associated enzymes	290
	11.2.2	Electron transport chain complexes I–V	294
	11.2.3	Additional electron transport chain and associated components	295
11.3	Posttranslational control of mitochondrial metabolism and function		299
	11.3.1	Phosphorylation – dephosphorylation	299
	11.3.2	Dithiol-disulfide interconversion	300
	11.3.3	Other oxidative modifications	301
	11.3.4	Supramolecular complexes	302
11.4	Integration of mitochondrial metabolism with other metabolic pathways		303
	11.4.1	Mitochondrial metabolism during photosynthesis	304
	11.4.2	Ascorbate biosynthesis	305
	11.4.3	Mitochondrial fatty acid synthesis	305
	11.4.4	The glyoxylate cycle and lipid respiration	306
11.5	Mitochondrial metabolism of reactive oxygen species		307
	11.5.1	Mitochondrial ROS generation	307
	11.5.2	Mechanisms to scavenge mitochondrial ROS	307
	11.5.3	Mechanisms to avoid mitochondrial ROS generation	309
	11.5.4	Signaling functions of mitochondrial ROS	311

	11.6	Additional stress-induced metabolic pathways associated with plant mitochondria	311
		11.6.1 GABA shunt	311
		11.6.2 Mitochondrial amino acid catabolism	312
		11.6.3 Formate dehydrogenase	312
		11.6.4 Mitochondrial aldehyde dehydrogenases	313
		11.6.5 Mitochondrial glycerol-3-phosphate dehydrogenase	314
		11.6.6 Nucleoside diphosphate kinase	314
		11.6.7 Root organic anion exudation	315
	11.7	Concluding remarks	315
	Acknowledgements		316
	References		316
12	**Photosynthetic carbon–nitrogen interactions: modelling inter-pathway control and signalling**		**325**
	CHRISTINE H. FOYER, GRAHAM NOCTOR and PAUL VERRIER		
	12.1	Introduction	325
	12.2	Integration of C and N metabolism in leaves	326
	12.3	Control of nitrate assimilation rates and the C/N interaction	329
	12.4	Pathway coordination and the C/N signal transduction network	331
	12.5	Modelling the C/N interaction	333
		12.5.1 Construction of a tentative model to explore the sensitivities in the GS and GOGAT reactions	335
	12.6	Conclusions and perspectives	342
	Acknowledgements		342
	References		342
13	**Control of sulfur uptake, assimilation and metabolism**		**348**
	MALCOLM J. HAWKESFORD, JONATHAN R. HOWARTH and PETER BUCHNER		
	13.1	Introduction	348
	13.2	Sulfate uptake and distribution	351
		13.2.1 Transcriptional regulation of transport	352
		13.2.2 Post-translation controls	353
	13.3	The assimilatory pathway – activation and reduction	354
		13.3.1 Cytosolic pathways	354
		13.3.2 Reductive assimilation in the plastid	354
		13.3.2.1 Sulfate activation by ATP sulfurylase	354
		13.3.2.2 Sulfate reduction by APS reductase	355
		13.3.3. Transcriptional regulation and coordination with C and N pathways	356
	13.4	Control of flux through the assimilatory pathway – cysteine synthesis	357
		13.4.1 The 'cysteine synthase' complex	358
		13.4.2 Metabolic control of the 'cysteine synthase' complex	358

	13.4.3 Control of SAT activity by cysteine feedback	359
	13.4.4 The role of O-acetylserine as an 'inducer' of gene expression	359
	13.4.5 A model for control of cysteine synthesis	361
13.5	Control of flux to the various sinks after cysteine biosynthesis	362
	13.5.1 Methionine biosynthesis	364
13.6	Summary	364
Acknowledgements		365
References		365

Index   **373**

Contributors

Dr. Luisana Avilan Departamento de Biologia, Facultad de Ciencias, Universidad de Los Andes, Merida 5101, Venezuela

Professor Steven G. Ball Unité de Glycobiologie Structurale et Fonctionnelle, UMR8576 CNRS/USTL, IFR 118, Université des Sciences et Technologies de Lille, 59655 Villeneuve d'Ascq, Cedex France

Dr Oliver E. Bläsing Max-Planck Institute of Molecular Plant Physiology, Am Mühlenberg 1, 14476 Golm, Germany

Dr Peter Buchner Crop Performance and Improvement Division, Rothamsted Research, Harpenden, Hertfordshire AL5 2JQ, UK

Dr Christophe d'Hulst Unité de Glycobiologie Structurale et Fonctionnelle, UMR8576 CNRS/USTL, IFR 118, Université des Sciences et Technologies de Lille, 59655 Villeneuve d'Ascq, Cedex France

Professor Oliver Fiehn UC Davis Genome Center, Davis, CA 95616, USA

Professor Christine H. Foyer Crop Performance and Improvement Division, Rothamsted, Harpenden, Herts AL5 2JQ, UK

Dr Yves Gibon Max-Planck Institute of Molecular Plant Physiology, Am Mühlenberg 1, 14476 Golm, Germany

Dr Brigitte Gontero CNRS-Universités Paris VI et Paris VII, Institut Jacques Monod, Tour 43, Laboratoire de Génétique et Membranes, 2 place Jussieu, 75 005 Paris, France

Dr Malcolm J. Hawkesford Crop Performance and Improvement Division, Rothamsted Research, Harpenden, Hertfordshire AL5 2JQ, UK

Dr Jonathan R. Howarth Crop Performance and Improvement Division, Rothamsted Research, Harpenden, Hertfordshire AL5 2JQ, UK

Dr Sandrine Lebreton CNRS-Universités Paris VI et Paris VII, Institut Jacques Monod, Tour 43, Laboratoire de Génétique et Membranes, 2 place Jussieu, 75 005 Paris, France

Dr Zhongyi Li CSIRO Plant Industry, GPO Box 1600, Canberra ACT 2601, Australia

Dr John Lunn Max-Planck-Institut für Molekulare Pflanzenphysiologie, Am Mühlenberg 1, 14424 Potsdam, Germany

Dr Elspeth MacRae Mt Albert Research Centre, HortResearch, Private Bag 92169 Auckland, New Zealand

Dr Allison E. McDonald Department of Life Sciences and Department of Botany, University of Toronto at Scarborough, Scarborough, ON M1C 1A4 Canada

Professor Greg B. G. Moorhead Department of Biological Sciences, University of Calgary, 2500 University Drive N.W., Calgary, Alberta, Canada T2N 1N4

Dr Santiago Mora-Garcia Instituto Leloir, Patricias Argentinas 435, C1405BWE Buenos Aires, Argentina

Dr Matthew K. Morell CSIRO Plant Industry, GPO Box 1600, Canberra ACT 2601, Australia.

Professor Hugh G. Nimmo Division of Biochemistry & Molecular Biology, Institute of Biomedical & Life Sciences, University of Glasgow, Glasgow G12 8QQ, UK

Professor Graham Noctor Institut de Biotechnologie des Plantes, UMR CNRS 8618, Université de Paris XI, 91405 Orsay cedex, France

Dr Sadiq Rahman CSIRO Plant Industry, GPO Box 1600, Canberra ACT 2601, Australia

Dr Ahmed Regina CSIRO Plant Industry, GPO Box 1600, Canberra ACT 2601, Australia

Dr Fabiana G. Stolowicz Instituto Leloir, Patricias Argentinas 435, C1405BWE Buenos Aires, Argentina

Dr Lee J. Sweetlove Department of Plant Sciences, University of Oxford, Oxford, UK

Professor Mechthild Tegeder School of Biological Sciences, Center for Reproductive Biology, Center for Integrated Biotechnology, Washington State University, Pullman, WA 99164-4236, USA

Dr George W. Templeton Department of Biological Sciences, University of Calgary, 2500 University Drive N.W., Calgary, Alberta, Canada T2N 1N4

Dr Oliver Thimm CNAP, Department of Biology, University of York, PO Box 373, YO10 5YW York, UK

Dr Hue T. Tran Department of Biological Sciences, University of Calgary, 2500 University Drive N.W., Calgary, Alberta, Canada T2N 1N4

Dr Bjorn Usadel Max-Planck Institute of Molecular Plant Physiology, Am Mühlenberg 1, 14476 Golm, Germany

Professor Greg C. Vanlerberghe Department of Life Sciences and Department of Botany, University of Toronto at Scarborough, Scarborough, ON M1C 1A4 Canada

Dr Paul Verrier Biomathematics and Bioinformatics Division, Rothamsted Research, Harpenden, Hertfordshire AL5 2JQ, UK

Professor Andreas P. M. Weber Department of Plant Biology, Michigan State University, East Lansing, MI 48824-1312, USA

Professor Ricardo A. Wolosiuk Instituto Leloir, Patricias Argentinas 435, C1405BWE Buenos Aires, Argentina

Preface

The ability to control the rates of metabolic processes in response to changes in the internal or external environment is an indispensable attribute of living cells that must have arisen with life's origin. This adaptability is necessary for conserving the stability of the intracellular environment which is, in turn, essential for maintaining an efficient functional state. The remarkable advances in molecular genetics that have occurred over the past several decades have somewhat eclipsed areas of traditional biochemistry such as protein chemistry, enzymology and metabolic control. With many genomes sequenced and others nearing completion, the next step is the less straightforward task of analysing the expression and function of gene products (proteins), as well as more thoroughly elucidating metabolism and its control. The task of completing the picture of all cellular proteins, their actions and reactions, is one of the biggest challenges facing life science researchers today. Although molecular biology has generated a host of impressive techniques (i.e., protein overexpression, site-directed mutagenesis, metabolic engineering, cDNA microarrays, etc.) for assessing various aspects of protein/enzyme structure-function and regulatory control, one cannot deduce the properties of a functional protein or the kinetic and regulatory properties of an enzyme solely from genetic information. Furthermore, recent genome sequencing projects have revealed a plethora of gene sequences that encode proteins having unknown functions, and many organisms whose genomes are currently being sequenced have not had their metabolism extensively studied. Where feasible, their metabolic phenotype is determined using annotated genome sequence data. Thus there appears to be a resurgence of interest in protein, enzymological and metabolic research for understanding biological processes in the post-genome era. Efficient approaches are needed for determining (a) the function of unknown gene products, (b) protein expression and corresponding metabolite levels in different cells and in different sub-cellular compartments under various developmental and environmental conditions, (c) covalent modifications of proteins in response to different stimuli, (d) protein:protein interactions, (e) the relationship between protein structure and protein function, (f) membrane transporter proteins that selectively translocate specific metabolites between different sub-cellular compartments and (g) the sophisticated mechanisms that serve to control the flux of metabolites through specific metabolic pathways *in vivo*.

Although biochemists frequently employ the terms 'regulation' and 'control' interchangeably, the need to discriminate between these terms has been emphasised by David Fell [1]. *Metabolic control* refers to adjusting the output of a metabolic pathway in response to an external signal. By contrast, *metabolic regulation* occurs when an organism maintains some variable which remains relatively constant over time, despite fluctuations in external conditions. Homeostasis is therefore a consequence of metabolic regulation, which itself may be a result of metabolic control. For example, the regulation of mammalian blood glucose is largely due to the secreted peptide hormones glucagon ('starved' signal) and insulin ('fed' signal) controlling intracellular metabolism within the liver. In this case, the concentration of blood glucose is regulated (kept constant) mainly by controlling (varying) fluxes of metabolic pathways (i.e., glycogen breakdown versus synthesis, glycolysis and gluconeogenesis) in hepatocytes. Regulation and control are properties of highly elaborate metabolic systems. An ongoing challenge is to link our knowledge of molecular, reductionist-based, enzyme control mechanisms to organismal-level explanations of metabolic regulation.

The advent of genomics, proteomics and metabolomics has revolutionised the study of plant development and is now having a significant impact on the study of plant metabolism and its control. In the last few years, significant advances have been made as enzyme gene families are elucidated, and new proteinaceous and allosteric regulators are identified. Enzyme activity is the major factor influencing the magnitude of metabolic fluxes in any cell. Metabolic control may occur at several levels, beginning with gene transcription and proceeding through various stages of protein synthesis and turnover. More rapid alterations in metabolic flux occur through activation and inhibition of pre-existing key enzymes along the major metabolic pathways, particularly by mechanisms such as reversible covalent modification and by the actions of allosteric effector molecules that reflect the cell's adenylate energy charge, oxidation/reduction potential and/or the accumulation of metabolic end products. Discoveries concerning plant metabolic control continue to be made at a rapid rate, particularly in the field of signal transduction. Each discovery adds to the view that plant signal transduction and metabolic control networks have remarkable intricacy. Although great advances have been made in our understanding of the mechanisms that contribute to the control of plant metabolism, our comprehension of why plant metabolic systems behave as they do *in vivo* is incomplete. However, the tools to address these questions are rapidly evolving, and advances in the theory of metabolic control and in computing power to analyse metabolism have kept pace with experimental developments. This holds great promise for those plant molecular geneticists who wish to reap a harvest via the process of metabolic engineering. A volume that reviews this progress and can point out the major research areas for the future, therefore, is very timely.

In this volume, a group of international specialists present their ideas and interpretations of specific subject matter relevant to plant primary metabolism and its control. Each chapter is written by an acknowledged expert or group of experts and provides an informed discussion on how the problem of metabolic control may be evaluated using the wide assortment of sophisticated techniques available to the modern researcher. The chapters are interrelated in order to provide the reader with an integrated view, reviewing information from the current literature and developing novel hypotheses based upon data acquired from extensive and diverse research activities.

For the purposes of this volume, primary metabolism is defined as the primary auxotrophic pathways in plants (CO_2 fixation and assimilation; sucrose and starch synthesis) and those in common with the primary pathways in mammalian cells (i.e., glycolysis and respiration). However, Part I of the volume (the first six chapters) is devoted to more generic aspects of metabolic control, with chapters on plant enzyme control by reversible covalent modification (i.e., protein-kinase mediated phosphorylation and dithiol-disulfide interconversions), metabolite transporters and the emerging roles of genomics, proteomics and metabolomics for informing the study of plant metabolic control. The chapters in Part I provide a basis for full appreciation of the information in the seven chapters of Part II. The latter focus on the control of specific pathways and enzymes of primary plant metabolism.

William C. Plaxton
Michael T. McManus

Reference

[1] D. Fell (1997) *Understanding the Control of Metabolism*. Portland, London.

1 Evaluation of the transcriptome and genome to inform the study of metabolic control in plants

Oliver Thimm, Oliver E. Bläsing, Björn Usadel and Yves Gibon

1.1 Introduction

New technologies often lead to major advances in biology. For plant biology, recently developed profiling methods, together with bioinformatics, have profoundly revolutionised the subject driving it to the whole system level. The accumulation of sequence data from various species (and the completion of nucleotide sequencing of several plant genomes) and remarkable advances in analytical and computational methods have enabled the development of a range of functional genomics approaches. Because many enzymes are highly conserved, identity searches allow the discovery of homologous enzymes, novel isoforms and pathways. Sequence data can also give information about the intracellular localisation of the encoded proteins. Furthermore, these genomic resources allow the development of high-throughput transcript profiling techniques and provide background knowledge for protein profiling. Although genetic maps of pathways from primary metabolism are now pretty well described, those from the secondary metabolism are still far from complete. However, our knowledge about how metabolic pathways are controlled in higher plants is expanding (see further chapters in this volume), and high-throughput transcript profiling provides the first insight into this control.

Parallel transcriptional analysis or 'transcriptomics' is believed to be one of the most important experimental approaches for discovering the function of genes [1]. The documentation and analysis of how genes respond to environmental or developmental challenges, as well as to genetic changes (e.g. knocking out, repressing or overexpressing genes), should allow the assignment of hypothetical functions. A strong theme that has emerged from microarray experiments is that groups of genes that are functionally related tend to be co-regulated at the transcriptional level [2, 3]. Aside from the fact that it may be a long time before the functional relationship is fully understood, finding such patterns of expression is definitely not trivial. Indeed, the major bottleneck in such experiments is no longer the generation of data, but the analysis and interpretation of the datasets produced. To meet this problem, a variety of bioinformatic tools have recently been developed in order to extract relevant information from large-scale datasets. However, many bioinformatic solutions are not intuitively accessible for biologists, because they imply data processing algorithms that can only be understood and applied with the assistance of statisticians and programmers. Moreover, bioinformatic tools cover mainly data management, processing and visualisation, whereas the development of tools for the integration of data and their interpretation is still in its infancy. With regard to the costs of profiling experiments, the output rarely meets the initial expectations or the intrinsic potential of the generated data.

In this chapter, we present the transcript profiling technologies that are available and discuss the different steps that are required in transcriptomics approaches, from planning of experiments to data visualisation. As an example, we present results obtained by comparing the transcriptome of an *Arabidopsis thaliana* starchless mutant with its corresponding wild type.

1.2 Transcript profiling technologies

Several methods have now been developed to measure steady-state levels of mRNA for hundreds to thousands of genes in parallel. DNA microarrays are the most commonly used tools to date, although other transcript profiling technologies based on nucleotide sequencing have been developed. These include serial analysis of gene expression (SAGE), massively parallel signature sequencing or using fragment sizing, differential display, and cDNA-amplified fragment length polymorphism analysis [4, 5].

The DNA microarray technique is based on nucleic acid hybridisation, a property initially discovered by Gillespie and Spiegelman [6], and is analogous to the Northern blot technique [7]. Indeed, this technique quantifies the highly specific hybridisation of an immobilised DNA strand (probe) with a nucleic acid strand, which is a labelled antisense copy of the target mRNA. Immobilised DNA should be in excess to allow a pseudo-first-order hybridisation kinetic, in order that the derived signal will be linearly related to the concentration of the labelled target. To date, two types of microarrays have evolved: DNA double-strand-based microarrays and oligonucleotide-based microarrays. DNA microarrays consist of DNA probes of various lengths spotted or ink-jetted onto nylon membranes or glass slides with a chemically modified surface. DNA probes can be full-length cDNAs, expressed sequence tag (EST) clones or amplified PCR fragments that encode the expressed part of a gene. Therefore, the length of the spotted DNA fragments ranges from a few hundred to thousands of base pairs.

Oligonucleotide-based arrays are made of oligonucleotides generated by standard synthesis and subsequently spotted, or by oligonucleotides directly synthesised on the solid support by photolithographic techniques. The typical length of oligonucleotides ranges from 20 to 80 base pairs. Probe density represents 20 000–30 000 spots per glass slide for DNA arrays and can be dramatically increased in the case of arrays prepared with photolithographic techniques ($>250 000$ oligonucleotides cm^{-2}). As a consequence, oligonucleotide arrays are best suited for the evaluation of the full transcriptome. Since the generation of oligonucleotide arrays relies on the quality of currently available sequence information, hybridisation targets can, however, represent outdated genome annotations, leading to either unspecific or missed hybridisation events. A disadvantage of cDNA microarrays is that they rely on collections of cDNA clones, for which quality and identity must be ensured over the whole microarray manufacturing process. In addition, gene families with closely related members could cause misleading cross-hybridisation problems. Prior to hybridisation, target mRNA is extracted from the plant material and reversely transcribed into cDNA, which is labelled with radioactivity or with fluorescent dyes. In the case of oligonucleotide-based arrays, cDNA is often used to synthesise labelled antisense RNA by linear amplification. Microarrays made of nylon membranes have to be hybridised with radioactive target cDNA, because labelling with fluorescent dyes is not applicable, due to the high background fluorescence of such membranes. In contrast, glass slides allow the hybridisation with fluorescent targets, the most commonly used being cDNAs labelled with Cy5 or Cy3 that allows the detection of different samples (e.g. control and treatment samples) hybridised to a single chip at different wavelengths. Treatment and control samples are combined in one hybridisation event, so that a ratio between the fluorescent signals of Cy5 and Cy3 can be established, reducing chip-to-chip variations. Finally, signals are detected with a phosphoimager or with a laser scanner attached to a confocal microscope.

Transcript profiling involves RNA preparation, reverse transcription, probe labelling, microarray manufacturing, hybridisation, signal detection and quantification. All these steps are prone to error, so that the results of microarray experiments are rarely comparable when generated by different expression profiling approaches or research groups, even when the same mRNA pool was used [8]. However, we anticipate that in the near future, companies will provide full service on extracted RNA, thus further reducing technical error.

In addition to the model Arabidopsis, major crop plants such as rice and sugar cane are also becoming a target for commercial expression profiling. Consequently, plant research will benefit from additional information on phylogenetic diversity. Also, real time RT-PCR [9] has emerged as a powerful alternative to microarrays, when subsets of genes are studied. Indeed, real time RT-PCR allows the precise quantification of transcript levels, with a much higher linearity range and sensitivity than microarray platforms [10].

1.3 Transcript profiling workflow

To ensure a high standard of publications, the editorial boards of 'The Plant Cell' and 'Plant Physiology' have recently modified their criteria of acceptance for manuscripts dealing with transcriptomics [11, 12]. Now, transcript profiling experiments have to be designed and described properly, based on the MIAME standard (Minimum Information about a Microarray Experiment, [13]). Datasets have to include an adequate number of replicates; the verification of gene subsets with alternative approaches such as real time PCR will not be regarded as compensatory for replicates. Data analysis should include appropriate statistical methods, so that arbitrary methods such as the calculation of fold changes (FC) to select differentially expressed gene are no longer accepted. Additionally, the entire dataset has to be deposited in public gene expression repositories, for example, TAIR (The Arabidopsis Information Resource, [14]).

1.3.1 Data generation

Profiling experiments are not merely expensive; they are also extremely time-consuming in terms of data analysis and interpretation. Furthermore, a dramatic increase in the number of parameters that can be measured does not mean that the quality of the information improves. Indeed, a poorly designed experiment is likely to lead to 'a flood of misleading or un-interpretable data, which will be even more difficult to identify and put aside than in the past' [15].

Preliminary experiments or literature searches may be very useful to help reveal differential expressions of genes in an accurate way. Of central importance is that many genes exhibit strong oscillations throughout a night and day cycle [16, 17], including a large proportion of the genes involved in metabolism [18, 19]. As a consequence, when comparing a mutant with its corresponding wild type for example, it will be very difficult to choose the critical time point, unless a precise time window has been established via preliminary experiments in which diagnostic markers are measured to evaluate the physiological and/or developmental state. Ideally, a 24-h time-course experiment would be performed in both genotypes, but this is expensive and requires sophisticated analyses. As an alternative, samples taken throughout the day and/or the night periods might be pooled, but considerable information will be lost. However, such approaches should at least give an estimate of average transcript levels (see example below). Another possibility is to grow plants in continuous light (http://www.Arabidopsis.org/info/expression/ATGenExpress.jsp), although this growth condition is artificial and can hide important but conditional phenotypic traits. For example, starchless mutants will only exhibit a carbon starvation phenotype, if they are grown in photoperiods of less than 16 h [19]. A minimum of three replicates from three independent experiments for each data point has been suggested, which are eventually combined with dye-swap replicates when cDNA arrays are being used [20]. Information about biological variation is not necessarily desired and can be reduced by pooling a consistent number of individuals.

Another important variable is the way in which the plant material is harvested. Whenever possible, sampling should be achieved at the location of the experiment and as quickly as possible. Indeed, any change such as a temperature shift, or a wound response provoked by a dissection, can exert dramatic effects on the transcriptome. As an example, our colleagues have observed anomalously high levels of transcripts encoding heat shock proteins in plants acclimated to low temperature (M.A. Hannah, personal communication). However, they finally established that this was due to the transfer of the plants to room temperature prior to the harvest. With these considerations in mind, it is apparent that performing transcript profiles from single cells [21] is particularly challenging and requires sophisticated techniques [22].

After sampling, the isolation of clean RNA is tedious work, depending on the tissue of interest. Several generic protocols are available for plant material, and kits including disposable cartridges can be purchased from various companies. However, in certain cases, dedicated protocols are necessary. For example, a method involving a hot borate buffer has been recommended for the extraction of high quality RNA from seeds [23]. To ensure quality, the integrity of RNA has to be checked by gel or capillary electrophoresis, in combination with UV spectrometry. In general, a final step consisting of the isolation of polyA$^+$ RNA is performed in order to reduce the technical noise due to background hybridisation.

1.3.2 Data management

Efficient data analysis depends on the individual organisation of raw and normalised experimental data. A crucial step for data handling is the unique naming and description of every chip file in any subsequent analyses. This specification should contain an intuitive abbreviation of the investigated genotype, whether a control or test experiment was carried out, an indication of the analysed time-point or the developmental stage in serial experiments and an index of replication. For each experiment an additional descriptive file should inform about further experimental details based on MIAME standards [13]. Database systems are the dedicated tools for data management, especially if multiple experiments, projects or users have to be organised [24–26]. Databases standardise data formats, use controlled vocabularies for experimental descriptions, define user-specific data access and facilitate the exchange of data. A well-structured and transparent data resource avoids experiment redundancy and allows data mining and large-scale data analysis (e.g. co-response analysis or data integration approaches [24, 25]).

A number of information management systems (IMS), including commercial software, have been developed in various research fields to cope with more and more complex datasets. These include medical research [27], microarray analysis in Arabidopsis [28], proteomics [29] or protein crystallography [30]. Classically, IMS allow project management, sample processing and tracking, data retrieving, sorting, visualisation and correlative approaches. Typically, such systems are centralised, often web-based, with restricted or conditional access. We are currently developing a flexible IMS, which is stand-alone, and involves various modules for the design of experiments, the collection and management of data and its subsequent analysis (J. Hannemann *et al.*, unpublished). This IMS will also allow the import of pre-existing data. Basically, the aim is to collect and link genotypic, environmental and phenotypic data in a 'decentral' mode, but with the possibility of collaborative sharing of data. The synchronisation of data from different research groups will be made possible by the edition of the nomenclatures. Indeed, the sharing of data is a critical issue for data mining, as this is the best way to increase replication and number of analytes, especially those requiring extensive equipment and know-how.

1.3.3 Data processing

Data processing is the most sensitive step in gene chip analysis, since this technology has still inherent methodological drawbacks that make it susceptible to technical variation and signal misinterpretation. Many authors have contributed on the current technological and statistical problems, and many bioinformatic tools have been developed to increase the accuracy and precision of gene chip analysis. Because of the strong effects of data processing strategies on the final results, the user should carefully choose his/her data analysis tools. Therefore, the processing strategy and its underlying algorithms should be understood in detail. This is hardly ever achieved by biologists, since an advanced mathematical and statistical knowledge is required. Additionally, the field of data processing is progressing so quickly that it is very difficult to keep up-to-date.

A number of algorithms have been recently developed to improve the performance of the standard microarray analysis software (Microarray Suite, MAS) released by Affymetrix™. MAS is a stand-alone platform allowing the processing of raw data in a predefined, but partially flexible workflow. Many of these algorithms are modular program packages using the open source statistical scripting language R [31]. Relevant packages are integrated into the Bioconductor project, which is dedicated to the analysis of genomic data [32]. Some of these tools work at a command-line level, which represents an additional hurdle for researchers without any programming experience. In contrast, R and the Bioconductor project are well documented and offer beginners comprehensive online help. In addition, many R courses are available for free, following the idea of an open source platform (e.g. http://compdiag.molgen.mpg.de/ngfn/pma2005.shtml). For the first time, researchers can assemble their own data analysis pipelines by downloading pre-programmed modules that can be found in the public domain. Furthermore, several Bioconductor packages have been integrated into user-friendly graphical interfaces (e.g. affylmGUI, http://bioinf.wehi.edu.au/affylmGUI/) to meet the demand of biologists.

The following sections will briefly introduce and compare the current strategies of data processing in terms of their impact on data interpretation.

1.3.3.1 Raw data handling

The whole genome Arabidopsis microarray (ATH1) from Affymetrix, uses nearly 23,000 probe sets with 11 gene-specific 25-mers (probes) that have been designed to match the last 600 bp of 3′ coding sequences. The expression level of each gene is assessed by summarising the signal intensities of the individual probes. To deal with unspecific binding, cross-hybridisation and background noise, a derivative oligomer was designed (mismatch: MM) for every specific oligomer (perfect match: PM), by changing the 13th base to its complementary (PM/MM couples are defined as probe pairs). The first version of the Affymetrix™ MAS software (4.0) used a robust average of PM-MM differences for expression measurement that includes background correction and deals with outlier probes [33]. By using this approach, about one-third of all PM-MM differences turned out to be negative, suggesting that MM oligomers are also subject to specific hybridisation [34, 35]. Additionally, it was observed that the inherent linear scale measure is not appropriate to measure expression [34, 36]. The Affymetrix algorithm was further optimised (latest MAS release 5.0), defining the signal of a gene as the robust average (Tukey Biweight) of log (PM – CT) [37]. Since this algorithm uses a log scale measure, the problem of negative PM-MM differences had to be addressed and CT values were introduced, where CT = MM, if MM < PM. In cases where MM > PM, CT values are adjusted to be smaller than PM. But this correction only masks the recurrent specific signal binding to MM probes. Therefore, alternative strategies have been suggested by several authors, chiefly RMA (log-scale robust multi-array analysis) method from Irizarry et al. [34], dCHIP from Li and Wong [36, 38], the PDNN model (positional-dependent-nearest-neighbour) from Zhang et al. [39] and the GCRMA approach from Wu et al. [40].

RMA includes a three-step procedure that corrects for background noise, normalises across multiple arrays and finally summarises probe-level values using PM information only. PM signals are supposed to contain specific binding, non-specific binding, background and optical noise. To correct the specific signal for background noise, RMA makes use of the assumption that the signal is exponentially distributed, whereas the background noise is normally distributed. Distribution parameters are estimated by PM values and the modelled background noise is subtracted from PM signal intensities to assess specific signal intensities. Li and Wong observed that PM value variation could be considerably higher within a probe set than the variance of certain PM across different arrays, indicating a strong probe affinity effect. A multiplicative model is used in dCHIP to estimate the probe affinity effect from a minimum number of 8–10 arrays, to remove outlier probe intensities and to calculate expression measures [36].

PDNN and GCRMA assess unspecific binding by referring to available oligomer sequence information. Zhang et al. [39] developed the PDNN model – a free energy model for the formation of RNA-DNA duplexes. It reflects the specific contribution of different parts of oligomers to the overall probe binding stability and assesses specific and non-specific binding without the use of MM signal information. The background correction used in GCRMA combines the physical aspects of Zhang et al. and the non-specific binding model of Naef and Magnasco [41]. Using available sequence information, GCRMA performs a background adjustment based on GC content and probe affinities (sum of position-dependent base effects) of PM and MM [40, 42].

1.3.3.2 Normalisation
Signal intensity is not only the result of mRNA expression level, but is also strongly affected by technical variability that includes all technical, chemical and human factors that are involved in sample preparation, hybridisation and image processing. The purpose of normalisation strategies is to reduce these effects and to enhance the comparability of specific signals across different arrays.

In addition to different expression measures, the software tools mentioned in Section 1.3.3.1 also make use of different normalisation strategies. Although the performances of the different strategies have been compared intensively, a common gold standard has not been defined to date. Generally, it can be stated that the choice of a strategy depends on the individual user's needs. Excellent examples of intelligible discussions dealing with gene expression measurements and normalisation strategies can be found in Saviozzi and Calogero [43] and O'Connell [44].

The MAS5.0 linear approach scales summarised probe set signal intensities. This results in the same average value for all arrays, but does not affect the correlation of the data [43, 45]. To deal with commonly observed non-linear relations between different arrays, Li and Wong proposed a non-linear scaling strategy that is used in dCHIP [36, 38]. Basically, arrays are normalised at a probe intensity level to a chosen baseline array with a median overall brightness. This normalisation is only carried out on non-differentially expressed genes (invariant set) which are defined in parallel with an iterative intensity ranking approach [38]. RMA and GCRMA use the quantile method with the aim to make the distribution of probe intensities (before summarisation) the same for all arrays analysed. The expression estimate for each gene is assessed with a robustly fitted (median polish) log-scale expression effect/probe effect model [44]. It has been shown that RMA expression measure has a higher precision than dCHIP and MAS5.0, when applied to replicates of the same experiment. RMA-derived probe values showed better correlation and smaller standard deviation (SD) especially at low gene expression levels [34, 46]. Comparing the performance on the basis of fold changes (FC: gene expression value of a control experiment/ gene expression value of a treatment experiment) in spike-in experiments, RMA shows higher sensitivity and specificity when compared with MAS5.0 and dCHIP.

Besides normalised expression values, MAS5.0 provides test statistics that allow a significant improvement of performance based on (i) gene-specific detection calls (either A = absent, M = marginal or P = present), indicating the level of expression and the quality of the probe set hybridisation and (ii) FC P-values derived from non-parametric test statistics. Scatter plot analyses of data normalised with MAS5.0 have shown that high SD at low expression levels mainly originate from genes that are classified as absent. To reduce signal variability and improve performance, genes called absent should be removed routinely from further analysis [43]. When MAS5.0 P-values are used for FC analysis, MAS5.0 performed as well as RMA and better than dCHIP [46]. Basically, MAS5.0 provides an accurate summary including background and MM correction, but sometimes the real expression level is underestimated, since MM can also contain specific signal intensities [34, 44]. Otherwise, RMA expression measures have a higher precision of expression measures, but are less accurate for low expression values [40, 44]. The normalisation with RMA and dCHIP showed a higher correlation and smaller SD when applied to replicates of the same experimented data [43]. Nevertheless, RMA normalisation should be applied carefully, since adjusting distributions across multiple arrays may lead to the removal of signals in the tails [34]. In particular, if many genes are differentially expressed or only a small portion of genes is expressed at all, the quantile normalisation may increase the number of false negatives. The normalisation across all datasets also complicates the interpretation of data, as is observed if additional arrays are integrated into an existing dataset, for example. The necessary re-normalisation across old and new data could lead to changed expression values in the original dataset and alter prior results. RMA FC estimates are about 10–20% compressed in comparison to FC calculated by MAS5.0 [46]. This is important to note, since many researchers still use a certain FC cut-off to restrict the number of differentially expressed genes for subsequent analysis.

1.3.4 Data analysis

As mentioned previously, the introduction of genome scale profiling experiments revolutionised plant biology in such a way that it is no longer possible to work on results expressed as means ± SD. Instead, a multitude of tools have been adapted or developed to help making sense out of the data.

1.3.4.1 Differential expression

One objective of microarray experiments is the identification of differentially expressed genes under different experimental conditions. Even after careful normalisation and filtering procedures, expression data are still noisy and statistical methods have to be applied to test whether changes in expression levels are significant [47, 48].

An early approach was the calculation of FC, and the definition of a general cut-off for a significant change in expression. Several authors agreed that a two-fold change in expression can reliably be detected by modern transcriptomic systems [49–51]. Although FC provide an intuitive value, they are not confirmed statistically, and the definition of a cut-off threshold remains arbitrary. High FC of low expressed genes should be interpreted carefully, since their signal variation across replicates is typically high. Complementary, conventional t-tests can be used, providing the probability whether a change in expression was detected by chance. As a result, even minor changes can be highly scored, irrespective of their biological relevance. Methods based on t-tests depend on strong parametric assumptions, which are often violated by the restricted number of replicates that are commonly used in microarray experiments [52].

The most widespread method to identify differential expression is the 'significance analysis of microarrays' (SAM), which assimilates a series of gene-specific t-tests [48]. SAM scores genes based on their change in expression, in relation to SD of replicates, without strong parametric assumptions [48, 52]. As a compensate for chip replicates, SAM uses repeated data permutations to generate controls and assess statistical significance [43]. Compared to the FC method, SAM reduces the false discovery rate (FDR, percentage of genes identified by chance) in one reported example from 73–84% to 12% [48].

The simple rank product (RP) method circumvents the previously mentioned parametric problems and provides both biological and statistical meaningful values [47]. RP is based on ranked FC, calculated pairwise from comparison control and treatment experiments. FC ranks (divided by the number of chip genes) are multiplied and averaged by a geometric mean to protect against outliers. A permutation-based procedure converts RP values to E-values that estimate the statistical significance of rank products. In comparison to current methods, the simple RP method outperformed the sophisticated SAM with a comparable FDR of 10% [47]. Moreover, the similarity of RP and FC results indicates a high biological relevance of RP values. The non-parametric nature of the approach makes RP particularly useful if only a low number of replicates are available.

1.3.4.2 Data mining
Data mining was initially defined as the 'non trivial extraction of implicit, previously unknown, and potentially useful information from data' [53]. Basically, the aim is to find patterns in large datasets by the use of various strategies, ranging from classical approaches such as cluster analysis [2] or principal component analysis (PCA) to machine learning approaches such as artificial neural networks (ANN), decision trees or support vector machines [3].

(i) *Making sense of in-house data.* Whereas the great advantage of multi-parallel platforms is the recording of thousands of different values at the same time, it is simply not feasible to browse through all these data. A very straightforward way to make use of such a dataset is to use a 'guilt by association' approach [54]. This consists of searching for genes that behave similarly with respect to a given gene already associated with a particular process, assuming that a functional relationship may exist. A further step to bring the data into a meaningful context is clustering, which consists of grouping genes and/or experiments that behave similarly. The idea is that the partitioning of the individual data points into groups will reveal new commonalities, or point out further potential group members. A number of tools suitable for clustering, commercial or free, stand-alone or Web based (TIGR Multiexperiment Viewer (TMEV), Genesis, TU Graz, R statistics environment) are available. Prior to clustering, a distance function has to be defined, typically an Euclidean distance or a correlation (for a discussion of different distance functions see, for example, D'Haeseleer *et al.* [55]). Most programs include hierarchical cluster analysis (HCA), self-organising maps (SOM) and k-means clustering. On the one hand, HCA builds cluster trees that are very similar to phylogenetic trees. A sensitive step here is to cut the tree at the right height to obtain groups. On the other hand, both k-means clustering and SOM require the pre-definition of the number of clusters to be made. This poses a challenge, since the number of expected clusters is not known, *a priori*. In fact, the multiple combinations of clustering methods and distance functions that can be chosen have strong consequences on the group composition. Unfortunately, there are no clear guidelines to deal with this problem [55], and depending on methods used, contradicting outputs might occur. However, a clearly defined question can help in evaluating clusters. For example, a procedure able to group well-identified genes properly will have the potential to associate further unknown genes to this group.

Another way to deal with multi-dimensional datasets is PCA. PCA aids in the explorative identification of key variables that best explain the differences in observations (experiments). PCA reduces dimensionality and preserves as much variance as possible. The method is well suited for the analysis of large datasets. However, the researcher has to identify the biological background of the key variables [56]. A similar method is correspondence analysis, which at the same time projects experiments and genes, so that genes and experiments that resemble each other group together. It has been used by Fellenberg *et al.* [57] to show that some yeast cell cycle experiments were probably wrongly classified in terms of cell cycle progression.

(ii) *Mining large databases.* After having extracted as much information as possible from a given microarray experiment, it is possible to extend data mining one step further. For example, it can be interesting to check for additional information about candidate genes in previous experiments performed by other research groups. Fortunately, many plant microarrays are deposited in public databases (e.g. NASC [58], TAIR [14], BarleyBase [59], GEO [60], EBI [61] and the Stanford Microarray Database [62]). A more straightforward approach is the mining for new information pertaining to genes of interest. Questions that are usually asked include where a gene is expressed, if it is co-expressed with other 'interesting' genes and whether such an association is context dependent. Several plant databases now offer expression values, as well as correlations over a broad range of experiments, following the 'guilt by association' approach. Currently, there seem to be two different approaches. One is to include all available experiments for calculations of distance measures and is featured by Genevestigator [26] and Expression Angler (http://bbc.botany.utoronto.ca/ntools/cgi-bin/ntools_expression_angler.cgi). The other approach, featured by CSB.DB [24], focuses on manually selected sets of experiments with similar biological contexts. Using the latter approach, the biologist can test the context dependence of gene associations, thus learning more about the nature of the associations.

A majority of data mining approaches rely on a context-dependent classification of experiments, and thus on a standardised description of the experiments. Efforts have been undertaken to establish standards for microarray experiments (MIAME) and ontologies for microarray experiments are being developed within the Microarray Gene Expression Data Society [63]. A context-dependent re-classification of experiments might be performed to answer specific questions and will require precise information about the experiments. Having done so, a new experiment might be classified as belonging to previously encountered groups. For example, the investigation of an expression profile of a mutant might match profiles conducted under a specific stress. The classification methods that are used most often are ANN, classification trees and support vector machines. Basically, these supervised machine-learning approaches rely on pre-defined classes obtained with training sets. One major problem is that many training samples are needed to obtain a high recognition rate. Even though various microarray platforms may deliver conflicting results (for example, see [64] for a comparison of Arabidopsis microarrays), it has been shown that ANN can be trained on multi-platform data to efficiently discriminate expression data derived from different tissues [65]. A rather novel approach is to take the functional classifications of genes relevant for experiment classification, thus looking at classes of genes that are affected, rather than single genes. Using this approach, Breitling *et al.* [66] could successfully put microarray experiments from three different organisms into a meaningful context.

Another approach is the extraction of *cis*-elements from genes that behave similarly. The assumption is made that if genes are transcribed together, they should also share some *cis*-elements. For example, Chen *et al.* [67] showed the enrichment of DRE and ABRE elements in cold responsive genes of Arabidopsis. To further increase the accuracy of these searches, it is often desirable to use available data as much as possible. Thus it is possible to run promoter searches with both co-expression information and sequence data from homologues of different species to obtain better predictions [68]. To bring all different information together, co-response networks can be built [69]. These can then be analysed for centrality and other network theoretical parameters. Genes having many connections to other genes (the so-called hubs) are believed to have key functions and might, therefore, constitute ideal targets for knock-out or overexpression experiments.

Finally, data mining should not be perceived as a 'magic solution that will pick out the gold from the mud of huge datasets'. It is likely that genes that are always co-responding will be found. These include subunits of a protein complex, such as the ribosomal proteins. However, in the case of regulatory pathways, a co-response is probably very often conditional, and its identification will require extensively documented experiments.

1.3.4.3 Functional categorisation

The aim of gene ontologies is the organisation, description and visualisation of biological knowledge [70]. Thus functional gene categories (FGC) play a major role in modern genomics. Transcript profiling analysis is highly facilitated with the use of FGC, since affected categories can quickly be identified and biologically interpreted. Furthermore, FGC allow an inter- and intra-species transfer of knowledge when combined with sequence similarity analysis. Necessary gene information is gathered by a combination of manual and electronic approaches. Despite the development of sophisticated text-mining strategies, manual interference is still indispensable, involving the extraction of textual information from scientific publications and the verification of electronic approaches (Fluck *et al.*, in preparation).

The most comprehensive and widespread ontology in use is GO (gene ontology) developed by the Gene Ontology Consortium, as a generic tool to classify biological genomes by three main categories, namely: (i) biological process (e.g. purine metabolism), (ii) molecular function (e.g. kinase activity) and (iii) cellular component (e.g. Golgi apparatus) [70]. Recently, the vocabulary of the Arabidopsis GO was extended to describe (iv) anatomy and (v) developmental stages [71]. GO has developed 17,593 terms that have been used to assign about 75% of the Arabidopsis genome to at least one category. These terms are organised in a complex hierarchical 'child to parent' relationship, where each 'child' may belong to multiple 'parents'. In parallel, the MIPS database has developed the FunCat system, which covers approximately 60% of annotated Arabidopsis genes using a restricted set of 28 main categories [72]. Recently an additionally FGC system has been developed, which is characterised by a high flexibility of category structure and low redundancy of functional gene assignments [73]. Thirty-four major FGC (BINs) and several hundred associated subBINs are used to group genes, either by broad functional and motif similarities (e.g. alcohol dehydrogenases or cytochrome P450s) or to assign individual genes to defined steps in biochemical pathways if sufficient biological evidence is available.

Obviously, the GO initiative represents the most comprehensive and accepted approach to organise and develop gene annotations. Therefore, a variety of bioinformatics tools have been developed to browse, query and edit GO terms, including AmiGO (http://www.godatabase.org/cgi-bin/amigo/go.cgi), COBrA (www.xspan.org) and DAG-Edit (http://www.geneontology.org/GO.tools.shtml). Furthermore, functional information can be used for bioinformatic-aided interpretation of transcriptomic data. GOToolBox, FuncAssociate, PathwayProcessor, PathMAPA, GiGA, MAPPFinder and ArrayXPath all use different statistical tests such as bootstrap analysis, Fisher Exact Test or different ranking methods to reveal significantly affected GO categories in transcriptomics datasets [74–79]. Equivalent tools are available for the FunCat and BIN system (Classification SuperViewer, http://bbc.botany.utoronto.ca and MapMan, http://gabi.rzpd.de/projects/MapMan/), which use Monte Carlo simulation or the non-parametric Wilcoxon Rank Sum Test for the identification of functional hotspots, respectively. The suitability of FGC for biological research strictly depends on assignment quality and category structure of the system itself, rather than on available bioinformatic support. Although the GO system benefits from comprehensiveness, its structure complicates an intuitive extraction of genes involved in a biological process of interest. The GO slim system, a simplified version of the original GO system that uses only 40 high-level GO terms [71], only slightly improves information access.

The FunCat system benefits from its simple and well-organised structure, but a detailed data analysis is severely hampered by obvious annotation mistakes, presumably introduced by electronic annotation procedures. Moreover, the high number of multiple gene entries in the FunCat and GO system negatively affects a statistical functional data analysis and data visualisation (see next section). The BIN system was designed to complement the GO system. It makes use of the wealth of biological knowledge stored in public databases (such as GO, TAIR, KEGG and AraCyc) that has been intensively re-organised and curated to serve biologist-friendly data visualisation and analysis [14, 73, 80, 81]. However, a functional-based transcriptomics analysis is the key for a deeper understanding of plant biology, especially when combined with sophisticated data mining strategies [24]. FGC analyses allow the combination of co-responding pathways or categories to new functional modules. Furthermore, this approach aids to functionally associate putative or unknown genes, when concerted expression patterns with known processes have been revealed.

1.3.5 Data visualisation

Visualisation tools translate efficiently abstract numeric values into accessible graphical information. In genomics, visual compression of information and reduction of dimensionality address different purposes. For example, scatter plot analyses are used to ensure data quality, in particular to investigate the effects of replication, normalisation or experimental treatment on data correlation of tens of thousands of individual data points [82]. Furthermore, well-known statistical analysis methods such as cluster and PCA analysis have been rediscovered for genomics applications, after being combined with powerful visualisation tools (e.g. TMEV). However, the development of visual pathway analysis tools represents an even more revolutionary advance for genomics data interpretation. Pathway analysis tools convert expression data into false colours that are subsequently mapped onto images of biochemical pathways or biological processes. Data mapping depends on functional classification of each gene *a priori*. This allows a spatial data arrangement onto maps, according to the assigned functional category. Thus pathway tools make use of biological knowledge to place transcriptomic data into their functional background. The visual integration of biological knowledge allows the user a detailed data interpretation, regardless of his/her area of expertise. With the use of visual pathway tools the major limitations of classic list approaches can be overcome. Typically, huge lists of differentially expressed genes are generated. To restrict the number of genes for manual inspection, data are filtered according to FC or P-values derived from statistical tests (e.g. SAM [48]). Thus, small expression changes, in particular, are excluded from further analysis and valuable information is lost. Even minor, but consistent changes in expression across pathways or pathway branches may indicate a concerted regulatory response [73].

During the last 3 years, a variety of pathway visualisation tools have been developed [73, 76–79, 83–86], of which MetNet, GiGA, Pathway Processor, PathMAPA and MAPMAN are applicable to Arabidopsis chip data. Although the concept of these pathway tools is very similar, the real potential of a pathway-aided analysis can only rarely be exploited. Suitable pathway analysis tools should enable: (i) a targeted analysis of expression changes across related pathways (e.g. photosynthesis, glycolysis or the citrate cycle), (ii) an untargeted discovery of functional modules (concerted response of unrelated pathways and processes) and (iii) functional association of unknown genes. The fulfilment of these claims depends primarily on the structure and quality of the ontology used and on the flexibility of the data display. Although a display of whole datasets quickly reveals a global trend in specific transcriptional responses, a detailed investigation at a single gene level is indispensable for the generation of biological hypothesis (e.g. the inspection of transcriptional responses of individual members of a gene family). This is only possible if the used ontology is of high quality.

As mentioned in Section 3.4.3, only MAPMAN makes use of a manually curated gene ontology that was especially developed for visual pathway analysis. Other tools use publicly available ontologies such as GO, KEGG and MIPS, which suffer from a non-intuitive structure, electronic annotation mistakes and redundant functional assignments [70, 87–89]. Untargeted discovery of functional modules and functional association of unknown genes depends on a flexible user-driven display, as only the abandonment of known pathway structures and a visual re-arrangement of the gene display can reveal new regulatory coherences. Therefore, static pathway maps (e.g. KEGG pathway maps) are only of limited value. They provide the user a wealth of proven biological knowledge, but hamper a creative data interpretation. Furthermore, suitable visualisation tools are characterised by flexible data input formats. This allows the display of results from upstream data analyses (e.g. cluster or promoter motif analysis), sub-cellular localisation or the visual integration of heterogeneous profiling techniques (see examples in Section 4).

1.4 What can we learn from transcript profiles performed in a starchless mutant?

An example of the application of transcript profiles is work with the starchless *pgm* mutant of Arabidopsis that lacks plastidic phosphoglucomutase, an enzyme that is essential for starch synthesis in the leaf [90]. In plants, normally a fraction of the photosynthate is exported during the day as sucrose from source leaves to support respiration and growth in the rest of the plant, but during the night, the entire plant becomes a net consumer of fixed carbon. Some photosynthate is stored in leaves as starch in the light and is re-mobilised at night to support leaf respiration, as well as the continued synthesis and export of sucrose. In the *pgm* mutant, sugars accumulate during the day, but are rapidly depleted in the first part of the night. A recurring phase of sugar starvation during the second part of the night leads to a severe growth impairment [91]. In the present experiment, transcript profiles were performed to evaluate the impact of these alterations in sugar metabolism on the expression of genes. Plants of both *pgm* and wild type (WT) genotypes were harvested every 4 h throughout a day and night cycle and three biological replicates were prepared in the case of the WT, but only one in the case of *pgm*. One replicate was made out of 15 pooled plants and total RNA was extracted with a TriZol protocol. After hybridisation on ATH1 microarrays, a quality check was carried out using the 'affy'-package of the R software environment. In general, all probes were found to contribute significantly to the expression signal. Subsequently, all wild-type and *pgm* data were normalised separately with the RMA method. The replication, checked by using scatter plots, pair-wise Pearson correlations and PCA, was found to be very good in the WT (not shown).

Average expression was calculated as the mean of six data points throughout the day and night cycle, thus mimicking a pooling of samples. As shown in Plate 1, the mutation provokes a general decrease in the expression of the genes involved in the light reactions of photosynthesis and in the synthesis of tetrapyrroles. Genes encoding nitrate reductase, and to a lesser extent other enzymes involved in the synthesis of amino acids, were also found to be down-regulated. In contrast, a quite large proportion of enzymes involved in the degradation of amino acids were found to be up-regulated. A naive interpretation of such results would be that photosynthesis and nitrogen assimilation are decreased in the mutant, while the recycling of amino acids is increased. Interestingly, one gene encoding asparagine synthetase is strongly induced in *pgm*, which also exhibits high asparagine levels, a typical response to carbon starvation [92, 93].

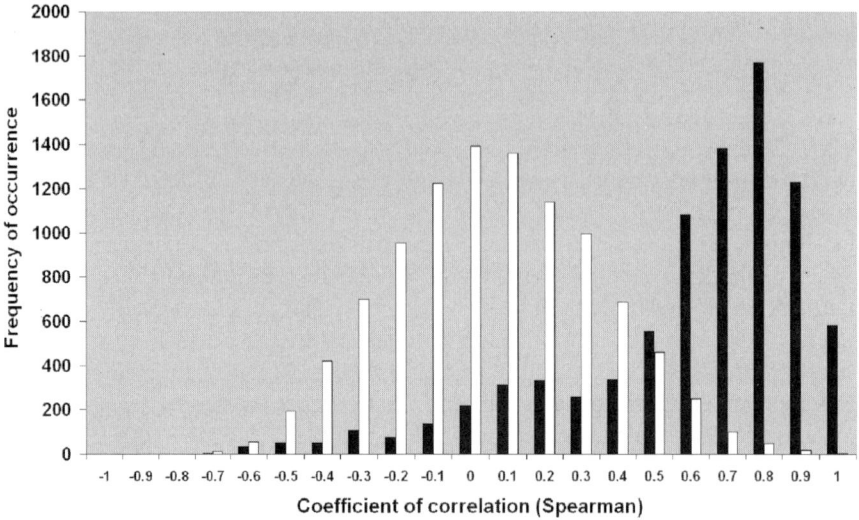

Figure 1.1 The CSB Database was queried with all genes from the Bins 'Photosynthesis Light reaction' and 'Tetrapyrrole Biosynthesis'. The Spearman correlation coefficients were grouped according to their values and their occurrences counted. A strong bias towards high correlation coefficients can be detected (black bars). For better comparison the database was also queried with randomised genes where a nearly normal distribution can be observed (white bars).

In fact, enzymes involved in the fixation of nitrogen actually had lower activities in the mutant, and glutamate dehydrogenase was strongly increased [22]. However, photosynthesis, estimated as the net carbon fixation, was found to be only slightly lower in the mutant [94].

A general decrease in the expression of genes related to photosynthesis nevertheless suggests a co-regulatory event associated with the altered carbon status in the mutant and thus constitutes a pattern. CSB.DB was then queried with all members of this BIN in the 'multigene query' mode, to check whether these genes correspond in general. As shown in Figure 1.1, these genes were found to be strongly co-regulated over a set of 51 microarrays obtained from various treatments. This suggests that these genes share a common regulatory pathway, for which various signals, such as alterations in the carbon status, are integrated upstream. Indeed, this rather simple example confirms that co-responding genes can be functionally related.

To compare amplitudes of diurnal changes in gene expression in the WT and in *pgm*, the log 2 ratios of amplitude calculated as the difference between maximum and minimum values, in the *pgm* mutant and in the wild type were established, and visualised with the MapMan program (Plate 2). As expected, the mutation provokes a global increase in the amplitude of changes in gene expression in metabolism (spots in blue). Interestingly, a few genes show markedly decreased amplitudes (spots in red) in *pgm*, namely the two genes encoding nitrate reductase, a cab gene, two genes involved in amino acid metabolism, and genes involved in starch degradation. Less marked was the apparent decrease in amplitude in many photosynthesis-associated genes.

An important question arising from these results was to check whether larger variations of steady state transcript levels were reflected by changes in the corresponding enzymes. For that purpose, 23 enzymes from the primary metabolism were determined throughout a diurnal cycle in both genotypes [19] and log 2 of the amplitude ratios was calculated. Plate 3 indicates that in *pgm*, the expression of most of the genes encoding the enzymes we measured had markedly larger amplitudes, but in most cases, this had no consequence on the corresponding activities. Furthermore, there is a trend of increased diurnal changes of the enzymes involved in sucrose and hexoses-P metabolism, particularly fructokinase and cytosolic fructose-1,6-bisphosphatase, but without being necessarily related to alterations in the expression of the corresponding genes (e.g. hexokinases). This example suggests that transcriptomics alone cannot capture the entire physiology of a complex organism. Consequently, it will be necessary to gather and integrate as much data as possible from other levels, i.e. proteins, metabolites and fluxes, to uncover all functions of the remaining 'unknown' genes. The controls on primary metabolism exerted at the post-translational level and by key metabolites are discussed in more detail in later chapters.

1.5 Conclusion/perspectives

Plant biology was literally revolutionised by the development of high-throughput expression profiling techniques, which have prefaced the post-genomic era. Because these powerful methods generate vast amounts of data, new challenges have appeared, triggering an intensive development of statistical approaches and programming solutions. Fortunately, these approaches are not organism specific, and thus plant scientists benefit from the progress made in other disciplines, such as medicine or microbiology. The necessity to efficiently manage large datasets spurred the development of numerous databases. Currently, it can be predicted that data will converge to a few centralised public databases, since some journals encourage the deposition of 'omics data in the public domain'. Many solutions are proposed to exploit the data obtained form microarray experiments. In most cases, algorithms aim to classify analytes and/or experiments. However, the availability of tens of classification tools, together with their various options, renders the choice of the ideal strategy difficult. Indeed, there is no gold standard, but we can hope that more user-friendly and self-explanatory programs will be developed. Moreover, biological education will hopefully include more advanced statistics in the future.

Despite the remarkable achievements of transcriptomics, we are still far from a systems understanding. We have, in fact, just begun the exploration of the plant system as a whole since post-transcriptional and post-translational regulation events are not taken into account in transcriptomic studies, despite being major players in the control of metabolism. Therefore, functional genomics will largely benefit from complementary approaches such as metabolite, protein or enzyme profiling. In contrast to transcript profiling, these techniques need still intensive manual input and tremendous expert knowledge. While the isolation of nucleic acids benefits from the similar chemical properties, metabolites, protein and enzymes differ highly in structure, physicochemical properties and turnover. Thus, only small fractions of the plant's metabolome, proteome and enzymome can be assessed in parallel [19, 94, 95], but already provides useful information when integrated with transcriptomics data [19, 96, 97].

In view of the rapid progress being made in the analytical aspects of metabolomics, proteomics and enzyme profiling, more sophisticated approaches for data integration and interpretation will be required. In particular, the planning of experiments and data management will be aided by statistics and information management systems. Moreover, we predict that public data mining will be one of the most important biological tools of the future, since profiling techniques will be still expensive and not affordable for every scientist. For example, data mining will be a premium source for hypothesis-driven research. Therefore, new algorithms and interfaces have to be developed that allow, for example, pattern matching queries across multiple databases and across heterogeneous datasets to reveal multi-level regulatory networks. A prerequisite is that databases use a standardised nomenclature. Several attempts have been made to impose a controlled vocabulary for the annotation of genes and the description of experiments [13, 71]. However, current genomics databases still suffer from low transparency, heterogeneous data processing strategies and different naming of public datasets [14, 58].

The term 'phenomics' has been introduced more recently and defines the high-throughput analysis of phenotypes [98]. It aims to understand the complex phenotypic consequences of genetic mutations or variations at the level of the organism [99], and probably poses the biggest challenge of bioinformatics to date. Phenomics extends our current understanding of data integration, since apart from the unification of all available genomics and post-genomics data, the concatenation of numeric and of textual information is necessary. Textual information originates from controlled vocabularies of phenotype descriptions of ecotypes and mutants and furthermore from biological knowledge that has been gathered from publications using text-mining approaches [71, 100]. Once these phenomics resources have been set up, it is possible to link genomic information with phenotypes across different species and to proceed once again in the understanding of whole plant systems.

Acknowledgements

We thank Mark Stitt for his kind and generous support, Alisdair Fernie and Ian Graham for constructive discussions of the manuscript and Jan Hanneman for his comments on information management systems. The authors are funded by the German Academic Exchange Service (DAAD), the Max Planck Society and the German Ministry for Research and Technology, in the framework of the German plant genomics program Genomanalyse im Biologischen System Pflanze (0312277A and 0313110).

References

1. C. Somerville and S. Somerville (1999) Plant functional genomics. *Science* **285**, 380–383.
2. M.B. Eisen, P.T. Spellman, P.O. Brown and D. Botstein (1998) Cluster analysis and display of genome-wide expression patterns. *Proceedings of the National Academy of Sciences of the United States of America* **95**, 14863–14868.
3. M.P. Brown, W.N. Grundy, D.Lin *et al.* (2000) Knowledge-based analysis of microarray gene expression data by using support vector machines. *Proceedings of the National Academy of Sciences of the United States of America* **97**, 262–267.
4. J. Donson, Y. Fang, G. Espiritu-Santo *et al.* (2002) Comprehensive gene expression analysis by transcript profiling. *Plant Molecular Biology* **48**, 75–97.
5. B.C. Meyers, T.H. Vu, S.S. Tej *et al.* (2004) Analysis of the transcriptional complexity of *Arabidopsis thaliana* by massively parallel signature sequencing. *Nature Biotechnology* **22**, 1006–1011.
6. D. Gillespie and S. Spiegelman (1965) A quantitative assay for DNA-RNA hybrids with DNA immobilized on a membrane. *Journal of Molecular Biology* **12**, 829–842.
7. J.C. Alwine, D.J. Kemp and G.R. Stark (1977) Method for detection of specific RNAs in agarose gels by transfer to diazobenzyloxymethyl-paper and hybridization with DNA probes. *Proceedings of the National Academy of Sciences of the United States of America* **74**, 5350–5354.
8. R.A. Irizarry (2004) Multiple lab comparison of microarray platforms. Department of Biostatistics Working Papers, Johns Hopkins University, Working Paper 71, pp. 1–18.
9. C.A. Heid, J. Stevens, K.J. Livak and P.M. Williams (1996) Real time quantitative PCR. *Genome Research* **6**, 986–994.
10. T. Czechowski, R.P. Bari, M. Stitt, W.R. Scheible and M.K. Udvardi (2004) Real-time RT-PCR profiling of over 1400 Arabidopsis transcription factors: unprecedented sensitivity reveals novel root- and shoot-specific genes. *Plant Journal* **38**, 366–379.
11. N. Raikhel and S. Somerville (2004) Modification of the data release policy for gene expression profiling experiments. *Plant Physiology* **135**, 1149.
12. R. Jorgensen (2004) Criteria for publication in The Plant Cell. *The Plant Cell* **16**, 1645–1646.
13. A. Brazma, P. Hingamp, J. Quackenbush *et al.* (2001) Minimum information about a microarray experiment (MIAME) – toward standards for microarray data. *Nature Genetics* **29**, 365–371.
14. S.Y. Rhee, W. Beavis, T.Z. Berardini *et al.* (2003) The Arabidopsis Information Resource (TAIR): a model organism database providing a centralized, curated gateway to Arabidopsis biology, research materials and community. *Nucleic Acids Research* **31**, 224–228.
15. M. Stitt and A.R. Fernie (2003) From measurements of metabolites to metabolomics: an 'on the fly' perspective illustrated by recent studies of carbon–nitrogen interactions. *Current Opinion in Biotechnology* **14**, 136–144.
16. S.L. Harmer, L.B. Hogenesch, M. Straume *et al.* (2000) Orchestrated transcription of key pathways in Arabidopsis by the circadian clock. *Science* **290**, 2110–2113.
17. R. Schaffer, J. Landgraf, M. Accerbi, V. Simon, M. Larson and E. Wisman (2001) Microarray analysis of diurnal and circadian-regulated genes in Arabidopsis. *The Plant Cell* **13**, 113–123.
18. S.M. Smith, D.C. Fulton, T. Chia *et al.* (2004) Diurnal changes in the transcriptome encoding enzymes of starch metabolism provide evidence for both transcriptional and posttranscriptional regulation of starch metabolism in Arabidopsis leaves. *Plant Physiology* **136**, 2687–2699.
19. Y. Gibon, O.E. Blaesing, J. Hannemann *et al.* (2004) A robot-based platform to measure multiple enzyme activities in Arabidopsis using a set of cycling assays: comparison of changes of enzyme activities and transcript levels during diurnal cycles and in prolonged darkness. *The Plant Cell* **16**, 3304–3325.
20. R. Alba, Z.J. Fei, P. Payton *et al.* (2004) ESTs, cDNA microarrays, and gene expression profiling: tools for dissecting plant physiology and development. *Plant Journal* **39**, 697–714.
21. N. Leonhardt, J.M. Kwak, N. Robert, D. Waner, G. Leonhardt and J.I. Schroeder (2004) Microarray expression analyses of Arabidopsis guard cells and isolation of a recessive abscisic acid hypersensitive protein phosphatase 2C mutant. *Plant Cell* **16**, 596–615.

22. J. Kehr (2003) Single cell technology. *Current Opinion in Plant Biology* **6**, 617–621.
23. S. de Folter, J. Busscher, L. Colombo, A. Losa and G.C. Angenent (2004) Transcript profiling of transcription factor genes during silique development in Arabidopsis. *Plant Molecular Biology* **56**, 351–366.
24. D. Steinhauser, B. Usadel, A. Luedemann, O. Thimm and J. Kopka (2004) CSB.DB: a comprehensive systems-biology database. *Bioinformatics* **20**, 3647–3651.
25. D. Steinhauser, B.H. Junker, A. Luedemann, J. Selbig and J. Kopka (2004) Hypothesis-driven approach to predict transcriptional units from gene expression data. *Bioinformatics* **20**, 1928–1939.
26. P. Zimmermann, M. Hirsch-Hoffmann, L. Hennig and W. Gruissem (2004) GENEVESTIGATOR. Arabidopsis microarray database and analysis toolbox. *Plant Physiology* **136**, 2621–2632.
27. M. Suistomaa, A. Kari, E. Ruokonen and J. Takala (2000) Sampling rate causes bias in APACHE II and SAPS II scores. *Intensive Care Medicine* **26**, 1773–1778.
28. H. Parkinson, U. Sarkans, M. Shojatalab *et al.* (2005) Array Express – a public repository for microarray gene expression data at the EBI. *Nucleic Acids Research* **33**, D553–D555.
29. N. Allet, N. Barrillat, T. Baussant *et al.* (2004) *In vitro* and *in silico* processes to identify differentially expressed proteins. *Proteomics* **4**, 2333–2351.
30. C.S. Goh, N. Lan, N. Echols *et al.* (2003) SPINE 2: a system for collaborative structural proteomics within a federated database framework. *Nucleic Acids Research* **31**(11), 2833–2838.
31. R Development Core Team (2004) R: A language and environment for statistical computing.
32. R.C. Gentleman, V.J. Carey, D.M. Bates *et al.* (2004) Bioconductor: open software development for computational biology and bioinformatics. *Genome Biology* **5**, R80.
33. Affymetrix (1999) Microarray Suite User Guide, Version 4. Affymetrix.
34. R.A. Irizarry, B. Hobbs, F. Collin *et al.* (2003) Exploration, normalization, and summaries of high density oligonucleotide array probe level data. *Biostatistics* **4**, 249–264.
35. F. Naef, C.R. Hacker, N. Patil and M. Magnasco (2002) Empirical characterization of the expression ratio noise structure in high-density oligonucleotide arrays. *Genome Biology* **3**, 18.01–18.11.
36. C. Li and W.H. Wong (2001) Model-based analysis of oligonucleotide arrays: expression index computation and outlier detection. *Proceedings of the National Academy of Sciences of the United States of America* **98**, 31–36.
37. K. Aggarwal and K.H. Lee (2003) Functional genomics and proteomics as a foundation for systems biology. *Brief Functional Genomics and Proteomics* **2**, 175–184.
38. C. Li and W. Wong (2001) Model-based analysis of oligonucleotide arrays: model validation, design issues and standard error application. *Genome Biology* **2**, 1–11.
39. L. Zhang, M.F. Miles and K.D. Aldape (2003) A model of molecular interactions on short oligonucleotide microarrays. *Nature Biotechnology* **21**, 818–821.
40. Z. Wu, R.A. Irizarry and R. Gentleman (2004) A model based background adjustment for oligonucleotide expression arrays. Department of Biostatistics Working Papers, Johns Hopkins University, Working Paper 1, pp. 1–26.
41. F. Naef and M.O. Magnasco (2003) Solving the riddle of the bright mismatches: labeling and effective binding in oligonucleotide arrays. *Physical Review E Statistical, Nonlinear, and Soft Matter Physics* **68**, 1–4.
42. Z. Wu and R.A. Irizarry (2004) Preprocessing of oligonucleotide array data. *Nature Biotechnology* **22**, 656–658.
43. S. Saviozzi and R.A. Calogero (2003) Microarray probe expression measures, data normalization and statistical validation. *Computational Functional Genomics* **4**, 442–446.
44. M. O'Connell (2003) Differential expression, class discovery and class prediction using S-PLUS and S+ArrayAnalyser. *Special Interest Group on Knowledge Discovery and Data Mining Explorations* **5**, 38–47.
45. Affymetrix (2003) Microarray Suite User Guide, Version 5. Affymetrix.
46. R.A. Irizarry, B.M. Bolstad, F. Collin, L.M. Cope, B. Hobbs and T.P. Speed (2003) Summaries of Affymetrix Gene Chip probe level data. *Nucleic Acids Research* **31**, e15.
47. R. Breitling, P. Armengaud, A. Amtmann and P. Herzyk (2004) Rank products: a simple, yet powerful, new method to detect differentially regulated genes in replicated microarray experiments. *Federation of European Biochemical Societies Letters* **573**, 83–92.

48. V.G. Tusher, R. Tibshirani and G. Chu (2001) Significance analysis of microarrays applied to the ionizing radiation response. *Proceedings of the National Academy of Sciences of the United States of America* **98**, 5116–5121.
49. R. Wang, K. Guegler, S.T. LaBrie and N.M. Crawford (2000) Genomic analysis of a nutrient response in Arabidopsis reveals diverse expression patterns and novel metabolic and potential regulatory genes induced by nitrate. *The Plant Cell* **12**, 1491–1509.
50. R. Schaffer, J. Landgraf, M. Accerbi, V.V. Simon, M. Larson and E. Wisman (2001) Microarray analysis of diurnal and circadian-regulated genes in Arabidopsis. *The Plant Cell* **13**, 113–123.
51. M. Seki, M. Narusaka, H. Abe *et al.* (2001) Monitoring the expression pattern of 1300 Arabidopsis genes under drought and cold stresses by using a full-length cDNA microarray. *The Plant Cell* **13**, 61–72.
52. W. Pan (2002) A comparative review of statistical methods for discovering differentially expressed genes in replicated microarray experiments. *Bioinformatics* **18**, 546–554.
53. W.J. Frawley, G. Piatetsky-Shapiro and C.J. Matheus (1991) Knowledge discovery in databases: an overview. In: *Knowledge Discovery in Databases*, (eds G. Piatetsky-Shapiro and W.J. Frawley), AAAI Press/MIT Press, Menlo Park, CA/Cambridge, MA. pp. 1–27.
54. X. Wen, S. Fuhrman, G.S. Michaels *et al.* (1998) Large-scale temporal gene expression mapping of central nervous system development. *Proceedings of the National Academy of Sciences of the United States of America* **95**, 334–339.
55. P. D'Haeseleer, S. Liang and R. Somogyi (2000) Genetic network inference: from co-expression clustering to reverse engineering. *Bioinformatics* **16**, 707–726.
56. S. Raychaudhuri, J.M. Stuart and R.B. Altman (2000) Principal components analysis to summarize microarray experiments: application to sporulation time series. *Pacific Symposium on Biocomputing*, 455–466.
57. K. Fellenberg, N.C. Hauser, B. Brors, A. Neutzner, J.D. Hoheisel and M. Vingron (2001) Correspondence analysis applied to microarray data. *Proceedings of the National Academy of Sciences of the United States of America* **98**, 10781–10786.
58. D.J. Craigon, N. James, J. Okyere, J. Higgins, J. Jotham and S. May (2004) NASCArrays: a repository for microarray data generated by NASC's transcriptomics service. *Nucleic Acids Research* **32**, 575–577.
59. L. Shen, J. Gong, R.A. Caldo *et al.* (2005) BarleyBase – an expression profiling database for plant genomics. *Nucleic Acids Research* **33**, 614–618.
60. T. Barrett, T.O. Suzek, D.B. Troup *et al.* (2005) NCBI GEO: mining millions of expression profiles – database and tools. *Nucleic Acids Research* **33**, 562–566.
61. H. Parkinson, U. Sarkans, M. Shojatalab *et al.* (2005) ArrayExpress – a public repository for microarray gene expression data at the EBI. *Nucleic Acids Research* **33**, 553–555.
62. C.A. Ball, I.A. Awad, J. Demeter *et al.* (2005) The Stanford Microarray Database accommodates additional microarray platforms and data formats. *Nucleic Acids Research* **33**, 580–582.
63. C. Ball, A. Brazma, H. Causton *et al.* (2004) An open letter on microarray data from the MGED Society. *Microbiology* **150**, 3522–3524.
64. J. Allemeersch, S. Durinck, R. Vanderhaeghen *et al.* (2005) Benchmarking the CATMA microarray. A novel tool for Arabidopsis transcriptome analysis. *Plant Physiology* **137**, 588–601.
65. G. Bloom, I.V. Yang, D. Boulware *et al.* (2004) Multi-platform, multi-site, microarray-based human tumor classification. *American Journal of Pathology* **164**, 9–16.
66. R. Breitling, A. Amtmann and P. Herzyk (2004) Iterative Group Analysis (iGA): a simple tool to enhance sensitivity and facilitate interpretation of microarray experiments. *BMC Bioinformatics* **5**, 34.
67. W. Chen, N.J. Provart, J. Glazebrook *et al.* (2002) Expression profile matrix of Arabidopsis transcription factor genes suggests their putative functions in response to environmental stresses. *The Plant Cell* **14**, 559–574.
68. T. Wang and G.D. Stormo (2003) Combining phylogenetic data with co-regulated genes to identify regulatory motifs. *Bioinformatics* **19**, 2369–2380.

69. V.J. Nikiforova, B. Gakiere, S. Kempa et al. (2004) Towards dissecting nutrient metabolism in plants: a systems biology case study on sulphur metabolism. *Journal of Experimental Botany* **55**, 1861–1870.
70. M. Ashburner, C.A. Ball, J.A. Blake et al. (2000) Gene ontology: tool for the unification of biology. The Gene Ontology Consortium. *Nature Genetics* **25**, 25–29.
71. T.Z. Berardini, S. Mundodi, L. Reiser et al. (2004) Functional annotation of the Arabidopsis genome using controlled vocabularies. *Plant Physiology* **135**, 745–755.
72. A. Ruepp, A. Zollner, D. Maier et al. (2004) The FunCat, a functional annotation scheme for systematic classification of proteins from whole genomes. *Nucleic Acids Research* **32**, 5539–5545.
73. O. Thimm, O. Blasing, Y. Gibon et al. (2004) MAPMAN: a user-driven tool to display genomics data sets onto diagrams of metabolic pathways and other biological processes. *Plant Journal* **37**, 914–939.
74. D. Martin, C. Brun, E. Remy, P. Mouren, D. Thieffry and B. Jacq (2004) GOToolBox: functional analysis of gene datasets based on Gene Ontology. *Genome Biology* **5**, R101.
75. G.F. Berriz, O.D. King, B. Bryant, C. Sander and F.P. Roth (2003) Characterizing gene sets with FuncAssociate. *Bioinformatics* **19**, 2502–2504.
76. P. Grosu, J.P. Townsend, D.L. Hartl and D. Cavalieri (2002) Pathway Processor: a tool for integrating whole-genome expression results into metabolic networks. *Genome Research* **12**, 1121–1126.
77. D. Pan, N. Sun, K.H. Cheung et al. (2003) BMC PathMAPA: a tool for displaying gene expression and performing statistical tests on metabolic pathways at multiple levels for Arabidopsis. *Boinformatics* **4**, 56.
78. S.W. Doniger, N. Salomonis, K.D. Dahlquist, K. Vranizan, S.C. Lawlor and B.R. Conklin (2003) MAPPFinder: using Gene Ontology and GenMAPP to create a global gene-expression profile from microarray data. *Genome Biology* **4**, R7.
79. H.J. Chung, M. Kim, C.H. Park, J. Kim and J.H. Kim (2004) ArrayXPath: mapping and visualizing microarray gene-expression data with integrated biological pathway resources using Scalable Vector Graphics. *Nucleic Acids Research* **32** (Web Server issue), W460–W464.
80. L.A. Mueller, P. Zhang and S.Y. Rhee (2003) AraCyc: a biochemical pathway database for Arabidopsis. *Plant Physiology* **132**, 453–460.
81. M. Kanehisa, S. Goto, S. Kawashima, Y. Okuno and M. Hattori (2004) The KEGG resource for deciphering the genome. *Nucleic Acids Research* **32**, D277–D280.
82. O. Thimm, B. Essigmann, S. Kloska, T. Altmann and T.J. Buckhout (2001) Response of Arabidopsis to iron deficiency stress as revealed by microarray analysis. *Plant Physiology* **127**, 1030–1043.
83. N. Moseyko and L.J. Feldman (2002) VIZARD: analysis of Affymetrix Arabidopsis GeneChip data. *Bioinformatics* **18**, 1264–1265.
84. Z. Hu, J. Mellor, J. Wu and C. DeLisi (2004) VisANT: an online visualization and analysis tool for biological interaction data. *BMC Bioinformatics* **5**, 17.
85. E.S. Wurtele, J. Li, L. Diao et al. (2003) MetNet: software to build and model the biogenetic lattice Arabidopsis. *Computational Functional Genomics* **4**, 239–245.
86. R. Breitling, A. Amtmann and P. Herzyk (2004) Graph-based iterative group analysis enhances microarray interpretation. *BMC Bioinformatics* **5**, 100.
87. H. Ogata, S. Goto, W. Fujibuchi and M. Kanehisa (1998) Computation with the KEGG pathway database. *Biosystems* **47**, 119–128.
88. H. Schoof, R. Ernst, V. Nazarov, L. Pfeifer, H.W. Mewes and K.F. Mayer (2004) MIPS *Arabidopsis thaliana* Database (MAtDB): an integrated biological knowledge resource for plant genomics. *Nucleic Acids Research* **32**, D373–D376.
89. H. Schoof, P. Zaccaria, H. Gundlach et al. (2002) MIPS *Arabidopsis thaliana* Database (MAtDB): an integrated biological knowledge resource based on the first complete plant genome. *Nucleic Acids Research* **30**, 91–93.
90. T. Caspar and B.G. Pickard (1989) Gravitropism in a starchless mutant of Arabidopsis: implications for the starch-statolith theory of gravity sensing. *Planta* **177**, 185–197.

91. Y. Gibon, O.E. Blasing, N. Palacios-Rojas *et al.* (2004) Adjustment of diurnal starch turnover to short days: depletion of sugar during the night leads to a temporary inhibition of carbohydrate utilization, accumulation of sugars and post-translational activation of ADP-glucose pyrophosphorylase in the following light period. *Plant Journal* **39**, 847–862.
92. R. Brouquisse, J.P. Gaudillere and P. Raymond (1998) Induction of a carbon-starvation-related proteolysis in whole maize plants submitted to light/dark cycles and to extended darkness. *Plant Physiology* **117**, 1281–1291.
93. C. Devaux, P. Baldet, J. Joubes *et al.* (2003) Physiological, biochemical and molecular analysis of sugar-starvation responses in tomato roots. *Journal of Experimental Botany* **54**, 1143–1151.
94. L.J. Sweetlove, R.L. Last and A.R. Fernie (2003) Predictive metabolic engineering: a goal for systems biology. *Plant Physiology* **132**, 420–425.
95. K.M. Oksman-Caldentey, D. Inze and M. Oresic (2004) Connecting genes to metabolites by a systems biology approach. *Proceedings of the National Academy of Sciences of the United States of America* **101**, 9949–9950.
96. M.Y. Hirai, M. Yano, D.B. Goodenowe *et al.* (2004) Integration of transcriptomics and metabolomics for understanding of global responses to nutritional stresses in *Arabidopsis thaliana*. *Proceedings of the National Academy of Sciences of the United States of America* **101**, 10205–10210.
97. D. Greenbaum, C. Colangelo, K. Williams and M. Gerstein (2003) Comparing protein abundance and mRNA expression levels on a genomic scale. *Genome Biology* **4**, 117.
98. D. Edwards and J. Batley (2004) Plant bioinformatics: from genome to phenome. *Trends in Biotechnology* **22**, 232–237.
99. B. Parvin, Q. Yang, G. Fontenay and M.H. Barcellos-Hoff (2002) BioSig: an imaging bioinformatic system for studying phenomics. *Computer* **35**, 65.
100. D. Hanisch, J. Fluck, H.T. Mevissen and R. Zimmer (2003) Playing biology's name game: identifying protein names in scientific text. *Pacific Symposium on Biocomputing*, 403–414.

2 The use of proteomics in the study of metabolic control

Lee J. Sweetlove

2.1 Introduction

Proteomics has emerged as one of the three central planks of the postgenomic landscape (the other two being transcriptomics and metabolomics) and there has been an explosion of interest and activity in the proteomic field of research. Proteomics can be defined as the systematic study of the complete protein content of an organism. Originally, the word was coined as a response to the emergent field of transcriptomics and the proteome tended to be thought of in terms of abundance of the protein products expressed by the genome [1]. That is, there was a direct linkage being made between transcription and protein abundance. As the field has developed, however, the original meaning of the term proteome has been extended to include post-transcriptional elements of the proteome: protein isoforms generated by alternative splicing and post-translational modifications, regulatory post-translational changes such as phosphorylation and the organization of proteins into complexes [2]. In some sense, any study of the properties or abundance of a protein constitutes proteomics and the word proteomics is often associated with research dealing with the behavior of a single protein or a small group of proteins. However, to most people, proteomic research is that which retains an element of systematic and global coverage of protein behavior in a cell or organism. Not withstanding the fact that technical considerations generally limit analysis to a small proportion of the total proteome, it is the intent of systematic coverage that is at the heart of proteomics. In this chapter, in which the use of proteomics to study the control of metabolic pathways and networks will be considered, proteomics will be viewed in this systematic light and a distinction will be made between proteomics and traditional studies of the properties and behavior of single proteins.

Unlike transcriptomics, which can be viewed as a mature technology, proteomics is still very much a developing method. While transcriptomics exploits sequence-specific hybridization to capture specific transcripts and gives genuinely global coverage, the lack of an equivalent capture methodology for proteins has limited the extent to which the proteome can be interrogated. Emerging approaches such as aptamers or antibodies may eventually lead to protein chips that could approach the complete proteome [3], but this approach is still in the very early stages of development. Most current proteomic research instead relies on a combination of protein fractionation depending on physicochemical properties and protein identification by mass spectrometry [4]. These methods generally give access to hundreds of proteins – a small proportion of the estimated total protein set. Despite these technical limitations, and the maturity of the transcriptomic approach, there is still much to be gained from the proteomic method in terms of investigating control of metabolic pathways and networks. While some elements of pathway flux, particularly developmentally programmed changes, are controlled at the transcriptional level [5, 6], it is thought that the majority of metabolic control mechanisms operate at the post-transcriptional level [7]. Protein abundance is a consequence not just of transcription of the appropriate gene, but also of the rate of translation of the resultant mRNA transcript [8] as well as the rate of protein turnover [9]. In addition, control of protein abundance is just one factor that controls enzyme activity – a whole host of post-translational regulatory inputs come into play to allow rapid modulation of enzyme activity in response to altered biochemical and physiological demands [10]. Proteomics is the only approach that will capture all these different levels of metabolic control and therefore, despite its current limitations, has the most potential to uncover novel aspects of mechanisms of metabolic network contr1ol.

In this chapter, the different ways in which proteomics has been used to investigate plant metabolism will be discussed. The chapter will begin with a very brief overview of the basic methodologies of proteomics and will then review a series of different proteomic approaches that have relevance to control of metabolism. The first of these is quantitative proteomics, which seeks to determine change in protein abundance in relation to different physiological or genetic perturbations. The second approach concerns the issue of protein localization, and the impact of the growing catalog of different organellar proteomes will be reviewed. Third, the issue of post-translational modifications (PTMs) of enzymes will be dealt with and the considerable potential that proteomics has to systematically identify post-translational modifications such as phosphorylation and thiol redox changes will be reviewed. Finally, the importance of protein–protein interactions and the ways in which proteomics can be used to probe protein complexes will be discussed. The chapter will conclude with a look at the future in terms of the emerging technologies that will drive the expansion of proteomics into a genuinely global approach. Throughout the chapter, the different proteomic approaches to the investigation of metabolic control will be illustrated using examples from the literature. Wherever possible, to limit the number of metabolic pathways being considered and to give a common theme throughout the chapter, the examples will be restricted to the metabolic pathways of central carbon metabolism and respiration (Figure 2.1). While this chapter will concentrate specifically on the application of proteomics to the control of primary plant metabolism, readers interested in more general aspects of plant proteomics should consult one of the many excellent reviews written on the subject [11–17].

2.2 Proteomic methodologies

The emergence of proteomics has been driven by the development of a number of specific technologies. In this section, an overview of the main experimental principles of the proteomic approach will be given. This overview will be necessarily brief as

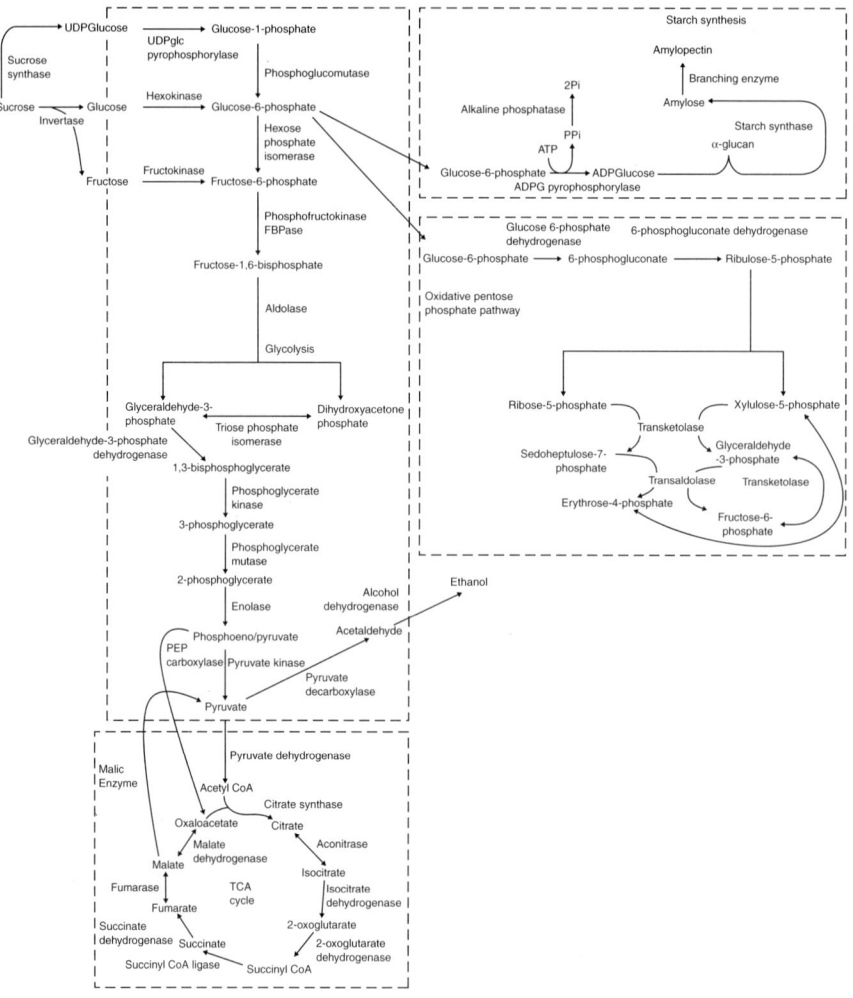

Figure 2.1 Pathways of primary carbon metabolism and respiration.

the main focus of this chapter is the application of proteomics, not the development of the technology behind it. If a more detailed description of technical aspects of proteomics is required, the reader is urged to consult one of the many reviews of this topic. The following reviews on generic aspects to proteomics methodologies are particularly well written and authoritative [4, 18–21]. In addition, the practical basis of proteomics in plants has recently been expertly reviewed [22].

2.2.1 Extraction of proteins from plant tissue

Effective and consistent extraction of proteins from plant tissue is one of the most critical and demanding steps of any proteomic experiment and yet protein extraction is frequently paid insufficient attention in the race to develop evermore sophisticated automated protein identification platforms and state-of-the-art mass spectrometers. The aim of protein extraction is to reproducibly extract, and maintain in a soluble state, the full protein complement of the tissue under consideration. Given the wide range of physicochemical properties of different proteins, this is inevitably an ideal that is rarely, if ever, achieved. Instead, protocols have been developed with a particular downstream application in mind [22]. Thus, there are specific methods designed to avoid disruption of protein complexes when fractionation of native proteins by blue-native gel electrophoresis (BN-PAGE) is desired [23]. Similarly, specific subsets of proteins can be preferentially extracted i.e. membrane proteins [24–26] and cell wall-bound proteins [27]. Besides paying attention to the physicochemical properties of the protein set to be extracted, one must also attempt to preserve the post-translational state of the protein. Thus, rapid acid extractions are required to maintain redox status of oxidizable amino acid side groups such as thiols [28]; the use of phosphatase inhibitors is necessary to preserve phosphorylation status and protease inhibitors should be used to prevent unwanted proteolysis. Whichever extraction procedure is used, it is often necessary to further clean up the sample to remove contaminants such as ribonucleotides, lipids, polysaccharides and secondary metabolites such as phenolics that interfere with subsequent protein fractionation steps. Indeed, plants can be considered the most challenging of all organisms in this respect, since they frequently contain high levels of these contaminating molecules [22]. Generally, protein precipitation is used as a convenient method both to remove such contaminants and also to concentrate the sample. The commonly used acetone/trichloroacetic acid method is extremely effective with tissues such as young leaves but can result in coextraction of polymeric compounds, particularly in more mature tissues where the content of cell wall polysaccharide and polyphenols is higher [29]. A more generically useful technique involves phase extraction in phenol followed by protein precipitation with methanol and ammonium acetate [30]. This method gives the highest quality protein extracts for subsequent gel electrophoresis and has proved capable of delivering high quality protein extracts from tissues with high polyphenolic or polysaccharide content [31, 32].

2.2.2 *Separation, display and quantification of proteins*

Once proteins have been extracted, the next challenge is to resolve and quantify as many of the different proteins within the extract as possible. By far and away the most accessible of the available technologies to do this is two-dimensional gel electrophoresis (2DGE). The boom in 2DGE has been driven by the commercialization of isoelectric gel electrophoresis strips with an immobilized pH gradient that allow highly reproducible separation of proteins according to charge. A large variety of such strips are available covering different pH ranges. Particularly important is the development of overlapping narrow-pH-range strips that allow for much greater resolution across the pH range [21]. Isoelectric focusing is generally used as the first dimension separation followed by sodium dodecyl sulfate gel electrophoresis (SDS-PAGE) as the second. Alternatively, BN-PAGE can be used for the first dimension [33]. The three methods have even been combined to provide a threedimensional separation [34]. An interesting alternative to multidimensional electrophoresis approaches is to use sucrose density gradient centrifugation to fractionate native protein complexes and then further resolve these protein fractions using standard SDS-PAGE [35].

Upon completion of the electrophoretic steps, the proteins need to be visualized. Ideally, the protein stain used should provide linear quantitation over several orders of magnitude and avoid modification of proteins in such a way that interferes with subsequent identification by mass spectrometry. Thus, it is generally preferable to avoid methods such as silver staining which oxidatively modify proteins and have a limited linear response range [21]. The most robust and accessible stain is colloidal coomassie blue [36], although more sensitive fluorescent stains are now available. A recent comparison of available protein stains revealed that the fluorescent dyes Sypro Ruby (Invitrogen Ltd) and Deep Purple (Amersham Biosciences Ltd) were approximately three times more sensitive than colloidal coomassie blue but were prone to saturation with the most abundant protein spots [37]. It is also worth bearing in mind that the use of fluorescent dyes necessitates picking spots 'blind' when it comes to extracting proteins from the gel for identification, and therefore requires an expensive robotic system linked to the fluorescent imaging system. Furthermore, the increased sensitivity of these stains often results in the detection of proteins of insufficient abundance to enable routine identification using mass spectrometry [15].

2.2.3 Identification of proteins by mass spectrometry

If the development of reproducible 2DGE approaches is considered to be one driver of the proteomic boom, the other has been the development of increasingly accurate and affordable mass spectrometers that can exploit genome sequence information to identify proteins. The basic approach is illustrated in Figure 2.2. Essentially, protein identification is achieved by digestion of the protein with a site-specific protease such as trypsin to generate a diagnostic set of peptides (known as a peptide mass fingerprint). The masses of the peptides are measured accurately with a mass spectrometer and matched to databases of predicted peptide mass fingerprints generated from translations of genomic sequences [18]. Generally, this type of analysis is done using relatively inexpensive time-of-flight or ion-trap mass spectrometers [19] with peptide ionization and vaporization being achieved with a laser (this method is generally referred to by the acronym, MALDI: matrix-assisted laser desorption ionization). This method is robust, tolerant of contaminants and readily automated. However, the results are sometimes ambiguous and the approach only works well with mixtures of proteins of extremely low complexity. Increasingly a more sophisticated approach is being adopted that uses two mass spectrometers in tandem (MS/MS) [4]. In this approach, individual peptides can be isolated from the spectrum and further fragmented by energetic collision with an inert gas to give a series of fragment ions.

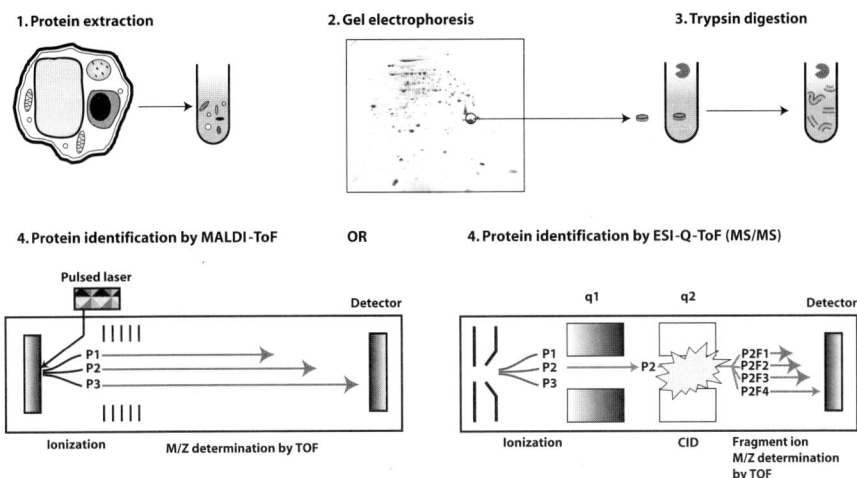

Figure 2.2 Identification of proteins by mass spectrometry. (1) Proteins are extracted from cellular material. (2) Proteins are fractionated by gel electrophoresis. (3) Proteins are excised from the gel and digested in-gel using trypsin or other sequence-specific proteases. (4) Peptides are analyzed by mass spectrometry to identify the protein. Using a single mass spectrometer (usually with a MALDI ionization source) the mass of each peptide can be determined to give a peptide mass fingerprint that is diagnostic of a particular protein. The resultant peptide mass list is searched against predicted peptide masses derived from protein databases. Alternatively a tandem mass spectrometer (MS/MS) can be used. Most commonly this is a hybrid quadrupole-time-of-flight (Q-ToF) spectrometer, although other combinations (i.e. ToF-ToF) are available. This allows a single peptide to be isolated and further fragmented through collision induced dissociation (CID). The masses of the fragment ions can be used to identify the parent peptide. This approach is useful if peptides from complex mixtures of proteins are to be analyzed.

Utilizing the same principle as peptide mass fingerprinting, the fragment ion mass pattern can be mapped back onto the genome sequence databases for protein identification. Because of the ability to isolate individual peptides and analyze them one at a time, relatively complex mixtures of peptides can be handled, allowing the identification of hundreds of proteins in a complex mixture. In addition, the method has a greater success rate at identifying proteins from nonsequenced species because some domains of proteins are more conserved from species to species. Peptides from these conserved domains often yield statistically significant identifications. Generally, when using tandem MS/MS, peptide ionization is by electrospray which is compatible with in-line liquid chromatography to provide further fractionation of the peptide sample prior to mass spectrometry. In addition, information about post-translational modifications of specific amino acid residues can also be divined [38]. While the tandem MS/MS method is undoubtedly more powerful, the hardware is significantly costlier than single mass spectrometer setups and requires specialized knowledge to operate. Nevertheless, it is clear that the majority of proteomic experiments will end up being performed using tandem MS/MS machines.

2.2.4 Gel-free proteomic approaches

A reliance on gel electrophoresis to display and quantify proteins represents a major bottleneck in the proteomic workflow since sophisticated image-analysis software is required and extraction of the required information is notoriously difficult to automate [39]. Furthermore, 2DGE that uses isoelectric focusing as a first dimension underrepresents hydrophobic proteins [24]. Therefore, much effort has been expended to develop alternative, more high-throughput and unbiased approaches to this problem. The ability of tandem mass spectrometers to analyze complex mixtures of peptides means that it is possible to analyze proteins directly in the mass spectrometer without prior separation by gel electrophoresis [4]. For very complex protein mixtures (i.e. a total cellular extract) it is still necessary to perform some prefractionation of the sample to avoid loss of resolution due to the presence of peptides of similar mass. Liquid chromatographic separations are well suited to this task and can either take the form of a single fractionation (generally using a C18 reverse-phase microbore LC column) [40] or a multidimensional chromatographic approach (sometimes referred to by the acronym MuDPIT – multidimensional protein identification technology) [41]. In fact, the latter technique is most commonly two dimensional, consisting of a strong anion exchange resin followed by C18 reverse-phase resin [42, 43]. This so-called shotgun approach to proteomics represents the current state-of-the-art in terms of throughput and can rapidly generate large catalogs of protein identity. However, quantitation within this approach is inherently complicated by variability in the efficiency of ionization of peptides in different samples. That is, the amount of a peptide peak observed in the mass spectrum can be as much influenced by ionization efficiency as by the actual amount of the peptide in the sample. A solution to this problem has emerged that is based on stable isotope labeling, the principle being that two peptides of identical chemical structure but different isotope composition are expected, according to stable isotope dilution theory, to generate the same response in a mass spectrometer [44]. Thus, an isotopically labeled reference sample can be combined with the test sample prior to mass spectrometry and this will result in a pair of peaks for each peptide that represents mass isotopomers. Since both peptides were ionized together, their relative abundances can be validly compared. There are a number of different ways of introducing isotope tags into protein populations. One method is to incubate cell cultures with isotopically labeled nutrients (i.e. ^{15}N amino acids) and compare to a control sample incubated with the equivalent light isotope (i.e. ^{14}N amino acids). This approach has been dubbed 'SILAC' for stable isotope labeling with amino acids in cell culture [45]. Alternatively, isotope labels can be attached chemically via isotope-coded affinity tags (ICATs) or similar agents [46, 47]. The tags contain a reactive group specific for particular amino acid functional groups (i.e. sulfydryls, amino groups, phosphate-ester groups and N-linked carbohydrates [4]), a linker with either light or heavy isotopes as well as an affinity tag (such as biotin) to allow purification of tagged peptides. The latter step can be useful if one is particularly interested in a particular class of proteins (i.e. cysteine-containing proteins as putative thioredoxin targets). In addition, it reduces the complexity of the mixture, facilitating subsequent mass spectrometry.

2.3 Cataloging protein localization

Over the past few years researchers in plant sciences have been applying the approaches of proteomics to a range of plant species (although the model plant species *Arabidopsis thaliana* and rice predominate due to the availability of complete genome sequence information). To date, the proteomic effort has been dominated by efforts to generate lists of proteins that are present in particular tissues, cell types or organelles. This concentration on the cataloging of proteins has in part been driven by the fact that this type of qualitative proteomics is most readily tackled by current technologies. However, there are useful biological reasons for establishing a catalog of protein localization [14]. In a metabolic context, genes that encode enzymes usually exist as small families with each member encoding different isozymes [48]. One view of the existence of such gene families is that it is representative of functional redundancy within the plant genome [49]. However, it is perhaps more likely that each isozyme fulfils specific roles in specific locations within the plant or within the cell [5]. Therefore, establishing the localization of members of enzyme families in a systematic fashion provides an essential molecular foundation upon which an understanding of metabolic control can be built. It is almost impossible to make an accurate assessment of how metabolic networks are constructed and controlled if we do not know which isozymes are present where.

2.3.1 Localizing proteins to different tissues

Studying the properties of enzymes localized in different tissues and subcellular compartments of plants is one of the mainstays of classical metabolic biochemistry. For example, an investigation into the regulatory properties of sucrose-phosphate synthase in different rice tissues allowed the identification of an isozyme specific to nonphotosynthetic tissues [50]. The proteomic approach now offers researchers the opportunity to undertake a much more systematic examination of which isozymes of enzymes accumulate in different plant tissues. This knowledge represents the first level of the information hierarchy required to establish the regulatory properties of the metabolic network present in a given location. This information could then be combined with studies of the properties of each gene product. Given that over expression of recombinant enzymes in heterologous systems such as bacteria is now routine and one-step purification of overexpressed enzymes can be achieved through the use of affinity tags, such information could potentially be generated in a relatively systematic and high-throughput manner.

To give a flavor of the type of information that can be gleaned from tissue-specific proteomic studies, the following section will consist of several examples from the literature. One of the earliest attempts to profile the proteome of different tissues was a study of green and etiolated shoots from rice [51]. Proteins were extracted from the two different tissues and fractionated using 2DGE. A small number of proteins were then identified using Edman degradation sequencing. This technique for protein identification is less efficient than the mass-spectrometry-based approaches that have superseded it; Edman degradation requires comparatively large amounts of protein and many proteins become modified during the electrophoresis steps such that they are 'blocked' from sequencing. Consequently, the dataset generated by Komatsu *et al.* [51] is rather limited and as such care has to be exercised in drawing conclusions from the list of proteins presented. Nevertheless, in the context of primary metabolism, two interesting observations were made. First, a plastidic aldolase was discovered in both green and etiolated tissues. Second, an isozyme of the glycolytic enzyme, phosphofructokinase, was also found to be present in both tissues. Other rice proteomic studies have focused on anther tissues to begin to establish which proteins are important in pollen production [52, 53]. Using 2DGE and either peptide mass fingerprinting or tandem MS/MS, Imin *et al.* [53] identified 53 proteins present in anther tissues at the young microspore stage. Among these proteins were several enzymes associated with respiratory pathways including triose-phosphate isomerase, phosphoglyceromutase, enolase, malate dehydrogenase (cytosolic), pyruvate dehydrogenase, aconitase, mitochondrial complex I subunits and mitochondrial ATP synthase subunits. The identification of these enzymes of glycolysis and respiration as particularly abundant in the anther tissues confirms the importance of energy supply for pollen formation [54]. Given this fact, it is perhaps unsurprising that mitochondrial genome mutations often lead to cytoplasmic male sterility [55]. However, when the very early stages of male gametophyte development were assessed, a different picture emerged. Here, respiratory proteins did not dominate and instead enzymes associated with sucrose breakdown and starch synthesis were found [52], corresponding to a known phase of starch accumulation in the developing pollen grain.

Besides analyzing proteome variations in tissues in different developmental contexts, attempts have also been made to look at specialized tissues such as phloem. Phloem is made up of sieve element cells and companion cells. The sieve element, although a living conduit, lacks a nucleus and many of the usual organelles and is supported metabolically by companion cells. Nevertheless, sieve element sap has been found to contain a considerable protein complement which probably originates in the companion cells and is transferred to the sieve element via plasmodesmatal connections [56]. Two recent studies have examined the nature of the proteome of extracted sieve element sap. The sieve element sap of cucumber and pumpkin phloem was found to contain a number of antioxidant enzymes [57]. A more comprehensive study of the sieve element proteome in *Ricinus communis* revealed not only antioxidant proteins, but also a number of enzymes, including phosphoglycerate mutase, phosphoglycerate kinase, enolase, malate dehydrogenase and UMP kinase [58]. The presence of these glycolytic enzymes raises the possibility that the sieve element is metabolically active in its own right. One possibility is that limited metabolism of sucrose occurs, possibly as a signalling or metabolite-sensing mechanism.

By far and away the most comprehensive study of a tissue-specific proteome of a single plant species was conducted in rice [59]. Proteins were extracted from leaves, roots and seeds and analyzed either by 2DGE followed by tandem MS/MS or by using the MuDPIT method. Over 2500 unique proteins were identified: 1002 leaf-specific proteins, 1350 root-specific proteins and 877 seed-specific proteins. In accordance with the distribution of functional classes of proteins encoded by the rice genome, metabolic proteins represented the second most abundant class of proteins in the dataset. Within this metabolic dataset, the main metabolic pathways of primary metabolism were fully represented (glycolysis/gluconeogenesis, citric acid cycle, oxidative pentose-phosphate pathway and most pathways of amino acid biosynthesis) and were found to occur within each of the three tissues analyzed. However, there were distinct tissue-specific patterns of the distribution of the isoforms of these enzymes. Some plastidial and cytosolic isozymes of glycolysis were found to be present in all tissues (aldolase, triose-phosphate isomerase, glyceraldehyde-3-phosphate dehydrogenase, phosphoglycerate kinase and phosphoglyceromutase) but the majority of metabolic enzymes were found to be tissue-specific in their localization. A case in point is ADP-glucose pyrophosphorylase (AGPase), a key enzyme of starch biosynthesis. As will be discussed in Chapter 10, the AGPase protein is a heterotetramer consisting of two small catalytic subunits and two large regulatory units [60]. The two isoforms of the small subunit were detected in both leaf and seed. However, two isoforms of the large subunit were detected exclusively in seed while the third was leaf specific. Given that the large subunit determines the allosteric properties of AGPase, this information allows one to make an assessment of the differential control of storage starch synthesis in a seed and transitory starch synthesis in a leaf.

While the various studies mentioned thus far represent pioneering efforts to establish the foundations of a plant proteome, it is clear that the coverage of the proteome generated by such studies is patchy and a long way from complete. Rice dominates in terms of species for which we have tissue-specific proteomic information and surprisingly there is very little information of this type for the other main model plant species, Arabidopsis, where most of the studies have focused on subcellular proteomes. Even in the relatively extensive rice dataset, only a few tissue types are covered and many have not been investigated. Furthermore, although tissue-specific proteomes are a useful starting point, very few tissues are homogeneous in terms of cell type. Ideally, further dissection of tissues into specific cell types is required. Such an approach has been initiated for barley seeds in which the proteomes from embryo, aleurone layer and endosperm have been investigated [61]. Clearly, there is much work to be done in this area of tissue-specific proteomics.

2.3.2 *Establishing subcellular protein localization: methodologies*

The vast majority of the research of plant proteomes to date has focused on the issue of establishing lists of proteins resident in specific organelles. So far, proteomes have been published with varying degrees of comprehensiveness for the chloroplast, amyloplast, mitochondrion, peroxisome, nucleus and vacuole as well

Table 2.1 Publicly available proteome databases

Database	Website URL
Plastid proteome database	http://ppdb.tc.cornell.edu
Arabidopsis mitochondrial proteome database	http://www.ampdb.bcs.uwa.edu.au/
Arabidopsis mitochondrial proteome project	http://www.gartenbau.uni-hannover.de/genetik/AMPP
Mitop2	http://ihg.gsf.de/mitop2/start.jsp
AraPerox (database of Arabidopsis peroxisomal proteins)	http://www.araperox.uni-goettingen.de/
Aramemnon database of membrane proteins	http://aramemnon.botanik.uni-koeln.de/
Protein GFP fusions	http://deepgreen.stanford.edu http://bioinf.scri.sari.ac.uk/cgi-bin/ProtLocDB/home
dbSubLoc	http://www.bioinfo.tsinghua.edu.cn/dbsubloc.html
Rice protein database	http://gene64.dna.affrc.go.jp/RPD

Note added in proof: Much of the information in these databases has been collated into a single database of Arabidopsis subcellular protein localization (www.suba.bcs.uwa.edu.au/).

as other subcellular locations such as the apoplast/cell wall. These subcellular proteomes are drawn exclusively from either Arabidopsis or rice. Many of the groups carrying out these investigations have deposited their protein sets in publicly available, searchable databases (Table 2.1).

There are essentially two ways of investigating subcellular localization of proteins. One involves introducing transgenes encoding proteins with tags (generally at the C-terminus of the protein to avoid interfering with protein targeting that involves an N-terminal pre-sequence) that can be visualized by microscopy. The most popular tag is green fluorescent protein (GFP) or one of its variants. The great advantage of this approach is that it can be done systematically by cloning a library of cDNAs into the appropriate vector and introducing the vectors into the organism under investigation. With a GFP tag, confocal microscopy can be used to visualize the tag at subcellular resolution. This approach has already been successfully used to systematically investigate the subcellular distribution of the yeast proteome [62, 63] and is now being applied to plants [64]. One of the main criticisms of this approach is inappropriate localization of the tagged protein, either due to the tag interfering with normal targeting or due to overexpression artifacts [65]. It should be possible to overcome the latter problem by using native promotor sequences [66]. However, the possibility of different targeting of the tagged proteins to the native protein remains and must be taken into account when interpreting the results of protein tagging experiments.

An alternative approach is to attempt to purify different organelles from the other cellular constituents and identify the proteins in the purified organelle fraction [67]. Although this approach does not offer the same throughput as the tagging method, it is less likely to generate artifactual protein localizations. However, there is an important caveat to this statement: the reliability of the organellar proteome generated in this way is entirely dependent upon the purity of the organelle preparation. Generally, organelles are purified according to their density using density gradient centrifugation. For organellar proteomics, it is usually necessary to modify and refine the method to ensure minimal contamination from other subcellular compartments [68]. However, even with the utmost care, some level of contamination is inevitable and given the increasing sensitivity of protein stains and mass spectrometers, these contaminants are more likely to be picked up. In future, it may be necessary to exploit other physicochemical properties of organelles as well as size/density to form the basis of purification. Promising possibilities include free-flow electrophoresis, which fractionates on the basis of charge [69] and immuno-affinity methods [70]. As the amount of final protein material required for identification becomes ever less with increasing sensitivity of protein identification technologies, the emphasis needs to shift away from quantity and toward quality [14].

Some would argue that the quest for an entirely pure organelle fraction is fruitless and whatever combination of purification techniques are used, some low level of contamination is unavoidable [71]. An alternative approach is to exploit quantitative information within an imperfectly pure organelle preparation. One method uses the distribution of known marker proteins throughout a density gradient [20, 71]. Cells are extracted and organelles are fractionated using a sucrose density gradient. Following centrifugation, the gradient is divided into a number of fractions and the proteins in each fraction are displayed and quantified using 2DGE. From this information, the abundance of known organelle marker enzymes across the density gradient can be plotted to give a profile of organelle distribution. Other proteins can be added to the organelle set if their abundance profile is statistically the same as the marker protein. A related approach uses ICATs to pinpoint contaminating proteins [72, 73]. The technique relies on a quantitative comparison of two organelle-enriched fractions, each of which contains cross-contaminants from the other organelle. Thus, if a quantitative comparison is made between an endoplasmic reticulum (ER)-enriched fraction contaminated with Golgi and a Golgi-enriched fraction contaminated with ER, then the genuine ER proteins will be more abundant in the former and genuine Golgi proteins will be more abundant in the latter.

2.3.3 Mitochondrial and chloroplast proteomes

Such careful quantitative approaches are relatively recent developments and require more complex experimental production lines. For this reason, nearly all the reported organellar proteome studies rely on a simple organelle purification. Moreover, the vast majority of the organelle studies to date concentrate on just two organelles – the chloroplast and the mitochondrion. At least part of the reason for this is the relative ease with which these two organelles can be isolated to a high degree of purity. However, perhaps the main motivating factor for the interest in these organelles is their central roles in carbon assimilation, energy metabolism, biosynthesis, redox homeostasis and programmed cell death. Put simply, chloroplasts and mitochondria are pivotal to the most fundamental biochemical reactions upon which plants depend. Increasingly, the breadth of processes that involve mitochondria and chloroplasts is being recognized and there is a growing interest in establishing a complete listing of the proteomes of these two organelles so that the full breadth of their physiological functions can be understood.

The very reasons that have made mitochondria and chloroplasts an attractive target for proteomics researchers had previously driven research into the function of these organelles using more classical approaches over the last few decades. In fact, they are the most intensively studied of all organelles and metabolism has historically been the main focus of attention. For this reason, despite extensive proteomic studies of mitochondria [74] and chloroplasts [75], there have been relatively few surprises relating to their metabolic proteomes. However, for both organelles, some important refinements of our appreciation of their metabolic capacity have emerged as a consequence of the proteomic investigations.

The main function of mitochondria is the production of ATP and biosynthetic precursors, and it is therefore to be expected that the enzymes associated with the citric acid cycle and the respiratory chain dominate the mitochondrial proteome. In Arabidopsis, a total of 30 proteins from the citric acid cycle and 78 from the respiratory chain have been identified [74]. The majority of these proteins are already well studied, but the proteomic investigations do offer some clarification. For example, the exact distribution of the products of the four aconitase genes in Arabidopsis was not clear. Aconitase occurs both in the cytosol and mitochondrial matrix. In yeast, a single gene encodes aconitase and an inefficient mitochondrial targeting method is believed to result in dual localization [76]. It was assumed that a similar situation might occur in plants. Proteomic studies have demonstrated that products of three of the four Arabidopsis aconitase genes reside in the mitochondrion (and moreover show evidence of post-translational modifications to produce more than three isoforms). In addition, blue-native gel electrophoresis has revealed that NAD-malic enzyme is present as a high molecular weight complex in rice mitochondria [77]. The complex corresponds to an oligomerization of the malic enzyme holoenzyme that has been observed *in vitro* and is thought to have regulatory consequences [78]. NAD-malic enzyme is a ubiquitous plant enzyme that represents an alternative metabolic route to pyruvate kinase for pyruvate formation. The presence of a high molecular weight malic enzyme complex may suggest that the enzyme is predominantly in its most active form in mitochondria from rice seedlings.

Clarification of the composition of the large protein complexes of the mitochondrial respiratory chain has also been possible, largely as a result of the use of blue-native gel electrophoresis to separate native proteins according to size [34, 77, 79–81]. In particular, complex I, which in eukaryotes is a huge complex of some 30–40 component proteins, has been studied in more detail [80]. Blue-native proteomics revealed 30 different complex I proteins in Arabidopsis and 24 in rice. A number of these proteins are highly conserved and are present in other eukaryotes. However, the study revealed a number of plant-specific proteins of complex I, including a series of ferripychelin-binding proteins as well as a number of small proteins of unknown function. Careful optimization of the extraction conditions used prior to blue-native gel electrophoresis has also revealed that there is a higher order of structural organization to the mitochondrial respiratory chain [23, 82].

Using mitochondria from a number of plant species, it was discovered that complex I (NADH dehydrogenase) and complex III (cytochrome c reductase) form a super complex with a variety of different stoichiometries. Further studies revealed that a super complex consisting of complex I, III and also complex IV (cytochrome c oxidase) exists in potato mitochondria in lower abundance. These complexes have been termed respirasomes. The fact that the component respiratory complexes can associate together with different stoichiometries suggests that the formation of super complexes could be an important mechanism that regulates the flow of electrons through mitochondrial respiratory chain.

The mitochondrial proteome has also revealed aspects of mitochondrial metabolism that were not previously well understood. For example, a study of pea leaf mitochondria noted the remarkable abundance of aldehyde dehydrogenases [83]. Some nine different aldehyde dehydrogenase proteins were observed, representing 7.5% of the total soluble mitochondrial protein. Aldehyde dehydrogenase can oxidize a broad range of aldehydes and could be involved in diverse functions from detoxification of acetaldehyde produced during fermentation [84] to the catabolism of amino acids [85]. These enzymes may also be important for normal pollen formation [86]. Another example of an underappreciated metabolic pathway that has been highlighted by proteomics is the GABA (γ-amino butyric acid) shunt, which allows the citric acid cycle to be by-passed from 2-oxoglutarate to succinate. Most of the enzymes that make up this pathway have now been identified [40, 68] and recently mutations in one of the enzymes have revealed the importance of the pathway for normal growth and development [87]. Proteomics has also proved to be useful in clarifying the exact distribution of isozymes between compartments. For example, the ascorbate-glutathione cycle has long been recognized as an important antioxidant pathway in chloroplasts and the nuclear genes encoding the chloroplast-targeted proteins have been identified [88]. However, despite biochemical evidence for the existence of the same enzymes in mitochondria [89], the genes that encode the mitochondrial isozymes have proved difficult to pin down. Mitochondrial proteomics provided a clue as to the identity of these genes by revealing that two enzymes of the ascorbate-glutathione cycle that were present in the mitochondrion were the same gene products as those present in chloroplasts [40]. Subsequent experiments demonstrated that the entire ascorbate-glutathione cycle is in fact dual-targeted between chloroplast and mitochondrion [90].

The proteome of the chloroplast has been just as intensively studied as that of the mitochondrion [75] and as is the case for the mitochondrion there are relatively few metabolic surprises. Investigations of the chloroplast proteome have paid more attention to the suborganellar compartmentation within the chloroplast and a series of papers have been published that focus on the proteomes of either the chloroplast envelope membrane [91–93] or the thylakoid (lumen or membrane-associated) [94–97]. Since most of the primary metabolic pathways of the chloroplast occur in the stroma, these studies reveal relatively little about metabolic control, although some new metabolic aspects are revealed. For example, a proteomic study of integral envelope membrane proteins identified a new phosphate (Pi)/proton transporter that was previously thought to be located in the plasma membrane [93].

The textbook version of chloroplast Pi stoichiometry has Pi entering the chloroplast in exchange for triose-phosphate via the triose-phosphate translocator. The identification of a Pi carrier that could lead to net uptake of Pi into the chloroplast driven by the envelope electrochemical potential provides new insight into the mechanisms used to maintain the stromal Pi concentration for ATP synthesis. A different study of the chloroplast envelope membrane proteome used a variety of extraction procedures to maximize proteome coverage [92]. This study identified several proteins involved in fatty acid metabolism and points to a more general involvement of envelope membrane proteins in lipid metabolism than had perhaps been previously recognized.

A recent study of the chloroplast proteome made no attempt to fractionate the different subcellular compartments of the chloroplast, but instead used a multidimensional chromatographic approach to fractionate a complete chloroplast protein extract [98]. Since this approach was unbiased in its coverage, it recovered proteins for each of the subcellular compartments including the stroma and is therefore more informative about a range of metabolic pathways. For example, the study established that a key enzyme of purine synthesis is located in the chloroplast and strongly supports a plastidic location for the synthesis of purines. Because of the depth of the coverage in this study and the reliable detection of the vast majority of known abundant proteins (such as the enzymes of the Calvin cycle) the authors argued that a failure to detect proteins suggests that they are present in very low abundance. This line of reasoning is used to identify the main metabolic activities of the chloroplast. Thus, it is argued that detection of all the enzymes of the Calvin cycle except for a ribose epimerase suggests that the main route for the regeneration of ribulose 5-phosphate is via sedoheptulose 1,7-bisphosphate and not via xylulose 5-phosphate (see Figure 2.3). Similarly, the failure to detect enzymes involved in aromatic amino acid synthesis leads the authors to suggest that the synthesis of aromatic amino acids is down-regulated in the light. While such arguments are beguiling, there is an important caveat to consider. Low abundance is one reason for a failure to detect a protein in a proteomic study of this kind, but there are several other factors that could contribute. For example, small proteins tend to be underrepresented in proteomic studies as they produce relatively few tryptic digestion products. In addition, some peptides do not ionize with high efficiency (and therefore are underrepresented in the mass spectrum) or produce low levels of ion fragments that are used to identify proteins in the MS/MS method. Therefore, the failure to identify a protein using MS/MS cannot unequivocally be attributed to low abundance.

Chloroplasts are, of course, just one of many types of plastids that occur in plants and the proteomes of other plastid types (amyloplast, chromoplast and leucoplast) have been almost entirely neglected. There is but one study in which the amyloplast proteome of wheat has been characterized [99]. Some 171 proteins were identified, either soluble or membrane-associated and the list included the expected enzymes of starch synthesis as well as a number of storage proteins such as glutenin and gliadin. However, the presence of a significant number of mitochondrial proteins highlights the well-known difficulties of obtaining amyloplasts that are uncontaminated by other organelles.

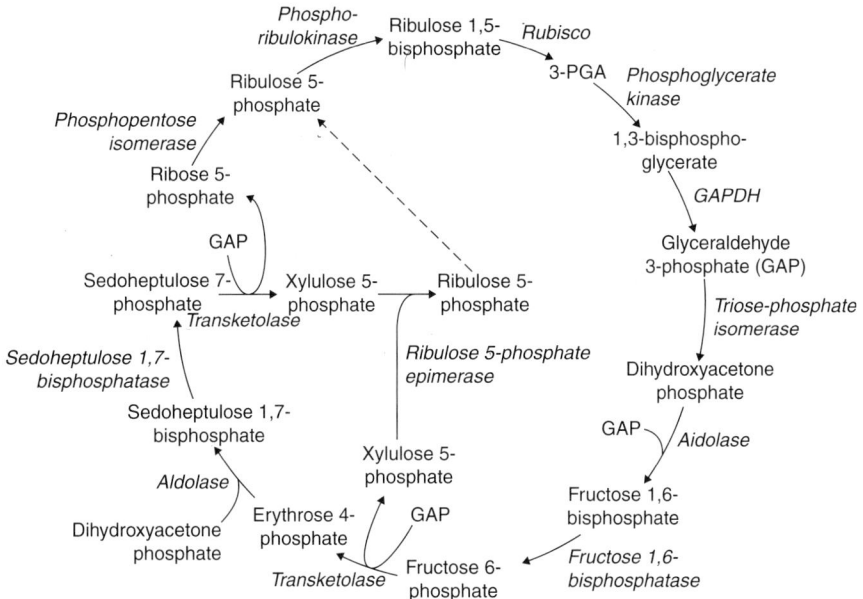

Figure 2.3 Simplified schematic of the Calvin cycle showing alternative routes for the regeneration of ribulose 5-phosphate.

2.3.4 Other subcellular proteomes

Although mitochondria and chloroplasts dominate the organellar proteome landscape, a number of studies have focused on other organelles and subcellular compartments. The peroxisome is a ubiquitous eukaryotic organelle and deserves extra attention in plants because of a diversification in its functional specialization. In addition to well-known roles in photorespiration and lipid mobilization, plant peroxisomes also have significant roles in nitrogen metabolism in root nodules [100], amino acid degradation [101] and synthesis of plant hormones such as jasmonic acid and auxin [102]. The extent of metabolic activity in plant peroxisomes is reflected by the large number of proteins that make up its proteome in comparison to peroxisomes from other eukaryotes [103]. In germinating oilseed cotyledons, the main function of peroxisomes is to catalyze the gluconoegenic mobilization of lipids to sugar via β-oxidation and glyoxylate cycle activity. Because of their functional specialization, such peroxisomes are known as glyoxysomes. An analysis of the proteome of glyoxysomes from etiolated Arabidopsis cotyledons identified the expected enzymes of β-oxidation as well as glyoxylate cycle enzymes such as malate synthase and isocitrate lyase [104]. More interestingly, however, the study also identified a NADP-specific isoform of isocitrate dehydrogenase. β-oxidation of fatty acids is dependent upon NADPH as a cofactor (the enzyme 2,4-dioenoyl CoA reductase requires stoichiometric amounts of NADPH for β-oxidation of fatty acids with double bonds at even positions [103]). Because the peroxisomal membrane is impermeable to NADPH, intraperoxisomal NADPH has to be generated using an aspartate/malate metabolite shuttle. The NADPH can be released from the reduced equivalent (malate) via conversion to isocitrate and then through the action of NADP-isocitrate dehydrogenase. The proteomic study clarifies exactly which member of the isocitrate dehydrogenase gene family in Arabidopsis encodes the peroxisomal targeted protein (At1g54 340) and improves our understanding of the NADPH generating system of the peroxisome.

The nucleus is another organelle that has been investigated at the proteome level [105, 106]. The majority of the proteins identified were unsurprisingly involved in maintenance and expression of nuclear DNA. However, besides these expected proteins the nuclear proteome contains some less obvious members, including a number of enzymes such as the glycolytic enzymes glyceraldehyde-3-phosphate dehydrogenase and phosphoglycerate kinase. The nuclear localization of certain glycolytic enzymes has also been observed in other eukaryotes [107] and has been previously observed in plants [108]. It is likely that rather than fulfilling a metabolic function, these proteins are performing secondary functions related to nuclear genome maintenance and expression [109–111]. The fact that the abundance of nuclear-localized glyceraldehyde-3-phosphate dehydrogenase and phosphoglycerate kinase increases in response to cold-stress may suggest a rapid regulatory mechanism of gene expression that involves translocation of glycolytic proteins from the cytosol to the nucleus [105, 107].

Another subcellular location that has been investigated using proteomics is the plant cell wall [27, 112]. Besides the expected proteins involved in cell wall structure, biosynthesis and degradation, these studies identified a large number of cytosolic or mitochondrial enzymes of primary carbon metabolism (such as glyceraldehyde-3-phosphate dehydrogenase, enolase, PEP carboxykinase, lactate dehydrogenase and citrate synthase). There are two views as to the presence of these cytosolic proteins in the cell wall proteome. One is that many cytoplasmic proteins bind with high affinity to the polysaccharide-rich cell wall residue during extraction and the presence of these protein should therefore be considered an artifact [22]. Alternatively, based on corroborating immunocytological data for some of these proteins (e.g. glyceraldehyde-3-phosphate dehydrogenase), the presence of these proteins can be viewed as another example of 'moonlighting' by enzymes of central metabolism. Additional experimentation is required to resolve this issue.

Finally, several studies have investigated the proteome of the tonoplast and the vacuole [113–115]. The identified proteins reveal the major metabolic activities of the vacuole from protein processing and degradation to ion transport and storage and antioxidant metabolism.

2.3.5 A stamp of authenticity for the subcellular protein postcode?

The case of the presence of unexpected proteins in an Arabidopsis cell wall fraction raises an important dilemma. Given the specter of contamination that haunts proteomic studies based on subcellular fractionation, how confidant can we be of the localization of proteins, particularly in cases where proteins are known to localize to other subcellular addresses? The issue is complicated by the fact that some organelles are more difficult to isolate to a high degree of purity than others and so the extent of contamination varies greatly between different studies. Furthermore, many proteins are legitimately localized in more than one subcellular compartment [90]. It is clear that to be more certain of the localization of a protein highlighted in a proteomic study, additional corroborating evidence is required. A good example of this is the unexpected discovery of the enzymes of glycolysis in the mitochondrial proteome of both plants [116] and animals [35]. The presence of these enzymes in the mitochondrial proteome was given credibility by additional experimentation in the plant system in which yellow-fluorescent protein tags were utilized to verify the mitochondrial localization *in vivo* [116]. Furthermore, protease-protection experiments indicated that the glycolytic enzymes were located on the outer face of the outer mitochondrial membrane [116]. This type of 'microcompartmentation' of enzymes could represent an important regulatory mechanism to control competing parallel demands on central metabolic pathways such as glycolysis: mitochondrially associated glycolysis could function exclusively to provide pyruvate for respiration, while cytosolic glycolysis could operate as a more branched pathway supplying carbon skeletons for amino acid biosynthesis and exchanging metabolites with the oxidative pentose-phosphate pathway.

An exemplary study of the yeast mitochondrial proteome points to the direction that the plant proteomic community could take to improve confidence in subcellular localization of proteins [117]. Besides conventional proteomic information, the study also drew on information from systematic screening of deletion mutants, mRNA abundance profiling, localization of tagged proteins, protein–protein interactions and computational predictions. These complementary approaches allowed the authors to identify an estimated 75% of the known yeast mitochondrial proteome. Furthermore, based on the number of pieces of corroborating evidence (and factors such as their specificity and coverage), each protein was given a score that provides a quantitative indication of the confidence of the localization.

2.4 Quantitative analyses of the proteome

The total set of expressed proteins is highly conditional – the proteome is dynamically regulated such that it is tailored to meet the physiological and biochemical demands placed on the system by different environmental and developmental variables. The extent to which the proteome is adapted to given conditions is emphasized by the extent to which the proteome varies in different ecotypes of Arabidopsis [118]. Understanding the relationship between changes in protein abundance and the metabolic status of the cell provides a valuable insight into the way in which metabolic networks are controlled. Control of the amount of an enzyme can be viewed as the first level of the regulatory hierarchy [119] with higher order levels consisting of post-translation regulatory changes and protein–protein interactions. Protein abundance is the result of a combination of gene expression, protein synthesis and protein turnover. The combined output of these regulatory mechanisms can be captured as changes in the abundance of proteins by making quantitative comparisons of the proteome as metabolism moves from one steady state to another. However, the experimental and technical demands required to execute such quantitative proteomics (see Section 2.2) have meant that relatively few quantitative proteomic studies have been attempted to date. Nevertheless, some quantitative proteomic studies are emerging and these demonstrate the value of this approach in terms of providing insight into metabolic control.

2.4.1 Examples of quantitative proteomics

One of the earliest attempts to map quantitative changes in the proteome investigated seed germination and priming in Arabidopsis [120]. Some 1300 seed proteins were resolved and quantified using 2DGE and 74 of these proteins were observed to significantly change in relative abundance (more than twofold) during seed imbibition and subsequent radicle protrusion. Some of these proteins were identified by mass spectrometry. Among the identified proteins were a number of enzymes of primary metabolism. It was found that PEP carboxylase and aconitase increased in abundance during imbibition whereas other citric acid cycle and glycolytic enzymes (citrate synthase, triose-phosphate isomerase, phosphoglycerate kinase, aldolase) remained unchanged. Given the importance of mitochondrial respiration for the germination process [121], these data may point to a key role for aconitase in the control of the citric acid cycle.

Proteomics has also been used to monitor changes in the pea leaf proteome during development [122]. In particular, the authors of this study were interested in relating proteomic changes to leaf nitrogen mobilization. The mobilization of leaf nitrogen occurs to support seed development and as a result total leaf nitrogen declines during leaf development. The work confirmed the importance of Rubisco as a nitrogen store in leaves: the amount of Rubisco (relative to total protein) decreased by as much as 60% during nitrogen mobilization and Rubisco degradation products were observed. A variety of other proteins were also reported to be increased in relative abundance. However, this observation should be treated with caution. Since Rubisco accounts for such a large proportion of leaf protein, a decrease in Rubisco will inevitably lead to an apparent increase in the relative abundance of other proteins without the absolute amounts of these proteins necessarily changing. Indeed, when calculated on a per-fresh-weight basis, these other proteins were constant in amount during leaf development.

One of the most rigorous quantitative proteomics studies conducted to date used greening maize as a system to study proteome changes during plastid biogenesis [123]. Plastids were isolated from five different developmental time points and four replicate sets of extracted protein were separated by 2DGE. Protein spots were automatically identified and quantified and gels assembled into match sets using proprietary software (PDQuest, Bio-Rad Laboratories). Normalized spot quantity across match sets was analyzed using a variety of hierarchical and non-hierarchical statistical methods to detect patterns of protein abundance profile during plastid development. One of the key findings from this study is that members of a given functional class of protein are generally coordinately regulated in expression. For example, changes in the abundance of enzymes of photosynthetic carbon assimilation were coordinated, showing a trend to increase during early development. However, the picture is not quite that simple because later in development different enzymes of carbon assimilation showed divergent abundance patterns.

Other quantitative studies of plant proteomes have concentrated not on development, but on the response to stress conditions. For example, 37 mitochondrial proteins were identified that changed in abundance in Arabidopsis cells exposed to oxidative stress [124]. These proteins fell into three classes. Nine proteins increased in abundance and represent potential candidates for components of mitochondrial antioxidant defenses. Twelve proteins were found to decrease in abundance, while 16 proteins that increased in abundance were found to be degradation products of smaller proteins. The identity of these latter two groups of proteins revealed the main proteins that are sensitive to oxidative damage as a result of the accumulation of reactive oxygen species. Enzymes of the citric acid cycle and respiratory chain complexes appear to be particularly vulnerable to oxidative damage and therefore mitochondrial oxidative stress will have an impact across the cell due to a limitation of ATP synthesis.

Quantitative proteomics has also been used to investigate the adaptive responses of plants to oxygen deficiency. A lack of oxygen inhibits mitochondrial ATP production and instead ATP is generated by fermentative pathways. Fermentation is fundamentally less efficient in producing ATP than mitochondrial oxidative phosphorylation. Furthermore, fermentation interferes with cytoplasmic pH regulation which can lead to a lethal acidification of the cytoplasm [125]. To survive oxygen deficiency, plants must therefore maintain tight control over cytoplasmic pH as well as marshal their metabolic pathways to be able to cope with a reduced supply of cellular ATP. An identification of the proteins that are present/increase in abundance during exposure to oxygen deficiency might therefore provide insight into the mechanisms by which plants survive this environmental stress condition. One such study identified 48 proteins that increased in abundance during a hypoxia pretreatment of maize roots that enhances tolerance of subsequent anoxia [126]. Besides proteins of the glycolytic and fermentative pathways involved in energy production during oxygen deficiency, proteins were identified across a wide range of functional categories including specific proteins associated with mRNA translation. A similar study looked at the response of the rice proteome to oxygen deprivation [127]. Two different types (cultivars?) of rice were used that were flooding-tolerant or flooding-intolerant (flooding leads to oxygen deprivation of submerged plant tissues). A number of enzymes involved in sugar mobilization and glycolysis were identified as key components of tolerance of oxygen deprivation due to flooding.

2.4.2 The use of high-throughput measurements of enzyme activity as a proxy for quantitative proteomics

When using a proteomic approach to investigate metabolism, ideally one would want to restrict the investigation to just those proteins (enzymes) involved in the metabolic pathways under consideration. However, by its very nature, proteomics is an unbiased approach and the proteins that will be identified and quantified will represent the full spectrum of protein classes in the plant cell. Although many enzymes are contained within the current proteome sets, there is no way to guarantee that all the enzymes of a given pathway are present [98]. Mark Stitt and colleagues [128] suggest an alternative approach that specifically targets enzymes. The idea is to use measurements of maximum catalytic activity of enzymes as a proxy for the metabolic proteome, assuming that there is a direct relationship between enzyme activity and protein abundance (although, clearly, enzyme kinetic properties can also be a factor). To enable enzyme activities to be measured on a sufficiently high-throughput to make the approach feasible, enzyme assay is automated using a robotic pipetting system. Furthermore, enzymes are assayed using extremely sensitive cycling assays. These assays require only small quantities of plant material and therefore maximize the number of enzymes that can be measured in a given sample.

Stitt and colleagues exploited this approach to investigate the relationship between the abundance of enzymes and their corresponding transcripts. Two biological scenarios in which transcript abundance and enzyme activity vary were assessed: first, changes were monitored during the diurnal cycle and second, during sugar starvation induced either by holding the plants in an extended dark period or by mutation of plastidic phosphoglucomutase. Some important findings were reported [128] and are summarized here. First, it is clear that there is a poor quantitative correlation with transcript abundance and the corresponding enzyme protein amount. Generally, the changes in protein are much smaller than the respective changes in transcript abundance. On average, protein abundance changes are damped by a factor of twofold or greater. Second, and more obviously, there is a time delay between changes in transcript abundance and corresponding enzymes. This reflects the fact that the processes of protein synthesis and turnover that regulate protein abundance are relatively slow in comparison to gene transcription. This fact is well appreciated and the time lag between transcript and protein abundance changes has been incorporated into recent mathematical models of bacterial metabolism [129]. One interpretation of the slow and damped response of protein abundance to transcript abundance changes is that it confers a combination of robustness and flexibility to the metabolic network. Robustness in that enzyme abundance only changes if there is a prolonged change in transcript. This ensures in a leaf, for example, in which changes in light, water potential and metabolite content occur on a regular, diurnal basis, excessive energy is not wasted in an inefficient process of repeated enzyme synthesis and turnover. Instead, post-transcriptional control of enzyme activity can balance the metabolic network against these changes in physiological status [130]. However, the regulatory network remains sufficiently flexible for the plant to be able to respond to its environment. Thus, while transcript abundance changes during the diurnal cycle have only a modest effect on enzyme amount within a single day, they do lead to marked changes in enzyme amount if the diurnal period is modified for several days.

2.5 The use of proteomics to investigate post-translational modification of proteins

Proteins, particularly enzymes, are extensively post-translationally modified to bring about rapid changes in their properties and localization. Post-translational modifications of enzymes that affect their kinetic properties are recognized to be one of the major mechanisms by which metabolic pathway flux is controlled. For example, the binding of effector molecules to allosteric sites allows feed-back and feed-forward regulatory mechanisms to operate that help maintain metabolic steady state. Alternatively, PTMs such as phosphorylation or the alteration of redox state of critical residues allow metabolic pathway flux to be modulated in response to environmental and biochemical stimuli. The presence of PTMs is apparent in most published 2DGE studies: multiple protein spots that are the product of the same gene are often observed. One explanation for the presence of these multiple spots is that they represent the products of alternatively spliced transcripts. However, usually the difference in apparent charge/size is sufficiently small that the presence of different PTMs is a more likely explanation. Despite the prevalence of post-translationally modified proteins and their demonstrable biological relevance, only in a very small number of cases have additional experiments been done to provide evidence for the presence of specific PTMs at specific amino acid residues [131]. Nevertheless, efforts are gathering pace to exploit the power of proteomics to begin to systematically map PTMs. There are two ways of tackling the systematic identification of PTMs. First, methods can be employed to visualize proteins that carry specific PTMs following gel electrophoresis. Although this method is within the technical reach of most laboratories, caution has to be exercised since gel electrophoresis can introduce artifactual protein modifications such as oxidation of methionine and modification of cysteine by acrylamide radicals [132] that could potentially mask *in vivo* PTMs [133]. The second approach uses mass spectrometry to identify the presence of specific PTMs at specific sites. This method is based on the observation of additional mass added to the amino acid by the PTM which can be pinpointed by allowing for variable mass additions during analysis of peptide fragment spectra. There are considerable technical challenges inherent in this approach, not least of which is the fact that many PTMs are not stable during mass spectrometry and trypsin digestion of proteins does not give coverage of the entire protein (because some tryptic peptides do not 'fly' well in the mass spectrometer).

To date, much of the systematic analysis of PTMs has focused on cataloging the 'phosphoproteome', although many other PTMs are accessible [38].

2.5.1 Systematic identification of phosphorylated proteins

Although the presence on twodimensional-gels (2D-gels) of multiple proteins that are the product of a single gene is indicative of PTMs, it gives no information as to the nature of the PTM. However, various methods exist to allow specific visualization of different PTMs. In the case of phosphorylation, a simple approach is to supply plant cells with [^{32}P]ATP. The radioactive ^{32}Pi isotope is incorporated into proteins during post-translational phosphorylation and can be visualized by autoradiography. This approach has been used to identify 14 phosphoproteins that are present in Arabidopsis mitochondria [134]. Prior to this study, the only mitochondrial protein that was known to be controlled by phosphorylation was the α-subunit of the pyruvate dehydrogenase complex and only a handful of other mitochondrial proteins had been shown to be phosphorylated. The majority of new mitochondrial phosphoproteins identified using the proteomic approach were either from the TCA cycle or the respiratory electron transport chain. In addition, several heat shock proteins were also shown to be phosphorylated as well as the mitochondrial antioxidant enzyme, superoxide dismutase. Further mass-spectrometric investigations pinpointed the phosphorylated residues within two phosphoproteins in potato mitochondria: pyruvate dehydrogenase complex and formate dehydrogenase [135].

An analogous approach to radio-labeling is to use antibodies or specific phosphoprotein stains to visualize phosphoproteins following gel electrophoresis. Both are now commercially available. However, at the time of writing, there are no documented uses of these approaches to study plant phosphoproteomes.

One of the potential problems with specific visualization of phosphoproteins following gel electrophoresis (be it by radio-labeling, antibodies or specific stains) is that the phosphoprotein display has to be related to proteins visible after staining the gel for total protein (so that the phosphorylated protein can be excised from the gel and identified by mass spectrometry). This can lead to difficulties if a low-abundance protein is strongly phosphorylated (i.e., at multiple sites). In extreme cases this can lead to a failure in finding a corresponding spot/band in the gel stained for total protein. Worse still, the low-abundance protein could be obscured by a nonphosphorylated high-abundance protein, leading to an erroneous assignment of phosphoprotein identity. To overcome these problems, an alternative approach first selects only those proteins that contain phosphorylated residues prior to fractionation and display using gel electrophoresis [136]. The approach uses an affinity chromatography step that exploits the interaction between Pi groups and metal ions immobilized on a chromatographic support. The method is known as immobilized metal-ion affinity chromatography (IMAC). The approach can be considered as one that enriches for phosphoproteins (since other negatively charged residues may also chelate metal ions) and therefore verification of the presence of phosphorylated residues by mass spectrometry is required. The IMAC method has been used to undertake the most comprehensive description of plant phosphoproteins to date [137]. The study focused on plasma membrane proteins from Arabidopsis and identified some 200 phosphoproteins. Proteins were identified using tandem MS/MS which also allowed identification of the precise sites of phosphorylation. The authors used this information to identify putative sequence motifs around phosphorylation sites that may help to predict the presence of phosphorylation sites in other proteins. The complete dataset has been made available as a searchable database (http://plantsp.sdsc.edu). From a metabolism point of view, it is interesting that several metabolite transporters were identified as phosphoproteins, raising the possibility that phosphorylation could regulate passage of metabolites across membrane boundaries. Ultimately, for a complete understanding of phosphorylation-based signal transduction cascades, protein phosphorylation needs to be observed on a time-dependent basis. In a landmark publication that demonstrates the full sophistication of the proteomic approach, Mann and colleagues used a combination of phosphotyrosine affinity purification and differential isotope labeling/quantitative mass spectrometry to undertake a comprehensive analysis of temporal changes in phosphorylation of proteins in human cells following treatment with a growth hormone (EGF) [138]. This study successfully captured the dynamics of not only all the previously known targets of the EGF receptor but also 81 signalling proteins and 31 novel effectors. This technical *tour de force* is an impressive demonstration of the power of proteomics to interrogate post-translational signalling pathways and lays down methods that could be adopted by the plant community to address similar issues in plant systems.

2.5.2 Systematic identification of protein redox modifications

Redox modifications of proteins have long been appreciated as an important mechanism by which enzyme activity is controlled (see Chapter 6 for further details). In redox-active subcellular compartments such as the mitochondrion and chloroplast, redox changes of regulatory residues link enzyme activity to electron transport processes. These redox changes are mediated by thioredoxin, a large family of thiol-containing proteins to which new members are still being added [139–141]. The control of the Calvin cycle by thioredoxin in response to electron transport in the thylakoid is one of the classic examples of post-translational redox control. However, recent proteomic studies have demonstrated that a whole host of other chloroplast enzymes are likely controlled by thioredoxin.

Two different approaches have been used to investigate novel thioredoxin protein targets in plant cells. In the first, the active site of thioredoxin is mutated such that the normally transient binding of the target protein is stabilized [142]. This means that the modified thioredoxin can be used as a capture agent in affinity chromatography and thioredoxin target proteins isolated in this way can be identified by mass spectrometry. Alternatively, purified thioredoxin/thioredoxin reductase can be used to reduce cellular proteins *in vitro*. Treated proteins are then separated by gel electrophoresis and stained with a fluorophore that is specific for reduced thiols [143]. Proteins that show evidence of having been reduced by thioredoxin can then be excised from the gel and identified by mass spectrometry. Using one or other of these two approaches, new targets for thioredoxins have been identified in the chloroplast [142, 144–146], mitochondrion [147] and other subcellular locations

[148, 149]. In addition, affinity proteomics has also been used to identify proteins that contain accessible thiols (but are not necessarily targets of thioredoxin) [28]. More details about the proteomic investigations into thioredoxin targets and their significance for redox control of metabolism can be found in Chapter 6.

In addition to changes in the redox state of protein thiols, a variety of other oxidative modifications can occur. Oxidative modifications can occur to most amino acids but among the most common are carbonylation of amine side groups, oxidation of methionine sulfur and oxidation of tryptophan [150]. Most of these modifications occur during oxidative stress conditions and can be seen as detrimental to normal protein function. There is a great deal of interest in mapping the precise effects of reactive oxygen species as well as identifying reliable molecular markers for oxidative damage. A proteomic approach that exploits the availability of antibodies to several of the known oxidized forms of amino acids could help to fulfill these objectives. Currently, little has been published on this aspect of plant proteomics but a number of studies have been initiated. For example, carbonyl groups can be identified by derivatization with dinitrophenyl hydrazine and detection of the conjugated dinitrophenyl group with specific antibodies. A preliminary study using this technique identified a number of proteins of rice leaf mitochondria that appear to be particularly sensitive to carbonyl group formation during mild oxidative stress conditions [151]. These include subunits of glycine decarboxylase, a photorespiratory enzyme that is known to be inactivated during stress conditions [152].

At this point, it is worth adding an important caveat for all proteomic analyses of PTMs: identification of a site of modification does not necessarily mean that modification at that site has a regulatory function. Additional experiments must be done to demonstrate that the PTM actually affects the functional properties of the protein in some way and can, therefore, be said to have biological relevance. While proteomics can rapidly generate lists of proteins that carry PTMs, this is only a prelude to the real task of identifying the regulatory significance of these changes.

2.6 The use of proteomics to investigate protein–protein interactions

In the context of metabolism, interactions between proteins have long been recognized as an important element of enzyme function. At the most basic level, proteins form stable interactions to build multisubunit holoenzymes. However, protein–protein interactions can also be more dynamic and have a regulatory element [10]. For example, the oligomeric state of some enzymes affects their activity [153]. Moreover, it is now widely accepted that sequential enzymes in a pathway may be physically associated with one another to form a metabolon [154]. Not only does this have the benefit of facilitating substrate channeling with associated kinetic advantages, but it also has a regulatory element. In complex, highly branched pathways, the formation of one endpoint metabolite over another can be gated by interaction of the appropriate enzymes to make a metabolon module [155]. This is known to occur in phenylpropanoid metabolism in plants and seems to be a mechanism that allows specification of one of a myriad of possible end products [156].

Besides these rather specific types of interaction, it is becoming apparent that interaction between proteins is a widespread phenomenon in eukaryotic cells and represents a major layer in the regulatory and signalling hierarchy. Several technologies have emerged in recent years that have the potential to put the protein 'interactome' within reach. These are yeast two-hybrid, tandem affinity purification (TAP) tags and use of fluorescent markers of interaction (either through fluorescence resonance energy transfer (FRET) or by the bringing together of a fluorophore split into two nonfluorescent modules).

Yeast two-hybrid has been around for many years and exploits the fact that activators of bacterial gene expression are modular (consisting of a DNA-binding domain and a gene expression activation domain). Constructs are made in which proteins to be tested are expressed with one of these two domains. If the two proteins interact, the domains are brought together and drive the expression of a reporter gene. The technique has the advantage of being scalable to address protein–protein interactions at the whole-proteome level. However, yeast two-hybrid is burdened by two main disadvantages: false positives are common and not all eukaryotic proteins that interact *in situ* will necessarily do so in the environment of the yeast nucleus [2].

Because of relatively high proportion of false positives and false negatives within the yeast two-hybrid approach, it is advisable to confirm any interaction with an alternative method. Ideally, this method should also be operable on a proteomic scale to provide an alternative systematic interactome set to the yeast two-hybrid, allowing for overlap in the two datasets to be identified [157]. Affinity chromatography is one method of assessing interactions: a bait protein is captured on an affinity resin and interacting proteins copurify. The problem in terms of scaling up this approach is the need to generate a specific affinity agent (either an antibody or the purified protein itself). However, the development of TAP tags has provided a generic protein domain that allows any tagged protein and its interacting partners to be purified using two serial affinity steps [158]. A variety of affinity domains have been used that are recognized by specific commercially available antibodies, allowing rapid affinity purification. Genome-scale analyses of protein–protein interactions in yeast have already begun in earnest and interactome sets based on affinity tags have been published [159, 160]. As of yet, relatively few examples exist in which TAP tags have been used to study protein–protein interactions in plants. However, the technique has been demonstrated to be feasible in plants [161] and several groups are currently using TAP tags to identify interacting proteins in plants.

The TAP tag method relies upon the ability of proteins to interact being maintained *in vitro*. Not only may this potentially lead to some interactions being missed, but it also presents a rather static picture. This overlooks crucial information. Protein–protein interactions are thought to be highly dynamic and it is the ability to bring proteins transiently together that confers the regulatory aspect of protein–protein interaction. Thus, the ideal method of studying protein–protein interactions would allow interactions to be investigated as they occur: *in vivo* and on a time-resolved basis. Methods using fluorescent tags are now emerging that satisfy these criteria. One method utilizes the phenomenon of FRET between compatible fluorophores. Essentially, if the two fluorophores are sufficiently close to one another, emission energy from one is transferred to excite the second. Thus, if blue fluorescent protein (BFP) and GFP are in sufficient proximity to one another, excitation of BFP will lead to green, not blue fluorescence due to FRET. Thus, fluorescent FRET pairs can be used to tag proteins and allow protein–protein interactions to be monitored using confocal microscopy *in vivo* and in real time. The disadvantage with the FRET approach is that it is technically demanding and simpler alternatives may be preferable. For example, fluorescent proteins can be split into two nonfluorescent modules. If these modules are used as protein tags, then interacting proteins will bring the two modules together and restore fluorescence. This technique, using a split yellow fluorescent protein (YFP), has been demonstrated to be feasible in plants [162]. Such techniques will undoubtedly become the method of choice for detecting protein–protein interactions between pairs of target proteins and have the potential to be scaled up to enable a systematic catalog of protein–protein interactions to be defined. However, the pairwise nature of the approach is a major disadvantage given that most protein complexes involve more than two component proteins [159]. Therefore, the ideal workflow would be to recognize all components of a complex using TAP tagging and then study interactions between component proteins *in vivo* using fluorescent tags.

2.7 Future perspectives

In this chapter, the various ways in which the proteomic approach can be utilized to study plant metabolism and metabolic control have been outlined. The ability of proteomics to address all aspects of the metabolic control architecture – from protein abundance, to protein localization (in specific tissues/cell types or at the subcellular level), to regulatory post-translational modifications to the composition of protein complexes – makes it an invaluable tool for attempts to reach a more comprehensive understanding of control of metabolic networks. However, despite its obvious potential, the ability of proteomics to make real inroads into our understanding of metabolic control is hamstrung by technical limitations.

The most obvious of these is the abundance threshold of proteomics. Whichever method is used to fractionate and identify proteins, current proteomic techniques have a clear bias toward more abundant proteins [15, 98]. Not only does this limit the scope of proteomics, but it may also lead a distorted view of the importance of a particular protein or group of proteins. For example, a characterization of different post-translational modifications (phosphorylation, oxidation and redox modification by thioredoxin) within the mitochondrial proteome reveals a similar set of proteins for each modification [74]. While there is no reason that proteins should not carry multiple post-translational modifications, it is interesting that what unites the proteins in this case is the fact that they are all abundant proteins and dominate the mitochondrial proteome. It is entirely possible that the reason that these particular proteins are the ones identified is because they are easy to visualize and identify because of their abundance. Thus, one should be wary of assuming that heat shock proteins are the most highly regulated proteins in the mitochondrial proteome just because they turn up in all the lists of proteins carrying PTMs.

There are two areas of development that address the abundance-threshold of proteomics. The first is the increasing sensitivity of mass spectrometers and the second is more reliable, routine methods to prefractionate and enrich protein samples. In combination, these two advances are moving us toward a point of being able to tackle smaller subcellular proteomes (such as those of the mitochondrion and chloroplast) in their entirety. The composite subcellular proteome modules could then be combined to give a complete cellular proteome. One remaining problem in this scenario is the lack of a cytosolic proteome. The cytosol presents a unique problem in that it is difficult to isolate a cytosolic fraction without contamination from proteins that are released from organelles ruptured during cell extraction. The cytosol may also have escaped a proteomic interrogation because it is perceived to be a less interesting compartment than the organelles. Whatever the reason, the lack of a plant cytosol proteome means that when analyzing gene families, cytosolic localization of the gene products has to be inferred by absence from organellar proteomes and the lack of a targeting presequence [48].

While improvements in mass spectrometry platforms and more sophisticated protein fractionation technologies will go a long way in improving the reach of proteomics, the capital investment required to establish such resources is high. Moreover, mass spectrometers capable of automated protein detection and identification require specialist knowledge to operate. Logistically, these issues can be solved by establishing centralized proteome resources that provide a service to users. Nevertheless, this dependence on capital expenditure and specialist knowledge limits the extent to which proteomics can become a standard part of the modern biologist's experimental arsenal. What ultimately is required is a simple platform to monitor the proteome that is analogous to microarray technology used to monitor the transcriptome. Indeed, it is the lack of an equivalent hybridization-based capture technology that has limited the development of proteomics. That said, the concept of a 'protein chip' is entirely feasible using either antibodies or aptamers (DNA or RNA molecules that have been selected from random pools based on their ability to bind to other molecules) as the affinity capture agent [163]. The main hurdles are the time and cost factors required to manufacture the required antibodies and aptamers. Nevertheless, it is clear that eventually protein chips will be commercialized and may provide a (relatively) cheap and simple alternative to the mass spectrometry-based approach. Moreover, although whole proteome chips may never be realized, the approach does at least offer a practical solution to the problem of how to monitor the abundance of defined groups of proteins [164]. From a metabolism standpoint, one could envisage a targeted protein chip on which are arrayed antibodies to the enzymes of primary carbon metabolism, for example. Protein arrays are also ideally suited to the task of systematically identifying protein binding partners [165].

While there remains much scope for improvement, it is clear that proteomics has become an essential tool to investigate the control of metabolism and promises to dramatically speed up the pace of progress over the next few years. We can look forward to a new era of metabolism in which the challenge will be to bring all the post-genomic information together in the form of testable models of the entire metabolic network of plant cells.

References

1. M. Tyers and M. Mann (2003) From genomics to proteomics. *Nature* **422**, 193–197.
2. E. Phizicky, P.I. Bastiaens, H. Zhu, M. Snyder and S. Fields (2003) Protein analysis on a proteomic scale. *Nature* **422**, 208–215.
3. A. Mirzabekov and A. Kolchinsky (2002) Emerging array-based technologies in proteomics. *Current Opinion in Chemical Biology* **6**, 70–75.
4. R. Aebersold and M. Mann (2003) Mass spectrometry-based proteomics. *Nature* **422**, 198–207.
5. S.A. Ruuska, T. Girke, C. Benning and J.B. Ohlrogge (2002) Contrapuntal networks of gene expression during Arabidopsis seed filling. *Plant Cell* **14**, 1191–1206.
6. E. Urbanczyk-Wochniak, A. Luedemann, J. Kopka *et al.* (2003) Parallel analysis of transcript and metabolic profiles: a new approach in systems biology. *EMBO Reports* **4**, 989–993.
7. B.H. ter Kuile and H.V. Westerhoff (2001) Transcriptome meets metabolome: hierarchical and metabolic regulation of the glycolytic pathway. *FEBS Letters* **500**, 169–171.
8. R. Kawaguchi and J. Bailey-Serres (2002) Regulation of translational initiation in plants. *Current Opinion in Plant Biology* **5**, 460–465.
9. M. Estelle (2001) Proteases and cellular regulation in plants. *Current Opinion in Plant Biology* **4**, 254–260.
10. W.C. Plaxton (1997) Metabolic regulation. In: *Plant Metabolism*, (eds D.T. Dennis, D.H. Turpin, D.D. Lefebvre and D.B. Layzell) Addison Wesley Longman, Harlow, UK, 2nd edn. pp. 50–68.
11. K.J. van Wijk (2001) Challenges and prospects of plant proteomics. *Plant Physiology* **126**, 501–508.
12. M. Rossignol (2001) Analysis of the plant proteome. *Current Opinion in Biotechnology* **12**, 131–134.
13. O.K. Park (2004) Proteomic studies in plants. *Journal of Biochemistry and Molecular Biology* **37**, 133–138.
14. A.H. Millar (2004) Location, location, location: surveying the intracellular real estate through proteomics in plants. *Functional Plant Biology* **31**, 563–571.
15. J.L. Heazlewood and A.H. Millar (2003) Integrated plant proteomics – putting the green genomes to work. *Functional Plant Biology* **30**, 471–482.
16. B. Kersten, L. Burkle, E.J. Kuhn *et al.* (2002) Large-scale plant proteomics. *Plant Molecular Biology* **48**, 133–141.
17. F.M. Canovas, E. Dumas-Gaudot, G. Recorbet, J. Jorrin, H.P. Mock and M. Rossignol (2004) Plant proteome analysis. *Proteomics* **4**, 285–298.
18. M. Mann, R. Hendrickson and A. Pandey (2001) Analysis of proteins and proteomes by mass spectrometry. *Annual Review of Biochemistry* **70**, 437–473.
19. R.D. Smith (2002) Trends in mass spectrometry instrumentation for proteomics. *Trends in Biotechnology* **20**, S3–S7.
20. H. Steen and M. Mann (2004) The ABC's (and XYZ's) of peptide sequencing. *Nature Reviews in Molecular and Cellular Biology* **5**, 699–711.
21. K.S. Lilley, A. Razzaq and P. Dupree (2002) Two-dimensional gel electrophoresis: recent advances in sample preparation, detection and quantitation. *Current Opinion in Chemical Biology* **6**, 46–50.
22. J.K. Rose, S. Bashir, J.J. Giovannoni, M.M. Jahn and R.S. Saravanan (2004) Tackling the plant proteome: practical approaches, hurdles and experimental tools. *Plant Journal* **39**, 715–733.

23. H. Eubel, J. Heinemeyer and H.P. Braun (2004) Identification and characterization of respirasomes in potato mitochondria. *Plant Physiology* **134**, 1450–1459.
24. V. Santoni, M. Molloy and T. Rabilloud (2000) Membrane proteins and proteomics: un amour impossible? *Electrophoresis* **21**, 1054–1070.
25. V. Santoni, T. Rabilloud, P. Doumas *et al.* (1999) Towards the recovery of hydrophobic proteins on two-dimensional electrophoresis gels. *Electrophoresis* **20**, 705–711.
26. M. Ferro, D. Seigneurin-Berny, N. Rolland *et al.* (2000) Organic solvent extraction as a versatile procedure to identify hydrophobic chloroplast membrane proteins. *Electrophoresis* **21**, 3517–3526.
27. S. Chivasa, B.K. Ndimba, W.J. Simon *et al.* (2002) Proteomic analysis of the *Arabidopsis thaliana* cell wall. *Electrophoresis* **23**, 1754–1765.
28. K. Lee, J. Lee, Y. Kim *et al.* (2004) Defining the plant disulfide proteome. *Electrophoresis* **25**, 532–541.
29. R.S. Saravanan and J.K. Rose (2004) A critical evaluation of sample extraction techniques for enhanced proteomic analysis of recalcitrant plant tissues. *Proteomics* **4**, 2522–2532.
30. Y. Meyer, J. Grosset, Y. Chartier and J.C. Cleyet-Marel (1988) Preparation by two-dimensional electrophoresis of proteins for antibody production: antibodies against proteins whose synthesis is reduced by auxin in tobacco mesophyll protoplasts. *Electrophoresis* **9**, 704–712.
31. W. Wang, M. Scali, R. Vignani *et al.* (2003) Protein extraction for two-dimensional electrophoresis from olive leaf, a plant tissue containing high levels of interfering compounds. *Electrophoresis* **24**, 2369–2375.
32. K. Vander Mijnsbrugge, H. Meyermans, M. Van Montagu, G. Bauw and W. Boerjan (2000) Wood formation in poplar: identification, characterization, and seasonal variation of xylem proteins. *Planta* **210**, 589–598.
33. H. Schagger (2001) Blue-native gels to isolate protein complexes from mitochondria. *Methods in Cell Biology* **65**, 231–244.
34. W. Werhahn and H.P. Braun (2002) Biochemical dissection of the mitochondrial proteome from *Arabidopsis thaliana* by three-dimensional gel electrophoresis. *Electrophoresis* **23**, 640–646.
35. S.W. Taylor, E. Fahy, B. Zhang *et al.* (2003) Characterization of the human heart mitochondrial proteome. *Nature Biotechnology* **21**, 281–286.
36. V. Neuhoff, N. Arold, D. Taube and W. Ehrhardt (1988) Improved staining of proteins in polyacrylamide gels including isoelectric focusing gels with clear background at nanogram sensitivity using Coomassie Brilliant Blue G-250 and R-250. *Electrophoresis* **9**, 255–262.
37. F. Chevalier, V. Rofidal, P. Vanova, A. Bergoin and M. Rossignol (2004) Proteomic capacity of recent fluorescent dyes for protein staining. *Phytochemistry* **65**, 1499–1506.
38. M. Mann and O.N. Jensen (2003) Proteomic analysis of post-translational modifications. *Nature Biotechnology* **21**, 255–261.
39. B. Kersten, T. Feilner, P. Angenendt, P. Giavalisco, W. Brenner and L. Burkle (2004) Proteomic approaches in plant biology. *Current Proteomics* **1**, 131–144.
40. J.L. Heazlewood, J.S. Tonti-Filippini, A.M. Gout, D.A. Day, J. Whelan and A.H. Millar (2004) Experimental analysis of the Arabidopsis mitochondrial proteome highlights signalling and regulatory components, provides assessment of targeting prediction programs, and indicates plant-specific mitochondrial proteins. *Plant Cell* **16**, 241–256.
41. J.P. Whitelegge (2002) Plant proteomics: BLASTing out of a MudPIT. *Proceedings of the National Academy of Sciences of the United States of America* **99**, 11564–11566.
42. D.A. Wolters, M.P. Washburn and J.R. Yates, 3rd (2001) An automated multidimensional protein identification technology for shotgun proteomics. *Analytical Chemistry* **73**, 5683–5690.
43. A.J. Link (2002) Multidimensional peptide separations in proteomics. *Trends in Biotechnology* **20**, S8–S13.
44. Y. Lu, P. Bottari, F. Turecek, R. Aebersold and M.H. Gelb (2004) Absolute quantification of specific proteins in complex mixtures using visible isotope-coded affinity tags. *Analytical Chemistry* **76**, 4104–4111.
45. S.E. Ong, B. Blagoev, I. Kratchmarova *et al.* (2002) Stable isotope labeling by amino acids in cell culture, SILAC, as a simple and accurate approach to expression proteomics. *Molecular and Cellular Proteomics* **1**, 376–386.

46. H. Zhang, W. Yan and R. Aebersold (2004) Chemical probes and tandem mass spectrometry: a strategy for the quantitative analysis of proteomes and subproteomes. *Current Opinion in Chemical Biology* **8**, 66–75.
47. M.R. Flory, T.J. Griffin, D. Martin and R. Aebersold (2002) Advances in quantitative proteomics using stable isotope tags. *Trends in Biotechnology* **20**, S23–S29.
48. P.G. Sappl, J.L. Heazlewood and A.H. Millar (2004) Untangling multi-gene families in plants by integrating proteomics into functional genomics. *Phytochemistry* **65**, 1517–1530.
49. The Arabidopsis genome initiative (2000) Analysis of the genome sequence of the flowering plant *Arabidopsis thaliana*. *Nature* **408**, 796–815.
50. G.C. Pagnussat, L. Curatti and G. Salerno (2000) Rice sucrose-phosphate synthase: identification of an isoform specific for heterotrophic tissues with distinct metabolic regulation from the mature leaf enzyme. *Physiologia Plantarum* **108**, 337–344.
51. S. Komatsu, A. Muhammad and R. Rakwal (1999) Separation and characterization of proteins from green and etiolated shoots of rice (*Oryza sativa* L.): towards a rice proteome. *Electrophoresis* **20**, 630–636.
52. T. Kerim, N. Imin, J.J. Weinman and B.G. Rolfe (2003) Proteome analysis of male gametophyte development in rice anthers. *Proteomics* **3**, 738–751.
53. N. Imin, T. Kerim, J.J. Weinman and B.G. Rolfe (2001) Characterisation of rice anther proteins expressed at the young microspore stage. *Proteomics* **1**, 1149–1161.
54. L.P. Taylor and P.K. Hepler (1997) Pollen germination and tube growth. *Annual Review of Plant Physiology and Plant Molecular Biology* **48**, 461–491.
55. P.S. Schnable and R.P. Wise (1998) The molecular basis of cytoplasmic male sterility and fertility restoration. *Trends in Plant Science* **3**, 175–180.
56. K.J. Oparka (2004) Getting the message across: how do plants exchange macromolecular complexes? *Trends in Plant Science* **9**, 33–41.
57. C. Walz, M. Juenger, M. Schad and J. Kehr (2002) Evidence for the presence and activity of a complete antioxidant defence system in mature sieve tubes. *Plant Journal* **31**, 189–197.
58. A. Barnes, J. Bale, C. Constantinidou, P. Ashton, A. Jones and J. Pritchard (2004) Determining protein identity from sieve element sap in *Ricinus communis* L. by quadrupole time of flight (Q-TOF) mass spectrometry. *Journal of Experimental Botany* **55**, 1473–1481.
59. A. Koller, M.P. Washburn, B.M. Lange *et al.* (2002) Proteomic survey of metabolic pathways in rice. *Proceedings of the National Academy of Sciences of the United States of America* **99**, 11969–11974.
60. M.A. Ballicora, Y. Fu, N.M. Nesbitt and J. Preiss (1994) ADP-glucose pyrophosphorylase form potato tubers. Site-directed mutagenesis studies of the regulatory sites. *Plant Physiology* **118**, 265–274.
61. C. Finnie, K. Maeda, O. Ostergaard, K.S. Bak-Jensen, J. Larsen and B. Svensson, (2004) Aspects of the barley seed proteome during development and germination. *Biochemical Society Transactions* **32**, 517–519.
62. W.K. Huh, J.V. Falvo, L.C. Gerke *et al.*, (2003) Global analysis of protein localization in budding yeast. *Nature* **425**, 686–691.
63. A. Kumar, S. Agarwal, J.A. Heyman *et al.* (2002) Subcellular localization of the yeast proteome. *Genes and Development* **16**, 707–719.
64. N.M. Escobar, S. Haupt, G. Thow, P. Boevink, S. Chapman and K. Oparka (2003) High-throughput viral expression of cDNA-green fluorescent protein fusions reveals novel subcellular addresses and identifies unique proteins that interact with plasmodesmata. *Plant Cell* **15**, 1507–1523.
65. A. Sickmann, J. Reinders, Y. Wagner *et al.* (2003) The proteome of *Saccharomyces cerevisiae* mitochondria. *Proceedings of the National Academy of Sciences of the United States of America* **100**, 13207–13212.
66. G.W. Tian, A. Mohanty, S.N. Chary *et al.* (2004) High-throughput fluorescent tagging of full-length Arabidopsis gene products in planta. *Plant Physiology* **135**, 25–38.
67. S. Taylor, E. Fahy and S. Ghosh (2003) Global organellar proteomics. *Trends in Biotechnology* **21**, 82–88.

68. A.H. Millar, L.J. Sweetlove, P. Giege and C.J. Leaver (2001) Analysis of the Arabidopsis mitochondrial proteome. *Plant Physiology* **127**, 1711–1727.
69. H. Zischka, G. Weber, P.J. Weber *et al.* (2003) Improved proteome analysis of *Saccharomyces cerevisiae* mitochondria by free-flow electrophoresis. *Proteomics* **3**, 906–916.
70. R.R. Burgess and N.E. Thompson (2002) Advances in gentle immunoaffinity chromatography. *Current Opinion in Biotechnology* **13**, 304–308.
71. T.A. Prime, D.J. Sherrier, P. Mahon, L.C. Packman and P. Dupree (2000) A proteomic analysis of organelles from *Arabidopsis thaliana*. *Electrophoresis* **21**, 3488–3499.
72. T.P. Dunkley, P. Dupree, R.B. Watson and K.S. Lilley (2004) The use of isotope-coded affinity tags (ICAT) to study organelle proteomes in *Arabidopsis thaliana*. *Biochemical Society Transactions* **32**, 520–523.
73. T.P. Dunkley, R. Watson, J.L. Griffin, P. Dupree and K.S. Lilley (2004) Localization of organelle proteins by isotope tagging (LOPIT). *Molecular and Cellular Proteomics* **3**, 1128–1134.
74. A.H.M. Millar, J.L. Heazlewood, B.K. Kristensen, H.P. Braun and I.M. Moller (2004) The plant mitochondrial proteome. *Trends in Plant Science* **10**, 36–43.
75. P. Jarvis (2004) Organellar proteomics: chloroplasts in the spotlight. *Current Biology* **18**, 317–319.
76. S.P. Gangloff, D. Marguet and G.J. Lauquin (1990) Molecular cloning of the yeast mitochondrial aconitase gene (ACO1) and evidence of a synergistic regulation of expression by glucose plus glutamate. *Molecular and Cellular Biology* **10**, 3551–3561.
77. J.L. Heazlewood, K.A. Howell, J. Whelan and A.H. Millar (2003) Towards an analysis of the rice mitochondrial proteome. *Plant Physiology* **132**, 230–242.
78. S.D. Grover and R.T. Wedding (1982) Kinetic ramifications of the association-dissociation behaviour of NAD-malic enzyme. *Plant Physiology* **70**, 1169–1172.
79. V. Kruft, H. Eubel, L. Jansch, W. Werhahn and H.P. Braun (2001) Proteomic approach to identify novel mitochondrial proteins in Arabidopsis. *Plant Physiology* **127**, 1694–1710.
80. J.L. Heazlewood, K.A. Howell and A.H. Millar (2003) Mitochondrial complex I from Arabidopsis and rice: orthologs of mammalian and fungal components coupled with plant-specific subunits. *Biochimica et Biophysica Acta* **1604**, 159–169.
81. P. Giegé, L.J. Sweetlove and C.J. Leaver (2003) Identification of mitochondrial protein complexes in Arabidopsis using two-dimensional blue-native polyacrylamide gel electrophoresis. *Plant Molecular Biology Reporter* **21**, 133–144.
82. H. Eubel, L. Jansch and H.P. Braun (2003) New insights into the respiratory chain of plant mitochondria. Supercomplexes and a unique composition of complex II. *Plant Physiology* **133**, 274–286.
83. J. Bardel, M. Louwagie, M. Jaquinod *et al.* (2002) A survey of the plant mitochondrial proteome in relation to development. *Proteomics* **2**, 880–898.
84. O. Kursteiner, I. Dupuis and C. Kuhlemeier (2003) The pyruvate decarboxylase1 gene of Arabidopsis is required during anoxia but not other environmental stresses. *Plant Physiology* **132**, 968–978.
85. G.W. Goodwin, P.M. Rougraff, E.J. Davis and R.A. Harris (1989) Purification and characterization of methylmalonate-semialdehyde dehydrogenase from rat liver. Identity to malonate-semialdehyde dehydrogenase. *Journal of Biological Chemistry* **264**, 14965–14971.
86. F. Liu, X. Cui, H.T. Horner, H. Weiner and P.S. Schnable (2001) Mitochondrial aldehyde dehydrogenase activity is required for male fertility in maize. *Plant Cell* **13**, 1063–1078.
87. N. Bouche, A. Fait, D. Bouchez, S.G. Moller and H. Fromm (2003) Mitochondrial succinic-semialdehyde dehydrogenase of the γ-aminobutyrate shunt is required to restrict levels of reactive oxygen intermediates in plants. *Proceedings of the National Academy of Sciences of the United States of America* **100**, 6843–6848.
88. G. Noctor and C.H. Foyer (1998) Ascorbate and glutathione: keeping active oxygen under control. *Annual Review of Plant Physiology and Plant Molecular Biology* **49**, 249–279.
89. A. Jimenez, J.A. Hernandez, L.A. delRio and F. Sevilla (1997) Evidence for the presence of the ascorbate-glutathione cycle in mitochondria and peroxisomes of pea leaves. *Plant Physiology* **114**, 275–284.

90. O. Chew, J. Whelan and A.H. Millar (2003) Molecular definition of the ascorbate-glutathione cycle in Arabidopsis mitochondria reveals dual targeting of antioxidant defenses in plants. *Journal of Biological Chemistry* **278**, 46869–46877.
91. J.E. Froehlich, C.G. Wilkerson, W.K. Ray *et al.* (2003) Proteomic study of the *Arabidopsis thaliana* chloroplastic envelope membrane utilizing alternatives to traditional two-dimensional electrophoresis. *Journal of Proteome Research* **2**, 413–425.
92. M. Ferro, D. Salvi, S. Brugiere *et al.* (2003) Proteomics of the chloroplast envelope membranes from *Arabidopsis thaliana*. *Molecular and Cellular Proteomics* **2**, 325–334.
93. M. Ferro, D. Salvi, H. Riviere-Rolland *et al.* (2002) Integral membrane proteins of the chloroplast envelope: identification and subcellular localization of new transporters. *Proceedings of the National Academy of Sciences of the United States of America* **99**, 11487–11492.
94. G. Friso, L. Giacomelli, A.J. Ytterberg *et al.* (2004) In-depth analysis of the thylakoid membrane proteome of *Arabidopsis thaliana* chloroplasts: new proteins, new functions, and a plastid proteome database. *Plant Cell* **16**, 478–499.
95. M. Schubert, U. Petersson, B. Haas, C. Funk, W. Schroder and T. Kieselbach (2002) Proteome map of the chloroplast lumen of *Arabidopsis thaliana*. *Journal of Biological Chemistry* **277**, 8354–8365.
96. J.B. Peltier, G. Friso, D.E. Kalume *et al.* (2000) Proteomics of the chloroplast: systematic identification and targeting analysis of lumenal and peripheral thylakoid proteins [see comments]. *Plant Cell* **12**, 319–341.
97. J.B. Peltier, O. Emanuelsson, D.E. Kalume *et al.* (2002) Central functions of the lumenal and peripheral thylakoid proteome of Arabidopsis determined by experimentation and genome-wide prediction. *Plant Cell* **14**, 211–236.
98. T. Kleffmann, D. Russenberger, A. von Zychlinski *et al.* (2004) The *Arabidopsis thaliana* chloroplast proteome reveals pathway abundance and novel protein functions. *Current Biology* **14**, 354–362.
99. N.L. Andon, S. Hollingworth, A. Koller, A.J. Greenland, J.R. Yates and P.A. Haynes (2002) Proteomic characterization of wheat amyloplasts using identification of proteins by tandem mass spectrometry. *Proteomics* **2**, 1156–1168.
100. D.P.S. Verma (2002) Peroxisome biogenesis in root nodules and assimilation of symbiotically-reduced nitrogen in tropical legumes. In: *Plant Peroxisomes: Biochemistry, Cell Biology and Biotechnological Applications,* (eds A. Baker and I.A. Graham), pp. 191–220. Kluwer Academic, Dordrecht.
101. B.K. Zolman, M. Monroe-Augustus, B. Thompson *et al.* (2001) *chy1*, an Arabidopsis mutant with impaired b-oxidation, is defective in a peroxisomal b-hydroxyisobutyryl-CoA hydrolase. *Journal of Biological Chemistry* **276**, 31037–31046.
102. I. Feussner and C. Wasternack (2002) The lipoxygenase pathway. *Annual Review of Plant Biology* **53**, 275–297.
103. S. Reumann, C. Ma, S. Lemke and L. Babujee (2004) AraPerox. A database of putative Arabidopsis proteins from plant peroxisomes. *Plant Physiology* **136**, 2587–2608.
104. Y. Fukao, M. Hayashi, I. Hara-Nishimura and M. Nishimura (2003) Novel glyoxysomal protein kinase, GPK1, identified by proteomic analysis of glyoxysomes in etiolated cotyledons of *Arabidopsis thaliana*. *Plant and Cell Physiology* **44**, 1002–1012.
105. M.S. Bae, E.J. Cho, E.Y. Choi and O.K. Park (2003) Analysis of the Arabidopsis nuclear proteome and its response to cold stress. *Plant Journal* **36**, 652–663.
106. T.T. Calikowski, T. Meulia and I. Meier (2003) A proteomic study of the Arabidopsis nuclear matrix. *Journal of Cellular Biochemistry* **90**, 361–378.
107. Z. Dastoor and J.L. Dreyer (2001) Potential role of nuclear translocation of glyceraldehyde-3-phosphate dehydrogenase in apoptosis and oxidative stress. *Journal of Cell Science* **114**, 1643–1653.
108. L.E. Anderson, X. Wang and J.T. Gibbons (1995) Three enzymes of carbon metabolism or their antigenic analogs in pea leaf nuclei. *Plant Physiology* **108**, 659–667.
109. C.J. Jeffery (1999) Moonlighting proteins. *Trends in Biochemical Science* **24**, 8–11.
110. C.J. Jeffery (2003) Moonlighting proteins: old proteins learning new tricks. *Trends in Genetics* **19**, 415–417.

111. M.A. Sirover (1999) New insights into an old protein: the functional diversity of mammalian glyceraldehyde-3-phosphate dehydrogenase. *Biochimica et Biophysica Acta* **1432**, 159–184.
112. A.R. Slabas, B. Ndimba, W.J. Simon and S. Chivasa (2004) Proteomic analysis of the Arabidopsis cell wall reveals unexpected proteins with new cellular locations. *Biochemical Society Transactions* **32**, 524–528.
113. T. Shimaoka, M. Ohnishi, T. Sazuka *et al.* (2004) Isolation of intact vacuoles and proteomic analysis of tonoplast from suspension-cultured cells of *Arabidopsis thaliana*. *Plant and Cell Physiology* **45**, 672–683.
114. W. Szponarski, N. Sommerer, J.C. Boyer, M. Rossignol and R. Gibrat (2004) Large-scale characterization of integral proteins from Arabidopsis vacuolar membrane by two-dimensional liquid chromatography. *Proteomics* **4**, 397–406.
115. C. Carter, S. Pan, J. Zouhar, E.L. Avila, T. Girke and N.V. Raikhel (2004) The vegetative vacuole proteome of *Arabidopsis thaliana* reveals predicted and unexpected proteins. *Plant Cell* **16**, 3285–3303.
116. P. Giege, J.L. Heazlewood, U. Roessner-Tunali *et al.* (2003) Enzymes of glycolysis are functionally associated with the mitochondrion in Arabidopsis cells. *Plant Cell* **15**, 2140–2151.
117. H. Prokisch, C. Scharfe, D.G. Camp, 2nd *et al.* (2004) Integrative analysis of the mitochondrial proteome in yeast. *PLoS Biology* **2**, 795–804.
118. F. Chevalier, O. Martin, V. Rofidal *et al.* (2004) Proteomic investigation of natural variation between Arabidopsis ecotypes. *Proteomics* **4**, 1372–1381.
119. Z.N. Oltvai and A.L. Barabasi (2002) Systems biology. Life's complexity pyramid. *Science* **298**, 763–764.
120. K. Gallardo, C. Job, S.P. Groot *et al.* (2001) Proteomic analysis of Arabidopsis seed germination and priming. *Plant Physiology* **126**, 835–848.
121. D.C. Logan, A.H. Millar, L.J. Sweetlove, S.A. Hill and C.J. Leaver (2001) Mitochondrial biogenesis during germination in maize embryos. *Plant Physiology* **125**, 662–672.
122. S. Schiltz, K. Gallardo, M. Huart, L. Negroni, N. Sommerer and J. Burstin (2004) Proteome reference maps of vegetative tissues in pea. An investigation of nitrogen mobilization from leaves during seed filling. *Plant Physiology* **135**, 2241–2260.
123. P.M. Lonosky, X. Zhang, V.G. Honavar, D.L. Dobbs, A. Fu and S.R. Rodermel (2004) A proteomic analysis of maize chloroplast biogenesis. *Plant Physiology* **134**, 560–574.
124. L.J. Sweetlove, J.L. Heazlewood, V. Herald *et al.* (2002) The impact of oxidative stress on Arabidopsis mitochondria. *Plant Journal* **32**, 891–904.
125. J.K.M. Roberts, J. Callis, O. Jardetzky, V. Walbot and M. Freeling (1984) Cytoplasmic acidosis as a determinant of flooding tolerance in plants. *Proceedings of the National Academy of Sciences of the United States of America* **81**, 6029–6033.
126. W.W. Chang, L. Huang, M. Shen, C. Webster, A.L. Burlingame and J.K. Roberts (2000) Patterns of protein synthesis and tolerance of anoxia in root tips of maize seedlings acclimated to a low-oxygen environment, and identification of proteins by mass spectrometry. *Plant Physiology* **122**, 295–318.
127. H. Dubey, G. Bhatia, S. Pasha and A. Grover (2003) Proteome maps of flood-tolerant FR 13A and flood-sensitive IR 54 rice types depicting proteins associated with O2-deprivation stress and recovery regimes. *Current Science* **84**, 83–89.
128. Y. Gibon, O.E. Blaesing, J. Hannemann *et al.* (2004) A robot-based platform to measure multiple enzyme activities in Arabidopsis using a set of cycling assays: comparison of changes of enzyme activities and transcript levels during diurnal cycles and in prolonged darkness. *Plant Cell* **16**, 3304–3325.
129. M.W. Covert and B.O. Palsson (2002) Transcriptional regulation in constraints-based metabolic models of *Escherichia coli*. *Journal of Biological Chemistry* **277**, 28058–28064.
130. J.H. Hendriks, A. Kolbe, Y. Gibon, M. Stitt and P. Geigenberger (2003) ADP-glucose pyrophosphorylase is activated by posttranslational redox-modification in response to light and to sugars in leaves of Arabidopsis and other plant species. *Plant Physiology* **133**, 838–849.
131. E. Wagner, S. Luche, L. Penna *et al.* (2002) A method for detection of overoxidation of cysteines: peroxiredoxins are oxidized *in vivo* at the active-site cysteine during oxidative stress. *Biochemical Journal* **366**, 777–785.

132. K.M. Swiderek, M.T. Davis and T.D. Lee (1998) The identification of peptide modifications derived from gel-separated proteins using electrospray triple quadrupole and ion trap analyses. *Electrophoresis* **19**, 989–997.
133. A. Sickmann, M. Mreyen and H.E. Meyer (2002) Identification of modified proteins by mass spectrometry. *IUBMB Life* **54**, 51–57.
134. N.V. Bykova, H. Egsgaard and I.M. Moller (2003) Identification of 14 new phosphoproteins involved in important plant mitochondrial processes. *FEBS Letters* **540**, 141–146.
135. N.V. Bykova, A. Stensballe, H. Egsgaard, O.N. Jensen and I.M. Moller (2003) Phosphorylation of formate dehydrogenase in potato tuber mitochondria. *Journal of Biological Chemistry* **278**, 26021–26030.
136. N. Imam-Sghiouar, R. Joubert-Caron and M. Caron (2005) Application of metal-chelate affinity chromatography to the study of the phosphoproteome. *Amino Acids* **28**, 105–109.
137. T.S. Nuhse, A. Stensballe, O.N. Jensen and S.C. Peck (2004) Phosphoproteomics of the Arabidopsis plasma membrane and a new phosphorylation site database. *Plant Cell*, **16**, 2394–2405.
138. B. Blagoev, S.E. Ong, I. Kratchmarova and M. Mann (2004) Temporal analysis of phosphotyrosine-dependent signalling networks by quantitative proteomics. *Nature Biotechnology* **22**, 1139–1145.
139. V. Collin, P. Lamkemeyer, M. Miginiac-Maslow *et al.* (2004) Characterization of plastidial thioredoxins from Arabidopsis belonging to the new y-type. *Plant Physiology* **136**, 4088–4095.
140. E. Gelhaye, N. Rouhier, J. Gerard *et al.* (2004) A specific form of thioredoxin h occurs in plant mitochondria and regulates the alternative oxidase. *Proceedings of the National Academy of Sciences of the United States of America* **101**, 14545–14550.
141. C. Laloi, N. Rayapuram, Y. Chartier, J.M. Grienenberger, G. Bonnard and Y. Meyer (2001) Identification and characterization of a mitochondrial thioredoxin system in plants. *Proceedings of the National Academy of Sciences of the United States of America* **98**, 14144–14149.
142. K. Motohashi, A. Kondoh, M.T. Stumpp and T. Hisabori (2001) Comprehensive survey of proteins targeted by chloroplast thioredoxin. *Proceedings of the National Academy of Sciences of the United States of America* **98**, 11224–11229.
143. H. Yano, S. Kuroda and B.B. Buchanan (2002) Disulfide proteome in the analysis of protein function and structure. *Proteomics* **2**, 1090–1096.
144. C. Marchand, P. Le Marechal, Y. Meyer, M. Miginiac-Maslow, E. Issakidis-Bourguet and P. Decottignies (2004) New targets of Arabidopsis thioredoxins revealed by proteomic analysis. *Proteomics* **4**, 2696–2706.
145. P. Rey, S. Cuine, F. Eymery *et al.* (2005) Analysis of the proteins targeted by CDSP32, a plastidic thioredoxin participating in oxidative stress responses. *Plant Journal* **41**, 31–42.
146. Y. Balmer, A. Koller, G. Del Val, W. Manieri, P. Schurmann and B.B. Buchanan (2003) Proteomics gives insight into the regulatory function of chloroplast thioredoxins. *Proceedings of the National Academy of Sciences of the United States of America* **100**, 370–375.
147. Y. Balmer, W.H. Vensel, C.K. Tanaka *et al.* (2004) Thioredoxin links redox to the regulation of fundamental processes of plant mitochondria. *Proceedings of the National Academy of Sciences of the United States of America* **101**, 2642–2647.
148. J.H. Wong, N. Cai, Y. Balmer *et al.* (2004) Thioredoxin targets of developing wheat seeds identified by complementary proteomic approaches. *Phytochemistry* **65**, 1629–1640.
149. H. Yano, J.H. Wong, Y.M. Lee, M.J. Cho and B.B. Buchanan (2001) A strategy for the identification of proteins targeted by thioredoxin. *Proceedings of the National Academy of Sciences of the United States of America* **98**, 4794–4799.
150. B. Halliwell and J.M.C. Gutteridge (1999) *Free Radicals in Biology and Medicine,* 3rd edn. Oxford University Press, Oxford.
151. B.K. Kristensen, P. Askerlund, N.V. Bykova, H. Egsgaard and I.M. Moller (2004) Identification of oxidised proteins in the matrix of rice leaf mitochondria by immunoprecipitation and two-dimensional liquid chromatography-tandem mass spectrometry. *Phytochemistry* **65**, 1839–1851.
152. N.L. Taylor, D.A. Day and A.H. Millar (2002) Environmental stress causes oxidative damage to plant mitochondria leading to inhibition of glycine decarboxylase. *Journal of Biological Chemistry* **277**, 42663–42668.

153. S.D. Grover and R.T. Wedding (1984) Modulation of the activity of NAD malic enzyme from *Solanum tuberosum* by changes in oligomeric state. *Archives of Biochemistry and Biophysics* **234**, 418–425.
154. P.A. Srere (1987) Complexes of sequential metabolic enzymes. *Annual Review of Biochemistry* **56**, 89–124.
155. J. Ovadi and P.A. Srere (2000) Macromolecular compartmentation and channeling. *International Review of Cytology* **192**, 255–280.
156. B.S.J. Winkel (2004) Metabolic channeling in plants. *Annual Review of Plant Biology* **55**, 85–107.
157. C. von Mering, R. Krause, B. Snel *et al.* (2002) Comparative assessment of large-scale data sets of protein-protein interactions. *Nature* **417**, 399–403.
158. A. Bauer and B. Kuster (2003) Affinity purification-mass spectrometry. Powerful tools for the characterization of protein complexes. *European Journal of Biochemistry* **270**, 570–578.
159. A.C. Gavin, M. Bosche, R. Krause *et al.* (2002) Functional organization of the yeast proteome by systematic analysis of protein complexes. *Nature* **415**, 141–147.
160. Y. Ho, A. Gruhler, A. Heilbut *et al.* (2002) Systematic identification of protein complexes in *Saccharomyces cerevisiae* by mass spectrometry. *Nature* **415**, 180–183.
161. J.S. Rohila, M. Chen, R. Cerny and M.E. Fromm (2004) Improved tandem affinity purification tag and methods for isolation of protein heterocomplexes from plants. *Plant Journal* **38**, 172–181.
162. M. Walter, C. Chaban, K. Schutze *et al.* (2004) Visualization of protein interactions in living plant cells using bimolecular fluorescence complementation. *Plant Journal* **40**, 428–438.
163. D.S. Wilson and S. Nock (2002) Functional protein microarrays. *Current Opinion in Chemical Biology* **6**, 81–85.
164. B. Schweitzer, S. Roberts, B. Grimwade *et al.* (2002) Multiplexed protein profiling on microarrays by rolling-circle amplification. *Nature Biotechnology* **20**, 359–365.
165. H. Zhu, M. Bilgin, R. Bangham *et al.* (2001) Global analysis of protein activities using proteome chips. *Science* **293**, 2101–2105.

3 Study of metabolic control in plants by metabolomics

Oliver Fiehn

3.1 Introduction

3.1.1 What is metabolomics?

The idea of 'metabolomics' has been coined and developed in the last decade to comprehensively study metabolism under genetic and environmental perturbations [1, 2]. However, the first papers involving metabolite profiling techniques were published well over 30 years ago, with the aim, at that time, of rapid medical diagnostics [3]. The underlying idea behind the use of metabolomics in plant biology today is to detect metabolic effects of genetic or environmental perturbation which may only distantly relate to known or presumed primary (enzymatic) alterations. Metabolomics, therefore, seeks to detect 'unexpected' events on a comprehensive scale, and it widely acknowledges the presence of novel metabolites with unknown chemical structure or biological function.

In this respect, it differs from classical control theory that has been applied more frequently to select well-known pathways or regulatory circuits with the objective to understand these pathways in a mathematical manner using well-defined models and assumptions. Usually, mathematical control models need to be supported by high level metabolite measurements such as flux data. Although some efforts have been reported to derive larger metabolic models from isotope calculations of protein hydrolysates, we are still far away from reaching the goal universal and global 'fluxome' [4] analysis, especially with regard to plant research. Metabolomics does not try to reach this goal. Its use in studying metabolism has so far been more of an observatory and confirmatory role. It aims less at directly deriving insights into the cellular organization of metabolism. Due to its power to detect broad classes of metabolites, including unknowns, metabolomics is best used for studying system properties (such as networks) and changes (control) of metabolite levels in disparate parts of metabolism.

In many studies involving genetic or environmental perturbations it appears that certain metabolic modules are in counter balance with others, such as sugars and amino acids (C/N balance), whereas other large modules such as lipid metabolism are less affected (or more tightly regulated) under these conditions. Hence, metabolomics may be best suited to identify which broader parts of metabolism are influenced in response to developmental, genetic or environmental changes. Data can then be used to generate novel hypotheses about potential cellular causes (changes in enzymatic or transport activities) that are responsible for such changes.

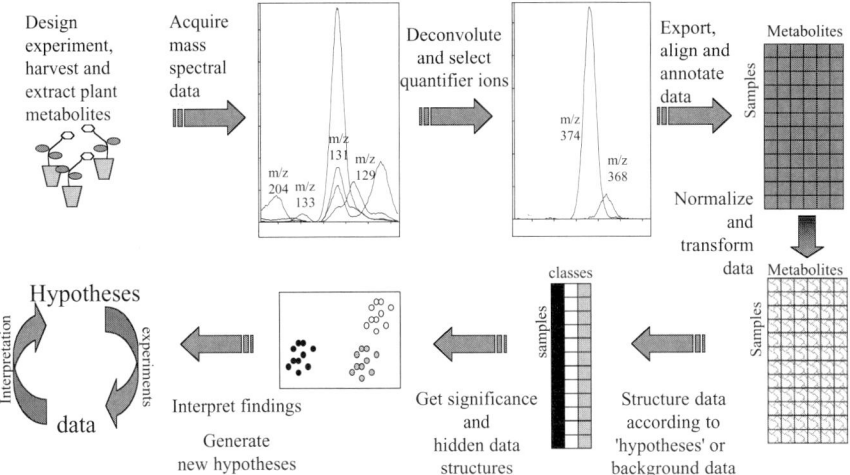

Figure 3.1 Flowchart of a plant metabolomic experiment.

Therefore, metabolomics will frequently generate more questions than answers, a concept that still needs to be embraced by classical hypothesis-driven research.

A general outline of this idea is depicted as Figure 3.1, which gives a flowchart of the way data are generated, annotated, transformed, structured and interpreted to gain novel hypotheses, before more experiments may verify these hypotheses. Starting from thoughts on system properties of metabolic networks, this chapter focuses on this flowchart, and specifically the problems associated with generating and annotating valid metabolite data. It adds a compilation of recent work in plant metabolomics to give an overview about the breadth and scope for which this technique is used in trying to understand plant metabolism.

3.1.2 Systemic properties in metabolic networks

Control and regulation are often used as synonyms, but this is actually not the case. As David Fell has pointed out in his famous book *Control of Metabolism* (1997 [5]), these two terms point to biochemical properties that are rather different in their respective meanings. Regulation is the ability of a complex system to maintain its basic properties (e.g. metabolite levels) independent of external factors that continuously try to push the system out of balance. A plant cell is exposed in short time intervals to many stochastic factors such as wind, light intensity differences, physical interferences or influx deviations of external transport metabolites. The system would become very unstable if each of these short-term pulses required immediate responses. There are a number of regulatory steps that inhibit metabolic overreactions, but instead introduce response lag times by using threshold systems, active transport steps, or reversibility of reactions. In total, these delay steps render the system to become 'robust' which is an important property to maintain the system at a given steady state. Complementary to such robust regulation of steady state levels is the necessity to alter metabolite levels depending on certain stress conditions or developmental needs. The responsible general system property is called 'flexibility'. System flexibility is a prerequisite of the capability to 'control' or alter defined steady states without affecting other parts of the system, depending on external or internal stimuli. Any system needs capabilities to react in a fast and coordinated manner on immediate needs and threats, even if the triggering signals for such needs are of low abundance and transient in nature. Examples might be heat shock, wounding responses or herbivore attacks. The glucosinolate–myrosinase system commonly found in plants of the order Brassicales is one such example: myrosinases are thioglucosidases capable of hydrolyzing glucosinolates upon nonspecific generalist herbivore attack, which leads to a release of a suite of compounds with cytotoxic or feeding deterrent effects. Other examples of 'control' can be found in classic physiology. In physiological terms, cold acclimation (by increased values in carbohydrates) or leaf senescence (altered ratios of catabolism versus anabolism) are examples of 'control' or 'system flexibility', whereas the tendency to keep metabolic fluxes in a narrow range under a given set of environmental parameters (the steady state) is an example for metabolic 'regulation' or system 'robustness'.

3.2 Metabolomic methods

3.2.1 Historic perspective of plant metabolite analysis

How do plant systems manage to keep these two fundamental properties in balance? In principle, the global nature of metabolomic surveys should be directly suitable to answer this question. Metabolomics aims at quantification and identification of all metabolites of a given biological system under defined conditions. Metabolomic data may thus be used to assess network properties such as metabolite connectivities, or changes in metabolic ratios, individual metabolite levels and pathways. When stable isotope tracers are applied, even changes in fluxes or flux ratios can be assessed up to a certain extent [6]. However, it is still a methodological challenge to acquire comprehensive metabolic data, given the large differences in metabolite size, lipophilicity, volatility, charge state and other physicochemical properties.

Classically, analytical chemistry and plant physiology have focused on analyses of a limited number of select metabolite targets. The history of such target analyses tells us how introducing new instruments or methods has opened windows of research opportunities and how methodological advances have changed the view of plant metabolite functions and diversity.

Analyzing metabolite levels for plant physiological (or clinical) purposes has proceeded over the last 230 years, at least. At the end of the 18th century, Scheele and Vanquelin examined single primary plant metabolites such as citric, malic, oxalic, gallic and tartaric acid. Around the same time, microscopy was introduced which enabled Markgraf to discover sucrose in sugar beet. In the 19th century, an array of analytical instruments was developed that fostered metabolite analysis, both for quantitative and qualitative purposes, including colorimeters, polarimeters, volumetric devices and photometers [7]. In that period, over 40 isolated plant metabolites were characterized by Berzelius. However, it was only during the last century that new techniques allowed the detection of the true richness of plant metabolomes, especially the so-called secondary metabolites.

The technological breakthrough did not come by a novel detection system but by better separation of metabolites. In 1906, the Russian botanist Mikhail Tswett invented chromatography by separating plant leaf extracts over powdered calcium carbonate and found chlorophyll pigments to be separated in several visible bonds [8]. The technique was eventually adopted and refined by Kuhn and Lederer in the 1930s who used it for carotenoid separations and purification of a large number of vitamins. This novel separation system boosted the number of compounds detected, particularly in combination with novel detectors which were based on a number of physical principles such as fluorescence emission, amperometry, light diffusion or light absorption in the visible and ultraviolet range. After the Second World War, the use of visible or UV absorbance, in particular, was widely adopted and allowed compound identification and comparison between laboratories based on spectral libraries [9].

3.2.2 Modern instrumentation in metabolite analysis

Today, virtually all these analytical methods are applied in physiological or medical research to separate, purify, detect and characterize compounds. However, it seems that for true metabolomics, only a select combination of methods is suitable. In general, a metabolomic method must be capable of detecting unambiguously, identifying and quantifying a large number of individual metabolites in a given sample. This requirement calls for high analytical resolution, universality (to detect metabolites irrespective of chemical substructures), selectivity (to acquire an analytical signal that is specific for a given metabolite), high dynamic range (to detect metabolites both at high and very low concentration), high precision (quantitative reproducibility), good accuracy (quantitative correctness) and high throughput (considering the need for statistically valid statements for a set of biological experiments). Metabolome number estimates range from about 500 (for prokaryotes) to many thousands of analytes (for vascular plants). There is no single method that can fulfil all the above mentioned requirements for so many analytes.

For this reason, the technology for true metabolomic approaches has only been enabled by breakthroughs in two areas, namely computer power and instrument engineering. This duo has fostered applications of the two dominant detection systems, nuclear magnetic resonance (NMR) and mass spectrometry (MS). An advantage of NMR is the linearity of quantitative responses on increasing metabolite concentrations, almost irrespective of the chemical compound class. Large signals in NMR can directly be interpreted as high level concentrations, whereas in MS, quantitative responses strongly rely on the ionization potential of each metabolite. Therefore, quantitation in MS is limited to relative abundances of a given metabolite between samples, or requires calibration curves if absolute comparisons of different metabolites are needed. Nevertheless, MS is generally more favored than NMR due to four reasons: (i) MS is advantageous with respect to the capability to resolve complex mixtures of compounds, (ii) for most compounds, MS is far more sensitive than NMR, (iii) due to the need for high end magnets, NMR instruments are usually far more expensive than most MS instruments and (iv) coupling of NMR to separation techniques such as liquid chromatography is far from straightforward. This leads to the main use of NMR as a tool for 'metabolite fingerprinting' in which complex, unresolved spectra are compared from tens to hundreds of samples. This allows the classification of differences in the global control of metabolite levels according to the underlying experimental design. Correspondingly, even experienced research consortia rarely identify more than 30 individual metabolites per NMR spectra of plant extracts [10, 11], whereas the use of chromatography-coupled MS leads to routine identifications of up to 150 metabolites per sample [16].

3.2.3 Sample preparation for metabolomics

There is no optimal way to prepare comprehensive plant extracts. Some compounds such as ADP/ATP have such high turnover rates that anything beyond freeze clamping may just be too slow to efficiently stop any of the residual postharvest enzyme activity. Other compounds such as plant hormones may have low turnover rates but are of such low abundance (in whole organs) that large plant biomasses need to be prepared. These two cases mark the opposite and contradictory ends of the range of extraction prerequisites. Another conflicting constraint is the ranges of lipophilicity versus hydrophilicity. For example, leaf waxes need very nonpolar solvents (such as hexane), whereas sugars can only be extracted with solvents of high polarity (such as water). The solubilization power of a given solvent mixture may be altered by additional modifications such as application of heat, microwaves, pressure or ultrasonication. However, the general problem remains of the contradiction between metabolome-wide comprehensiveness and quantitative completeness (recovery) of the extraction method. *Arabidopsis thaliana* leaves have been shown to serve as an example of how to systematically maximize comprehensiveness and reproducibility in sample preparation procedures [12]. However, this study was restricted to primary metabolites that are detectable by gas chromatography-based methods.

In general, therefore, there are only two possible solutions for unbiased and comprehensive metabolomics. Either extractions are performed with sequential steps of solvent polarity (which may cause miscibility and precipitation problems when combining the resultant extracts) or compromises are accepted. The general notion is that it is unavoidable to take compromises in metabolomics. Even if these are accepted, it is mandatory to test and report the limits of the chosen method of sample preparation. For example, many protocols suggest the use of methanol during extraction. However, even slight amounts of methanol at ambient temperature and physiological pH cause chlorophyll to decompose by autoxidation, demetallation and methylation [13, 14]. Consequently, if methanol extraction is performed, a range of porphyrins and other chlorophyll allomerization products are unavoidable and are detected by LC/MS. Other metabolites can potentially get altered in analogous reactions under comparably mild conditions. Apart from autoxidation, there are a number of other factors contributing to the formation of artifacts or loss of compounds. For example, thawing of biomass must be carefully avoided as long as proteins are not fully precipitated (for complete enzyme inactivation). Some enzymes such as hydrolases or phosphatases are still active even in methanolic solutions at ambient temperatures.

The large losses in compounds that have been observed have been compared in a direct comparison of two published plant extraction methods: the 70°C hot MeOH:H_2O (4:1 v/v) protocol [15] and the -15°C cold $CHCl_3$:MeOH:H_2O (2:5:2, v/v/v) strategy [16]; see Figure 3.2. Glucose-6-phosphate and other compounds with high turnover rates were barely detectable using the hot enzyme inactivation/extraction method, whereas recovery was high using the cold protein precipitation method. A likely reason for this striking difference is that frozen plant material might not reach 70°C in the first seconds after addition of the methanolic mixture, and needs time to heat up. This time frame needed for heating the extraction slurry of ground-frozen Arabidopsis leaves and methanolic solvent may then last long enough to reduce the already low abundance levels of hexose phosphates

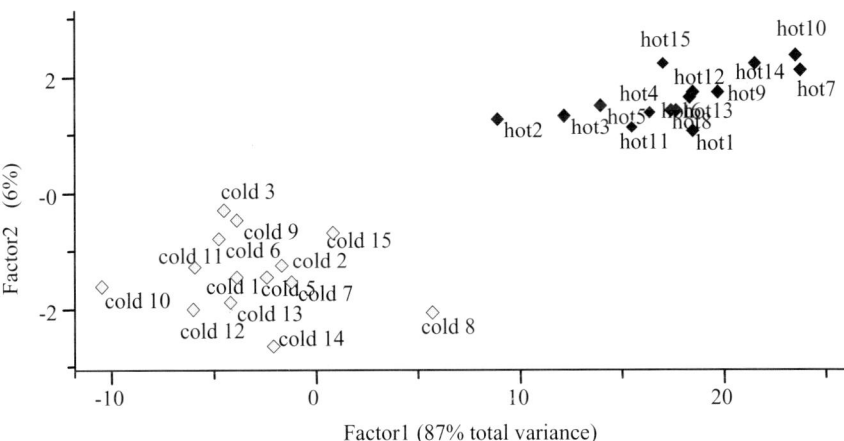

Figure 3.2 Principle component analysis of a direct comparison of the +70°C (hot) and the −15°C cold methanolic extraction method (cold) (unpublished results). Each method was applied 15 times on identical homogenized, deep frozen Arabidopsis leaf material. For each replicate, 20 mg FW material was taken. Data acquisition was performed by GC-TOF mass spectrometry. Principle component vector 1 separated the two methods, explaining 87% of the total variance of the data set. Investigation of the vector loading scores and *t*-test analysis resulted in identification of phosphorylated intermediates as most important discriminatory metabolites with clearly less being recovered from the +70°C (hot) extracts.

and other phosphorylated intermediates to below the detection limit. Conversely, during cold extraction, enzymes are kept inactive at all times and in addition, the simultaneous presence of chloroform ensures immediate protein precipitation. Some protocols favor lyophilization instead of using fresh (frozen) plant material. However, even this procedure bears risks as many ligands directly interact with proteins or are tightly attached to cell walls or membranes. The degree of this interaction may be even higher in lyophilized material, which may cause losses in extraction of subsequent certain caged metabolites. Unfortunately, no comprehensive study has yet been published that compares these protocols.

Other physicochemical factors pose even larger risks in reducing metabolite recovery. Catecholamines, which are genuine metabolites in potato leaves, decompose when exposed to light for longer than 15–30 min [17]. Even more severe is the effect of oxygen that may be dissolved in the extraction solutions. Minute amounts of O_2 will suffice to oxidize cysteine, glutathione, tocopherol or ascorbate. Correspondingly, all extraction solvents and storage containers must be carefully degassed with argon or other noble gases.

Another important step that potentially leads to metabolite losses is solvent volume reduction. Concentration of solvents to complete dryness will inevitably cause losses of volatile and semivolatile components (such as terpenes). In addition, other compounds, such as complex lipids, may face reduced recoveries. For example, lipids may precipitate on surfaces during sample preparation or fractionation, with limited potential to get resolubilized (depending on the actual solvents and conditions). To conclude, therefore, published reports about the number of 'detected metabolites' in metabolomics need to be taken with caution, particularly if reports do not specify (a) the precise extraction method, sample preparation conditions and comparisons to method blank controls and (b) the number and names of unambiguously identified compounds and how confidence in metabolite annotation was achieved.

3.2.4   Metabolome coverage

3.2.4.1   The quest for combining sensitivity and selectivity
There are many challenges and open questions in plant metabolome analysis. For example, the simple question as to how large a plant metabolome is for a given species in a set of typical environments is still unanswered. There is growing evidence that the size of metabolomes cannot simply be computed using reconstructed pathways from genome annotations. On the one hand, many genes (and enzymes) as yet lack clear functional annotation, but on the other hand, enzymes may be far less specific than classically anticipated [18]. For example, deletion of a single amino acid residue may already lead to different enzyme substrate specificity and lead to new products [19]. Therefore, gene annotations may easily be misled by the sole reliance on gene sequence homology or the degree of amino acid identity.

The alternative possibility of using instrumental analytics to tackle metabolome size is equally difficult. An important reason for this is that both NMR and MS still lack the resolution, sensitivity or universality necessary to detect and identify all components of a given sample. Developments for improved resolution may remedy this problem, for example, by exploiting more than one physicochemical property for separation prior to detection. An obvious possibility is to utilize gas phase ion mobility [20] in addition to classical chromatography. However, this technique has not yet been applied to plant samples. Other choices may include combining different chromatographic techniques such as coupling liquid chromatography to capillary zone electrophoresis (LC × CE; lipophilicity versus charge) or using gas chromatography with different column polarities (GC × GC, volatility versus lipophilicity) [21]. Even classical high pressure liquid chromatography (HPLC) (LC in short) has undergone dramatic improvements in performance, from the 4.6 mm i.d. columns that were used in the 1980s, to the 0.2 mm i.d. capillary columns used in the 1990s, to today's monolithic columns [27] (Figure 3.3) or ultrahigh pressure HPLC that leads to increased chromatographic resolution with over 100 000 theoretical plates.

Other advances in instrumentation to try to improve universality includes ionization for MS, which is achieved by introducing coupled interfaces (ESI × APCI,

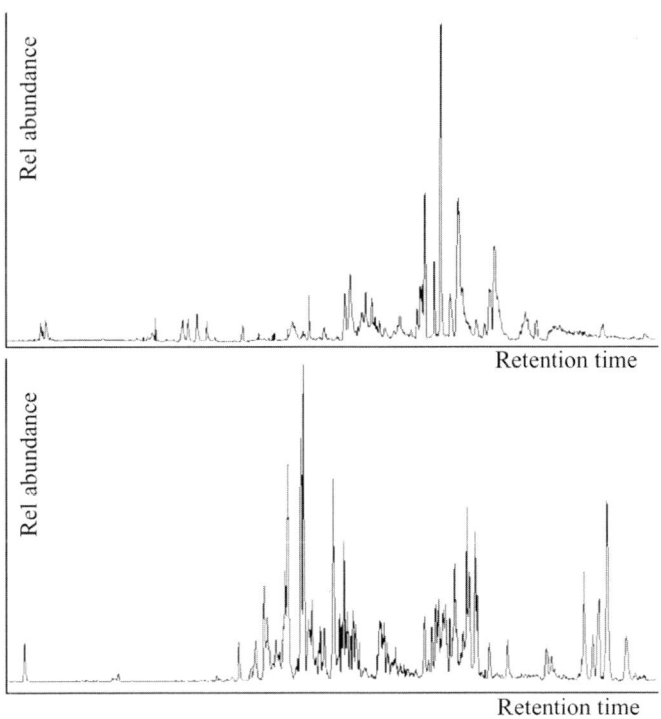

Figure 3.3 Chromatographic resolution in LC/MS. Upper panel: injection of 100 µl *Arabidopsis thaliana* leaf extract (i-propanol:water 2:1) onto a classical 4.6 mm C18 reverse phase column (20 000 theoretical plates/m, 1 ml/min flow rate.) Lower panel: injection of 0.2 µl of the same Arabidopsis extract onto a monolithic capillary C18 column (100 000 theoretical plates/m, 0.01 ml/min flow rate).

electrospray and chemical ionization) or characteristics of different types of mass spectrometers (linear ion trap, capable of tandem mass spectrometry 'in time' and 'in space'). However, it seems obvious that improvements in resolution for a single method will not cause a quantum leap in metabolome coverage. To date, uses of COSY and TOCSY two-dimensional NMR methods [22] have not succeeded in significantly increasing the number of identified compounds in metabolomic surveys. Approaches to use high-end Fourier-transform ion cyclotron mass spectrometers (FT-MS) are equally limited in the potential to target full metabolome surveys [59]. Despite this ultimate mass spectrometric resolution of $R > 500\,000$, isomeric compounds (such as glucose and fructose) cannot be distinguished due to their identical elemental composition, and even mass accuracy in FT-MS of <1 ppm does not allow unambiguous calculation of elemental compositions above approximately 400–500 Da without additional information [23, 24]. In addition, each metabolite subjected to electrospray/mass spectrometry usually gives rise to 3–5 further signals apart from isotopic ions. This is typically caused by adduct formation or in-source fragmentation ions. Therefore, the number of detected ions (mass signals or m/z values) must not be mistaken with the number of detected metabolites. Consequently, publications that report hundreds to thousands of mass signals based on MS-based metabolite fingerprinting [25] or LC/MS profiles [26, 27] do not give a reasonable estimate of metabolome size or coverage. Many of these ions certainly will account for novel and unknown compounds; however, others may simply be catabolic products which have been produced during postharvest biological processes or chemical by-products that occurred during sample preparation (or simply chemical artifacts due to impurities in solvents and plastic ware).

3.2.4.2 Cellular and subcellular metabolomics

A further reason why the complement of all metabolites present for a given plant, say Arabidopsis or rice, is not known is the lack of spatial resolution. It is a truism that a plant consists of many organs, and that each organ may include many tissue types and each tissue type may comprise various cell types. All published reports so far support the notion that different tissue types comprise varying metabolomes. Different biological roles of individual cell types support the further expectation of detecting striking differences on the low-level spatial resolution, e.g. between trichome and epidermis cells, or between parenchyma and bundle sheath cells. Lastly, intracellular organization of metabolism is also highly structured into compartments, each of which serves specific functions that lead to large metabolic differences. One example is a report of metabolite profiling of isolated chloroplasts and subfractions including the envelope, the stroma and the thylakoids in a study on the activity of three 13-lipoxygenases under stress conditions [28]. In this study, barley leaves treated with methyl jasmonate resulted in a remarkable increase of linolenic acid, free 9- and 13-hydroperoxy linolenic acid and the corresponding hydroxy- and aldehyde derivatives. The subcellular fractionation confirmed that these compounds were preferentially accumulated in the envelope and the stroma, therefore directly linking the localization of the products and substrates with the corresponding lipoxygenases.

Apart from lack of spatial resolution, the number and concentrations of metabolites are controlled in response to plant development and environmental stimuli. On the other hand, the very nature of this flexible and unsteady metabolome state may serve as a valuable source of information of the physiological condition and the underlying regulatory network if carefully designed physiological (and genetic) plant experiments are carried out. Unfortunately, given the challenges of spatial and temporal resolution, today's analytical methods still seem to be inadequate with respect to acquiring the full complement of metabolites at the required sensitivity and for multiple biological snapshots.

3.2.4.3 Compound identification

In order to come closer to high metabolomic coverage, there are basically two approaches. One is to work toward a comprehensive analysis of all known and previously described metabolites by combining and unifying extraction, fractionation and improved detection methods. The metabolome coverage would increase in a stepwise manner by building blocks of certain compound classes until large overview analyses have been achieved. The advantage of this approach is that exact quantifications would enable comparison of results across experiments and databases. A clear disadvantage is that most known metabolites are simply not commercially available. As an example, from the general metabolite list given in the LIGAND database in KEGG [29], only about 1000 compounds can be purchased. Furthermore, this approach would always be biased by past knowledge and disregard the potential importance and impact of novel metabolites. The alternative approach is to embark on comprehensive structural elucidation by *de novo* analysis of (NMR and MS) spectra, called dereplication [30]. In the past 10–15 years, some progress has been made in computational methods and the establishment of metabolite spectra databases. However, many (secondary) metabolites are structurally so complicated that current methods are incapable of performing an automatic structural dereplication. Instead, detailed interpretations by experienced natural product chemists are needed. On-line acquisition of UV, NMR and MS spectra after separation of compounds by liquid chromatography enables rapid gathering of the necessary structural information which potentially leads to a partial or a complete on-line *de novo* structure determination of natural products [31]. However, there is always a lesser level of confidence in metabolite annotations if these are based on spectral interpretation rather than comparison with authentic standards. It is hardly imaginable that dereplication reaches the level of acceptance like gene sequencing or peptide mapping. Consequently, an increase in metabolome coverage can only be achieved by running both approaches simultaneously, i.e. building blocks of metabolite profiles of known standards, and *de novo* compound identification. It is mandatory to specify in metabolomic reports how exact a given metabolite annotation is. For example, whether or not the applied method is able to distinguish isobaric and/or isomeric compounds, or if even higher confidence levels are reached that allow distinguishing chiral compounds such as D/L-isomers, allomers, enantiomers or diastereomers.

In addition, the choice of the data acquisition method is dictated by the biological question, as unbiased metabolomics is inadequate for many research projects. If, for example, biological hypotheses are narrowed to selected pathways or a small number of metabolic elementary modes, there is no need to use metabolomics. Instead, 'metabolite profiling' methods (that look for a limited set of pre-defined analytes) or classical target analyses can be applied. Conversely, target methods are unsuitable to answer questions about systemic control of network properties for which metabolomics is appropriate. In comparison to metabolomics, the focus on a selection of 'known' metabolites by metabolite profiling or target analysis disables an unbiased search for novel phytochemicals which may bear important physiological relevance (for example, as signalling molecules).

3.2.5 Quality control

Once a certain protocol for plant metabolomics has been developed, and is established for a research project, it is a prerequisite to monitor the quality of metabolite identification and quantification over the whole project period, preferentially over years. This is essentially the difference between method development and method validation. Developing a protocol basically means a proof-of-principle that a certain analytical objective can be fulfilled. However, a method is only validated if these objectives are strictly defined, and if the exact parameters and conditions are given as to how to achieve and monitor these results, e.g. by quality control charts. In metabolomics, this is relevant to both quantification and compound identification. For example, for each method the relative standard deviations need to be given, and it also needs to be specified how these limits are controlled and monitored on a routine (daily) basis. In addition to validation by relative error ranges, some compounds may be quantified in absolute concentrations using reference compounds and comparison to control methods to give quantification accuracies. Use of absolute instead of relative quantification refers to the question under study. For example, for some biological studies, determination of nanomolar concentrations may be essential for calculating turnover rates or crop nutritional quality. For other studies, such as functional genomics, assessment of metabolic control by x-fold values may be sufficient.

Once the validation criteria are clearly laid out, metabolomic protocols become 'standard operating procedures' (SOPs). SOPs are commonly seen for industrial processes including agro-biotechnology companies, but for academic laboratories, combining larger genomic programs with exploratory metabolomics, such SOPs are becoming more and more essential. In any case, high quality reports on metabolomics or metabolite profiling should always include detail as to how many of the claimed metabolite signals were detectable in all the samples of a given biological experiment, what level of variance was found for these variables within the experiment and how method blanks were used to ensure that detected peaks were not artificially formed during the process. Without such specifications, the reported results and sometimes even the biological interpretations may become questionable.

3.3 Metabolomic databases

Publishing metabolite data is straightforward in classical target analysis. The experimental sections in peer-reviewed scientific journals usually refer to established and widely accepted methods, and data can be presented as average values or even as individual results for each sample, when appropriate. For metabolomics, publishing data is not so straightforward. Metabolomic results are usually data-rich, but poor in information. If only x-fold changes of metabolites are published with respect to the controls, then the data may contain only limited information for comparison with other experiments or conditions. Instead, metabolite levels must be deposited in a publicly accessible way that allows reusing the data under different aspects by giving the SOPs of sample preparation, data acquisition, data processing and the corresponding results. In 2004, a variety of reports have highlighted the importance of providing such information, among them being a general architecture for metabolomic databases ArMet [32] and considerations about the minimal information of a metabolomic experiment, MIAMet [33]. These considerations have only partially materialized in publicly available plant metabolomic databases [34]. For a range of compounds, agro-biotechnology companies have published validated metabolite data of crop nutritional value [35]. However, for fundamental research, no equivalent is known that is as comprehensive and validated. The basic reason for this lack is that there are very many aspects and parameters that need to be associated with 'metabolite levels' in order to turn these into informative and interpretable patterns that are useful for external researchers. Biochemical properties and cellular relationships can be mapped onto software platforms that can be interrogated in order to enhance the interpretability of data [36], but the very details of biological experiments and data acquisition are hard to capture in a standardized way. Some progress can be reviewed in a public forum that originated from a biomedical, pharmaceutical and toxicological background, *Standard Metabolic Reporting Structures* (SMRS) led by the Imperial College, London, UK [37]. Some of the associated problems are common for all databases that are reporting metabolite levels. The metabolites need to be named by unique identifiers in a consistent and traceable way to allow data exchange between different databases, for example, for system biology applications. Astonishingly enough, there are no such (publicly accessible) repositories of unique metabolite names.

The most comprehensive repository may be CAS, the Chemical Abstracts Service of the American Chemical Society, that includes information about millions of compounds, among them being biogenic metabolites. However, this service comes with high charges and it does not contain links to genomic databases. Furthermore, some compound identifiers in CAS have been changed or erased over time, and some components have multiple entries. Despite widespread popularity among (plant) biologists, databases such as Goto's LIGAND repository [29] cannot serve as the authoritative resource for metabolite identifiers because many compounds, especially lipids, are not well covered. Efforts have been launched recently to compile comprehensive metabolite lists such as MetaCyc [38] and INCHI [39], and it is therefore very likely that the problem of consistent metabolite annotations will soon be solved.

In addition, the underlying plant biology experiments need to be described in detail to allow reuse of the metabolite results. This is a serious problem that seems to be very hard to tackle in an appropriate way. For publishing experiments in a peer-reviewed plant journal, it is expected to explain experimental details in both the 'materials and methods' section and the flow text. However, publication of data in database repositories cannot follow the same path. Any unstructured flow text description of biological study designs is insufficient. The concept of publicly available databases is that results can be queried and downloaded for comparative studies. This concept requires, therefore, a logical and consistent structure of entering information about the underlying experimental design and details. So far, unfortunately, there is no consensus in the plant community on vocabularies and items that are mandatory for describing a given experiment. One reason is that experimental designs are at the heart of a study and are therefore very different and hard to describe, and to capture in a fixed database structure. The other reason is that, so far, the biological community has relied on the peer-review system to ensure that sufficient information is given to enable reuse of data or the repeat of a study. However, there is usually no peer-review system associated with database entries. Metabolomics research groups may learn from Web-based entry forms that have been developed for describing transcript microarray data using a study annotator [40] that supports quantitative data with a structured ontology on the relationships and properties of various study designs and experimental details.

For general acceptance of metabolomic databases, a consensus needs to be sought how to name and structure plant biological experiments with respect to terms, structural hierarchies, ontologies and controlled vocabularies. Related efforts have resulted in compulsory repositories such as for naming species (in NCBI [41], for Arabidopsis germplasm in *the Arabidopsis information resource* TAIR [42] or for naming plant organs in Plantontology.org [43]). In the meantime, metabolomic databases need to describe, but not prescribe, which experimental details need to be given.

3.4 Pathways, clusters and networks: applications of plant metabolomics

Recent studies in plant biology involving the four different types of metabolite analysis (target analysis, metabolite profiling, metabolomics, metabolite fingerprinting) may be classified into five broad research areas, some of which are overlapping: confirming the effects of bioengineering plant metabolic enzymes (Section 3.4.1), studying plant biochemistry including the connectivity of pathways (Section 3.4.2), observing and cataloguing physiological effects during developmental or

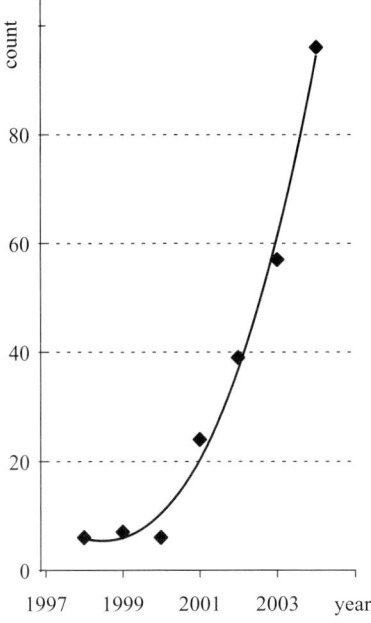

Figure 3.4 Count of hits in the literature database of the Institute for Scientific Information (ISI). Search words relating to metabolomics were restricted to the plant specific literature. 71% of the counts were original reports, and 20% were classified as review articles. 77 authors were found with three or more articles.

environmental transitions (Section 3.4.3), designing and utilizing better analytical methods (Section 3.4.4) and applying profiling methods in food science (Section 3.4.5). A literature survey using metabolomics/metabolite profiling terminology in the ISI database in May 2005 resulted in over 3000 hits, and 296 hits were still found when the search was restricted to plant-related search terms (Figure 3.4). Consequently, only select publications are presented here to serve as an overview of how these techniques may be applied to plant metabolism research.

3.4.1 Bioengineering of metabolism

In bioengineering, metabolite analysis is typically restricted to only a few target compounds or select pathways are profiled to conclude the effectiveness of a certain treatment. An example here is the selection of plant lines with high secondary metabolite levels, such as ginsenosides, for which 993 EMS mutant lines were tested by LC/MS and LC/UV [44]. If metabolite profiling or target analysis is used, hundreds of lines involving thousands of analyses can easily be scrutinized to guide the bioengineering efforts because both sample purification and instrumental methods can easily be optimized for this task. Sometimes more directed efforts in bioengineering are put forth by overexpressing novel genes into plant systems, such as for formation of ketocarotenoids in tomato and tobacco [45]. Metabolite profiling can then demonstrate the effectiveness of this transformation, and in addition, the level of formation of side products or substrates and catabolites of the primary products (such as hydroxylated intermediates). An example of how metabolite profiling may unravel unexpected effects of plant bioengineering is the transformation of stilbene sythase genes from grape to tomato plants [46]. This transgenic overexpression resulted in accumulation of trans-resveratrol and trans-resveratrol-glucopyranoside, but also of soluble antioxidants such as ascorbate and glutathione. In contrast, membrane-bound antioxidants such as tocopherol and lycopene were not affected.

Another important field of research is geared toward more complex traits such as protein content or optimized carbon partitioning and fluxes. Metabolite profiling of transgenic bean plants which expressed a *Corynebacterium glutamicum* phosphoenolpyruvate carboxylase (PEPC) in a seed-specific manner [47] showed that metabolic fluxes shifted from sugars and starch into organic acids and free amino acids. This ultimately led to a gain of 20% more protein per gram seed dry weight and an increase of total seed dry weight of more than 20%. This report also shows that there is some overlap between bioengineering (complex) traits, efforts toward the biochemical relationships and the use of metabolite profiling to verify these. Consequently, a major use of metabolomics is found in plant biochemistry.

3.4.2 Plant biochemistry

3.4.2.1 Pathway analysis
Interestingly, potentially conflicting data on the biological role of PEPC were reported for Arabidopsis lines with up to 75% reduced PEPC activity [48]. In this report, it was found by ^{1}H-NMR fingerprinting that levels of various primary metabolites were indeed affected by reduced PEPC activity but without having an impact on overall plant growth. The authors concluded that this finding supported the idea that PEPC had little impact on anplerotic carbon fluxes and was just the opposite to what was found with *C. glutamicum* PEPC in transgenic beans [47]. These reports may serve to illustrate that even major conclusions may differ depending on the experimental setup (e.g. underexpression of an endogenous enzyme-coding gene vs overexpression of a transgene) and, obviously, on the plant species that is being investigated.

A particularly nice example for a successful plant biochemical study involving metabolomics is a report on malate synthase involved in Arabidopsis seedling growth [49]. In this report, it was shown that this glyoxylate shunt enzyme is partially dispensible for lipid utilization and gluconeogenisis, in contrast to bacteria. This study may also serve as an example that each enzyme needs to be studied individually to determine its functions and biological roles, *in vivo,* and that homology searches and comparisons alone carry high risks of misleading conclusions. Another excellent study compared the effects of inducible activation of cytosolic yeast invertase with constitutive gene overexpression in growing potato tubers [50]. By this approach, the primary effects of increased sucrose mobilization caused by overexpression could be distinguished from pleiotropic effects such as a switch from starch synthesis to respiration. This switch was only observed to occur in constitutive gene overexpression during plant development, but not in short-term inducible overexpression. Such experiments are extremely important to better understand the control of metabolism *in vivo,* and to distinguish causes from mere observations of end-point effects.

Other studies have focused on secondary plant biochemistry [51]. In this research field, an elegant study reported the metabolic and transcriptional characterization of phenylpropanoid biosynthesis in Arabidopsis. Metabolomic approaches could show that despite absence of phenotypic alterations, specific functions of the Arabidopsis phenylalanine lyases PAL1 and PAL2 were elaborated [52]. The authors showed that PAL1 is the primary factor involved in the formation of phenylpropanoids, as well as more complex and less understood alterations in other metabolic modules.

Further reports apply plant metabolomics by integrating classical methods such as isotope labeling or by extending data mining toward network analysis. Isotope labeling with ^{13}C tracers is particularly useful for ^{13}C-NMR analysis. Consequently, this technique has been used for plant biochemistry. A study on rice coleoptiles under anaerobic conditions revealed that glutamine and malate pools were generated from multiple turns of the TCA cycle and that there was a high contribution of the glyoxylate shunt toward malate formation under these conditions [53]. Primary metabolism was also the focus of a study investigating potato plants with underexpression of sucrose synthase II that was found to be primarily localized in vascular tissues [54]. Largely different effects of underexpression of this enzyme isoform were found for source and sink tissues, with major effects on control of sugar alcohol metabolism in leaves and of control of amino acid metabolism in tubers. Besides classical and multivariate statistics, these effects were revealed by directly ranking differences in metabolic network connectivities.

Comparative correlation analysis has also been used to study control of primary metabolism responding to elicitation by methyl jasmonate, yeast elicitor or UV light [55]. In this study, glycine, serine and threonine pathways were found to be perturbed and induction of threonine aldolase activity was suggested from these data.

3.4.2.2 *Flux measurements*

The result of metabolomic analyses is a series of measurements of metabolite levels, that is, snapshots of metabolism. However, this is just one way to study the control of metabolism. Metabolic snapshot data are usually not sufficient to directly derive enzyme activities and hierarchical structures of pathways and, even more importantly, the dynamics of carbon partitioning between the different organs or the different metabolic cycles. A range of possible scenarios may explain a finding of altered metabolite levels: anabolic reactions might be faster, the catabolic fate might be different or transport activities may have been changed. Conversely, even if metabolite levels remain unchanged in an experiment, the underlying enzyme activities may still be different. If, for example, both anabolic and catabolic reactions are altered in the same way and intensity, the steady state levels of a given metabolite should not change (although fluxes are increased), Therefore, metabolite snapshot data should be complemented by flux data.

Theoretically, it should be possible to derive fluxes from snapshot measurements if we had the ability to measure true concentrations of all substrates and products at faster time intervals, assuming we would know the total network structure. However, all these constraints are not fulfilled at the current state of metabolomic practice, as outlined above. Even if snapshots were taken in a time series, and even if all substrates and products of a 'pathway' were covered, today we are still unable to detect the flow of products back into the pathway (either by reversible reactions or via other routes through the metabolic network) or find potential new side fluxes out of (or into) the pathway by unforeseen additional enzymatic activities. Here, the use of labeled compounds is most appropriate, either by radioactive labeling or by stable isotope tracers (e.g. isotopomer analysis). These techniques, however, are also restricted in use by the need to feed in labeled substrates which (a) may not be taken up quickly enough and (b) are then quickly diluted through the metabolic network. Therefore, only short distances or small parts of the total metabolic network can be imputed that have reasonably high metabolic turnover rates or that lead to and from strong carbon sinks such as starch. A global view on all metabolic fluxes (a 'fluxome' [4]) is still out of reach by current techniques, even if fluxes are inferred from other metabolic sinks such as proteins. For vascular plants, a highly suitable way to analyze fluxes is to use NMR-based techniques [56]. Using current methods, a combination of metabolomic snapshot data at high number of biological replicates (to get the breadth of metabolic networks at high statistical significance levels) and flux measurements (on select and important pathways [57]) therefore seems to be the most practical solution to reach a more complete picture of metabolic control and regulation.

3.4.3 *Physiological studies*

Physiological adaptation to environmental stress has been the focus of several studies on Arabidopsis plants. By comparing Ws-2 and Cvi-1 ecotypes to Arabidopsis lines overexpressing CBF transcription factor genes, it has been shown that CBF overexpression configured the metabolome of Arabidopsis in a way that resembles cold acclimation treatments [58], proving the major contribution of this gene family in the cold response pathway. Two other studies compared plants under sulfur stress. By using both mass-spectrometry-based metabolite fingerprints and gene transcription levels [59], it was confirmed that an immediate response on sulfur starvation is a decrease in glucosinolate biosynthesis. Another report combined data from GC/MS and LC/MS measurements upon sulfur deprivation [60] and found that the metabolic system was rebalancing not only sulfur metabolism but also partitioning of carbon and nitrogen in a time dependent manner. This occurred mainly by accumulation along the O-acetylserine-serine-glycine pathway which led to storage of nitrogen in glutamine and allantoin pools. Apart from nitrogen shuffling, further effects were found in lipid catabolism, purine metabolism and enhanced photorespiration.

Other reports have focused on metabolic effects upon environmental perturbations. For example, responses to heat stress, drought stress and a combination of heat and drought stress were compared in Arabidopsis by metabolite profiling [61]. It was suggested that sucrose and other sugars replace proline as the major osmoprotectant under combined drought and heat stress, whereas proline levels were controlled in response to drought stress alone. Another important area of very active research has been the study of metabolic responses to pests and pathogens. Interestingly, an infestation of tomato plants with spider mites *(Tetranychus urticae)* caused a delay of 4 days between increased levels of terpene biosynthetic transcripts and the emission of volatile terpenoids [62]. This work sheds light on the time-decoupling of control layers in metabolic responses, which also questions the validity of approaches using direct transcript–metabolite correlations for generating hypothesis on the functional annotation of metabolic genes [63].

In order to distinguish cause and effects, most reports involve studying the time dependence of metabolic responses instead of mere data associations or multivariate clustering. A good example of studying the hormonal effects on control of metabolism was an investigation of a 48-day time course upon elicitation of Arabidopsis roots with salicylic acid, jasmonic acid, chitosan and two fungal cell wall elicitors [64]. Upon treatment, 289 secondary metabolite peaks were profiled by LC/MS of which 10 peaks were confirmed by NMR structural elucidation to be compounds exhibiting antimicrobial activity at concentrations detected in the root exudates. Investigating metabolic relationships has been the focus of a physiological study on sink–source transitions in developing aspen leaves [65]. Besides confirming anticipated changes in sugar and amino acid metabolism, the study also revealed that control of nitrogen storage (determined by altered asparagine concentrations) was sequestered by changes in malate concentration and transaminase activity in this developmental time course.

3.4.4 Plant metabolomic methods

Novel insights into the control of metabolism go hand in hand with improved methods. For this reason, various research groups validated the usefulness of novel methods or improved protocols by using model plant experiments. Among these methods is the use of vapor phase extraction for phytohormone analysis upon pathogen infection of Arabidopsis plants [66], fractionation methods applied to rice plants to reduce metabolite complexity prior to instrumental analysis [67], or application of capillary electrophoresis coupled to mass spectrometry for rice leaf analysis [68], a method that was primarily used previously for microbial metabolomics. Volatile compounds responding to infection with pathogens are best monitored by gas chromatography/mass spectrometry, for example using head space sampling [69]. For example, when applied to onion bulbs that were inoculated with *Erwinia carotovora ssp. carotovora*, *Fusarium oxysporum* or *Botrytis allii* [70], 259 compounds were detectable in the headspace of which 25 were found to be specific to one or more pathogens. For phytohormones, an elegant and simple novel extraction procedure has been proposed [71]: plant tissues were extracted in mixtures of n-propanol and dichloromethane followed by methylation using trimethylsilylcliazomethane. The reaction products were heated, collected on a polymeric absorbent and analyzed by GC/NCI-MS. This approach has been applied to Arabidopsis, tobacco and tomato plants, enabling to quantify signalling crosstalk interactions at the level of synthesis and accumulation of phytohormones.

Most methods used for studying metabolic control involve invasive techniques. A new idea has recently been proposed to enable detecting metabolites *in vivo* and in real time at subcellular resolution by using protein-based nanosensors based on FRET fluorescence-based microscopy. The prototypes of these sensors have been shown to work in yeast and in mammalian cell cultures [72, 73]. An extension to multiple metabolites and detection in plant cells would offer extreme versatility and direct insights into metabolic control, e.g. following time courses after glucose pulses or inhibition of specific enzymes. Analyzing time-related metabolomic data has also been the focus of a study of growth related gradients in poplar trees by magic angle spinning NMR analysis [74]. When investigating trees with underexpression of the PttMYB76 gene involved in the phenylpropanoid and lignification pathways, growth-related metabolic gradients were detected in the plant internode direction. Factors affecting NMR spectra have been investigated for potato and tomato samples [75]. This study emphasizes that, as with any method, great care must be taken to control method parameters in order to allow robust assessments of metabolite levels over hundreds of samples.

Methods based on gas chromatography/mass spectrometry have evolved in two directions. Two methods have been published that avoid detailing individual metabolites, but rather compare full spectra sections in order to align and compare hundreds of chromatograms, followed by multivariate analysis and retrospective investigation of differences related to the plant experimental designs [76, 77]. However, apart from looking at differences in the control of metabolism, it is equally interesting to note which metabolites are tightly regulated at a defined steady state level. That is, which metabolites are not altered between experiments. Hence, it seems a favorable option to quantify levels for each individual compound that is detectable in the metabolomic experiment. For GC/MS, LC/MS or CE/MS approaches, this involves individual peak detection with subsequent mass spectral deconvolution of overlapping peaks (Figure 3.5) and peak alignment by retention indices, in order to be able to compare data between experiments, laboratories or metabolomic databases. Examples for this strategy is the compilation of known and unknown metabolites from *Lotus japonicus* nodules, roots, leaves and flowers [78] or the investigation of metabolites from Arabidopsis leaves [16].

3.4.5 Food science

The last typical field of application of metabolomics in control of metabolism is food science. Food quality is easily and rapidly deteriorated by a number of different pests, and therefore it must be tightly monitored to prevent major losses. One typical report on such efforts is the use of GC/MS to profile volatile metabolites in apple to diagnose fungal infections [79]. In this report, four different fungi were applied to infect apples and the responses to these infections were classified by 20 significant biomarkers out of a total of 498 detected different volatile metabolite peaks. Such methods enable a

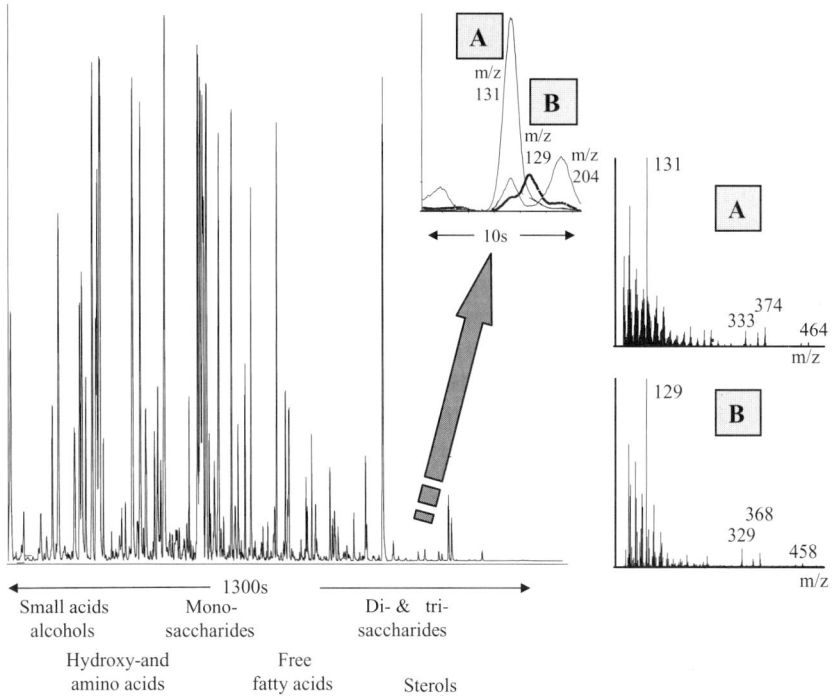

Figure 3.5 Deconvolution of overlapping peak analysis of an Arabidopsis leaf extract by GC-TOF. Left panel: primary plant metabolites are separated within 1300 s. Due to the complexity of metabolomic extracts, co-elution of compounds is inevitable. Right panels: Mass spectra of d6-cholesterol (A) and cholesterol (B) are deconvoluted by automatic algorithms (ChromaTOF 3.0), enabling unambiguous metabolite annotations.

rapid survey during food storage by observing metabolic effects for disease diagnostics rather than trying to understand the biochemical or physiological control mechanisms. A more recent but important branch of research tries to improve nutritional quality or metabolic traits in foods by genetic breeding using the analysis of quantitative trait loci (QTL). Metabolic effects were assessed as a result of the introgression of a 9 cM region of the wild tomato species, *Lycopersicon pennelli*, into a cultivated tomato line (*Lycopersicon esculentum* IL9-2-5) [80]. Metabolite contents in ripe fruits were found to have increased sucrose and glucose levels that were due to altered kinetic properties of a fruit apoplastic invertase. A few other metabolic perturbations were found, including aspartate and alanine biosynthesis.

In food quality control, concerns have arisen that genetic modifications may result in potentially harmful or undesirable metabolite alterations. In order to study the substantial equivalence of genetically modified (GM) potato tubers to classical cultivars, 40 GM lines, modified in primary carbon metabolism, glycoprotein processing or polyamine and ethylene metabolism, were analyzed by NMR and LC-UV [81]. Differences in average metabolite levels were less than threefold, which was

3.5 Outlook

Most of the published work in plant metabolite analysis is, so far, either classical hypothesis-driven target analysis or multitarget metabolite profiling, that is restricted to usually below 100 identified compounds. Although considerable hypothesis-driven research can be undertaken using these methods, the prospects of using truly unbiased metabolomics are alluring. Two major bottlenecks need to be tackled. The first is that too many metabolic peaks remain unidentified. This raises concerns that many of these may not genuinely reflect control of metabolic states but rather arise from insignificant enzymatic side reactions, or, even worse, are indeed chemical artifacts produced during sample preparation. In principle, this argument is hard to rebut, especially as long as there are no consistent metabolomic databases and no major efforts for rapid identification *de novo* of unknown compounds. Various metabolomic databases are or will be made public in the near future. However, it is doubtful how many known metabolites these databases will include. The second major drawback is a gap in the interpretation of metabolic snapshot data. Very often, the general finding of metabolomic studies is that a large number of compounds have been altered in response to a given experimental treatment. Such observations remain mere physiological descriptions if other levels of information are not integrated to result in a comprehensive picture of plant biology (for example, spatial and temporal resolution of such metabolic snapshot data). A further way to improve the interpretation of snapshot data is to refer these to a 'plant physiology and plant biochemistry knowledge database' that may be inferred from both theoretical considerations (such as metabolic control analysis) and text mining approaches. The ultimate aim of all these efforts must remain an understanding of events and effects in plant physiology, which can be tested by constructing data models and predict metabolic alterations under (genotype x environment) conditions that were not tested before [82]. This level of understanding of plant metabolism is still out of reach, but with modern methods in dissecting plants on the molecular and cellular levels it is not impossible anymore!

References

1. O. Fiehn (2002) Metabolomics – the link between genotypes and phenotypes. *Plant Molecular Biology*, **48**, 155–171.
2. S. Oliver (1997) Yeast as a navigational aid in genome analysis. *Microbiology* **143**, 1483–1487.
3. E.C. Horning and M.G. Horning (1971) Human metabolic profiles obtained by GC and GC/MS. *Journal of Chromatographic Science* **9**, 129–140.
4. N. Zamboni and U. Sauer (2004) Model-independent fluxome profiling from H-2 and C-13 experiments for metabolic variant discrimination. *Genome Biology* **5**, Art. No. R99 (http:// genomebiology.com/).

5. D. Fell (1997) *Understanding the Control of Metabolism.* Portland Press, London.
6. J. Kikuchi, K. Shinozaki and T. Hirayama (2004) Stable isotope labeling of *Arabidopsis thaliana* for an NMR-based metabolomics approach. *Plant and Cell Physiology* **45**, 1099–1104.
7. J. Buettner (2005) Impacts of laboratory methodology on medical thinking in the 19th century. *Clinical Chemistry and Laboratory Medicine* **38**, 57–63.
8. M. Tswett (1906) Physical chemical studies on chlorophyll adsorptions. *Berichte der Deutschen botanischen Gesellschaft* **24**, 316–323.
9. R.A. Friedel and M. Orchin (1951) *Ultraviolet Spectra of Aromatic Compounds.* John Wiley & Sons, New York.
10. P. Krishnan, N.J. Kruger and R.G. Ratcliffe (2005) Metabolite fingerprinting and profiling in plants using NMR. *Journal of Experimental Botany* **56**, 255–265.
11. N.J.C. Bailey, M. Oven, E. Holmes, M.H. Zenk and J.K. Nicholson (2004) An NMR-based metabolomic approach to the analysis of the effects of xenobiotics on endogenous metabolite levels in plants. *Spectroscopy – An International Journal* **18**, 279–287 Sp. Iss. SI.
12. J. Gullberg, P. Jonsson, A. Nordstrom, M. Sjostrom and T. Moritz (2004) Design of experiments: an efficient strategy to identify factors influencing extraction and derivatization of *Arabidopsis thaliana* samples in metabolomic studies with gas chromatography/mass spectrometry. *Analytical Biochemistry* **331**, 283–295.
13. R.G. Brereton, A. Rahmani, Y.Z. Liang and O.M. Kvalheim (1994) Investigation of the allomerization reaction of Chlorophyll-alpha. Use of diode-array HPLC, mass spectrometry and chemometric factor analysis for the detection of early products. *Photochemistry and Photobiology* **59**, 99–110.
14. K. Hyvarinen, J. Helaja, P. Kuronen, I. Kilpelainen and P.H. Hynninen (1995) H-1 and C-13 NMR-spectra of the methanolic allomerization products of 13(2)(R)-Chlorophyll-alpha. *Magnetic Resonance in Chemistry* **33**, 646–656.
15. O. Fiehn, J. Kopka, P. Doermann, T. Altmann, R. Trethewey and L. Willmitzer (2000) Metabolic profiling for plant functional genomics. *Nature Biotechnology* **18**, 1157–1161.
16. W. Weckwerth, K. Wenzel and O. Fiehn (2004) Process for the integrated extraction, identification and quantification of metabolites, proteins and RNA to reveal their co-regulation in biochemical networks. *Proteomics* **4**, 78–83.
17. J. Szopa, G. Wilczynski, O. Fiehn, A.Wenczel and L. Willmitzer (2001) Identification and quantification of catecholamines in potato plants (*Solarium tuberosum*) by GC-MS. *Phytochemistry* **58**, 315–320.
18. W. Schwab (2003) Metabolome diversity: too few genes, too many metabolites?. *Phytochemistry* **62**, 837–849.
19. C. Lindermayr, J. Fliegmann and J. Ebel (2003) Deletion of a single amino acid residue from different 4-coumarate: CoA ligases from soybean results in the generation of new substrate specificities. *Journal of Biological Chemistry* **278**, 2781–2786.
20. J.M. Clark, K.A. Daum and J.H. Kalivas (2003) Demonstrated potential of ion mobility spectrometry for detection of adulterated perfumes and plant speciation. *Analytical Letters* **36**, 215–244.
21. J.L. Hope, B.J. Prazen, E.J. Nilsson, M.E. Lidstrom and R.E. Synovec (2005) Comprehensive two-dimensional gas chromatography with time-of-flight mass spectrometry detection: analysis of amino acid and organic acid trimethylsilyl derivatives, with application to the analysis of metabolites in rye grass samples. *Talanta* **65**, 380–388.
22. H.R. Tang, Y.L. Wang, J.K. Nicholson and J.C. Lindon (2004) Use of relaxation-edited one-dimensional and two dimensional nuclear magnetic resonance spectroscopy to improve detection of small metabolites in blood plasma. *Analytical Biochemistry* **325**, 260–272.
23. C.A. Hughey, R.P. Rodgers and A.G. Marshall (2002) Resolution of 11000 compositionally distinct components in a single electrospray ionization Fourier transform ion cyclotron resonance mass spectrum of crude oil. *Analytical Chemistry* **74**, 4145–4149.
24. A.C. Stenson, A.G. Marshall and W.T. Cooper (2003) Exact masses and chemical formulas of individual Suwannee River fulvic acids from ultrahigh resolution electrospray ionization Fourier transform ion cyclotron resonance mass spectra. *Analytical Chemistry* **75**, 1275–1284.

25. S.J. Murch, H.P. Rupasinghe, D. Goodenowe and P.K. Saxena (2004) A metabolomic analysis of medicinal diversity in Huang-qin (Scutellaria baicalensis Georgi) genotypes: discovery of novel compounds. *Plant Cell Reports* **23**, 419–425.
26. E. von Roepenack-Lahaye, T. Degenkolb, M. Zerjeski, *et al.* (2003) Profiling of Arabidopsis secondary metabolites by capillary liquid chromatography coupled to electrospray ionization quadrupole time-of-flight mass spectrometry. *Plant Physiology* **134**, 548–559.
27. V.V. Tolstikov, A. Lommen, K. Nakanishi, N. Tanaka and O. Fiehn (2003) Monolithic silica-based capillary reversed-phase liquid chromatography/electrospray mass spectrometry for plant metabolomics. *Analytical Chemistry* **75**, 6737–6740.
28. A. Bachmann, B. Hause, H. Maucher *et al.* (2002) Jasmonate-induced lipid peroxidation in barley leaves initiated by distinct 13-LOX forms of chloroplasts. *Biological Chemistry* **383**, 1645–1657.
29. S. Goto, T. Nishioka and M. Kanehisa (2000) LIGAND: chemical database of enzyme reactions. *Nucleic Acids Research* **28**, 380–382.
30. C. Steinbeck (2001) The automation of natural product structure elucidation. *Current Opinion in Drug Discovery & Development* **14**, 338–342.
31. J.L. Wolfender, K. Ndjoko and K. Hostettmann (2003) Liquid chromatography with ultraviolet absorbance-mass spectrometric detection and with nuclear magnetic resonance spectroscopy: a powerful combination for the on-line structural investigation of plant metabolites. *Journal of Chromatography A* **1000**, 437–455.
32. H. Jenkins, N. Hardy, M. Beckmann *et al.* (2004) A proposed framework for the description of plant metabolomics experiments and their results. *Nature Biotechnology* **22**, 1601–1605.
33. R.J. Bino, R.D. Hall, O. Fiehn *et al.* (2004) Potential of metabolomics as a functional genomics tool. *Trends in Plant Science* **9**, 418–425.
34. J. Kopka, N. Schauer, S. Krueger *et al.* (2005) GMD@CSB.DB: the Golm metabolome database. *Bioinformatics* **21**, 1635–1638.
35. W.P. Ridley, R.D. Shillito, I. Coats *et al.* (2004) Development of the international life sciences institute crop composition database. *Journal of Food Composition and Analysis* **17**, 423–438.
36. E.S. Wurtele, J. Li, L.X. Diao *et al.* (2003) MetNet: software to build and model the biogenetic lattice of Arabidopsis. *Comparative and Functional Genomics* **4**, 239–245.
37. J. Lindon (chair). Author group from organisations: Imperial College (London), Nestlé, Aventis, Novo Nordisk, Bristo-Myers Squibb, TNO, Umetrics, University of Wales (Aberystwyth), GlaxoSmithKline, Unilever, Pfizer, European Bioinformatics Institute, The Medical Research Council, Bruker, Roche, Eli Lilly (2005) The standard metabolic reporting structure – an open standard for reporting metabolic data. http://www.smrsgroup.org/, March 09, 2005.
38. W.A. Warr (ed.) (2004) IUPAC Project Meetings: Extensible Markup Language (XML) Data dictionaries and chemical identifier. http://www.warr.com/inchi.pdf NIST, Gaithersburg, MA, USA, June 2004.
39. P. Murray-Rust, H. Rzepa and S.E. Stein (2005) The INChI as an LSID for molecules in life-science. http://lists.w3.org/Archives/Public/public-swls-ws/2004Sep/att-0026/inchi.html, March 9, 2005.
40. E. Manduchi, G.R. Grant, H. He *et al.* (2004) RAD and the RAD study-annotator: an approach to collection, organization and exchange of all relevant information for high-throughput gene expression studies. *Bioinformatics* **20**, 452–459.
41. D.L. Wheeler, T. Barrett, D.A. Benson *et al.* (2005) Database resources of the National Center for Biotechnology Information. *Nucleic Acids Research* **33**, D39–D45 (Special Issue).
42. S.Y. Rhee, W. Beavis, T.Z. Berardini *et al.* (2003) The Arabidopsis Information Resource (TAIR): a model organism database providing a centralized, curated gateway to Arabidopsis biology, research materials and community. *Nucleic Acids Research* **31**, 224–228.
43. Plant Ontology Consortium (POC) (2002) The plant ontology™ consortium and plant ontologies. *Comparative and Functional Genomics* **3**(2), 137–142.
44. S.S. Woo, J.S. Song, J.Y. Lee *et al.* (2004) Selection of high ginsenoside producing ginseng hairy root lines using targeted metabolic analysis. *Phytochemistry* **65**, 2751–2761.
45. L. Ralley, E.M.A. Enfissi, N. Misawa, W. Schuch, P.M. Bramley and P.D. Fraser (2004) Metabolic engineering of ketocarotenoid formation in higher plants. *Plant Journal* **39**, 477–486.

46. G. Giovinazzo, L. D'Amico, A. Paradiso, R. Bollini, F. Sparvoli and L. DeGara (2005) Antioxidant metabolite profiles in tomato fruit constitutively expressing the grapevine stilbene synthase gene. *Plant Biotechnology Journal* **3**, 57–69.
47. H. Rolletschek, L. Borisjuk, R. Radchuk *et al.* (2004) Seed-specific expression of a bacterial phosphoenolpyruvate carboxylase in *Vicia narbonensis* increases protein content and improves carbon economy. *Plant Biotechnology Journal* **2**, 211–219.
48. A. Moing, M. Maucourt, C. Renaud *et al.* (2004) Quantitative metabolic profiling by 1-dimensional H-1-NMR analyses: application to plant genetics and functional genomics. *Functional Plant Biology* **31**, 889–902.
49. J.E. Cornah, V. Germain, J.L. Ward, M.H. Beale and S.M. Smith (2004) Lipid utilization, gluconeogenesis, and seedling growth in Arabidopsis mutants lacking the glyoxylate cycle enzyme malate synthase. *Journal of Biological Chemistry* **279**, 42916–42923.
50. B.H. Junker, R. Wuttke, A. Tiessen *et al.* (2004) Temporally regulated expression of a yeast invertase in potato tubers allows dissection of the complex metabolic phenotype obtained following its constitutive expression. *Plant Molecular Biology* **56**, 91–110.
51. K. Morino, F. Matsuda, H. Miyazawa, A. Sukegawa, H. Miyagawa and K. Wakasa (2005) Metabolic profiling of tryptophan-overproducing rice calli that express a feedback-insensitive alpha subunit of anthranilate synthase. *Plant and Cell Physiology* **46**, 514–521.
52. A. Rohde, K. Morreel, J. Ralph *et al.* (2004) Molecular phenotyping of the pal1 and pal2 mutants of *Arabidopsis thaliana* reveals far-reaching consequences on and carbohydrate metabolism. *Plant Cell* **16**, 2749–2771.
53. T.W.M. Fan, A.N. Lane and R.A. Higashi (2003) *In vivo* and *in vitro* metabolomic analysis of anaerobic rice coleoptiles revealed unexpected pathways. *Russian Journal of Plant Physiology* **50**, 787–793.
54. W. Weckwerth, M.E. Loureiro, K. Wenzel and O. Fiehn (2004) Differential metabolic networks unravel the effects of silent plant phenotypes. *Proceedings of the National Academy of Sciences of the United States of America* **101**, 7809–7814.
55. C.D. Broeckling, D.V. Huhman, M.A. Farag *et al.* (2005) Metabolic profiling of Medicago truncatula cell cultures reveals the effects of biotic and abiotic elicitors on metabolism. *Journal of Experimental Botany* **56**, 323–336.
56. R.G. Ratcliffe and Y. Shachar-Hill (2005) Revealing metabolic phenotypes in plants: inputs from NMR analysis. *Biology Reviews* **80**, 27–43.
57. A.R. Fernie, P. Geigenberger and M. Stitt (2005) Flux an important, but neglected, component of functional glenomics. *Current Opinion in Plant Biology* **8**, 174–182.
58. D. Cook, S. Fowler, O. Fiehn and M.F. Tomashow (2004) A prominent role for the CBF cold response pathway in configuring the low-temperature metabolome of Arabidopsis. *Proceedings of the National Academy of Sciences of the United States of America* **101**, 15243–15248.
59. M.Y. Hirai, M. Yano, D.B. Goodenowe *et al.* (2004) Integration of transcriptomics and metabolomics for understanding of global responses to nutritional stresses in *Arabidopsis thaliana*. *Proceedings of the National Academy of Sciences of the United States of America* **101**, 10205–10210.
60. V.J. Nikiforova, J. Kopka, V. Tolstikov *et al.* (2005) Systems rebalancing of metabolism in response to sulfur deprivation, as revealed by metabolome analysis of Arabidopsis plants. *Plant Physiology* **138**, 304–318.
61. L. Rizhsky, H.J. Liang, J. Shuman, V. Shulaev, S. Davletova and R. Mittler (2004) When defense pathways collide. The response of Arabidopsis to a combination of drought and heat stress. *Plant Physiology* **134**, 1683–1696.
62. M.R. Kant, K. Ament, M.W. Sabelis, M.A. Haring and R.C. Schuurink (2004) Differential timing of spider mite-induced direct and indirect defenses in tomato plants. *Plant Physiology* **135**, 483–495.
63. E. Urbanczyk-Wochniak, A. Luedemann, J. Kopka *et al.* (2003) Parallel analysis of transcript and metabolic profiles: a new approach in systems biology. *EMBO Reports* **4**, 989–993.
64. T.S. Walker, H.P. Bais, K.M. Halligan, F.R. Stermitz and J.M. Vivanco (2003) Metabolic profiling of root exudates of *Arabidopsis thaliana*. *Journal of Agricultural and Food Chemistry* **51**, 2548–2554.

65. M.L. Jeong, H.Y. Jiang, H.S. Chen, C.J. Tsai and S.A. Harding (2004) Metabolic profiling of the sink-to-source transition in developing leaves of quaking aspen. *Plant Physiology* **136**, 3364–3375.
66. E.A. Schmelz, J. Engelberth, J.H. Tumlinson, A. Block and H.T. Alborn (2003) The use of vapor phase extraction in metabolic profiling of phytohormones and other metabolites. *Plant Journal* **39**, 790–808.
67. T. Frenzel, A. Miller and K.H. Engel (2002) Metabolite profiling – a fractionation method for analysis of major and minor compounds in rice grains. *Cereal Chemistry* **79**, 215–221.
68. S. Sato, T. Soga, T. Nishioka and M. Tomita (2004) Simultaneous determination of the main metabolites in rice leaves using capillary electrophoresis mass spectrometry and capillary electrophoresis diode array detection. *Plant Journal* **40**, 151–163.
69. L.H. Lui, A. Vikram, Y. Abu-Nada, A.C. Kushalappa, G.S.V. Raghavan and K. Al-Mughrabi (2005) Volatile metabolic profiling for discrimination of potato tubers inoculated with dry and soft rot pathogens. *American Journal of Potato Research* **82**, 1–8.
70. B. Prithiviraj, A. Vikram, A.C. Kushalappa and V. Yaylayan (2004) Volatile metabolite profiling for the discrimination of onion bulbs infected by Erwinia carotovora ssp carotovora, Fusarium oxysporum and Botrytis allii. *European Journal of Plant Pathology* **110**, 371–377.
71. E.A. Schmelz, J. Engelberth, H.T. Alborn *et al.* (2003) Simultaneous analysis of phytohormones, phytotoxins, and volatile organic compounds in plants. *Proceedings of the National Academy of Sciences of the United States of America* **100**, 10552–10557.
72. S. Okumoto, K. Deuschle, M. Fehr *et al.* (2004) Genetically encoded sensors for ions and metabolites. *Soil Science and Plant Nutrition* **50**, 947–953.
73. M. Fehr, D.W. Ehrhardt, S. Lalonde and W.B. Frommer (2004) Minimally invasive dynamic imaging of ions and metabolites in living cells. *Current Opinion in Plant Biology* **7**, 345–351.
74. S. Wiklund, M. Karlsson, H. Antti *et al.* (2005) A new metabonomic strategy for analysing the growth process of the poplar tree. *Plant Biotechnology Journal* **3**, 353–362.
75. M. Defernez and I.J. Colquhoun (2003) Factors affecting the robustness of metabolite fingerprinting using H-1 NMR spectra. *Phytochemistry* **62**, 1009–1017.
76. P. Jonsson, J. Gullberg, A. Nordstrom *et al.* A strategy for identifying differences in large series of metabolomic samples analyzed by GC/MS. *Analytical Chemistry* **204**, 1738–1745.
77. A.L. Duran, J. Yang, L.J. Wang and L.W. Sumner (2003) Metabolomics spectral formatting, alignment and conversion tools (MSFACTs). *Bioinformatics* **19**, 2283–2293.
78. G.G. Desbrosses, J. Kopka and M.K. Udvardi (2005) Lotus japonicus metabolic profiling. Development of gas chromatography-mass spectrometry resources for the study of plant-microbe interactions. *Plant Physiology* **137**, 1302–1318.
79. A. Vikram, B. Prithiviraj, H. Hamzehzarghani and A. Kushalappa (2004) Volatile metabolite profiling to discriminate diseases of McIntosh apple inoculated with fungal pathogens. *Journal of the Science of Food and Agriculture* **84**, 1333–1340.
80. C. Baxter, F. Carrari, A. Bauke *et al.* (2005) Fruit carbohydrate metabolism in an introgression line of tomato with increased fruit soluble solids. *Plant and Cell Physiology* **46**, 425–437.
81. M. Defernez, Y.M. Gunning, A.J. Parr, L.V.T. Shepherd, H.V. Davies and I.J. Colquhoun (2004) NMR and HPLC-UV profiling of potatoes with genetic modifications to metabolic pathways. *Journal of Agricultural and Food Chemistry* **52**, 6075–6085.
82. L.V. Lejay, D.E. Shasha, P.M. Palenchar *et al.* (2004) Adaptive combinatorial design to explore large experimental spaces: approach and validation. *Systems Biology* **2**, 1–7.

4 Metabolite transporters in the control of plant primary metabolism

Mechthild Tegeder and Andreas P. M. Weber

4.1 Introduction

A fundamental process in the physiology of plants is the selective partitioning of organic metabolites among different organelles, cells, tissues and organs. Various transport mechanisms exist to accommodate the vectorial transport of metabolites, and these mechanisms are coordinated and regulated at different levels to achieve normal physiological functions in the whole plant. Transporters are involved in basic metabolic pathways, partitioning of metabolites within and between cells, and intermediate and long-distance transport between tissues and organs, respectively. While at the most basic level, plants assimilate inorganic carbon and nitrogen into reduced compounds required for plant growth, at the molecular level the variety of produced metabolites is large and the pathways that they feed into are complex and interconnected. These pathways are also often partitioned between organelles, cells or even tissues and organs. Thus, transporters are critical for sustaining the complexity of biosynthesis/catabolism and growth, and through their potential to affect the availability of substrates or products they are in a position to regulate metabolism and growth. An understanding of the types of transporters present in plants, their location and kinetic properties are necessary in describing metabolic fluxes and their control, as well as basic partitioning of nutrients between growth and storage.

Plant cells are highly compartmentalized, which is a reflection of how metabolic pathways have evolved and are partitioned at the sub-cellular level. The compartmentation of metabolic pathways augments options for control, permits the simultaneous operation of pathways that compete for the same substrates and helps avoid futile cycles. Metabolite transporters play critical roles in connecting the parallel and interdependent biosynthetic and catabolic pathways and thus represent the integrating elements in these metabolic networks, similar to interchanges in road networks. In addition, in vascular plants, long-distance transport is critical for the allocation of organic carbon and nitrogen compounds from their location of synthesis in so-called 'source' organs to developing or reproductive plant organs ('sinks') that rely heavily on import of the organic compounds for growth and development. This chapter will provide a brief overview of carbon and nitrogen assimilation and then focus on intracellular transport processes in plant cells. In addition, the characteristics of source and sink organs and principles of source to sink translocation of assimilates will be discussed. Particular emphasis will be given on recent insights derived from forward and reverse genetic approaches and *in vitro* studies of transporter function.

4.2 Photoassimilation and assimilate transport in source cells

Chloroplasts are the sole sites of photosynthetic carbon assimilation and the predominant sites of nitrogen assimilation in plant cells. Triose phosphates (TPs), the net products of carbon assimilation by the reductive pentose-phosphate pathway (RPPP), serve as the principle precursor for all other biosynthetic reactions in plants. For example, recently assimilated carbon can be allocated to starch and sucrose biosynthesis, nitrogen and sulfur metabolism, fatty acid biosynthesis, cell wall biosynthesis, secondary metabolism and a plethora of other metabolic pathways. Carbon allocation to these pathways is controlled at multiple levels, such as transcription, translation and post-translational and allosteric control of enzyme activity, subcellular compartmentation and the distribution of specific pathways between different plant tissues. In addition, environmental factors such as temperature, light intensity and water supply need to be integrated with developmental programs, nutrient status, effects of a variety of biotic and abiotic stresses and source–sink interactions.

4.2.1 Carbon assimilation by the reductive pentose-phosphate pathway (Calvin cycle)

Inorganic carbon dioxide is assimilated into organic C compounds in the chloroplast stroma by the reductive pentose-phosphate pathway (Calvin cycle). As outlined by Gontero *et al.* (Chapter 7), the carbon dioxide acceptor molecule ribulose 1,5-bisphosphate (RubP) is carboxylated by ribulose 1,5-bisphosphate carboxylase/oxygenase (Rubisco), yielding an unstable C6 intermediate that rapidly hydrolyzes into two molecules of 3-phosphoglyceric acid (3-PGA). 3-PGA represents the first stable intermediate of carbon fixation and is reduced to glyceraldehyde-3-phosphate (GAP) by the consecutive actions of phosphoglycerate kinase and NADPH-dependent glyceraldehyde-phosphate dehydrogenase. One ATP and one NADPH are consumed during the reduction of one 3-PGA. Hence, the carboxylation of one RubP and the subsequent reduction of the resulting two 3-PGA to GAP requires two ATP and two NADPH. GAP is freely interconvertible with dihydroxyacetone 3-phosphate (DAP) by the activity of triose-phosphate isomerase. GAP and DAP represent the actual end products of the reducing phase of the reductive pentose-phosphate pathway (RPPP). One out of six synthesized triose phosphates can be withdrawn from the Calvin cycle for sucrose or starch biosynthesis whereas five out of six TPs have to enter the regenerative phase of the RPPP to produce the CO_2 acceptor RubP.

4.2.2 The plastidic triose-phosphate pool – a metabolic crossway

The triose phosphates that are not required for the regeneration of RubP represent substrates for the first metabolic 'crossroad' in the allocation of recently assimilated CO_2 to different metabolic fates. TPs can either be retained inside the chloroplast to enter plastid-localized pathways, or they can be exported to the cytosol. Because chloroplasts are bound by a double membrane system (the chloroplast envelope), solutes cannot freely permeate between the chloroplast stroma and the surrounding cytosol. The plastidial metabolism is interfaced with cytosolic metabolism by solute transporters that reside in the inner chloroplast envelope membrane [1, 2]. These transporter proteins catalyze the specific exchange of solutes between plastid and cytosol. The export of the triose phosphates GAP and DAP to the cytosol is mediated by the triose phosphate/phosphate translocator (TPT) [3]. This transporter catalyzes the strict counter-exchange of phosphorylated C3 compounds such as triose phosphates and 3-PGA with plastidic phosphate (Pi), but does not accept phospho*enol*pyruvate (PEP), pentose phosphates or hexose phosphates (see [2, 4, 5] for recent reviews on the Pi translocator family). The strict one-to-one stoichiometry of the counter-exchange is important for the maintenance of Pi homeostasis in the stroma because Pi is required for the biosynthesis of ATP from ADP and Pi in the light reaction of photosynthesis [6]. If export of Pi from the plastid stroma in the form of TPs would not be balanced by counter-exchange with Pi (or phosphorylated carbon compounds), this would cause a depletion of the plastidial Pi pool that would eventually lead to inhibition of photosynthetic electron transport [7] and ultimately to damage of the photosynthetic machinery. As outlined in the next section, the coupling of TP export to the cytosol to the import of Pi is also important for governing the allocation of TPs to either plastidial or cytosolic metabolism.

4.2.2.1 *Communication between the starch and sucrose biosynthetic pathways via TPT*

At the end of the dark period, when the transitory starch pool has been depleted and no net photosynthesis occurs, foliar soluble sugar levels are low, as is the carbon status of sink tissues. Upon onset of light, carbon assimilation by the RPPP sets in and triose phosphates are produced. The two predominant metabolic fates of recently produced TPs are starch biosynthesis in the plastid stroma and sucrose biosynthesis in the cytosol (Figure 4.1) – a mechanism is required that regulates the partitioning of carbon between these competing pathways in response to the metabolic needs of the plant. During the early part of the photoperiod, the demand of sink tissues for carbon is high. Driven by a relatively high TP-to-Pi ratio in the stroma and a reverse ratio in the cytosol early in the light phase, the predominant portion of TPs is exported to the cytosol, thus fueling sucrose biosynthesis in source tissues and its export to the sinks. Pi bound in TPs is released during sucrose biosynthesis and thus becomes available as counter-exchange substrate for the export of additional TPs from the stroma (Figure 4.1). Hence, as long as the rate of sucrose biosynthesis keeps up with the rate of CO_2 assimilation, there will be sufficient Pi available in the cytosol to ensure the efficient export of TP from the stroma. However, sink demand decreases later during the day, causing a drop in the rate of sucrose export and, in response, also of its biosynthesis (see below for a more detailed discussion of the concept of 'sink strength'). Consequently, the release of Pi from TPs slows down, and more and more Pi becomes 'trapped' in organic compounds, which depletes the free Pi pool. Under such conditions, Pi is no longer available in the cytosol to drive the export of TPs from the

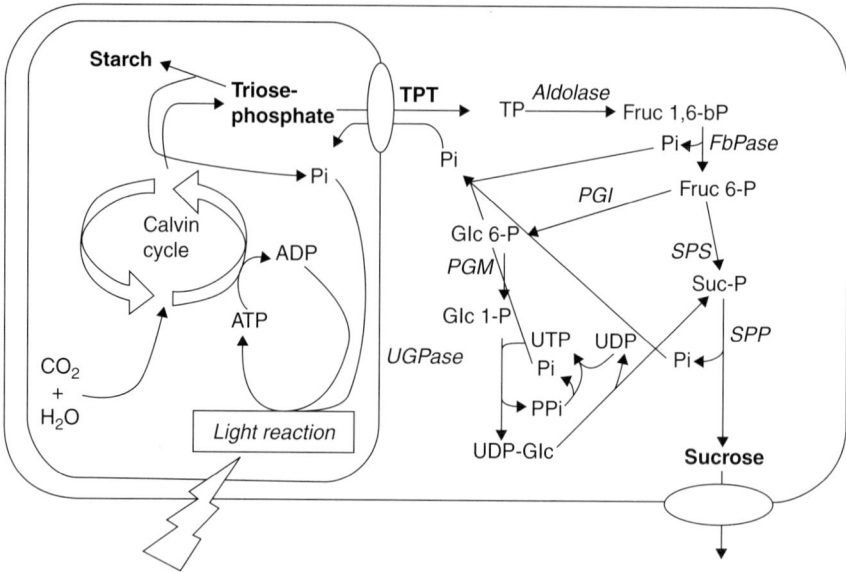

Figure 4.1 Simplified schematic representation of the connection between cytosolic sucrose and plastidic starch metabolism, emphasizing the central role of the triose-phosphate pool and the plastid envelope membrane triose-phosphate/phosphate translocator (TPT). Triose phosphates are the end products of CO_2 assimilation by the Calvin cycle. They can be either exported to the cytosol in counter-exchange with Pi by TPT, or they can be allocated to biosynthesis of transitory starch. Please note that for the sake of simplicity exact stoichiometries and the complete set of enzymes involved in both pathways are not given and the regeneration of UTP from UDP and PPi is abbreviated. Abbreviations: Fru 1,6-bP, fructose 1,6-bisphosphate; Fruc 6-P, fructose 6-phosphate; FbPase, fructose 1,6-bisphosphate phosphatase; SPP, sucrose-phosphate phosphatase; Glc 1-P, glucose 1-phosphate; Glc 6-P, glucose 6-phosphate; PGI, phosphoglucoisomerase; PGM, phosphoglucomutase; Pi, *ortho*-phosphate; PPi, pyrophosphate; SPS, sucrose-phosphate synthase; TP, triose phosphates; Suc-P, sucrose phosphate; TPT, triose-phosphate/phosphate translocator; UGPase, UDP-glucose pyrophosphorylase.

stroma and the concomitant Pi-depletion of the stroma impedes photophosphorylation because of substrate limitation. The resulting decrease of ATP levels in the stroma slows down the conversion of 3-PGA to TPs, leading to an increase of the steady-state 3-PGA pool. In summary, a reduced cytosolic consumption of TPs by sucrose-biosynthesis is communicated to the plastid via the TPT as a decrease in the Pi and an increase in the 3-PGA levels. This change in stromal metabolite levels relieves the Pi inhibition of ADP-glucose pyrophosphorylase (AGPase), the first committed step of starch biosynthesis, and, at the same time, allosterically activates this enzyme by 3-PGA [8, 9]. In addition to allosteric control, AGPase is post-translationally activated by redox modification via the thioredoxin system [10], potentially in response to redox or carbon-status signals (see Chapters 6 and 10 for additional details). The onset of starch biosynthesis releases Pi from the plastidic organic-phosphate pool and fuels photophosphorylation with fresh substrate, thus partially uncoupling the plastid stroma from Pi supply from the cytosol.

The central role of TPT as the communication pathway between sucrose and starch biosynthesis was emphasized by transgenic plants in which TPT expression was either knocked down by antisense repression or knocked out by a T-DNA insertion [11–17]. Antisense repression of the TPT transcript in transgenic potato plants to almost undetectable levels and a corresponding reduction of TPT activity by more than 30% showed that a decreased TPT transport capacity could be compensated by increased allocation of recently assimilated carbon into the transitory starch pool and a corresponding decrease in soluble sugar levels during the light period. However, during the following dark period, starch breakdown and sucrose export from leaves was significantly higher than in wild type plants. Reduced TPT activity thus resulted in an altered carbon partitioning between the plastidic starch and cytosolic sucrose pools during the photoperiod (more starch, less sucrose) and shifting the export of reduced carbon from source to sink tissues partially to the dark phase [15].

Similar to the above-described transgenic potato lines, tobacco plants in which TPT activity was reduced by approximately 70% using antisense also showed an increased rate of starch biosynthesis during the day. However, the actual starch accumulation was similar to the wild type [11]. This discrepancy could be explained by an induction of starch breakdown in the light, leading to soluble sugar and starch levels that were similar to the wild type. Obviously, recently assimilated carbon, instead of being exported to the cytosol in the form of triose phosphates, initially enters the transitory starch pool, from which it is rapidly released, most likely in the form of maltose and glucose (Glc) that are subsequently exported to the cytosol by specific transporters [1, 18–21]. The bottleneck in the 'day path' of carbon export from chloroplasts generated by reduced TPT activity is thus bypassed by increased flux of carbon through the starch pool, thereby converting TP into Glc and maltose that can leave the stroma independent of TPT. Further studies using these antisense plants and TPT overexpression lines showed that TPT activity does not limit photosynthetic carbon dioxide assimilation under ambient conditions; however, maximal rates of carbon dioxide assimilation at saturating CO_2 and light conditions are severely limited by TPT capacity in the wild type and increased assimilation rates under saturating conditions can be achieved by overexpression of TPT [12, 13].

The phenotype of Arabidopsis T-DNA insertion mutants in the TPT gene (*tpt-1*) is similar to that of the transgenic tobacco plants described above. The lack in TPT activity in *tpt-1* was compensated by an increased flux of recently assimilated carbon through the starch pool (i.e., simultaneous biosynthesis and degradation) in the light. However, simultaneous inhibition of starch biosynthesis in *tpt-1* by a genetic cross to a mutant deficient in AGPase activity caused a dramatic dwarfed phenotype of the resulting double mutant. This demonstrates the importance of starch biosynthesis as the major salvage pathway for compensating a defective TPT [14]. It is important to note that simultaneous biosynthesis and degradation of starch in the light does not occur in wild type plants under standard growth conditions [22]. The release of carbon from the transitory starch pool in the light seemingly represents a stress response, possibly triggered by metabolic signals, which arise as a result of Pi limitation of photosynthesis [17].

The regulatory network consisting of TP, 3-PGA, Pi, AGPase and TPT presented here is simplified because it does not include, for example, the intricate control of carbon flux in the cytosol by allosteric regulation of fructose 1,6-bisphosphatase (FbPase) and PPi-dependent phosphofructokinase by fructose 2,6-bisphosphate (F2,6bP) levels [23], and the allosteric and post-translational control of sucrose-phosphate synthase [24, 25]. In a discussion of the role of transporters in the control of primary metabolic pathways in photosynthetic source cells, however, it is important to emphasize the critical role of the chloroplast envelope membrane in separating the plastidic and cytosolic hexose-phosphate pools. If these pools were connected by a hexose-phosphate transporter in the envelope (i.e., a glucose 6-phosphate translocator), the control over the flux of TPs into the hexose-phosphate pool that is exerted by cytosolic FbPase could be bypassed by plastidic FbPase, which is not subject to control by F2, 6bP, thus 'short circuiting' a central control point in cytosolic carbon metabolism. In addition, during the dark period, a possibility for direct exchange of hexose-phosphates across the chloroplast envelope membrane could potentially lead to futile cycling between sucrose biosynthesis from starch breakdown products in the cytosol and the oxidative pentose-phosphate pathway in the plastid stroma. As further discussed below, recent work from Flügge's laboratory has provided convincing *in vivo* evidence for the absence of significant hexose-phosphate transporter activity in photosynthetic tissues of Arabidopsis [26]. Transgenic plants overexpressing the plastidic glucose 6-phosphate/phosphate translocator (GPT) in photosynthetic tissues would represent an interesting tool to further dissect the control of carbon flux and partitioning in photosynthetic cells.

4.2.3 Allocation of recently assimilated carbon to other pathways

Although starch and sucrose biosynthesis represent the major sink for recently assimilated carbon dioxide, a significant portion of reduced carbon is required as precursor for a large number of other primary and secondary metabolic pathways. Reduced carbon can either be directly withdrawn from the regenerative phase of the Calvin cycle, for example in the form of erythrose 4-phosphate (E4P) to fuel the plastid-localized shikimic acid pathway, or it is exported to the cytosol as TP to be converted by, for example, the glycolytic pathway to phospho*enol*pyruvate or pyruvate that serve as precursors for mitochondrial respiration as well as the biosynthesis of organic acids, amino acids, etc. Owing to space constraints, a comprehensive treatise of all possible interactions between the various metabolic pathways in multiple cellular compartments is not possible in this review. We will, therefore, illustrate the principles of interaction between pathways in different compartments with selected examples from amino acid, isoprenoid and nitrogen metabolism.

4.3 Nitrogen assimilation

The assimilation of inorganic nitrogen into organic compounds such as glutamate is an essential biochemical pathway in plant cells. The principle form of nitrogen that is converted into organic N-containing compounds is ammonia. Ammonia can be either taken up directly from the soil, supplied to the plant by symbiotic, nitrogen-fixing microorganisms or it can be generated by reduction of nitrate (see Plate 4). Nitrate is reduced to nitrite in the cytosol by assimilatory NADH-dependent nitrate reductase. Nitrite is then imported into the plastid stroma, either by diffusion or by an active transport process, where it is reduced to ammonia by nitrite reductase. Ammonia is assimilated into the organic form by the joint action of glutamine synthetase (GS) and ferredoxin or NADH-dependent glutamate synthase (Fd/NADH-GOGAT). This process consumes two electrons and one molecule of ATP. The major pathway for ammonia assimilation is the plastid located GS/GOGAT reaction cycle [27–29].

The reaction can be summarized as follows:

$$2\text{-oxoglutarate} + \text{glutamate} + \text{ATP} + 2\,\text{Fd}_{red} + \text{NH}_4^+ \rightarrow$$
$$2\,\text{glutamate} + \text{ADP} + \text{P}_i + 2\,\text{Fd}_{ox}$$

One glutamate can be withdrawn from the reaction cycle to serve as the principal amino group donor in plant metabolism, whereas the second molecule reenters the cycle to serve as an acceptor for ammonia in the GS-catalyzed reaction. In summary, the net reaction of the GS/GOGAT cycle produces one molecule of glutamate from one molecule of each of 2-oxoglutarate (2-OG) and ammonia.

As shown in Figure 4.2, *de novo* glutamate biosynthesis by GS and GOGAT requires the precursor 2-oxoglutarate. 2-OG is synthesized from isocitrate by isocitrate

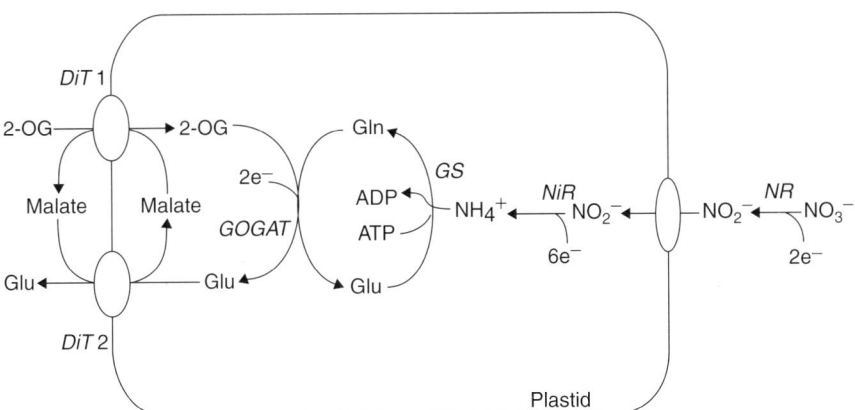

Figure 4.2 Connection of plastidic ammonia assimilation with cytosolic carbon metabolism by the two-translocator system for the transport of dicarboxylic acids and glutamate in the plastid envelope membrane. 2-OG cannot be synthesized inside the plastid stroma and therefore needs to be imported from the cytosol by DiT1. The end product of ammonia assimilation, Glu, is exported to the cytosol by DiT2. Malate cycles across both transporters, thus connecting them to a two-translocator system. The pathway for nitrite transport into plastids is unknown. Abbreviations: DiT1, 2-oxoglutarate/malate translocator; DiT2, glutamate/malate translocator; Glu, glutamate; Gln, glutamine; GOGAT, glutamine:oxoglutarate aminotransferase (glutamate synthase); GS, glutamine synthetase; NiR, nitrite reductase; NR, nitrate reductase; 2-OG, 2-oxoglutarate.

dehydrogenase (IDH). Although a plastidic IDH isozyme has been reported [30], it is generally accepted that 2-OG is synthesized in the cytosol and/or the mitochondria [31–33]. Hence, 2-OG has to be imported into the stroma from the cytosol. The uptake of 2-OG into the plastid stroma is catalyzed by a two-translocator system that is located in the inner plastid envelope membrane [34]. 2-OG is imported in counter-exchange with malate by a 2-oxoglutarate/malate translocator (DiT1). After conversion of 2-OG into glutamate by GS/GOGAT, glutamate is exported to the cytosol in counter-exchange with malate by a glutamate/malate translocator (DiT2). In summary, 2-OG is exchanged for glutamate by two malate-coupled translocators without net malate transport (Figure 4.2). DiT1 was the first component of the two-translocator system that was identified at the molecular level [35, 36], and recently also DiT2 from Arabidopsis [37] and other plant species [35, 38, 39] was reported. Similar to the coupling of plastidic starch metabolism with cytosolic sucrose metabolism by the triose-phosphate/phosphate translocator [40], plastidic dicarboxylate translocators couple plastidic and cytosolic C and N metabolism [35].

Surprisingly, it was recently found that reconstituted DiT1 was able to catalyze the counter-exchange of oxaloacetate (OAA) with malate *in vitro* [38, 41], raising the question of whether DiT1 might also have the function of an OAA/malate exchanger, which is an essential component of the redox shuttle (malate valve) between the plastid stroma and the cytosol [42–44]. It seems unlikely that DiT1 acts as OAA/malate exchanger *in vivo* because OAA-transport by DiT1 is strongly inhibited by malate [38]. The malate-sensitivity of OAA transport by DiT1 explains the need for a specific OAA/malate exchanger that was previously reported by Hatch *et al.* in isolated chloroplasts of spinach and maize [45]. Importantly, the OAA/malate exchanger is competitively inhibited by relatively low concentrations of 2-OG (K_i = 0.42 mM; [45]), raising the question whether the rate of OAA import into the chloroplast by the OAA/malate exchanger might be regulated by cytosolic 2-OG levels, thus coupling the export of reducing equivalents to the cytosol via the malate valve to the rate of 2-OG consumption by plastidic ammonia assimilation [46].

Nitrogen assimilation requires the interaction of three cellular compartments: plastid stroma, cytosol and mitochondria. The conversion of TP to 2-OG involves the glycolytic pathway (conversion of two molecules of TP to: (i) one molecule of PEP (by phosphoglycerate kinase, phosphoglycerate mutase and enolase) and (ii) one molecule of pyruvate (by above mentioned enzymes and pyruvate kinase), respectively), carboxylation of PEP to OAA by PEP-carboxylase and conversion of pyruvate to acetyl-CoA by mitochondrial pyruvate dehydrogenase (PDH). The subsequent reduction of OAA to malate and its conversion to 2-OG requires a partial tricarboxylic acid cycle and hence mitochondrial metabolism (see Chapter 12 as well as refs. [1, 2, 5, 35] and [47] for recent reviews). In leaves of C_3-type plants, approximately 90% of the capacity of the ammonia assimilatory machinery is used for reassimilation of ammonia that is released from the photorespiratory carbon cycle [48, 49]. During photorespiration, ammonia is generated in the mitochondrial matrix by glycine decarboxylase (GDC) [50, 51] and it was previously assumed that the plastidial isozyme of glutamine synthetase 2 (GS2) represents the major pathway for reassimilation of photorespiratory ammonia [52, 53]. The recent surprising discovery of dual targeting of glutamine synthetase 2 (GS2) to chloroplasts and mitochondria [54] opens the possibility of ammonia assimilation by GS in the mitochondrial matrix, directly at the site of ammonia generation by GDC. However, to maintain the nitrogen stoichiometry of the photorespiratory pathway, glutamine needs to be converted to glutamate, which serves as amino donor for the peroxisome-localized glyoxylate amino transferase reaction. The conversion of glutamine and 2-OG to glutamate is catalyzed by Fd-GOGAT (see above) and the current understanding is that Fd-GOGAT (and NADH-GOGAT) is exclusively localized in the plastid stroma [55, 56]. Hence, glutamine needs to be shuttled back to chloroplasts by a yet uncharacterized transport mechanism. Potentially, a glutamate/glutamine shuttle could operate between mitochondria and plastids. A glutamine/glutamate antiporter has been purified from rat liver mitochondria [57] and a similar transport system was also described in isolated chloroplasts [58]. The corresponding genes, however, are unknown. Overall, the photorespiratory pathway is highly compartmentalized, involving plastids, cytosol, peroxisomes and mitochondria. Metabolite transporters are critical to maintain the high net fluxes between these organelles [47]; however, to date only one transporter involved in this pathway, the plastidial glutamate/malate transporter DiT2, has been identified at the molecular level [38].

4.4 Amino acid and isoprenoid metabolism

Chloroplasts are the major intracellular site of amino acid biosynthesis in plants. For example, the minor amino acids Val, Arg, Ile, His, Leu, Trp, Tyr, Met, Phe, Cys and Lys and the major amino acid Glu are predominantly or exclusively synthesized in the plastid stroma. Enzymes involved in Thr biosynthesis carry plastid-targeting signals and are thus also likely to be localized in plastids [59, 60]. In addition, chloroplasts are the sole site of *de novo* glutamate biosynthesis from 2-OG by the various GS and GOGAT isozymes [49]. Asp, Gln, Ser and Gly can be synthesized in multiple compartments, whereas asparagine synthase, at least in Arabidopsis, seems to be confined to the cytosol [61]. The sole or predominant localization of some amino acid biosynthetic pathways in the plastid stroma requires (i) import systems for carbon precursors and cofactors of amino acid biosynthetic pathways and (ii) export systems for amino acids that are required in other cellular compartments for protein biosynthesis or as amino group donors for other biosynthetic pathways. In addition, amino uptake systems are needed for those amino acids that are not synthesized inside the plastid (i.e., proline) but are necessary for protein biosynthesis in the plastid stroma. Similarly, mitochondria also rely on amino acid uptake systems to provide mitochondrial protein biosynthesis with precursors. With the exception of the plastidic glutamate/malate transporter DiT2, which serves as the major Glu export pathway from plastids [38], none of the plastidic amino acid transporters has been unequivocally identified at the molecular level, whereas a mitochondrial arginine/ornithine transporter has been reported recently [62]. A comprehensive treatise of amino acid metabolism, its control and its compartmentation is beyond the scope of this chapter and the reader is referred to several excellent reviews covering this topic [49, 63–70]. We will focus here on recent findings relating to the compartmentation of amino acid biosynthesis and their metabolism in plant cells.

4.4.1 Methionine and S-adenosylmethionine metabolism

Although the Arabidopsis genome encodes three isozymes of methionine synthase that have different subcellular localizations, including cytosol and plastids [71], the *de novo* biosynthesis of the methionine precursor homocysteine from cysteine by cystathionine synthetase and cystathionine lyase is confined to the plastid stroma [72]; hence plastids are the unique site of *de novo* methionine biosynthesis in plant cells [71]. Importantly, Met not only serves as building block for protein biosynthesis, but is also a component of the activated methyl donor S-adenosylmethionine (SAM). SAM is the sole methyl donor in metabolism that is required in a wide range of transmethylation reactions by SAM-dependent methyltransferases such as the chlorophyll, tocopherol, plastoquinone, lipid and nucleic acid biosyntheses. SAM is synthesized from Met by SAM synthetase, an enzyme that is exclusively localized in the cytosol of plant cells [72, 73]. The end product of the methylation reaction, S-adenosyl homocysteine is recycled to Met via Hyc by AdoHyc hydrolase, methionine synthase and SAM synthase. Since SAM is required as a methyl donor in biosynthetic reactions in both mitochondria and chloroplasts, uptake systems for SAM are needed in both organelles. Such SAM transport systems have been recently described in yeast and human mitochondria [74, 75] and a transport system with similar kinetic properties and substrate specificity was identified in isolated spinach chloroplasts [71]. Both, mitochondria and chloroplasts take up SAM from the cytosol and export AdoHyc for regeneration to SAM in the cytosol. The corresponding carriers seem to catalyze both SAM uniport and SAM/AdoHyc counter-exchange.

4.4.2 Shikimic acid pathway and aromatic amino acid biosynthesis

The aromatic amino acids Trp, Tyr and Phe are derived from the shikimic acid pathway that is confined to the plastid stroma [76]. Chloroplasts, unlike most nongreen plastids, lack the activities of enolase and/or phosphoglyceromutase [77–79]; hence triose phosphates cannot be converted to PEP in the plastid stroma. Therefore, chloroplasts (and likely also other plastid subtypes) require a transporter for the import of PEP into the plastid stroma. A plastid envelope membrane transporter that catalyzes the counter-exchange of PEP with Pi, but that does not accept TPs or hexose phosphates as substrates (phospho*enol*pyruvate/phosphate translocator, PPT) was recently identified from cauliflower, maize and Arabidopsis [80, 81]. The Arabidopsis genome harbors two PPT genes (*PPT1*, *PPT2*) that both encode functional PPT proteins [82]. A knockout mutant in PPT1, *cue1*, displays a reticulate phenotype in which interveinal regions of the leaves are visibly pale, whereas paraveinal regions are green. This phenotype is accompanied by aberrantly shaped mesophyll cells containing abnormal chloroplasts, but bundle sheath cells and chloroplasts appear like the wild type [80]. It was hypothesized that the phenotype is due to reduced import of PEP into plastids, causing substrate limitation of the shikimate pathway, and possibly also of isopentenyl biosynthesis. This hypothesis was supported by metabolic analysis, demonstrating that metabolites derived from the shikimate pathway, such as phenylpropanoids, were severely reduced in *cue1* [80, 83]. In addition, constitutive overexpression of pyruvate:phosphate dikinase (PPDK) in the stroma of *cue1* abrogated the reticulate phenotype of mutant [83]. PPDK can generate PEP from pyruvate that can still be taken up from the cytosol, thereby bypassing the defective PPT1 and generating an alternative source for PEP in the plastid stroma. This approach also indicates that Arabidopsis mesophyll chloroplasts have a substantial capacity for pyruvate transport across the envelope membrane. However, the overexpression of PPDK did not alleviate all aspects of the *cue1* phenotype, indicating that the phenotype cannot exclusively be attributed to a reduction in PEP transport capacity [83].

The second precursor required in addition to PEP for the shikimate pathway is E4P, which can be withdrawn from the pentose-phosphate cycle. E4P is produced together with xylulose 5-phosphate (Xul 5P) from fructose 6-phosphate (F6P) and GAP in a reversible reaction catalyzed by transketolase (TK). Hence TK represents an important branch point in metabolism: the substrate F6P is the precursor for starch biosynthesis, GAP can be exported to the cytosol by TPT, E4P can serve as precursor for the shikimate pathway but also for the production of Xul 5P. TK is thus also critical for the regeneration of the CO_2 acceptor RubP and thus for the continued operation of the Calvin cycle.

The important role of TK in both photosynthetic and phenylpropanoid metabolism was demonstrated by antisense repression of TK in transgenic tobacco plants [84]. A relatively small decrease of TK activity (20–40%) inhibited the regeneration of RubP and carbon assimilation and caused marked changes in the partitioning of recently assimilated carbon between sucrose and starch biosynthesis. Reduced TK activity also caused pronounced decreases in aromatic amino acids, intermediates and end products of phenylpropanoid metabolism such as Tyr, Phe, Trp, caffeic acid, tocopherol and hydroxycinnamic acids [84]. In addition, a massive decrease in lignin content was observed if TK activity was reduced by 40% or more.

The shikimate pathway represents a striking example of coordination and interaction between plastidic and cytosolic metabolism and the role of solute transporters in this process. In photosynthetic tissues, carbon precursors are derived from the Calvin cycle. Whereas E4P can be directly withdrawn from the RPPP, PEP is synthesized in the cytosol from TP that has been exported from the chloroplast by TPT, and is then imported into the plastid by PPT. This 'detour' through parts of the cytosolic glycolytic pathway couples the production and use of PEP in the cytosol to its metabolism in plastids. Since multiple metabolic routes 'draw' from and 'feed' into the PEP pool, the connection between pathways in multiple cellular compartments at the level of a central metabolite is essential for cross-pathway and cross-compartment control of metabolism.

4.4.3 Isoprenoid synthesis via the deoxy-xylulose 5-phosphate pathway

Another example for a plastidic pathway that requires interaction of cytosolic and plastidial metabolism is the deoxy-xylulose 5-phosphate (DOXP) pathway for the biosynthesis of isopentenyldiphosphate (IPP), also known as the 2-methylerythritol-4-phosphate pathway [85, 86]. Most if not all plastid-synthesized isoprenoids such as carotenoids, tocopherol, isoprene and the phytol side chain of chlorophyll are derived from the DOXP pathway and not, as previously assumed, via the mevalonate (MVA) pathway. The MVA pathway is localized in the cytosol and uses acetyl-CoA as a precursor for the biosynthesis of the isoprenoid precursor IPP, whereas the plastidic DOXP pathway is fueled by pyruvate and GAP [86]. Whereas GAP can be withdrawn from the RPPP, either pyruvate or PEP has to be imported from the cytosol because photosynthetic plastids lack the glycolytic sequence from GAP to PEP [77–79]. Pyruvate transporters have been characterized in a number of C_4 [87, 88] and C_3 [89] plants. Additional evidence for efficient uptake of pyruvate into chloroplasts from C_3 plants comes from the observation that isolated spinach chloroplasts are able to synthesize pyruvate-derived amino acids upon external supply of pyruvate [90]. Moreover, as outlined above, the PPT-deficient mutant *cue1* could be complemented by overexpression of plastidic PPDK, demonstrating that pyruvate uptake into chloroplasts also occurs *in vivo* [91]. Another possible route for pyruvate import into chloroplast is uptake of PEP and conversion of PEP and ADP to pyruvate and ATP by pyruvate kinase.

The compartmental separation of the DOXP and MVA pathways for IPP biosynthesis is not absolute; considerable evidence suggests crosstalk between both pathways [92]. A recent study, in fact, demonstrated that intermediates of the DOXP and MVA pathways such as DOXP and IPP could be exchanged between the chloroplast stroma and the cytosol [93]. DOXP can be exchanged with Pi and 3-PGA, but not with glucose 6-phosphate and is actively accumulated in the plastid stroma. The transport of DOXP is preferentially mediated by the xylulose 5-phosphate/phosphate translocator (XPT) that was previously shown to accept triose phosphates, Xul 5-P, Ru 5-P and Ery 4-P as counter-exchange substrates for inorganic phosphate, whereas Rib 5-P and hexose phosphates are not transported by this carrier [94]. IPP is taken up into isolated chloroplasts at a lower rate than DOXP and it is not actively accumulated. IPP is transported by a transport system that is distinct from the plastidic Pi translocators – neither reconstituted recombinants GPT, TPT or XPT were able to catalyze the uptake of IPP into proteoliposomes. However, when total chloroplast envelopes were reconstituted into liposomes preloaded with Pi or phosphorylated carbon compounds, uptake of radiolabeled IPP was observed. Using back-exchange experiments, it was further shown that IPP transport by reconstituted chloroplast envelope membranes occurs unidirectionally, most likely by facilitated diffusion, but presence of phosphorylated compounds or Pi on the *trans*-side of the membrane is required as activators or positive regulators of the IPP transporter [94].

4.5 Sucrose and amino acid loading into the phloem for long-distance transport

In most plant species amino acids and sucrose represent the major transport form of organic nitrogen and carbon, respectively, and the overview on mechanisms of source to sink translocation of metabolites will focus on these.

Long-distance transport of amino acids between organs occurs generally via the phloem pathway and the xylem in the transpiration stream (Plate 4). Translocation of amino acids from photosynthetically active organs and storage sources to sinks is primarily by the phloem, though phloem–xylem exchange of amino acids can occur in both directions [95–97]. In contrast, sucrose, and some other sugars and sugar alcohols are translocated between source and sink only by the phloem [98–102]. In source leaves, following assimilation of CO_2 and inorganic nitrogen, and transient storage of organic carbon and nitrogen compounds, sucrose and amino acids are loaded into the phloem of the minor veins for long-distance transport to developing sink organs [102–104]. To accommodate the phloem translocation mechanism, sucrose and amino acids must be concentrated in the sieve tube sap, which provides an efficient means for transport of large amounts of sucrose and amino nitrogen to sinks such as apices, newly developing tissues and particularly strong sinks such as reproductive organs and seeds and vegetative storage organs that accumulate large amounts of protein, starch, or lipids [95, 105–107]. Plasma membrane localized transporters are key components of this translocation process both at the source and sink ends of the system (Plate 4).

4.5.1 Mobilization of stored carbon and nitrogen

Accumulation or transport of assimilates are dynamic processes that can respond rapidly to changing environmental, physiological and developmental conditions. These processes are equally applicable to green leaves and storage organs such as rhizomes, tubers and bulbs, which function as vegetative reproductive propagules, and seeds, where remobilization of starch, lipid, or protein reserves fuels growth of new shoots or plants. In many plant species large pools of starch can accumulate in the plastids of green leaves when conditions are optimal for photoassimilation of CO_2 and sucrose synthesis exceeds its export from the source leaf [108, 109]. During darkness or with increasing sink demand the starch can be converted to sucrose to be loaded into the phloem for long-distance transport [109], thus maintaining growth. In a similar manner, controlled proteolysis is essential in allowing plants to change their protein content during development, to adapt to new environmental conditions, to reutilize amino acids that would otherwise be lost during senescence of an organ [110–112] and during regrowth from vegetative propagules. Studies on a wide range of organisms showed that protein turnover is common to all plants and leaf protein degradation has a large number of functions in physiology and development. Leaf proteins are degraded when carbon and nitrogen are required to support growth of sink organs, and this process increases during leaf development from mature, green leaves to senescent leaves. In green cells most of the protein is located in chloroplasts, so these organelles are the major source of the organic nitrogen salvaged from senescing tissues. The most abundant of these proteins is Rubisco, located in the stroma of the chloroplast. In leaves of many leguminous plants nitrogen is also accumulated as vegetative storage protein in paraveinal mesophyll cell vacuoles whose amount is directly related to source–sink dynamics [113–115]. Several studies have demonstrated that the process of leaf senescence measured either as yellowing or declining rates of photosynthesis is broadly correlated with decreases in the amounts of Rubisco and other proteins involved in photosynthetic carbon reduction. The enzymes responsible for degradation of proteins during senescence have not been fully identified yet. However, senescence associated genes (SAGs) such as *SAG12* from Arabidopsis encoding for cysteine proteases are suggested to play a central role in the degradation of leaf proteins [112, 116, 117]. The resulting amino acids (and peptides) are available for long-distance transport to sink tissues.

4.5.2 Mechanisms of phloem loading and involvement of transporter proteins

For long-distance transport sucrose and amino acids have to be exported out of mature and the senescing leaf chlorenchyma cells and loaded into the phloem. In principle, there are two basic mechanisms by which the phloem can be loaded with assimilates. The first is referred to as the symplasmic path, which relies on direct movement between cells via plasmodesmata. Plasmodesmata are intercellular structures providing a potential symplasmic continuity between adjacent cells [118]. In the symplasmic loading mechanism, plasmodesmata connect the mesophyll cells and elements of the phloem such as bundle sheath cells, phloem parenchyma cells, companion cells and sieve elements (see Plate 4). The second mechanism is referred to as apoplastic phloem loading, in which sugars and amino acids are first exported into the phloem apoplast from nonphloem cells by an unknown process. The organic N and C compounds are then taken up into the sieve element/companion cell complex of the phloem by an energy-requiring plasma membrane transport step. Plasmodesmata are essentially absent between companion cell/sieve element complex and the mesophyll for this system, or when present, they might be 'closed' [119, 120]. Recent studies by Turgeon and Medville [121] suggest that only plants that transport oligosaccharides larger than sucrose are symplasmic loaders, and all other plant species follow phloem loading via the apoplast independent of the frequency of plasmodesmata.

Both apoplastic and symplasmic loading result in an increase of solutes in the sieve element, which is important for generating the driving force for phloem translocation [122]. Our current understanding of the mechanism that drives transport of sucrose and amino acids in the phloem is based on the osmotic pressure-flow hypothesis of Münch [122] and the pioneering molecular model of Giaquinta [123, 124] for sucrose uptake and translocation. It is well established that sucrose loading occurs via plasma membrane sucrose/proton symporters energized by proton pumping ATPases [98–100, 102, 103, 125, 126] transporting sucrose against a concentration gradient into the sieve element/companion cell complex (SE/CC). In a similar mechanistic fashion, recent studies indicate amino acid loading into the phloem is mediated by H^+-coupled amino acid symporters [101, 104, 127–129].

For sugar transport, many plants accumulate and translocate primarily sucrose, which requires high affinity and possible additional low affinity transporters. In contrast the amino acids are a heterogeneous chemical group and their transport is likely more complicated than that for sucrose. The relative concentrations of some specific amino acids in the phloem and the mesophyll cytosol of spinach and barley were shown to be similar, suggesting passive diffusion of amino acid into the phloem [130, 131]. However, studies with the same and other plant species also show that some amino acids are in higher concentrations in the phloem than in the cytosol of mesophyll cells, thus indicating that certain amino acids must be actively loaded into the SE/CC [130–133]. It has also been shown that a concentration gradient of amino acid can be found between the apoplast and the phloem [132], which indicates that transport systems are involved in the uptake of various amino acids from the apoplast into the phloem. This is supported by physiological studies with plasma membrane vesicles isolated from mature leaf tissue that have demonstrated the existence of proton-coupled (symport) amino acid transporters [134–138]. However, transport studies with membrane vesicles also give evidence that low-level facilitated diffusion mechanisms might coexist with the symporter-mediated accumulation [139]. In addition, similar composition of amino acids in the phloem and the mesophyll cells points to an uptake system which might not be highly selective. Physiological studies with leaf membrane vesicles support the hypothesis of a low selective uptake system for amino acids [135]. Nevertheless, more selective, high affinity systems for phloem loading of specific amino acids cannot be excluded.

4.5.3 Sucrose transporters

Sucrose transporters are among the most studied transport systems in plants and are thus the best characterized at the physiological and molecular levels. In many plants organic carbon is preferentially translocated in the form of sucrose, though some plants may transport large amounts of polyols and raffinose series sugars as well [100, 121]. It is well established that for long-distance translocation sucrose is taken up from the leaf apoplast and imported into the sieve element/companion cell complex of the phloem by H^+ coupled symporters (see Plate 4). So-called SUT/SUC transporters have been found to be responsible for this import step [101]. The physiology, expression, localization and function of the sucrose transporters have been extensively studied and a number of comprehensive reviews have been published [99, 101–103, 107, 140–143]. However, there is still much to be learned about these transporters. For example, with respect to phloem loading mechanisms differences have been found in localization of sucrose transporters from Arabidopsis (SUC2, [144]) and solanaceous species (SUT1, [145, 146]). While SUC2 was localized to the CC plasma membrane of Arabidopsis, SUT1 of tobacco and potato were detected in the SE membrane of leaves, petioles and stem and hypothesized to function in CC or SE loading, respectively. This might be interpreted as species differences in phloem loading systems but additional, undiscovered or uncharacterized SUT/SUC transporters might be present as indicated by recent studies with *Plantago major* [147] which may expand our concepts of the role of sucrose transporters in distinct cell types of the phloem.

4.5.4 Amino acid transporters

Considering the number of different amino acids and their analogs that might potentially be transported, one might hypothesize that a range of amino acid transporters with differing affinities must exist. In fact, the reality lies somewhere in between, with specific, selective and broad-spectrum transporters having been identified. Molecular studies on amino acid transport have mostly concentrated on Arabidopsis, where a minimum of 53 amino acid transporters are predicted to be present based on DNA sequences in the genome [148]. The AtAAPs (amino acid permeases) with eight members are the most extensively investigated amino acid transporters. They were characterized to be broad specificity proton symporters and to transport mainly neutral (and acidic) amino acids with low affinity [149]. Northern-blot analysis and promoter-reporter gene fusion studies revealed that the AAPs have distinct expression patterns, are developmentally regulated and are expressed in specific tissues indicating that the various transporters fulfill specific functions within the plant [104, 150–152]. Other transporters include the AtProTs, transporters for compatible solutes including the amino acid proline [153–155], AtANT1, an aromatic-neutral amino acid transporter [156], AtCATs, cationic amino acid transporters [157, 158] and AtLHTs. AtLHT1 was characterized to be a lysine and histidine transporter [159]. However, when analyzing the substrate specificity of another member of the family, AtLHT2, it was found that this protein transports uncharged and negatively charged amino acids with high affinity [160]. There is still little known about the actual cellular location of most of these transporters. While some of the transporters might be located in the plasma membrane where they are involved in import of amino acids into the cell, others could play a role in metabolite transfer processes across organelle membranes. Promoter-GUS studies in Arabidopsis, tobacco and oil seed rape have localized AAP transporters to the vasculature and vascular parenchyma [151, 161] and in the case of *AtAAP2* and *AtAAP3* to the phloem, suggesting a role in apoplastic phloem loading of amino acids [150, 151]. Additional studies and strategies such as employment of green fluorescent protein (GFP) reporter or development of viable transporter-specific antibodies will help resolve some of these questions. Unfortunately, while immunolocalization can be definitive, developing antibodies to membrane bound transporters is known to be notoriously difficult.

4.5.5 Genetic modification of phloem loading with assimilates

Antisense and reverse genetic studies have provided direct evidence that leaf phloem sucrose transporters are critical to sucrose accumulation by the phloem and that their presence is required for normal plant growth and development and for reproductive success [162–164]. These experiments showed that SUT1/SUC2 T-DNA insertion mutants and transgenic plants were characterized by accumulation of leaf carbohydrates and reduced export of sucrose out of the source leaf, with a concomitant adverse effect on sink development. In another study, overexpression of SUT1 under control of the constitutive CaMV 35S promoter in potato led to a reduced level of sucrose in source leaves and increased amounts in tubers, respectively [165]. However, little change was observed in photosynthetic rate in leaves and starch content in tubers. This lack of effects on metabolism could be dependent on the suitability of the promoter used but could also indicate that in some cases both metabolic and transport activities might need to be altered to drive processes downstream of translocation in sink tissue.

The relative contribution of an individual amino acid transporter to physiological parameters of amino acid translocation is less understood. As reviewed above, a number of putative plant amino acid transporters with varying properties have been identified; however, only one study directly demonstrates a physiological role for such a transporter in the plant [166]. Using potato plants, it was found that *antisense* inhibition of a leaf specific amino acid permease (*StAAP1*) under control of the constitutive CaMV 35S promoter leads to the reduction of the pool of free amino acids in potato tubers. These experiments prove a function of amino acid permeases in long-distance transport and also confirm that amino acid transport to, and accumulation in, sink organs can be manipulated by molecular genetic engineering in principal, and in particular at the source end. Studies on overexpression, knockout or knockdown of specific amino acid transporters in various organs will help to dissect the role of the members of the different amino acid transporter families in the physiology of the plant.

4.6 Phloem unloading in sinks and assimilate transport to developing seeds

Uptake, transport and partitioning of organic carbon and nitrogen between different plant parts depend on developmental and physiological status of the plant as well as on more complex concepts of 'sink strength'. Sink strength as a measurable term has been discussed controversially, since activity and sink size are dependent on photosynthesis, competition between sinks, the transport pathways and the demand by the rest of the plant [167, 168]. In addition, expression and control of transporters are important factors. Transport and transfer processes contributing to nitrogen and carbon flow from source to sink do not function independently, but are highly regulated and integrated events, both at the physiological and molecular levels [109, 169–171]. The availability of powerful molecular techniques and molecular genetic information on an increasing number of plant species provides the tools that have allowed us to make progress on identifying transporters involved in transport and transfer processes at the cellular, tissue and whole plant levels.

With respect to source to sink transport, the amount of organic nitrogen or carbon in sinks depends on the amounts of assimilates translocated in the phloem sap, indicating that phloem loading in source leaves is a key factor in the distribution process. Sucrose has been identified as an important signal molecule that influences carbon and nitrogen partitioning [172–174]. Similarly, the regulatory effect of nitrogen was shown by Crafts-Brandner *et al.* [175] who found that N deficiency enhanced the rate of leaf senescence and leads to a decline in amounts of protein in the leaf and probably to changes in amino acid transport activities. Leaf protein degradation can also be repressed by removal of developing sink organs (fruits, pods) indicating that sinks might be involved in the signalling process [111]. A well-studied example of this dynamic is depodded soybean, where it was shown that in old leaves Rubisco is still degraded but the mobilized N is apparently refixed in the paraveinal mesophyll as vegetative storage protein resulting in little change in total leaf protein [176, 177]. Thus, sink limitation can affect transport properties of the leaf [114, 178, 179].

4.6.1 *Assimilate distribution and transport in seed coats*

Seeds represent strong sinks for sugars and amino-N, and large amounts of organic carbon and nitrogen are translocated in support of metabolism and storage of proteins and starch or lipids in seed sinks [128, 180]. Transport and accumulation of sugars and nitrogen involves coordination of a set of separate processes in the maternal (seed coat) and filial (embryo/endosperm) seed tissues, because the developing embryo and endosperm are symplasmically isolated from the surrounding maternal tissue (see Plate 4). Thus, there is a requirement for membrane passage of imported sucrose and amino acids at the interface between these two tissues [181]. Phloem unloading of the organic compounds in the seed coats is considered to be symplasmic [102, 182–185]. Following symplasmic passage via the seed coat parenchyma cells, assimilates are exported into the apoplastic cavity from where they are subsequently absorbed by the growing embryo. While a number of experiments demonstrated that large amounts of amino acids and sugars are released into the apoplastic space between seed coat and embryo, the mechanism by which they are released is not understood. de Jong and Wolswinkel [186] proposed that release of amino acids from pea seed coats occurs by a facilitated membrane transport mechanism, probably through nonselective pores [187]. A similar system was postulated for efflux of sucrose from pea seed coats [188]. This is in contrast to studies done on *Vicia faba* and *Phaseolus vulgaris* [189, 190] where efflux experiments have demonstrated that sucrose/H^+ antiport is responsible for release of sucrose from the seed coat into the apoplast prior to loading into the developing cotyledons by a symporter. In these systems, sucrose release from the seed coat seems to be regulated by cell turgor that directly acts on activity of a sucrose/H^+ antiporter [191].

4.6.2 *Uptake of sucrose and amino acids by the developing embryo*

After export from the seed coat to the apoplastic cavity, assimilates are absorbed by the growing embryo. Besides sucrose, a mixture of 20 amino acids was found in the seed apoplast [192–194]. Glutamine, asparagine, alanine and threonine were the predominant amino acids, although this was dependent on seed developmental stage. The basic mechanisms of sucrose and amino acid absorption by the embryo have been studied extensively with leguminous seeds such as pea and soybean [188, 195, 196]. For amino acids these studies have demonstrated the presence of two different transport systems, a saturable and a nonsaturable one. The nonsaturable system appears to be of primary importance during early seed development. In soybean an additional saturable system is present but its activity is low compared to the non-saturable system and its contribution to amino acid uptake is insignificant during the early developmental stage. However, it is interesting to note that the soybean saturable system can be derepressed by N and C starvation, a mechanism that probably involves synthesis of new carriers [196]. Studies on valine absorption by pea cotyledons have shown similar results [195]. Uptake of valine during early cotyledon development was strictly dependent on the external amino acid concentration over the whole concentration range, indicating a diffusion-like mechanism. As seed development progresses, the passive transport pathway is supplemented by a saturable, and probably active, transport system. This saturable system appeared to operate as an H^+/symporter, and its activity increased rapidly up to the latest stage of seed development and was only slightly reduced by low osmolarity [195].

4.6.3 Specialized sites of import

In some plant species and tissues so-called transfer cells are located in strategic positions where they promote high membrane fluxes of solute between the apoplast and symplasm. Transfer cells have a specialized wall-membrane apparatus comprising an invaginated secondary ingrowth wall and an associated plasma membrane enriched in proteins that support solute exchange [183, 197]. In the seeds of many grain legumes, including pea, transfer cells develop at the interface between maternal and filial tissue. More specifically, transfer cells line the inner surface of the seed coat and the juxtaposed outer surface of the enclosed cotyledons [198–200]. For example in faba bean and pea cotyledons import and accumulation of solutes is facilitated by the development of a transfer cell complex at their outer epidermis surface [198, 199, 201, 202]. Within the epidermal transfer cells, wall ingrowths are polarized to the outer walls that face the seed coat, which release the assimilate into the apoplast for import into the cotyledons [183, 198, 199]. It is only recently that the existence and function of proton-coupled co-transport within transfer cells was demonstrated [199, 201, 203–206].

4.6.4 Sucrose and amino acid import into developing embryos/cotyledons

Sucrose transporters have been identified in a number of sink tissues including flowers and seeds of dicot and some monocot plant species [99, 101, 102]. Most of the studies have concentrated on sucrose import into legume seeds [199, 201, 203–209]. SUT protein and RNA have been localized to the epidermal transfer cell layers of *Pisum sativum* and *V. faba* cotyledons, where they are responsible for the high rates of uptake of sucrose from the seed apoplast [204–206]. In addition, some of the apoplastic sucrose might be cleaved by a plasma membrane bound invertase followed by import of fructose and glucose via monosaccharide transporters [99, 101, 206]. While expression of SUT transporters within the storage parenchyma has been found, they are suggested to play a function in retrieving sucrose leaking into apoplast from these storage cells, which are active in starch synthesis [199].

We have a poor understanding of the role of amino acid transporters in seed development compared to better-studied sucrose transporters. So far only a few genes coding for amino acid transporters have been shown to be expressed in seeds of Arabidopsis [210, 211], *P. sativum* [212] and *V. faba* [213–215]. All these transporters belong to the AAP family. Promoter-GUS analysis in Arabidopsis revealed *AtAAP1* and *AtAAP8* expression in seed coat and developing embryo. RNA *in situ* hybridization studies demonstrated that the legume *AAP*s were located in the storage parenchyma of pea and faba bean cotyledons, respectively, indicating a function in amino acid retrieval from the apoplast for protein synthesis and accumulation [213–216] (see Plate 4). However, *PsAAP1* is also highly expressed in the abaxial epidermal transfer cells of pea cotyledons [216] where it is hypothesized to import neutral and acidic amino acid from the apoplastic cavity of the seed.

4.6.5 *Genetic modification of assimilate transport in seeds*

Given the differing affinities of members of the amino acid transporter families, there is the possibility of targeting accumulation of a specific class of amino acids or overall amino acids through molecular genetic techniques. While there are no published studies on manipulation of amino acid transport processes aimed at effects on seeds, the potential importance of such manipulations is indicated by work with a sucrose transporter. It was demonstrated that SUT activity can be functionally overexpressed in storage parenchyma of developing pea cotyledons by using the pea vicilin promoter [217]. Heterologous expression of *StSUT1* in this tissue could significantly enhance the sucrose uptake capacity of the storage parenchyma cells and also increased cotyledon growth rate. However, influx in whole cotyledons through the outer surface of the epidermal transfer cell layer was only increased by ca 20%. From these results we might conclude that cellular location of transporter activity is probably a key determinant in genetic manipulation of import processes into seeds. Use of a promoter that targets transporter expression and function to the outer cell layer of developing embryos/cotyledons might help in enhancing metabolite uptake and lead to increased accumulation of storage compounds.

4.7 Assimilate transport and metabolism in sink cells

Storage sinks receive carbon precursors from source tissues that are then converted to and stored as starch, oil and proteins that accumulate in specific compartments of sink cells. Fatty acid, amino acid and starch biosynthesis are all associated with plastids; hence we will emphasize the role of plastids in sink cell metabolism. Carbon transport into nongreen plastids has been recently reviewed [218]; therefore, we will focus on recent developments since this last review.

4.7.1 The role of hexose-phosphate import into nongreen plastids

Glc 6-P is the sole precursor for starch biosynthesis in Arabidopsis (and likely also in other plant species). This was demonstrated by the starch-free phenotype of a mutant in the plastidic isozyme of phosphoglucomutase (PGM; conversion of Glc 6-P to Glc 1-P). Since Glc 1-P and Glc 6-P are freely interconvertible in the cytosol of plant cells due to the activity of cytosolic PGM, the cytosol contains both Glc 1-P and Glc 6-P. If Glc 1-P could be taken up into chloroplasts, a deficiency in the plastidic PGM isozyme should not lead to a starch-free phenotype because the defect would be bypassed by uptake of Glc 1-P. Hence, the starch-free phenotype of the plastidic PGM mutant unequivocally demonstrates that no pathway exists for the uptake of Glc 1-P into plastids [219, 220]. A mutant in the plastidic isozyme of phosphoglucoisomerase (conversion of Fru 6-P to Glu 6-P) contains starch in nongreen tissues whereas photosynthetically active tissues are starch free [221]. This finding indicates that mesophyll chloroplasts do not have significant Glc 6-P transport capacity that could bypass the interrupted link between the Calvin cycle and starch biosynthesis, whereas nongreen plastids are able to take up Glc 6-P from the cytosol and are thus not dependent on plastidic PGI activity. Glc 6-P is taken up into nongreen plastids by the GPT [222]. GPT has the broadest substrate specificity within the plastidic Pi translocator family [223]. It is able to exchange Pi for TPs, Glc 6-P, the pentose phosphates ribulose 5-phosphate, ribose 5-phosphate and xylulose 5-phosphate, as well as PEP, 3-PGA and E4P. The genome of Arabidopsis harbors two paralogous GPT genes, *AtGPT1* and *AtGPT2* [4]. Expression of the corresponding cDNAs in yeast cells and reconstitution into liposomes demonstrated that both genes encode functional Glc 6-P translocators [26]. Constitutive expression of either gene under control of the CaMV 35S promoter in an Arabidopsis mutant deficient in plastidic phosphoglucomutase [221] rescued the low-starch phenotype in photosynthetic tissues of this mutant, demonstrating that: (i) both transporters function as Glc 6-P translocators in intact plants and that (ii) Glc 6-P transport across the envelope membrane of Arabidopsis plants with wild type expression levels of GPTs is insignificant [26]. Homozygous knockout mutants in *GPT2* did not display an obvious phenotype, whereas homozygous knockout mutants in *GPT1* could not be obtained [26]. Arabidopsis plants hemizygous for T-DNA insertions in *GPT1* showed a distorted segregation ratio, reduced male and female transmission efficiency and an arrest in pollen and ovule development, indicating a critical role of *GPT1* during gametogenesis [26]. The female gametophyte development in the mutant was arrested at a stage before the polar nuclei fuse and pollen development was associated with less lipid body formation and a disintegration of the membrane system. In addition, an effect on the developing sporophyte was indicated by the complete lack of homozygous adult mutant plants and the occasional occurrence of aborted embryos [26]. Because of this severe phenotype, the physiological role of *GPT1* in intact plants is difficult to assess. It will require inducible RNAi repression or inducible expression of *GPT1* in a *gpt1* background to obtain further insight. The situation becomes even more complex, taking into account the recent identification of a novel plastidic (i.e., stroma-localized) hexokinase in tobacco [224] and of similar proteins in *Physcomitrella patens* [225, 226] and Arabidopsis (I. Krassovskaya and A.P.M. Weber, unpublished). Hexokinase activity has also been detected in the plastid stroma of developing castor bean seeds [227]. Since plastids are able to take up glucose [21, 228, 229] and ATP [230, 231] from the cytosol, they should, in theory, be independent of glucose 6-phosphate import from the cytosol. Obviously, this is not the case, as indicated by the severe phenotype of *gpt1*. A possible answer to this problem is nonoverlapping gene expression patterns and nonredundant functions of both pathways for supplying plastids with glucose 6-phosphate. Detailed studies of the tissue-specific and developmental expression patterns will be required to address this problem.

4.7.2 The role of ATP-transport into nongreen plastids

The adenine nucleotide translocator (NTT) was the first plastidic metabolite translocator that was identified in isolated chloroplasts, using a silicon–oil filtration–centrifugation technique [230]. The corresponding cDNA was identified by Neuhaus' group [231, 232]. Since nongreen plastids cannot produce ATP by photophosphorylation, it was proposed that NTT is required to supply nonphotosynthetic plastids (and also chloroplasts during the dark) with ATP to drive biosynthetic reactions such as protein, RNA, starch and fatty acid biosynthesis. Alternatively, ATP could be generated in nongreen plastids by glycolytic kinases (3-PGA kinase and pyruvate kinase) via substrate level phosphorylation.

4.7.3 Knockout of NTTs in Arabidopsis

The Arabidopsis genome encodes two plastidic ATP/ADP translocators, *AtNTT1* and *AtNTT2* [233]. Neuhaus's group has recently reported T-DNA knockout mutants and RNAi knockdown lines for both NTTs in Arabidopsis [234]. Surprisingly, it was found that NTTs are not required for a complete life cycle of Arabidopsis plants (i.e., a double-knockout was viable and able to produce seed). Nevertheless, the plants showed retarded development, a reduced ability to generate primary roots and delayed chlorophyll accumulation in seedlings. The phenotype was more severe under short-day conditions and alleviated under long-day conditions, indicating a substantial role of NTT in supplying chloroplasts with ATP at night. Seed weight and lipid contents were reduced in *AtNTT2* but not in *AtNTT1* knockout lines. Obviously, Arabidopsis embryo plastids are able to generate ATP from endogenous sources, possibly substrate-level phosphorylation of ADP by phosphoglycerate kinase and/or pyruvate kinase [234]. Hence, other metabolic routes can at least partially compensate the block in ATP import due to the knockout of NTT activity.

4.7.4 Antisense repression and overexpression of NTTs in potato

Transgenic potato plants showing antisense repression of the endogenous *NTT* and constitutive overexpression of *AtNTT1* have been analyzed in detail [235, 236].

Tuber size was massively decreased and tuber morphology was severely altered in antisense lines. The total amount of tuber starch was reduced and its composition changed, showing a reduced amylose to amylopectin ratio. Free sugar levels, UDP-Glc and hexose phosphates were increased in tubers from antisense plants, whereas PEP, isocitrate, adenine and uridine nucleotides, and inorganic pyrophosphate levels were slightly decreased. Potato plants overexpressing *AtNTT1* showed a marked increase in starch content, and the amylose to amylopectin ratio was higher than in the wild type [235]. Adenine and uridine nucleotides, and inorganic pyrophosphate levels were elevated in tubers from sense plants, whereas soluble sugars remained unaltered. The ADP-Glc content was reduced by 50% in antisense tubers, whereas its content was increased up to twofold in sense tubers. These results suggested a close interaction between plastidial adenylate transport and starch biosynthesis, indicating that ADP-Glc pyrophosphorylase (and thus starch biosynthesis) is ATP-limited in wild type potato plants [236].

4.7.5 A novel role for Rubisco in developing oilseeds

Oil is a major reserve material in many plant seeds. The major source for seed oil biosynthesis is sucrose, which is imported from source tissues via the phloem and is converted to the fatty acid precursor acetyl-CoA by glycolysis and pyruvate dehydrogenase. The C3 compound pyruvate is converted by PDH into the C2 compound acetyl-CoA and one molecule of CO_2. Hence, for every two carbon atoms converted to acetyl-CoA, one carbon atom (i.e., one-third of the carbon) would be lost as CO_2, therefore fatty acid biosynthesis was thought to be a quite wasteful process. Using metabolic flux and elementary flux mode analysis, Schwender *et al.* [237] recently demonstrated that oil production in *Brassica* seeds is significantly more efficient than predicted from the paper biochemistry – a loss of only one carbon atom per three carbon atoms converted to oil is reported (i.e., 40% less loss than previously assumed). Key to this unexpected finding is a previously undescribed role of Rubisco in the conversion of carbohydrates to oil. In this metabolic route, Rubisco does not act in the self-regenerating Calvin cycle, but it carboxylates ribulose 1,5-*bis*phosphate that is synthesized *de novo* from F6P and TPs using CO_2 that is produced by the PDH reaction, thus recovering part of the CO_2 previously believed to be lost in the process (see [237] and supplementary data for a discussion of the role of PEPC in reassimilation of CO_2 produced by PDH). In contrast to the Calvin cycle, the resulting 3-PGA is converted to pyruvate (and subsequently to acetyl-CoA) and not used to regenerate the CO_2 acceptor. Part of the ATP required for the activation of ribose 1-phosphate (Rib 1P) to RubP by phosphoribulokinase is generated by the light reaction in green seeds of *Brassica*. Likely, this metabolic scheme applies to most other 'green' seeds but not to oilseeds that do not have functional photosystems, such as sunflower seeds [237, 238]. The generation of ATP by photophosphorylation in green seeds may also explain the relatively small effect of a knockout in plastidic ATP/ADP translocators as outlined above.

4.8 Concluding remarks

The localization of biochemical and catabolic pathways has been resolved for many organic compounds, and it is clear that these pathways often involve or require transfer of intermediates between various compartments. Similarly, end products used as major organic nutrients for growth and reproduction need to be transported from cells, which synthesize them, to cells at sites of long-distance transport, and then into cells at sites of utilization. Membrane bound transporters are required in these processes and given the number of transported substrates, it is likely that a large and diverse population of transporters is operating in a tissue or even an individual cell in support of the normal physiology of a plant. Information on some transporters involved in short-, intermediate- and long-distance transport has become available within the last 15 years, but a very large number of transporters are yet to be characterized. Similarly, while a few specific transporters connecting metabolic pathways within a cell have been identified, many more must exist and their identification and characterization are critical to understanding the control of metabolic fluxes. In recent years much progress has been made on identification and characterization of metabolite transporters using plant genome information and heterologous expression systems such as *Saccharomyces cerevisiae* and *Xenopus* oocytes. However, for most of these transport proteins, especially the amino acid transporters, the location of activity, the control of expression and their physiological function are still unknown. While localization of expression of a transporter in a particular cell or tissue provides a sound basis for predicting an important function within that site, it does not provide the type of information needed to assess the overall physiological contribution to metabolite transport and transfer processes. This requires a combination of physiological and molecular genetic approaches. Identification of transporters with specific transport function (and their promoters) at critical points of metabolite distribution and partitioning is also needed for targeted manipulation of transport processes for various aspects of crop improvement, such as nutrient content, enhanced partitioning to harvestable products, manipulation of growth dynamics, etc. Identification of transporters and characterization of their kinetic parameters will be invaluable to our basic understanding of plant function and to attempts to modify specific aspects where such transporters are key regulators.

Acknowledgements

This work was supported by grants IOB 0135344, IOB 0448506 (M.T.) and MCB 0348074 (A.P.M.W.) from the National Science Foundation, grant 2001-35318-10990 (M.T.) from the National Research Initiative of the USDA Cooperative State Research, Education and Extension Service and the Nation and grant DE-FG02-04ER15562 (A.P.M.W) from the U.S. Department of Energy, which is greatly appreciated. We are grateful to Jörg Schwender for critical reading of the chapters on ATP transport into plastids and the novel role of Rubisco in seeds.

References

1. A.P.M. Weber (2004) Solute transporters as connecting elements between cytosol and plastid stroma. *Current Opinion in Plant Biology* **7**, 247–253.
2. A.P.M. Weber, R. Schwacke and U.I. Flügge (2005) Solute transporters of the plastid envelope membrane. *Annual Review of Plant Biology* **56**, 99–131.
3. U.I. Flügge (1999) Phosphate translocators in plastids. *Annual Review of Plant Physiology and Plant Molecular Biology* **50**, 27–45.
4. S. Knappe, U.I. Flügge and K. Fischer (2003) Analysis of the plastidic phosphate translocator gene family in Arabidopsis and identification of new phosphate translocator-homologous transporters, classified by their putative substrate-binding site. *Plant Physiology* **131**, 1178–1190.
5. A.P.M. Weber, J. Schneidereit and L.M. Voll (2004) Using mutants to probe the *in vivo* function of plastid envelope membrane metabolite transporters. *Journal of Experimental Botany* **55**, 1231–1244.
6. U.I. Flügge (1995) Phosphate translocation in the regulation of photosynthesis. *Journal of Experimental Botany* **46**, 1317–1323.
7. T.D. Sharkey (1985) O_2-insensitive photosynthesis in C-3 plants – its occurrence and a possible explanation. *Plant Physiology* **78**, 71–75.
8. S.K. Hwang, P.R. Salamone and T.W. Okita (2005) Allosteric regulation of the higher plant ADP-glucose pyrophosphorylase is a product of synergy between the two subunits. *FEBS Letters* **579**, 983–990.
9. J.M. Cross, M. Clancy, J.R. Shaw *et al.* (2004) Both subunits of ADP-glucose pyrophosphorylase are regulatory. *Plant Physiology* **135**, 137–144.
10. J.H. Hendriks, A. Kolbe, Y. Gibon, M. Stitt and P. Geigenberger (2003) ADP-glucose pyrophosphorylase is activated by posttranslational redox-modification in response to light and to sugars in leaves of Arabidopsis and other plant species. *Plant Physiology* **133**, 838–849.
11. R.E. Häusler, N.H. Schlieben, B. Schulz and U.I. Flügge (1998) Compensation of decreased triose phosphate/phosphate translocator activity by accelerated starch turnover and glucose transport in transgenic tobacco. *Planta* **204**, 366–376.
12. R.E. Häusler, N.H. Schlieben and U.I. Flügge (2000) Control of carbon partitioning and photosynthesis by the triose phosphate/phosphate translocator in transgenic tobacco plants (*Nicotiana tabacum*). II. Assessment of control coefficients of the triose phosphate/phosphate translocator. *Planta* **210**, 383–390.
13. R.E. Häusler, N.H. Schlieben, P. Nicolay, K. Fischer, K.L. Fischer and U.I. Flügge (2000) Control of carbon partitioning and photosynthesis by the triose phosphate/phosphate translocator in transgenic tobacco plants (*Nicotiana tabacum* L.). I. Comparative physiological analysis of tobacco plants with antisense repression and overexpression of the triose phosphate/phosphate translocator. *Planta* **210**, 371–382.
14. A. Schneider, R.E. Häusler, U. Kolukisaoglu *et al.* (2002) An *Arabidopsis thaliana* knock-out mutant of the chloroplast triose phosphate/phosphate translocator is severely compromised only when starch synthesis, but not starch mobilisation is abolished. *Plant Journal* **32**, 685–699.
15. D. Heineke, A. Kruse, U.I. Flügge *et al.* (1994) Effect of antisense repression of the chloroplast triose-phosphate translocator on photosynthetic metabolism in transgenic potato plants. *Planta* **193**, 174–180.
16. J.W. Riesmeier, U.I. Flügge, B. Schulz *et al.* (1993) Antisense repression of the chloroplast triose phosphate translocator affects carbon partitioning in transgenic potato plants. *Proceedings of the National Academy of Sciences of the United States of America* **90**, 6160–6164.
17. R.G. Walters, D.G. Ibrahim, P. Horton and N.J. Kruger (2004) A mutant of Arabidopsis lacking the triose-phosphate/phosphate translocator reveals metabolic regulation of starch breakdown in the light. *Plant Physiology* **135**, 891–906.
18. T. Niittyla, G. Messerli, M. Trevisan, J. Chen, A.M. Smith and S.C. Zeeman (2004) A previously unknown maltose transporter essential for starch degradation in leaves. *Science* **303**, 87–89.

19. S.C. Zeeman, S.M. Smith and A.M. Smith (2004) The breakdown of starch in leaves. *New Phytologist* **163**, 247–261.
20. S.E. Weise, A.P.M. Weber and T.D. Sharkey (2004) Maltose is the major form of carbon exported from the chloroplast at night. *Planta* **218**, 474–482.
21. A. Weber, J.C. Servaites, D.R. Geiger *et al.* (2000) Identification, purification, and molecular cloning of a putative plastidic glucose translocator. *Plant Cell* **12**, 787–801.
22. S.C. Zeeman, A. Tiessen, E. Pilling, K.L. Kato, A.M. Donald and A.M. Smith (2002) Starch synthesis in Arabidopsis. Granule synthesis, composition, and structure. *Plant Physiology* **129**, 516–529.
23. M. Stitt (1990) Fructose-2,6-bisphosphate as a regulatory metabolite in plants. *Annual Review of Plant Physiology and Plant Molecular Biology* **41**, 153–185.
24. S.C. Huber and J.L. Huber (1996) Role and regulation of sucrose-phosphate synthase in higher plants. *Annual Review of Plant Physiology and Plant Molecular Biology* **47**, 431–444.
25. J.E. Lunn and E. MacRae (2003) Complexities in the synthesis of sucrose. *Current Opinion in Plant Biology* **6**, 208–214.
26. P. Niewiadomski, S. Knappe, S. Geimer *et al.* (2005) The Arabidopsis plastidic glucose 6-phosphate/phosphate translocator GPT1 is essential for pollen maturation and embryo sac development. *Plant Cell* **17**, 760–775.
27. B.J. Miflin and P.J. Lea (1980) Ammonia assimilation. In: *The Biochemistry of Plants,* (ed. B.J. Miflin), Academic Press, London, pp. 169–202.
28. R.J. Ireland and P.J. Lea (1999) The enzymes of glutamine, glutamate, asparagine, and aspartate metabolism. In: *Plant Amino Acids. Biochemistry and Biotechnology,* (ed. B.K. Singh), Dekker, New York, pp. 49–109.
29. B.J. Miflin and P.J. Lea (1976) The pathway of nitrogen assimilation in plants. *Phytochemistry* **15**, 873–885.
30. S. Galvez, E. Bismuth, C. Sarda and P. Gadal (1994) Purification and characterization of chloroplastic NADP-isocitrate dehydrogenase from mixotrophic tobacco cells (comparison with the cytosolic isoenzyme). *Plant Physiology* **105**, 593–600.
31. M. Hodges (2002) Enzyme redundancy and the importance of 2-oxoglutarate in plant ammonium assimilation. *Journal of Experimental Botany* **53**, 905–916.
32. M. Lancien, P. Gadal and M. Hodges (2000) Enzyme redundancy and the importance of 2-oxoglutarate in higher plant ammonium assimilation. *Plant Physiology* **123**, 817–824.
33. M. Lancien, M.R.S. Ferrario, Y. Roux *et al.* (1999) Simultaneous expression of NAD-dependent isocitrate dehydrogenase and other Krebs cycle genes after nitrate resupply to short-term nitrogen-starved tobacco. *Plant Physiology* **120**, 717–726.
34. K.C. Woo, U.I. Flügge and H.W. Heldt (1987) A two-translocator model for the transport of 2-oxoglutarate and glutamate in chloroplasts during ammonia assimilation in the light. *Plant Physiology* **84**, 624–632.
35. A. Weber and U.I. Flügge (2002) Interaction of cytosolic and plastidic nitrogen metabolism in plants. *Journal of Experimental Botany* **53**, 865–874.
36. A. Weber, E. Menzlaff, B. Arbinger, M. Gutensohn, C. Eckerskorn and U.I. Flügge (1995) The 2-oxoglutarate/malate translocator of chloroplast envelope membranes: molecular cloning of a transporter containing a 12-helix motif and expression of the functional protein in yeast cells. *Biochemistry* **34**, 2621–2627.
37. M. Taniguchi, Y. Taniguchi, M. Kawasaki *et al.* (2002) Identifying and characterizing plastidic 2-oxoglutarate/malate and dicarboxylate transporters in *Arabidopsis thaliana*. *Plant and Cell Physiology* **43**, 706–717.
38. P. Renné, U. Dreßen, U. Hebbeker *et al.* (2003) The Arabidopsis mutant *dct* is deficient in the plastidic glutamate/malate translocator DiT2. *Plant Journal* **35**, 316–331.
39. Y. Taniguchi, J. Nagasaki, M. Kawasaki, H. Miyake, T. Sugiyama and M. Taniguchi (2004) Differentiation of dicarboxylate transporters in mesophyll and bundle sheath chloroplasts of maize. *Plant and Cell Physiology* **45**, 187–200.
40. U.I. Flügge (1999) Phosphate translocators in plastids. *Annual Review of Plant Physiology and Plant Molecular Biology* **50**, 27–45.

41. M. Taniguchi, Y. Taniguchi, M. Kawasaki *et al.* (2002) Identifying and characterizing plastidic 2-oxoglutarate/malate and dicarboxylate transporters in *Arabidopsis thaliana. Plant and Cell Physiology* **43**, 706–717.
42. D. Heineke, B. Riens, H. Grosse *et al.* (1991) Redox transfer across the inner chloroplast envelope membrane. *Plant Physiology* **95**, 1131–1137.
43. L.E. Fridlyand, J.E. Backhausen and R. Scheibe (1998) Flux control of the malate valve in leaf cells. *Archives of Biochemistry and Biophysics* **349**, 290–298.
44. R. Scheibe (2004) Malate valves to balance cellular energy supply. *Physiologia Plantarum* **120**, 21–26.
45. M.D. Hatch, L. Dröscher, U.I. Flügge and H.W. Heldt (1984) A specific translocator for oxaloacetate transport in chloroplasts. *FEBS Letters* **178**, 15–19.
46. A.P.M. Weber and W.M. Kaiser (2005) Rapid modulation of nitrate reduction in leaves by redox-coupling of plastidic and cytosolic metabolism. In: *Photosynthesis: Fundamental Aspects to Global Perspectives,* (eds A. van der Est and D. Bruce), Allen Press, Lawrence, KS, pp. 810–812.
47. A.P.M. Weber (2005) Synthesis, export, and partitioning of the end products of photosynthesis. In: *The Structure and Function of Plastids,* (eds R.R. Wise and J.K. Hoober), Kluwer Academics, Dordrecht, pp. 273–292.
48. B. Hirel and P.J. Lea (2001) Ammonia assimilation. In: *Plant Nitrogen,* (eds P.J. Lea and J.F. Morot-Gaudry), Springer-Verlag, Berlin, pp. 79–100.
49. G.M. Coruzzi (2003) Primary N-assimilation into amino acids in Arabidopsis. In: *The Arabidopsis Book,* (eds C.R. Somerville and E.M. Meyerowitz), American Society of Plant Biologists, Rockville, MD, pp. 1–17.
50. R. Douce, J. Bourguignon, M. Neuburger and F. Rebeille (2001) The glycine decarboxylase system: a fascinating complex. *Trends in Plant Science* **6**, 167–176.
51. R. Douce and M. Neuburger (1999) Biochemical dissection of photorespiration. *Current Opinion in Plant Biology* **2**, 214–222.
52. R.M. Wallsgrove, A.J. Keys, I.F. Bird, M.J. Cornelius, P.J. Lea and B.J. Miflin (1980) The location of glutamine-synthetase in leaf-cells and its role in the reassimilation of ammonia released in photo-respiration. *Journal of Experimental Botany* **31**, 1005–1017.
53. R.D. Blackwell, A.J.S. Murray and P.J. Lea (1987) Inhibition of photosynthesis in barley with decreased levels of chloroplastic glutamine synthetase activity. *Journal of Experimental Botany* **38**, 1799–1809.
54. M. Taira, U. Valtersson, B. Burkhardt and R.A. Ludwig (2004) *Arabidopsis thaliana* GLN2-encoded glutamine synthetase is dual targeted to leaf mitochondria and chloroplasts. *Plant Cell* **16**, 2048–2058.
55. M. Lancien, M. Martin, M.H. Hsieh, T. Leustek, H. Goodman and G.M. Coruzzi (2002) Arabidopsis glt1-T mutant defines a role of NADH-GOGAT in the non-photorespiratory ammonium assimilatory pathway. *Plant Journal* **29**, 347–358.
56. K.T. Coschigano, R. Melo-Oliveira, J. Lim and G.M. Coruzzi (1998) Arabidopsis gls mutants and distinct Fd-GOGAT genes: implications for photorespiration and primary nitrogen assimilation. *Plant Cell* **10**, 741–752.
57. C. Indiveri, G. Abruzzo, I. Stipani and F. Palmieri (1998) Identification and purification of the reconstitutively active glutamine carrier from rat kidney mitochondria. *Biochemical Journal* **333**(2), 285–290.
58. J. Yu and K.C. Woo (1988) Glutamine transport and the role of the glutamine translocator in chloroplasts. *Plant Physiology* **88**, 1048–1054.
59. G. Curien, R. Dumas, S. Ravanel and R. Douce (1996) Characterization of an *Arabidopsis thaliana* cDNA encoding an S-adenosylmethionine-sensitive threonine synthase. Threonine synthase from higher plants. *FEBS Letters*, **390**, 85–90.
60. M. Lee and T. Leustek (1999) Identification of the gene encoding homoserine kinase from *Arabidopsis thaliana* and characterization of the recombinant enzyme derived from the gene. *Archives in Biochemistry and Biophysics* **372**, 135–142.
61. H.M. Lam, M.H. Hsieh and G. Coruzzi (1998) Reciprocal regulation of distinct asparagine synthetase genes by light and metabolites in *Arabidopsis thaliana. Plant Journal* **16**, 345–353.

62. E. Catoni, M. Desimone, M. Hilpert et al. (2003) Expression pattern of a nuclear encoded mitochondrial arginine-ornithine translocator gene from Arabidopsis. *BMC Plant Biology* **3**, 1.
63. G. Galili and R. Höfgen (2002) Metabolic engineering of amino acids and storage proteins in plants. *Metabolic Engineering* **4**, 3–11.
64. G. Galili (2002) New insights into the regulation and functional significance of lysine metabolism in plants. *Annual Review of Plant Biology* **53**, 27–43.
65. B.K. Singh (1999) Biosynthesis of valine, leucine, and isoleucine. In: *Plant Amino Acids: Biochemistry and Biotechnology*, (ed. B.K. Singh), Marcel Dekker, New York, NY, pp. 227–247.
66. B.F. Matthews (1999) Lysine, threonine, and methionine biosynthesis. In: *Plant Amino Acids: Biochemistry and Biotechnology*, (ed. B.K. Singh), Marcel Dekker, New York, NY, pp. 205–225.
67. K.J. Denby and R.L. Last (1999) Diverse regulatory mechanisms of amino acid biosynthesis in plants. *Genetic Engineering (NY)* **21**, 173–189.
68. E.R. Radwanski and R.L. Last (1995) Tryptophan biosynthesis and metabolism: biochemical and molecular genetics. *Plant Cell* **7**, 921–934.
69. H. Hesse, V. Nikiforova, B. Gakiere and R. Hoefgen (2004) Molecular analysis and control of cysteine biosynthesis: integration of nitrogen and sulphur metabolism. *Journal of Experimental Botany* **55**, 1283–1292.
70. R. Hell, R. Jost, O. Berkowitz and M. Wirtz (2002) Molecular and biochemical analysis of the enzymes of cysteine biosynthesis in the plant *Arabidopsis thaliana*. *Amino Acids* **22**, 245–257.
71. S. Ravanel, M.A. Block, P. Rippert et al. (2004) Methionine metabolism in plants: chloroplasts are autonomous for *de novo* methionine synthesis and can import S-adenosylmethionine from the cytosol. *Journal of Biological Chemistry* **279**, 22548–22557.
72. S. Ravanel, B. Gakiere, D. Job and R. Douce (1998) The specific features of methionine biosynthesis and metabolism in plants. *Proceedings of the National Academy of Sciences of the United States of America* **95**, 7805–7812.
73. A.D. Hanson and S. Roje (2001) One-carbon metabolism in higher plants. *Annual Review of Plant Biology* **52**, 119–137.
74. G. Agrimi, M.A. Di Noia, C.M. Marobbio, G. Fiermonte, F.M. Lasorsa and F. Palmieri (2004) Identification of the human mitochondrial S-adenosylmethionine transporter: bacterial expression, reconstitution, functional characterization and tissue distribution. *Biochemical Journal* **379**(1), 183–190.
75. C.M. Marobbio, G. Agrimi, F.M. Lasorsa and F. Palmieri (2003) Identification and functional reconstitution of yeast mitochondrial carrier for S-adenosylmethionine. *EMBO Journal* **22**, 5975–5982.
76. J. Schmid and N. Amrhein (1995) Molecular organization of the shikimate pathway in higher plants. *Phytochemistry* **39**, 737–749.
77. P. Bagge and C. Larsson (1986) Biosynthesis of aromatic amino acids by highly purified spinach chloroplasts – compartmentation and regulation of the reactions. *Physiologia Plantarum* **68**, 641–647.
78. D. Van der Straeten, R.A. Rodrigues-Pousada, H.M. Goodman and M. van Montagu (1991) Plant enolase: gene structure, expression, and evolution. *Plant Cell* **3**, 719–735.
79. S. Borchert, J. Harborth, D. Schünemann, P. Hoferichter and H.W. Heldt (1993) Studies of the enzymic capacities and transport properties of pea root plastids. *Plant Physiology* **101**, 303–312.
80. S.J. Streatfield, A. Weber, E.A. Kinsman et al. (1999) The phosphoenolpyruvate/phosphate translocator is required for phenolic metabolism, palisade cell development, and plastid-dependent nuclear gene expression. *Plant Cell* **11**, 1609–1621.
81. K. Fischer, N. Kammerer, M. Gutensohn et al. (1997) A new class of plastidic phosphate translocators: a putative link between primary and secondary metabolism by the phosphoenolpyruvate/phosphate antiporter. *Plant Cell* **9**, 453–462.
82. S. Knappe, T. Löttgert, A. Schneider, L. Voll, U.I. Flügge and K. Fischer (2003) Characterization of two functional phosphoenolpyruvate/phosphate translocator (PPT) genes in Arabidopsis – AtPPT1 may be involved in the provision of signals for correct mesophyll development. *Plant Journal* **36**, 411–420.

83. L. Voll, R.E. Häusler, R. Hecker et al. (2003) The phenotype of the Arabidopsis *cue1* mutant is not simply caused by a general restriction of the shikimate pathway. *Plant Journal* **36**, 301–317.
84. S. Henkes, U. Sonnewald, R. Badur, R. Flachmann and M. Stitt (2001) A small decrease of plastid transketolase activity in antisense tobacco transformants has dramatic effects on photosynthesis and phenylpropanoid metabolism. *Plant Cell* **13**, 535–551.
85. J. Schwender, M. Seemann, H.K. Lichtenthaler and M. Rohmer (1996) Biosynthesis of isoprenoids (carotenoids, sterols, prenyl side-chain of chlorophylls and plastochinone) via a novel pyruvate/glyceraldehyde 3-phosphate non-mevalonate pathway in the green algae *Scenedesmus obliquus*. *Biochemical Journal* **316**, 73–80.
86. H.K. Lichtenthaler, J. Schwender, A. Disch and M. Rohmer (1997) Biosynthesis of isoprenoids in higher plant chloroplasts proceeds via a mevalonate-independent pathway. *FEBS Letters* **400**, 271–274.
87. S.C. Huber and G.E. Edwards (1977) Transport in C4 mesophyll chloroplasts: characterization of the pyruvate carrier. *Biochimica Biophysica Acta* **462**, 583–602.
88. N. Aoki, J. Ohnishi and R. Kanai (1992) 2 different mechanisms for transport of pyruvate into mesophyll chloroplasts of C4 plants-a comparative study. *Plant and Cell Physiology* **33**, 805–809.
89. M.O. Proudlove and D.A. Thurman (1981) The uptake of 2-oxoglutarate and pyruvate by isolated pea chloroplasts. *New Phytologist* **88**, 255–264.
90. D. Schulze-Siebert, D. Heineke, H. Scharf and G. Schulz (1984) Pyruvate-derived amino acids in spinach chloroplasts: synthesis and regulation during photosynthetic carbon metabolism. *Plant Physiology* **76**, 465–471.
91. L. Voll, R.E. Häusler, R. Hecker et al. (2003) The phenotype of the Arabidopsis *cue1* mutant is not simply caused by a general restriction of the shikimate pathway. *Plant Journal* **36**, 301–317.
92. A. Hemmerlin, J.F. Hoeffler, O. Meyer et al. (2003) Cross-talk between the cytosolic mevalonate and the plastidial methylerythritol phosphate pathways in tobacco bright yellow-2 cells. *Journal of Biological Chemistry* **278**, 26666–26676.
93. U.I. Flügge and W. Gao (2005) Transport of isoprenoid intermediates across chloroplast envelope membranes. *Plant Biology (Stuttg)* **7**, 91–97.
94. M. Eicks, V. Maurino, S. Knappe, U.I. Flügge and K. Fischer (2002) The plastidic pentose phosphate translocator represents a link between the cytosolic and the plastidic pentose phosphate pathways in plants. *Plant Physiology* **128**, 512–522.
95. J.S. Pate (1980) Transport and partitioning of nitrogenous solutes. *Annual Review of Plant Physiology* **31**, 79–99.
96. J.S. Pate, P.J. Sharkey and O.A.M. Lewis (1975) Xylem to phloem transfer of solutes in fruiting shoots of legumes, studied by a phloem bleeding technique. *Planta* **122**, 11–26.
97. A.J.E. van Bel (1984) Quantification of xylem-to-phloem transfer of amino acids by use of inuline (^{14}C) carboxylic acid as xylem transport marker. *Plant Science Letters* **15**, 285–291.
98. J.M. Ward, C. Kühn, M. Tegeder and W.B. Frommer (1997) Sucrose transport in higher plants. *International Review of Cytology* **178**, 41–71.
99. L.E. Williams, R. Lemoine and N. Sauer (2000) Sugar transporters in higher plants – a diversity of roles and complex regulation. *Trends in Plant Science* **5**, 283–290.
100. N. Noiraud, L. Maurousset and R. Lemoine (2001) Transport of polyols in higher plants. *Plant Physiology and Biochemistry* **39**, 717–728.
101. S. Lalonde, D. Wipf and W.B. Frommer (2004) Transport mechanisms for organic forms of carbon and nitrogen between source and sink. *Annual Review of Plant Biology* **55**, 341–372.
102. S. Lalonde, M. Tegeder, M. Throne-Holst, W.B. Frommer and J.W. Patrick (2003) Phloem loading and unloading of sugars and amino acids. *Plant Cell and Environment* **26**, 37–56.
103. D.R. Bush (1999) Sugar transporters in plant biology. *Current Opinion in Plant Biology* **2**, 187–191.
104. W.N. Fischer, B. Andre, D. Rentsch et al. (1998) Amino acid transport in plants. *Trends in Plant Science* **3**, 188–195.
105. J.S. Pate, P.J. Sharkey and C.A. Atkin (1977) Nutrition of a developing legume fruit. *Plant Physiology* **59**, 506–510.

106. W.B. Frommer, B. Hirner, C. Kühn *et al.* (1996) Sugar transport in higher plants. In: *Membranes: Specialized Functions in Plants,* (eds M. Smallwood, J.D. Knox and D.J. Bowles), BIOS Sci., Oxford, pp. 319–335.
107. R. Lemoine (2000) Sucrose transporters in plants: update on function and structure. *Biochimica Biophysica Acta* **1465**, 246–262.
108. R.N. Trethewey and T. ap Rees (1994) A mutant of *Arabidopsis thaliana* lacking the ability to transport glucose across the chloroplast envelope. *Biochemical Journal* **301**, 449–454.
109. M.J. Paul and C.H. Foyer (2001) Sink regulation of photosynthesis. *Journal of Experimental Botany* **52**, 1383–1400.
110. S. Gan and R.M. Amasino (1995) Inhibition of leaf senescence by autoregulated production of cytokinin. *Science* **270**, 1986–1988.
111. S. Gan and R.M. Amasino (1997) Making sense of senescence (molecular genetic regulation and manipulation of leaf senescence). *Plant Physiology* **113**, 313–319.
112. V. Buchanan-Wollaston (1997) The molecular biology of leaf senscence. *Journal of Experimental Botany* **48**, 181–199.
113. T.J. Tranbarger, V.R. Franceschi, D.F. Hildebrand and H.D. Grimes (1991) The soybean 94-kilodalton vegetative storage protein is a lipoxygenase that is localized in paraveinal mesophyll cell vacuoles. *Plant Cell* **3**, 973–987.
114. S.F. Klauer, V.R. Franceschi, M.S.B. Ku and D.Z. Zhang (1996) Identification and localization of vegetative storage proteins in legume leaves. *American Journal of Botany* **83**, 1–10.
115. A.J. Lansing and V.R. Franceschi (2000) The paraveinal mesophyll: a specialized path for intermediary transfer of assimilates in legume leaves. *Australian Journal of Plant Physiology* **27**, 757–767.
116. V. Buchanan-Wollaston, S. Earl, E. Harrison *et al.* (2003) The molecular analysis of leaf senescence – a genomics approach. *Plant Biotechnology Journal* **1**, 3–22.
117. B.F. Quirino, Y.S. Noh, E. Himelblau and R.M. Amasino (2000) Molecular aspects of leaf senescence. *Trends in Plant Science* **5**, 278–282.
118. W.J. Lucas and S. Wolf (1993) Plasmodesmata: the intercellular organelles of green plants. *Trends in Cell Biology* **3**, 308–315.
119. G.A. Thompson and A. Schulz (1999) Macromolecular trafficking in the phloem. *Trends in Plant Science* **4**, 354–360.
120. W.J. Lucas and J.Y. Lee (2004) Plasmodesmata as a supracellular control network in plants. *Nature Reviews Molecular Cell Biology* **5**, 712–726.
121. R. Turgeon and R. Medville (2004) Phloem loading. A reevaluation of the relationship between plasmodesmatal frequencies and loading strategies. *Plant Physiology* **136**, 3795–3803.
122. K.J. Oparka and R. Turgeon (1999) Sieve elements and companion cells-traffic control centers of the phloem. *Plant Cell* **11**, 739–750.
123. R. Giaquinta (1977) Possible role of pH gradient and membrane atpase in loading of sucrose into sieve tubes. *Nature* **267**, 369–370.
124. R. Giaquinta (1977) Possible role of protons and membrane ATPase in phloem loading of sucrose. *Plant Physiology* **59**, 750–755.
125. D.R. Bush (1992) The proton-sucrose symport. *Photosynthesis Research* **32**, 155–165.
126. D.R. Bush (1993) Proton-coupled sugar and amino-acid transporters in plants. *Annual Review of Plant Physiology and Plant Molecular Biology* **44**, 513–542.
127. A. Ortiz-Lopez, H. Chang and D.R. Bush (2000) Amino acid transporters in plants. *Biochimica Biophysica Acta* **1465**, 275–280.
128. S. Delrot, C. Rochat, M. Tegeder and W.B. Frommer (2001) Amino acid transport. In: *Plant Nitrogen,* (eds P.J. Lea and J.F. Morot-Gaudry), Springer-Verlag, Heidelberg, pp. 213–235.
129. L.E. Williams and A.J. Miller (2001) Transporters responsible for the uptake and partitioning of nitrogenous solutes. *Annual Review of Plant Physiology and Plant Molecular Biology* **52**, 659–688.
130. H. Winter, G. Lohaus and H.W. Heldt (1992) Phloem transport of amino-acids in relation to their cytosolic levels in barley leaves. *Plant Physiology* **99**, 996–1004.
131. G. Lohaus, M. Burba and H.W. Heldt (1994) Comparison of the contents of sucrose and amino-acids in the leaves, phloem sap and taproots of high and low sugar-producing hybrids of sugar-beet (*Beta vulgaris* L.). *Journal of Experimental Botany* **45**, 1097–1101.

132. G. Lohaus, H. Winter, B. Riens and H.W. Heldt (1995) Further studies of the phloem loading process in leaves of barley and spinach. The comparison of metabolite concentrations in the apoplastic compartment with those in the cytosolic compartment and in the sieve tubes. *Botanica Acta* **108**, 270–275.
133. B. Riens, G. Lohaus, D. Heineke and H.W. Heldt (1991) Amino-acid and sucrose content determined in the cytosolic, chloroplastic, and vacuolar compartments and in the phloem sap of spinach leaves. *Plant Physiology* **97**, 227–233.
134. Z.C. Li and D.R. Bush (1992) Structural determinants in substrate recognition by proton-amino acid symports in plasma membrane vesicles isolated from sugar beet leaves. *Archives of Biochemistry and Biophysics* **294**, 519–526.
135. Z.C. Li and D.R. Bush (1990) Delta-pH-dependent amino-acid-transport into plasma-membrane vesicles isolated from sugar-beet leaves. 1. Evidence for carrier-mediated, electrogenic flux through multiple transport-systems. *Plant Physiology* **94**, 268–277.
136. Z.C. Li and D.R. Bush (1991) Delta-pH-dependent amino-acid-transport into plasma-membrane vesicles isolated from sugar-beet (*Beta vulgaris* L.) leaves. 2. Evidence for multiple aliphatic, neutral amino-acid symports. *Plant Physiology* **96**, 1338–1344.
137. L.E. Williams, S.J. Nelson and J.L. Hall (1992) Characterization of solute proton cotransport in plasma-membrane vesicles from *Ricinus* cotyledons, and a comparison with other tissues. *Planta* **186**, 541–550.
138. K. Weston, J.L. Hall and L.E. Williams (1995) Characterization of amino-acid-transport in *Ricinus communis* roots using isolated membrane-vesicles. *Planta* **196**, 166–173.
139. L.E. Williams, S.J. Nelson and J.L. Hall (1990) Characterization of solute transport in plasma-membrane vesicles isolated from cotyledons of *Ricinus communis* L. 2. Evidence for a proton-coupled mechanism for sucrose and amino-acid-uptake. *Planta* **182**, 540–545.
140. S. Lalonde, E. Boles, H. Hellmann *et al.* (1999) The dual function of sugar carriers. Transport and sugar sensing. *Plant Cell* **11**, 707–726.
141. S. Delrot, R. Atanassova and L. Maurousset. (2000) Regulation of sugar, aminoacid and peptide plant membrane transporters. *Biochimica Biophysica Acta* **1465**, 281–306.
142. S. Delrot and J.L. Bonnemain (1981) Involvement of protons as a substrate for the sucrose carrier during phloem loading in *Vicia faba* leaves. *Plant Physiology* **67**, 560–564.
143. N. Sauer (1997) Sieve elements and companion cells – extreme division of labour. *Trends in Plant Science* **2**, 285–286.
144. R. Stadler and N. Sauer (1996) The *Arabidopsis thaliana AtSUC2* gene is specifically expressed in companion cells. *Botanica Acta* **109**, 299–306.
145. C. Kühn, V.R. Franceschi, A. Schulz, R. Lemoine and W.B. Frommer (1997) Macromolecular trafficking indicated by localization and turnover of sucrose transporters in enucleate sieve elements. *Science* **275**, 1298–1300.
146. L. Barker, C. Kuhn, A. Weise *et al.* (2000) SUT2, a putative sucrose sensor in sieve elements. *Plant Cell* **12**, 1153–1164.
147. R. Stadler, J. Brandner, A. Schulz, M. Gahrtz and N. Sauer (1995) Phloem loading by the PmSUC2 sucrose carrier from *Plantago major* occurs into companion cells. *Plant Cell* **7**, 1545–1554.
148. D. Wipf, U. Ludewig, M. Tegeder, D. Rentsch, W. Koch and W.B. Frommer (2002) Conservation of amino acid transporters in fungi, plants and animals. *Trends in Biochemical Sciences* **27**, 139–147.
149. W.N. Fischer, D.D. Loo, W. Koch *et al.* (2002) Low and high affinity amino acid H+-cotransporters for cellular import of neutral and charged amino acids. *Plant Journal* **29**, 717–731.
150. B. Hirner, W.N. Fischer, D. Rentsch, M. Kwart and W.B. Frommer (1998) Developmental control of H+/amino acid permease gene expression during seed development of Arabidopsis. *Plant Journal* **14**, 535–544.
151. S. Okumoto, R. Schmidt, M. Tegeder *et al.* (2002) High affinity amino acid transporters specifically expressed in xylem parenchyma and developing seeds of Arabidopsis. *Journal of Biological Chemistry* **277**, 45338–45346.
152. W.N. Fischer, M. Kwart, S. Hummel and W.B. Frommer (1995) Substrate specificity and expression profile of amino acid transporters (AAPs) in Arabidopsis. *Journal of Biological Chemistry* **270**, 16315–16320.

153. D. Rentsch, B. Hirner, E. Schmelzer and W.B. Frommer (1996) Salt stress-induced proline transporters and salt stress-repressed broad specificity amino acid permeases identified by suppression of a yeast amino acid permease-targeting mutant. *Plant Cell* **8**, 1437–1446.
154. R. Schwacke, S. Grallath, K.E. Breitkreuz *et al.* (1999) LeProT1, a transporter for proline, glycine betaine, and gamma-amino butyric acid in tomato pollen. *Plant Cell* **11**, 377–392.
155. S. Grallath, T. Weimar, A. Meyer *et al.* (2005) The AtProT family. Compatible solute transporters with similar substrate specificity but differential expression patterns. *Plant Physiology* **137**, 117–126.
156. L. Chen, A. Ortiz-Lopez, A. Jung and D.R. Bush (2001) ANT1, an aromatic and neutral amino acid transporter in Arabidopsis. *Plant Physiology* **125**, 1813–1820.
157. W.B. Frommer, S. Hummel, M. Unseld and O. Ninnemann (1995) Seed and vascular expression of a high-affinity transporter for cationic amino acids in Arabidopsis. *Proceedings of the National Academy of Sciences of the United States of America* **92**, 12036–12040.
158. Y.H. Su, W.B. Frommer and U. Ludewig (2004) Molecular and functional characterization of a family of amino acid transporters from Arabidopsis. *Plant Physiology* **136**, 3104–3113.
159. L. Chen and D.R. Bush (1997) LHT1, a lysine- and histidine-specific amino acid transporter in Arabidopsis. *Plant Physiology* **115**, 1127–1134.
160. Y.H. Lee and M. Tegeder (2004) Selective expression of a novel high-affinity transport system for acidic and neutral amino acids in the tapetum cells of Arabidopsis flowers. *Plant Journal* **40**, 60–74.
161. J. Tilsner, N. Kassner, C. Struck and G. Lohaus (2005) Amino acid contents and transport in oilseed rape (*Brassica napus* L.) under different nitrogen conditions. *Planta* **221**, 328.
162. J.W. Riesmeier, L. Willmitzer and W.B. Frommer (1994) Evidence for an essential role of the sucrose transporter in phloem loading and assimilate partitioning. *EMBO Journal* **13**, 1–7.
163. C. Kühn, W.P. Quick, J.W. Riesmeier, U. Sonnewald and W.B. Frommer (1996) Companion cell-specific inhibition of the potato sucrose transporter SUT1. *Plant, Cell and Environment* **19**, 1115–1123.
164. R. Lemoine, C. Kuhn, N. Thiele, S. Delrot and W.B. Frommer (1996) Antisense inhibition of the sucrose transporter in potato: effects on amount and activity. *Plant, Cell and Environment* **19**, 1124–1131.
165. G. Leggewie, A. Kolbe, R. Lemoine *et al.* (2003) Overexpression of the sucrose transporter SoSUT1 in potato results in alterations in leaf carbon partitioning and in tuber metabolism but has little impact on tuber morphology. *Planta* **217**, 158–167.
166. W. Koch, M. Kwart, M. Laubner *et al.* (2003) Reduced amino acid content in transgenic potato tubers due to antisense inhibition of the leaf H+/amino acid symporter StAAP1. *Plant Journal* **33**, 211–220.
167. D.R. Geiger and W.J. Shieh (1993) Sink strength – learning to measure, measuring to learn. *Plant, Cell and Environment* **16**, 1017–1018.
168. M. Stitt (1993) Sink strength – integrated systems need integrating approaches. *Plant, Cell and Environment* **16**, 1041–1043.
169. K. Herbers and U. Sonnewald (1998) Molecular determinants of sink strength. *Current Opinion in Plant Biology* **1**, 207–216.
170. C. Atkins (2000) Biochemical aspects of assimilate transfers along the phloem path: N-solutes in lupins. *Australian Journal of Plant Physiology* **27**, 531–537.
171. D.R. Geiger, J.C. Servaites and M.A. Fuchs (2000) Role of starch in carbon translocation and partitioning at the plant level. *Australian Journal of Plant Physiology* **27**, 571–582.
172. T. Roitsch (1999) Source-sink regulation by sugar and stress. *Current Opinion in Plant Biology* **2**, 198–206.
173. U. Wobus and H. Weber (1999) Sugars as signal molecules in plant seed development. *Biological Chemistry* **380**, 937–944.
174. S.I. Gibson (2005) Control of plant development and gene expression by sugar signalling. *Current Opinion in Plant Biology* **8**, 93–102.
175. S.J. Crafts-Brandner, R. Holzer and U. Feller (1998) Influence of nitrogen deficiency on senescence and the amounts of RNA and proteins in wheat leaves. *Physiologia Plantarum* **102**, 192–200.

176. V.R. Franceschi, V.A. Wittenbach and R.T. Giaquinta (1983) Paraveinal mesophyll of soybean leaves in relation to assimilate transfer and compartmentation. III. Immunohistochemical localization of specific glycopeptides in the vacuole after depodding. *Plant Physiology* **72**, 586–589.
177. V.A. Wittenbach, V.R. Franceschi and R.T. Giaquinta (1984) Soybean leaf storage proteins. In: *Current Topics in Plant Biochemistry and Physiology,* (eds D.D. Randall, D.G. Blevins, R.L. Larson and B.J. Rupp), pp. 19–30. University of Missouri, Columbia.
178. V.A. Wittenbach (1982) Effect of pod removal on leaf senescence in soybeans. *Plant Physiology* **70**, 1544–1548.
179. V.A. Wittenbach (1983) Effect of pod removal on leaf photosynthesis and soluble-protein composition of field-grown soybeans. *Plant Physiology* **73**, 121–124.
180. H. Weber, L. Borisjuk and U. Wobus (1997) Sugar import and metabolism during seed development. *Trends in Plant Science* **2**, 169–174.
181. J.H. Thorne (1985) Phloem unloading of C-assimilates and N-assimilates in developing seeds. *Annual Review of Plant Physiology and Plant Molecular Biology* **36**, 317–343.
182. C.E. Offler and J.W. Patrick (1984) Cellular structures, plasma-membrane surface-areas and plasmodesmatal frequencies of seed coats of *Phaseolus vulgaris* L. in relation to photosynthate transfer. *Australian Journal of Plant Physiology* **11**, 79–99.
183. C.E. Offler, S.M. Nerlich and J.W. Patrick (1989) Pathway of photosynthate transfer in the developing seed of *Vicia faba* L. Transfer in relation to seed anatomy. *Journal of Experimental Botany* **40**, 769–780.
184. C.E. Offler and J.W. Patrick (1993) Pathway of photosynthate transfer in the developing seed of *Vicia faba* L. – a structural assessment of the role of transfer cells in unloading from the seed coat. *Journal of Experimental Botany* **44**, 711–724.
185. M.A. Grusak and P.E.H. Minchin (1988) Seed coat unloading in *Pisum sativum* – osmotic effects in attached versus excised empty ovules. *Journal of Experimental Botany* **39**, 543–559.
186. A. de Jong and P. Wolswinkel (1995) Differences in release of endogenous sugars and amino-acids from attached and detached seed coats of developing pea-seeds. *Physiologia Plantarum* **94**, 78–86.
187. A. de Jong, J.W. Koerselman-Kooij, J.A.M.J. Schuurmans and A.C. Borstlap (1997) The mechanism of amino acid efflux from seed coats of developing pea seeds as revealed by uptake experiments. *Plant Physiology* **114**, 731–736.
188. A. de Jong, J.W. Koerselman-Kooij, J.A.M.J. Schuurmans and A.C. Borstlap (1996) Characterization of the uptake of sucrose and glucose by isolated seed coat halves of developing pea seeds. Evidence that a sugar facilitator with diffusional kinetics is involved in seed coat unloading. *Planta* **199**, 486–492.
189. S. Fieuw and J.W. Patrick (1993) Mechanism of photosynthate efflux from *Vicia faba* L. seed coats. *Journal of Experimental Botany* **44**, 65–74.
190. N.A. Walker, J.W. Patrick, W.H. Zhang and S. Fieuw (1995) Efflux of photosynthate and acid from developing seed coats of *Phaseolus vulgaris* L. – a chemiosmotic analysis of pump-driven efflux. *Journal of Experimental Botany* **46**, 539–549.
191. N.A. Walker, W.H. Zhang, G. Harrington, N. Holdaway and J.W. Patrick (2000) Effluxes of solutes from developing seed coats of *Phaseolus vulgaris* L. and *Vicia faba* L.: locating the effect of turgor in a coupled chemiosmotic system. *Journal of Experimental Botany* **51**, 1047–1055.
192. P. Wolswinkel and H. Deruiter (1985) Amino-acid release from the seed coat of developing seeds of *Vicia faba* and *Pisum sativum*. *Annals of Botany* **55**, 283–287.
193. C. Rochat and J.P. Boutin (1991) Metabolism of phloem-borne amino-acids in maternal tissues of fruit of nodulated or nitrate-fed pea-plants (*Pisum sativum* L.). *Journal of Experimental Botany* **42**, 207–214.
194. F.C. Lanfermeijer, M.A. Vanoene and A.C. Borstlap (1992) Compartmental analysis of amino-acid release from attached and detached pea seed coats. *Planta* **187**, 75–82.
195. F.C. Lanfermeijer, J.W. Koerselman-Kooij and A.C. Borstlap (1990) Changing kinetics of L-valine uptake by immature pea cotyledons during development – an unsaturable pathway is supplemented by a saturable system. *Planta* **181**, 576–582.

196. A.B. Bennett and R.M. Spanswick (1983) Derepression of amino acid-H+ co-transport in developing soybean embryos. *Plant Physiology* **72**, 781–786.
197. B.E.S. Gunning and J.S. Pate (1969) Transfer cells, plant cells with wall ingrowth, specialized in relation to short-distance transport of solutes – their occurrence, structure, and development. *Protoplasma* **68**, 107–133.
198. C.E. Offler, E. Liet and E.G. Sutton (1997) Transfer cell induction in cotyledons of *Vicia faba* L. *Protoplasma* **200**, 51–64.
199. M. Tegeder, X.D. Wang, W.B. Frommer, C.E. Offler and J.W. Patrick (1999) Sucrose transport into developing seeds of *Pisum sativum* L. *Plant Journal* **18**, 151–161.
200. J.T. van Dongen, A.M.H. Ammerlaan, M. Wouterlood, A.C. van Aelst and A.C. Borstlap (2003) Structure of the developing pea seed coat and the post-phloem transport pathway of nutrients. *Annals of Botany* **91**, 729–737.
201. R. McDonald, S. Fieuw and J.W. Patrick (1996) Sugar uptake by the dermal transfer cells of developing cotyledons of *Vicia faba* L. Experimental systems and general transport properties. *Planta* **198**, 54–63.
202. J.L. Bonnemain, S. Bourquin, S. Renault, C.E. Offler and D.G. Fisher (1991) Transfer cells: structure and physiology. In: *Phloem Transport and Assimilate Compartmentation,* (eds J.L. Bonnemain, S. Delrot, W.J. Lucas and J. Dainty), Quest Editions, Nantes, France, pp. 178–186.
203. X.D. Wang, G. Harrington, J.W. Patrick, C.E. Offler and S. Fieuw (1995) Cellular pathway of photosynthate transport in coats of developing seed of *Vicia faba* L. and *Phaseolus vulgaris* L. II. Principal cellular site(s) of efflux. *Journal of Experimental Botany* **46**, 49–63.
204. G.N. Harrington, Y. Nussbaumer, X.D. Wang *et al.* (1997) Spatial and temporal expression of sucrose transport-related genes in developing cotyledons of *Vicia faba* L. *Protoplasma* **200**, 35–50.
205. G.N. Harrington, Y. Nussbaumer, X.D. Wang, M. Tegeder, V.R. Franceschi, W.B. Frommer, J.W. Patrick and C.E. Offler (1997) Spatial and temporal expression of sucrose transport-related genes in developing cotyledons of *Vicia faba* L. *Protoplasma.* **200**, 35–50.
206. H. Weber, L. Borisjuk, U. Heim, N. Sauer and U. Wobus (1997) A role for sugar transporters during seed development: molecular characterization of a hexose and a sucrose carrier in fava bean seeds. *Plant Cell* **9**, 895–908.
207. J.W. Patrick (1997) Phloem unloading: sieve element unloading and post-sieve element transport. *Annual Review of Plant Physiology and Plant Molecular Biology* **48**, 191–222.
208. L. Borisjuk, H. Rolletschek, U. Wobus and H. Weber (2003) Differentiation of legume cotyledons as related to metabolic gradients and assimilate transport into seeds. *Journal of Experimental Botany* **54**, 503–512.
209. H. Weber, U. Heim, S. Golombek, L. Borisjuk and U. Wobus (1998) Assimilate uptake and the regulation of seed development. *Seed Science Research* **8**, 331–345.
210. W.N. Fischer, M. Kwart, S. Hummel and W.B. Frommer (1995) Substrate-specificity and expression profile of amino-acid transporters (AAPs) in Arabidopsis. *Journal of Biological Chemistry* **270**, 16315–16320.
211. S. Okumoto, R. Schmidt, M. Tegeder *et al.* (2002) High affinity amino acid transporters specifically expressed in xylem parenchyma and developing seeds of Arabidopsis. *Journal of Biological Chemistry* **277**, 45338–45346.
212. M. Tegeder, C.E. Offler, W.B. Frommer and J.W. Patrick (2000) Amino acid transporters are localized to transfer cells of developing pea seeds. *Plant Physiology* **122**, 319–325.
213. F. Montamat, L. Maurousset, M. Tegeder, W. Frommer and S. Delrot (1999) Cloning and expression of amino acid transporters from broad bean. *Plant Molecular Biology* **41**, 259–268.
214. M. Miranda, L. Borisjuk, A. Tewes *et al.* (2001) Amino acid permeases in developing seeds of *Vicia faba* L.: expression precedes storage protein synthesis and is regulated by amino acid supply. *Plant Journal* **28**, 61–71.
215. M. Miranda, L. Borisjuk, A. Tewes *et al.* (2003) Peptide and amino acid transporters are differentially regulated during seed development and germination in faba bean. *Plant Physiology* **132**, 1950–1960.

216. M. Tegeder, C.E. Offler, W.B. Frommer and J.W. Patrick (2000) Amino acid transporters are localized to transfer cells of developing pea seeds. *Plant Physiology* **122**, 319–325.
217. E. Rosche, D. Blackmore, M. Tegeder *et al*. (2002) Seed-specific overexpression of a potato sucrose transporter increases sucrose uptake and growth rates of developing pea cotyledons. *Plant Journal* **30**, 165–175.
218. K. Fischer and A. Weber (2002) Transport of carbon in non-green plastids. *Trends in Plant Science* **7**, 345.
219. H. Kofler, R.E. Häusler, B. Schulz, F. Gröner, U.I. Flügge and A. Weber (2000) Molecular characterization of a new mutant allele of the plastid phosphoglucomutase in Arabidopsis, and complementation of the mutant with the wild-type cDNA. *Molecular and General Genetics* **263**, 978–986.
220. T. Caspar, S.C. Huber and C. Somerville (1985) Alterations in growth, photosynthesis, and respiration in a starchless mutant of *Arabidopsis thaliana* (L.) deficient in chloroplast phosphoglucomutase activity. *Plant Physiology* **79**, 11–17.
221. T.S. Yu, W.L. Lue, S.M. Wang and J. Chen (2000) Mutation of Arabidopsis plastid phosphoglucose isomerase affects leaf starch synthesis and floral initiation. *Plant Physiology* **123**, 319–326.
222. B. Kammerer, K. Fischer, B. Hilpert *et al*. (1998) Molecular characterization of a carbon transporter in plastids from heterotrophic tissues: the glucose-6-phosphate/phosphate antiporter. *Plant Cell* **10**, 105–117.
223. U.I. Flügge, R.E. Häusler, F. Ludewig and K. Fischer (2003) Functional genomics of phosphate antiport systems of plastids. *Physiologia Plantarum* **118**, 475–482.
224. J.O. Giese, K. Herbers, M. Hoffmann, R.B. Klosgen and U. Sonnewald (2005) Isolation and functional characterization of a novel plastidic hexokinase from *Nicotiana tabacum*. *FEBS Letters* **579**, 827–831.
225. M. Thelander, T. Olsson and H. Ronne (2005) Effect of the energy supply on filamentous growth and development in *Physcomitrella patens*. *Journal of Experimental Botany* **56**, 653–662.
226. T. Olsson, M. Thelander and H. Ronne (2003) A novel type of chloroplast stromal hexokinase is the major glucose-phosphorylating enzyme in the moss *Physcomitrella patens*. *Journal of Biological Chemistry* **278**, 44439–44447.
227. J.A. Miernyk and D.T. Dennis (1983) Mitochondrial, plastid and cytosolic ioszymes of hexokinase from developing endosperm of *Ricinus communis*. *Archives of Biochemistry and Biophysics* **226**, 458–468.
228. G. Schäfer, U. Heber and H.W. Heldt (1977) Glucose transport into intact spinach chloroplasts. *Plant Physiology* **60**, 286–289.
229. J.C. Servaites and D.R. Geiger (2002) Kinetic characteristics of chloroplast glucose transport. *Journal of Experimental Botany* **53**, 1581–1591.
230. H.W. Heldt (1969) Adenine nucleotide translocation in spinach chloroplasts. *FEBS Letters* **5**, 11–14.
231. H.E. Neuhaus, E. Thom, T. Möhlmann, M. Steup and K. Kampfenkel (1997) Characterization of a novel eukaryotic ATP/ADP translocator located in the plastid envelope of *Arabidopsis thaliana* L. *Plant Journal* **11**, 73–82.
232. K.H. Kampfenkel, T. Möhlmann, O. Batz, M. van Montague, D. Inzé and H.E. Neuhaus (1995) Molecular characterization of an *Arabidopsis thaliana* cDNA encoding a novel putative adenylate translocator of higher plants. *FEBS Letters* **374**, 351–355.
233. T. Möhlmann, J. Tjaden, C. Schwoppe, H.H. Winkler, K. Kampfenkel and H.E. Neuhaus (1998) Occurrence of two plastidic ATP/ADP transporters in *Arabidopsis thaliana* L. – molecular characterisation and comparative structural analysis of similar ATP/ADP translocators from plastids and *Rickettsia prowazekii*. *European Journal of Biochemistry* **252**, 353–359.
234. J. Reiser, N. Linka, L. Lemke, W. Jeblick and H.E. Neuhaus (2004) Molecular physiological analysis of the two plastidic ATP/ADP transporters from Arabidopsis. *Plant Physiology* **136**, 3524–3536.

235. J. Tjaden, T. Möhlmann, K. Kampfenkel, G. Henrich and H.E. Neuhaus (1998) Altered plastidic ATP/ADP-transporter activity influences potato (*Solanum tuberosum* L.) tuber morphology, yield and composition of tuber starch. *Plant Journal* **16**, 531–540.
236. P. Geigenberger, C. Stamme, J. Tjaden *et al*. (2001) Tuber physiology and properties of starch from tubers of transgenic potato plants with altered plastidic adenylate transporter activity. *Plant Physiology* **125**, 1667–1678.
237. J. Schwender, F. Goffman, J.B. Ohlrogge and Y. Shachar-Hill (2004) Rubisco without the Calvin cycle improves the carbon efficiency of developing green seeds. *Nature* **432**, 779–782.
238. S.A. Ruuska, J. Schwender and J.B. Ohlrogge (2004) The capacity of green oilseeds to utilize photosynthesis to drive biosynthetic processes. *Plant Physiology* **136**, 2700–2709.

5 Role of protein kinases, phosphatases and 14-3-3 proteins in the control of primary plant metabolism

Greg B. G. Moorhead, George W. Templeton and Hue T. Tran

5.1 Introduction

The covalent modification of proteins is a fundamental regulatory event in cells, controlling essentially every cellular process examined. It is known that proteins can be modified on particular amino acids by a wide array of novel functional groups. A few of these modifications include acetylation, hydroxylation, methylation, farnesylation, ubiquitination, adenylylation, uridylylation and sumolyation, but it is thought that protein phosphorylation is the most prevalent means of protein covalent modification [1]. Early work on mammalian cells demonstrated that approximately one of every three proteins is modified by phosphorylation on serine, threonine and tyrosine residues, and this value is believed to be true for other eukaryotic organisms as well. This biochemical observation is now supported by genomic information that has revealed that the protein kinases, the enzymes responsible for the addition of phosphate to proteins, constitute one of the largest gene families of eukaryotes. They encode for more than 1.7, 2.1, 2.2 and 4.0% of the human, *Saccharomyces cerevisiae*, *Caenorhabditis elegans* and Arabidopsis genes, respectively [2–5]. These numbers are based on the 'eukaryotic protein kinases' (ePKs) that likely all evolved from a single catalytic domain and include serine/threonine and tyrosine kinases. A number of 'atypical protein kinases' also exist and these likely evolved independently from the ePKs [4]. The phosphoinositide-3-kinase-like kinase (PIKK) family of enzymes also phosphorylates serine and threonine residues. This is a relatively small group of enzymes, most of which function in DNA-damage sensing and repair. This family of enzymes is also conserved in plants and will be discussed. The protein phosphatase catalytic subunits are responsible for the enzymatic removal of phosphate from proteins. They constitute a smaller group of genes compared to the protein kinases, but are likely as prevalent as holoenzymes because certain catalytic subunits (such as protein phosphatase one) complex with a large number of other proteins to form many catalytic and regulatory subunit complexes. It is believed that only a small population of protein phosphatase regulatory subunits has thus far been identified in any organism.

The history of protein phosphorylation is long, being intimately linked with the story of modern biochemistry, particularly the study of mammalian glycogen metabolism. This research field can be traced to the late 1930s when Carl and Gerty

Cori demonstrated that glycogen phosphorylase resided in mammalian cells as two distinct forms with differing properties. These two forms were interconvertible and designated phosphorylase *a* and *b*. Nearly 20 years later, Edmond Fischer and Edwin Krebs were able to show that the interconversion of phosphorylase *b* and *a* was a protein phosphorylation event [6]. Fischer and Krebs were awarded the Nobel Prize in Medicine and Physiology in 1992 in recognition of their lifetime of work in protein phosphorylation. Since most of the early studies of protein phosphorylation were linked to metabolic enzymes, it was thought that the phosphorylation of proteins was primarily an event that caused conformational changes in enzymes and this regulated enzymatic activity and substrate specificity. It is now known that the phosphorylation of proteins controls not only enzymatic activity, but also can generate specific docking sites for other proteins, controls the shuttling of proteins between cellular compartments and regulates proteolytic degradation [1]. In fact, the generation of specific phosphorylation-dependent docking or interaction sites may be the most common function of protein phosphorylation.

5.2 Protein kinases

Protein kinases catalyze the transfer of the γ-phosphate of ATP to the serine, threonine, tyrosine or histidine residue of a substrate protein. The ePKs all share a conserved catalytic domain of approximately 280 amino acids. This domain is divided into 12 subdomains that are highly conserved with less conserved regions in between. The region between subdomains VII and VIII contains the signature sequences DFG and APE. These are the hallmark residues of the activation or T-loop. Phosphorylation of a serine, threonine or tyrosine in this loop positions the two lobes of the kinase to allow substrate binding and phosphate transfer [3]. Detailed studies of individual protein kinases have shown that substrates are selected, in part, based on the amino acids surrounding the phosphorylatable residue, thus small peptides are generally excellent substrates for protein kinases. The use of oriented peptide libraries has been utilized to define the preferred substrate motif for several protein kinases [7].

Genomics has told us in a humbling fashion that plants, with about 1019 ePKs (in this case, Arabidopsis), have about twice as many protein kinases as humans (518 ePKs). Cataloguing the protein kinase complement of a number of model organisms has also provided some interesting knowledge regarding plant protein kinases. First, nearly two-thirds of plant protein kinases are predicted to be serine/threonine receptor kinases [3]. Of particular interest, it is now known that the calcium-dependent protein kinases (CDPKs) are unique to plants and a particular subgroup of protists, while the phosphoenolpyruvate carboxylase (PEPC) kinases and the PEPC kinase-related kinases are unique to plants. There are 2 PEPC kinases, two PEPC kinase-related kinases, 34 CDPKs, 8 CDPK-related kinases and 38 members of the sucrose-nonfermenting related kinase group in the Arabidopsis genome [5, 8]. As mentioned above, most protein kinases are activated by phosphorylation of their activation or T-loops. Most of the CDPKs and all of the

CDPK-related kinases have an aspartate or glutamate at this position, which will partially mimic a phosphorylated residue. This suggests that these kinases do not get activated by an upstream protein kinase and may respond solely to calcium signals (but perhaps other signals as well). The PEPC kinases and PEPC kinase-related kinases have a glycine at this amino acid position.

Phosphorylation on tyrosine residues comprises only a small proportion of the total cellular protein phosphorylation and is utilized primarily in multicellular eukaryotes (protein tyrosine phosphorylation does occur to some extent in prokaryotes and single cell eukaryotes). Although only a small percentage of the phosphoproteome resides on tyrosine residues (for instance in mammals), it does not mean that the roles played by tyrosine-phosphorylated proteins are any less important than those played by serine/threonine phospho-proteins. The number of studies of tyrosine kinases is proportionally much greater than that of tyrosine phosphatases and this is most likely for historical reasons. The first tyrosine kinase (mammalian) was sequenced in 1980, and we now know there are 90 protein tyrosine kinases in the human genome. Arabidopsis has 53 protein- tyrosine kinases with few described functions. The phospho-tyrosine binding domain Src homology two (SH2) is prevalent in mammals where tyrosine phosphorylation is also relatively common. It is interesting that three SH2 domains containing proteins with no known function have been identified in plants [9] and comparison to other SH2 domains shows that all the critical residues for binding phospho-tyrosine are present. We should also note that plant phytochromes have serine/threonine protein kinase activity, but will not be discussed here. We refer readers to recent articles [10].

5.2.1 Phosphoinositide 3-kinase-like kinases

The PIKK family of enzymes is intimately linked with regulation in response to a wide variety of stress conditions, such as nutrient deprivation and DNA damaging conditions. The PIKK family is made up exclusively of extremely large proteins, ranging from 2500 to 3900 amino acids in length, but are relatively simple in design. These proteins are, as the name suggests, characterized by the similarity of their catalytic domain to that of the phosphoinositide kinases, though no PIKK member has ever been shown to have catalytic activity toward lipids. Other identifiable features of these proteins include their FAT and FATC domains (named for three members of the family, FRAP, ATM and TRRAP), which are essential for activity and are believed to be responsible for exposing the catalytic domain [11]. The rest of these large proteins consist of HEAT repeats (90–95%), which take the form of antiparallel α-helices that stack on top of one another, forming a large surface thought to be responsible for protein–protein interactions [12, 13]. Arabidopsis has close homologs to three of the five known catalytically active members of the PIKK family, ATM (ataxia-telangiectasia mutated), ATR (ataxia-telangiectasia and RAD3 related) and TOR (target of rapamycin). ATM and ATR are known to be involved in DNA damage, repair, translation and cell cycle progression in mammals and yeast, while TOR has been shown to be involved with nutrient and hormone signalling in several organisms including plants [14, 15].

5.3 Protein phosphatases

The recognition that mammalian glycogen phosphorylase was converted to the active *a* form by the ATP-dependent addition of a phosphate group immediately implied the existence of an *a* to *b* converting enzyme, or a protein phosphatase. Like the field of protein phosphorylation, knowledge of protein phosphatases emerged from studies of mammalian glycogen metabolism. In 1983, a series of seminal papers from the Cohen laboratory demonstrated that essentially all of the serine and threonine protein phosphatase activities in cells could be ascribed to four classes based on biochemical properties: type 1 and types 2A, 2B and 2C [16]. Two heat stable inhibitor proteins referred to as inhibitor 1 (I1) and inhibitor 2 (I2) could inhibit type one enzyme (PP1) activity. Type two enzymes were resistant to the inhibitor proteins I1 and I2 and had various metal ion requirements as listed here: Protein phosphatase 2A (PP2A) (none), PP2B (Ca^{2+}) and Protein phosphatase 2C (PP2C) (Mg^{2+}). PP1, PP2A and PP2C activities were later found in plant tissues [17] and consistent with a lack of PP2B activity, no PP2B genes have been identified in plant genomes [18]. PP2B is a Ca^{2+}-calmodulin-dependent protein phosphatase, but due to the lack of PP2B in plants, will not be discussed in detail here. Purification of mammalian forms of each of these enzymes (PP1, PP2A and PP2C) and subsequent cloning unveiled that protein phosphatase catalytic subunits are amongst the most conserved enzymes known. This also allowed for a molecular search for additional phosphatase catalytic subunits using degenerate PCR. This revealed the existence of the so-called novel protein phosphatases, which are named PP4, PP5, PP6 and PP7 [19]. This also demonstrated that PP1, PP2A and PP2B catalytic subunits are from the same gene family, which now also includes the novel protein phosphatases. Collectively, they are referred to as the PPP family. The magnesium-dependent PP2C enzymes likely evolved independently and are designated the PPM family.

The sequencing of the first tyrosine kinase around 1980 and the purification and cloning of the first protein tyrosine phosphatase 10 years later ushered in the era of tyrosine phosphorylation. A detailed analysis of the human genome suggests that there are a comparable number of tyrosine phosphatases (107; of this total, several are catalytically inactive and others are specific for mRNA and inositol phospholipids, but belong to this gene family) and tyrosine kinases (90; five are catalytically inactive) [20]. The Arabidopsis genome encodes for 53 predicted tyrosine kinases, and the last bioinformatic analysis of plant tyrosine phosphatases listed 21 genes [18]. Recently defined novel tyrosine phosphatases [20, 21] likely means this group has expanded the plant collection and this warrants a bioinformatic update.

The most striking difference between the tyrosine phosphatases and the serine/threonine phosphatases is that most tyrosine phosphatases appear to have additional modular domains that function to regulate and/or target the enzyme to another molecule or compartment. The serine/threonine catalytic subunits, on the other hand, specifically bind other proteins that target or regulate them in a similar fashion (PP1 being the best example). Some of the novel serine/threonine phosphatases do have N- or C-terminal extensions beyond the catalytic domain that likely play targeting or regulatory roles (see PP5 and the Kelch-domain phosphatase below). Similarly, the atypical dual specificity phosphatases have no domains outside the minimum catalytic domain and it is likely that they have additional, as yet undiscovered, targeting or regulatory subunits.

5.3.1 Protein phosphatase 1

Protein phosphatase 1 (PP1) is a widely expressed, highly conserved enzyme, found in every eukaryotic organism examined. In plants, PP1 has been implicated in various cellular processes, which include signal transduction, regulation of membrane channels, cell cycle control and development [22–27]. Native PP1 generally exists as a heterodimer consisting of an exceptionally well conserved catalytic subunit and a variable targeting or regulatory subunit. This regulatory subunit controls PP1 activity *in vivo* by targeting it to a particular subcellular location and confers substrate specificity to the enzyme complex. The functions of PP1 are thus contributed to the different PP1 regulatory subunits, but to date, there have been no PP1 regulatory subunits identified or characterized from higher plants. In mammalian systems, a single PP1 enzyme can dephosphorylate multiple substrates. This is because the regulatory subunits bind to PP1 in a mutually exclusive manner, thereby targeting the same catalytic subunit to various substrates and subcellular locations.

A PP1 catalytic subunit binding motif was found by screening a peptide library with PP1 [28]. It was shown that binding to PP1 is conferred to a protein containing the peptide motif (R/K)(V/I)X(F/W), where X denotes any amino acid other than large hydrophobic residues. Also, the motif is often preceded by 2–5 basic residues and followed by one acidic residue. Using sequence alignments and site-directed mutagenesis experiments, this consensus sequence has since been further defined as $(R/K)X_{0-1}(V/I)\{P\}(F/W)$, where $\{P\}$ is any residue other than proline and X is defined as above [29]. The complex of human PP1 with the RVXF peptide derived from the muscle glycogen PP1 targeting subunit (G_M) revealed the precise means of how the sidechains of the highly conserved V and F residues of this motif interact with the catalytic subunit [30]. Because of the high degree of conservation of PP1 amongst eukaryotes, it is likely the same regulatory mechanism is used in plant systems. Indeed, the PP1 residues that bind the RVXF motif are completely conserved in all plant PP1 enzymes.

After a long wait, the first crystal structure of PP1 associated with a regulatory subunit was published in 2004 (Figure 5.1, generated with MOLMOL [31]) [32]. The PP1-MYPT1 complex from smooth muscle shows how a regulatory subunit modifies the catalytic cleft of the enzyme to accommodate the substrate (myosin) and how other regions of the targeting subunit likely aid in the interaction of this complex with the substrate. It is also notable that the C-terminus, which is highly variable in all PP1s, and highly disordered in monomeric structures, was visible in the complex interacting with the ankyrin repeat of MYPT1.

Although the RVXF motif remains the primary means of interaction between the regulatory subunits and PP1, it has been suggested that secondary interactions

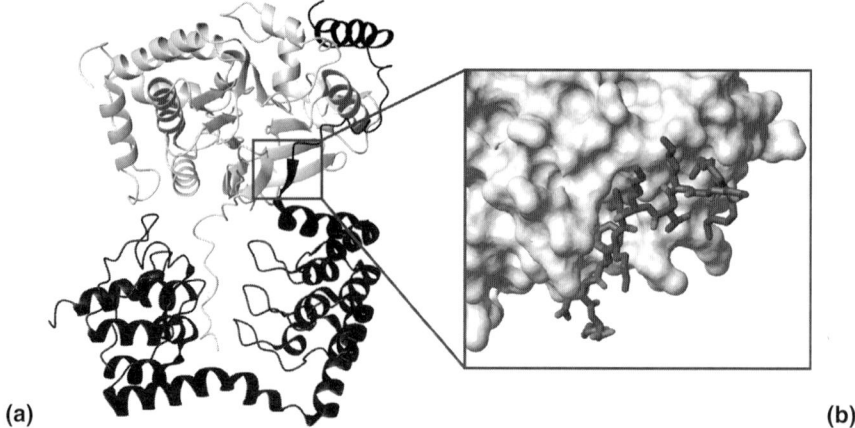

Figure 5.1 The PP1/MYPT1 complex. (a) PP1 (gray) was co-crystallized with a fragment (1-299) of MYPT1 (black) [PDB 1S70]. MYPT1 is a smooth muscle PP1 regulatory subunit, which targets PP1 to myosin to dephosphorylate myosin light chains. Nearly all PP1 regulatory subunits characterized so far bind PP1 through a short four amino acid sequence referred to as the RVXF/W motif. (b) The **RVXF/W** motif of MYPT1 (^{31}RKK**TKVKF**DGA^{41}; stick representation) in its binding pocket on the surface of PP1. Structures were constructed in MOLMOL [31].

with PP1 exist to stabilize binding and to provide crucial contacts for modulating the activity and/or substrate specificity of PP1. Studies carried out on known PP1 binding proteins show that the basic region preceding the RVXF motif appears to strengthen the binding and/or modulates the activity of PP1, whereas the RVXF motif itself appears to have no effect on activity [33, 34]. Because these secondary interactions alone are not sufficient to initiate formation of complexes, the notion of the RVXF motif acting mainly as an anchor, which allows the strongest association with PP1, is starting to gain acceptance. Undoubtedly, each novel PP1 binding protein has evolved to aid in the dephosphorylation of the appropriate substrate by assisting in binding the substrate and/or altering the PP1 structure to have increased activity against the phospho-target.

5.3.2 Protein phosphatase 2A

PP2A plays a role in numerous plant processes, including hormone signal transduction, metabolism, gene expression and development [23–26]. Native PP2A is found either as a heterodimer, consisting of the catalytic subunit (PP2Ac) and the regulatory A subunit, or as a heterotrimer, involving PP2Ac, the regulatory A subunit and a variable B regulatory subunit [35]. The 65 kDa A subunit, otherwise known as PR65 or the scaffolding protein, consists of 15 tandem HEAT repeats which it uses to link PP2Ac to different B subunits.

The catalytic subunit of PP2A is highly conserved, with a 79–82% identity amongst plant, human and yeast amino acid sequences [17]. Plant PP2Ac has been localized to various subcellular compartments, such as the nucleus [26] and cytosol, but has not been found in chloroplasts [17, 36]. PP2Ac is anchored constitutively to the A subunit and methylation of the leucine residue at the C-terminus of PP2Ac (as demonstrated in mammals and yeast) controls the binding of various B subunits to the core heterodimer. Its association with the regulatory B subunits controls the activity and substrate specificity of PP2A, although examples exist where the PP2A core is bound to proteins other than the B regulatory subunits [37].

At least 20 B regulatory subunits, ranging from 54–130 kDa, have been identified from mammalian systems. Using biochemistry and molecular biology, four distinct, unrelated groups have been identified which are the 55 kDa B group, the 52–74 kDa B' group, the 72–130 kDa B" group and the putative 93–110 kDa B''' group [35, 38–40]. A recent survey of the Arabidopsis genome revealed at least six PP2Ac genes, three A regulatory subunit genes, and two B, nine B' and five B" regulatory subunit genes [41]. Because of the numerous variable regulatory subunits, a large combination of PP2A complexes is possible, which certainly contributes to the diversity of PP2A functions.

A few novel PP2A regulatory B subunits have been identified in higher plants. Most recent was the isolation of the TONNEAU2 (TON2) gene, where researchers showed not only that TON2 encodes a putative novel regulatory subunit of PP2A, but also that it was involved in the control of cytoskeletal organization in plants [41]. The TON2 gene, which is expressed ubiquitously in Arabidopsis, encodes a 55 kDa protein that is highly conserved in higher plants. The C-terminal end of the protein displays significant similarity to the C-terminal region of the human PR72 protein, which is a representative member of the B" group of regulatory subunits. Also, using the yeast two hybrid system, the physical interaction between Arabidopsis TON2 protein and the A subunit of PP2A was verified. Taken together, these results justified the classification of the TON2 protein as a novel B" regulatory subunit. It is hypothesized that the function of the TON2-PP2A complex involves the control of the cortical cytoskeleton, as mutations in the TON2 gene resulted in abnormalities in the organization of this subcellular component. Information obtained from plant genomes will aid in the identification of novel regulatory subunits and the dissection of the unique functions of PP2A in all aspects of plant growth and development.

5.3.3 *Protein phosphatase 2C*

PP2C, although the largest phosphatase gene family in Arabidopsis, is one of the least well characterized serine/threonine phosphatases. A recent analysis of the Arabidopsis genome revealed 69 genes encoding putative, functional PP2C enzymes [18]. This is a considerably large number, as there are no more than 15 PP2Cs identified in the human genome [42]. But when one takes into account the number of kinase genes ($\sim$1019 Arabidopsis kinase genes compared to only $\sim$518 protein kinase genes in humans) it seems logical that there are this many PP2C genes to oppose the action of the protein kinases. Also, the large number of PP2C genes is needed to support the sophisticated signalling pathways that exist in the plant stress response mechanism.

PP2C genes in Arabidopsis have been divided into 10 different groups (A–J) based on amino acid sequence alignments [42]. Of these 10 groups, only three have representative genes with described functions. Group A consists of nine genes, including ABI1 and ABI2 (abscisic acid insensitive 1 and 2), which are genes associated with the abscisic acid signalling pathway. Group B (six genes) contains genes which show homology to alfalfa MP2C (Medicago sativa phosphatase 2C), which is a negative regulator of the stress activated MAPK pathway. Group C (seven genes) consists of the POL-type phosphatases (POLTERGIEST-type phosphatases), known for their role in flower development. Also, KAPP (kinase-associated protein phosphatase) was isolated as a binding partner of RLK5 (receptor-like kinase 5) and because it shares no similarity with the other 10 PP2C gene groups, it is the only Arabidopsis gene belonging to the KAPP gene cluster. As so few PP2C genes have known functions, it is not surprising that only one physical substrate of PP2C has been identified in plants.

The PP2C genes of Arabidopsis all share a common catalytic core consisting of 11 highly conserved subdomains [42]. Interestingly, the catalytic core is usually found at the C-terminal region but in a few cases, the PP2C genes begin with the catalytic core. The C-terminal region of PP2C has been implicated in conferring substrate specificity because many domains involved in protein–protein interactions, such as the forkhead associated (FHA) domain, have been described in this region. The variable N-terminal extensions of some PP2C genes may also allow interactions with specific substrates, as the uniqueness of the N-terminal extensions are well suited as binding sites for specific substrates or as attachment sites for different signalling complexes. A recent study by Scheible *et al.* showed a possible link between PP2C and nitrate signalling [43]. The experiment involved Arabidopsis seedlings grown for 7 days in liquid media containing complete nitrogen, then growing these plants for 2 days in low nitrogen liquid media. Nitrate was then added and seedlings harvested 30 min or 3 h later. RNA expression levels in the seedling tissue were measured using microarray technology and it was found that, in response to nitrate readdition, four PP2C genes were induced. In particular, one PP2C gene (At4g32950) containing a putative MAPK docking site was most highly induced. These results suggest a putative role for PP2C in the nitrate-signalling pathway. With the abundance of Arabidopsis PP2C genes, it will be a daunting task to ascribe specific functions to each gene. As genetic, molecular and biochemical tools become available, they will help in the elucidation of PP2C functions *in vivo* and will be invaluable in the identification of targets of the PP2C enzymes.

5.3.4 Novel protein phosphatases

The Arabidopsis genome encodes many novel protein phosphatases, including two genes for PP4, one gene each for PP5, PP6 and PP7 and one gene for an 'ancient' phosphatase most related to a bacterial protein phosphatase [18]. It was interesting to note that the one Arabidopsis gene encoding the bacterial-like phosphatase contains a putative chloroplast transit peptide. Also, the Arabidopsis genome has four genes encoding phosphatases with large N-terminal extensions. These were identified recently as BSU1 and BSU1-like proteins [44]. BSU1 is a novel nuclear protein phosphatase involved in the brassinolide signalling pathway in plants. Although PP4, PP5, PP6 and PP7 were identified years ago, very little research has been dedicated to these phosphatases in plants. There are no genes encoding either PPY or PPZ phosphatases in Arabidopsis although these phosphatases have been found in other eukaryotes such as *Drosophila melanogaster* and *S. cerevisiae*.

Much research has been done on PP5, as in mammals it functions in many signalling pathways that control growth arrest, apoptosis and DNA damage due to ionizing radiation [45–47]. Although its function is unknown, the Arabidopsis homolog of PP5 consists of all the structural features that characterize all PP5 phosphatases. These characteristics include three TPR (tetratricopeptide repeats) motifs in the N-terminal domain followed by a variable region and a phosphatase domain in the C-terminal region. The TPR motif is important as it allows for protein–protein interactions, thus it likely targets PP5 to specific substrates and subcellular localizations [48]. Also, it has been shown that the proteolytic removal of either the TPR domain or the C-terminal region of PP5 increases its catalytic activity, suggesting that these regions function in an autoinhibitory manner [49, 50]. In plants, PP5 catalytic activity can be stimulated by long chain fatty acids implicating they may have a role in the regulation of PP5 *in vivo*. Using the solved structures, the superimposition of the phosphatase domain of auto-inhibited human PP5 and PP2B, and PP1 complexed to calyculin A, it was found that all three structures had very similar features and characteristics. Thus it was hypothesized that the mechanism by which the TPR domain inhibits PP5 is similar to that of PP2B autoinhibition by its C-terminal domain (CTD) and PP1 inhibition by toxins such as calyculin A and microcystin [51]. Specifically, inhibition of phosphatase activity is carried out by a Glu residue that binds to a catalytic arginine residue found in the phosphatase, thereby effectively blocking access to the catalytic center. In human PP5, this Glu (Glu76) residue is found in the TPR domain and blocks phosphatase activity by interacting with the catalytic Arg275 residue. Similar glutamate and arginine residues, interacting in an inhibitory manner, have been identified in PP1-microcystin complexes (Arg96 of the PP1 catalytic site, which is equivalent to Arg275 of PP5) and autoinhibited PP2B (Glu481 in the C-terminal domain inhibits the catalytic Arg254, which is equivalent to Arg400 of PP5).

A novel Arabidopsis nuclear protein phosphatase, BSU1, was recently identified [44]. Three orthologs of BSU1 were identified in Arabidopsis and subsequently named BSL1, BSL2, and BSL3 for BSU1-like. Using these sequences we found that these four phosphatases are the same enzymes identified by Kerk *et al.* [18], which they defined as protein phosphatases with long N-terminal extensions. The BSU1 protein contains an N-terminal Kelch-repeat domain, a connecting middle region, followed by a C-terminal phosphatase domain. The Kelch domain has been implicated in conferring substrate specificity and defining subcellular localizations for BSU1, as Kelch domains, like TPR domains are involved in protein–protein interactions [52]. BSU1 was found to be a functional phosphatase, with catalytic activity similar to both PP1 and PP2A, when tested with various well-known PP1 and PP2A inhibitors. Also, it was shown that the Kelch domain behaved independently of the phosphatase domain as the Kelch domain did not interfere with the phosphatase activity in the C-terminal domain. Overall, BSU1 is implicated in the dephosphorylation of BES1 in the nucleus, thereby opposing the action of BIN2, which is a glycogen synthase kinase-3 homolog. Together, BSU1 and BIN2 are responsible for regulating the phosphorylation state of BES1 and thus they respond to the varying levels of the plant steroid hormone brassinolide. This study is a prime example of the discovery and characterization of novel protein phosphatases from Arabidopsis and it is hoped that many more findings of similar magnitude will be made in the near future.

5.3.5 *The tyrosine and dual specificity protein phosphatases*

Since the first purification and cloning of a tyrosine phosphatase over 15 years ago in mammals, and the identification of the dual specificity protein phosphatases (of which the MAPK phosphatases are the most famous), we now know there are four protein tyrosine phosphatase families. Classification is derived from work done primarily in mammalian systems, and the four protein tyrosine phosphatase families are based on the catalytic domains of these enzymes [20].

5.3.5.1 *Class I cysteine-based protein tyrosine phosphatases*

This is by far the largest family, with all enzymes likely evolving from a common ancestral gene, and includes the tyrosine specific enzymes and the group called dual specificity enzymes. The dual specificity enzymes are given this name based on the ability of the MAPK phosphatases to dephosphorylate tyrosine and threonine in the MAPK activation loop. This group also includes the atypical dual specificity enzymes, Slingshots, PRLs, the cdc14s (which dephosphorylate the activation loop of cyclin-dependent kinases), the PTENs and myotubularins. The last two subgroups specifically have evolved not to dephosphorylate proteins, but the D-3 position of inositol phospholipids. The Arabidopsis genome appears to encode a single tyrosine specific enzyme and about 20 dual specificity protein phosphatases.

5.3.5.2 *Class II cysteine-based protein tyrosine phosphatase*

In humans this class is represented by a single enzyme, which has a low molecular mass and activity specific for phosphotyrosine. It is more ancient than the Class I PTPs with representatives across archaea, eubacteria and eukaryotes, including plants. The Arabidopsis gene is predicted to have a mass of ~20 kDa and is 37% identical (48% similar) to the human enzyme.

5.3.5.3 *Class III cysteine-based protein tyrosine phosphatases*

This family constitutes the group of enzymes known as the CDC25 phosphatases and is responsible for dephosphorylating the cyclin-dependent kinases on neighboring tyrosine and threonine residues in their N-terminal domain. In yeast a single CDC25 performs this role, while humans have three CDC25s that perform dephosphorylation events at different points of the cell cycle. Once dephosphorylated at the N-terminus, the cyclin-dependent kinases can become fully activated and drive the cell cycle. If CDC25 is phosphorylated, for instance in response to DNA damage, the generated 14-3-3 binding motif functions to allow 14-3-3s to bind and shuttle CDC25 out of the nucleus. Yeast CDC25 has been shown to activate cell division in plants, suggesting a plant homolog performs the equivalent role [53]. No classic CDC25 homologs have been found in the Arabidopsis or rice genomes, although recently a small tyrosine phosphatase was identified in Arabidopsis that is functionally the CDC25 equivalent [53, 54].

5.3.5.4 Class IV protein tyrosine phosphatases
This class was only defined in 2003 when the protein of the *Drosophila* mutant Eyes Absent displayed tyrosine phosphatase activity [21]. To date, only four proteins of this large family of hydrolases have been characterized to have tyrosine phosphatase activity and thus the possibility of expansion of this class exists. It is thought that the cysteine-based class I, II and III enzymes evolved independently, but structurally they display a common ancestral fold. On the other hand, the class IV enzymes have a completely different catalytic mechanism, which utilizes a key aspartic acid residue and a cation [20, 55]. We found evidence for at least two homologs in Arabidopsis with the best matches to the human Eyes Absent protein being 37 and 31% identical.

5.3.6 RNA polymerase II phosphatases-FCP1 and SCP
The initiation of transcription by RNA polymerase II (RNAPII) involves a large complex of proteins containing several general transcription factors (TFIIA, TFIIB, TFIID, TFIIE, TFIIF, and TFIIH). At the time of initiation, RNAPII becomes highly phosphorylated on its CTD in a heptapeptide repeat [56]. This repeat has the consensus YSPTSPS (where serines 2 and 5 get phosphorylated) and is present 29, 42, 52 and 34 times respectively in yeast, Drosophila, mammal and Arabidopsis RNAPII CTD. Just after or during the termination of transcription, the CTD of RNAPII must be dephosphorylated to allow another round of preinitiation complex assembly. The original purification of the CTD phosphatase from HeLa cell extracts yielded a 150 kDa protein whose activity was Mg^{2+} dependent and is now designated FCP1 (TFIIF associated CTD phosphatase) [56, 57]. The presence of a BRCA1 carboxy-terminal (BRCT) domain was demonstrated through cloning of FCP1 and a functional BRCT is necessary for FCP1 to dephosphorylate CTD. FCP1 protein is associated with the complex through the 74 kDa subunit of TFIIF with the highly conserved last 82 amino acids of each protein interacting with the other. The p74 subunit of TFIIF greatly stimulates the phosphatase activity of FCP1 when phosphorylated CTD is used as substrate. The FCP1 protein is a CTD phosphatase, yet until recently it did not have any apparent homology to any other characterized protein phosphatase and thus represented a novel, specialized phosphatase domain. Further characterization of this domain has revealed a conserved

DXDX(T/V) motif necessary for catalysis that is present in members of the superfamily of phosphohydrolases and phosphotransferases and is now recognized as the same catalytic motif found in class IV tyrosine phosphatases [58]. Three proteins closely related to the catalytic region of FCP1, but lacking a BRCT domain, were identified in the human databases. One of these small CTD phosphatases (SCPs) preferentially dephosphorylates serine 5 of the CTD repeat and is also activated by p74 [59].

Searching the Arabidopsis genome predicted the presence of four FCP1-like proteins based on the catalytic region [53]. Two are the larger BRCT-domain containing enzymes, such as human FCP1, and two are the shorter SCP-like enzymes. The two SCP-like enzymes have just been characterized and, like their human counterpart, they preferentially dephosphorylate serine 5 of the Arabidopsis CTD repeat sequence.

5.3.7 Histidine phosphatases

The demonstration of phosphorylation on histidine residues in eukaryotes occurred over 40 years ago and it is believed that as much as 6% of the total phosphoproteome in eukaryotes resides on histidine residues [60]. Very few eukaryotic proteins are known that are phosphorylated on histidine; on the other hand, histidine phosphorylation is very well studied in prokaryotes due to the prevalence of two component systems, and well characterized histidine kinases and phosphatases. Very recently a small (14 kDa) phosphohistidine phosphatase was purified from mammalian tissue and its first substrate was identified as ATP-citrate lyase [61]. To date, no histidine phosphatases have been reported in plants. We have blasted this mammalian histidine phosphatase against the Arabidopsis genome and found no obvious homologs.

5.4 A multitude of phosphospecific binding modules

5.4.1 Phosphospecific binding modules

It is likely that the most significant observation in the last 20 years of signal transduction research is the realization that the specific interaction of proteins is responsible for much of the specificity and regulation of signal transduction events. The phosphorylation of a specific amino acid and the residues surrounding this site can provide a specific docking domain for other protein modules. The 14-3-3 proteins were the first proteins shown to specifically bind a target protein in a phosphoserine or threonine dependent fashion. To date, such phospho-dependent binding modules include 14-3-3 proteins, leucine rich repeats, FHA, WW, MH2, WD40, polobox and BRCT domains for serine and threonine phosphorylation and SH2 and protein tyrosine binding domains for phosphotyrosine [62, 63]. It is likely that additional, and perhaps many more phosphospecific binding modules exist and await discovery. A recently developed method to address this question was put forth by the Yaffe laboratory [64]. Their approach to identify new phosphospecific binding modules is based on the observation that protein kinases phosphorylate their substrates at discrete motifs. For instance, PKA phosphorylates preferentially at RRXSØ sequences (where Ø is any large hydrophobic residue and S is the phosphorylated serine) and cyclin-dependent kinases modify at SP motifs. The DNA damage response protein kinase ATM belongs to the PIKK family of protein kinases. ATM kinases phosphorylate proteins at a SQ motif, and it was predicted that part of the DNA damage response signalling cascade was based on generating phospho-SQ-binding sites on ATM substrates. Their strategy was to develop a phosphopeptide library biased toward this motif and couple this, plus the nonphosphorylated version of the same library, to beads and screen for the ability of *in vitro* translated polypeptides (from a cDNA library) to preferentially bind the phospho-SQ peptides [64]. After screening ~100,000 translated proteins of various lengths, several proteins or fragments thereof were flagged for their ability to bind selectively to the phospho, but not the dephospho-SQ library peptides. A fragment of the protein PTIP was identified as a strong phospho-SQ interactor. This fragment contained the four PTIP BRCT domains and the last two were shown to bind strongly to this phospho-SQ library. An oriented peptide library was then used to define the optimal phospho-binding motifs for the BRCT domains of PTIP and the breast cancer susceptibility gene, BRCA1. These results indicated that individual BRCT domains (functioning in pairs) likely have evolved in the context of each protein to yield specific binding motifs. The BRCA1 and PTIP BRCT domains both bind phospho-SQ peptides, but the surrounding amino acid sequences were different. The use of the oriented peptide library also demonstrated that BRCT domains are best defined as phosphoserine binding domains because there was in fact no strong selection for peptides with glutamine (Q) at the plus one position. The BRCT domain is conserved in more than 30 proteins of the human genome. A majority of these proteins are involved in the DNA damage response. Interestingly, the RNA polymerase CTD phosphatase FCP1 has a BRCT domain, and it is conserved in the two Arabidopsis FCP-like proteins [58]. No doubt, additional phospho-specific binding domains will be retrieved using the technique described. As yet, this sort of methodology has not been applied to plant systems.

5.4.2 *14-3-3 proteins*

14-3-3s were discovered in 1967 during an examination of abundant brain proteins and given this peculiar title based on their chromatography column elution fraction and position after starch gel electrophoresis [65]. 14-3-3s resurfaced in the literature many years later, initially as activators of tryptophan and tyrosine hydroxylases [66, 67], then as regulators of protein kinase C and the protein kinase Raf-1 [68–70]. The key experiment brought to light that binding of 14-3-3 to its target protein was dependent on the phosphorylation of that protein [71]. Since this discovery in 1996, 14-3-3s have emerged as regulators of a multitude of phosphorylation-dependent cellular events. Several proteomic studies to identify the 14-3-3 interactome in plants and humans have identified more than 350 proteins that reside in 14-3-3 complexes, although it is likely that not all directly bind 14-3-3s [72–76]. Some of the seminal work that defined 14-3-3 function was performed in plant systems and will be discussed below. Because much of the detailed knowledge of protein phosphorylation events linked to primary metabolism in plants involves 14-3-3 proteins, we will first provide a framework for understanding 14-3-3 protein function before discussing specific examples of 14-3-3-dependent processes in plants. We also refer readers to several recent excellent reviews on 14-3-3 proteins [77–79].

5.4.2.1 14-3-3 structures and function

The 14-3-3s have been found in every eukaryotic genome sequenced with no 14-3-3 or 14-3-3-like proteins present in any prokaryotes. 14-3-3s are acidic, approximately 30 kDa proteins that form homo- and heterodimers. The number of 14-3-3 isoforms varies from organism to organism with 12 in Arabidopsis, 7 in mammals and 2 in yeast, *C. elegans* and *D. melanogaster*. Their importance is best evidenced by yeast knockout studies of both 14-3-3 genes, as this yields a lethal phenotype.

The first 14-3-3 crystal structure demonstrated that the 14-3-3 dimer forms a cup or saddle shape with the most conserved residues on the interior of the cup or underside of the saddle. Each subunit has been shown to bind a defined phosphoserine or threonine peptide in an extended antiparallel orientation [63, 80]. It is the amino acids of the phospho-peptide binding region that are mostly high conserved across species with the residues equivalent to Lys^{49}, Arg^{56}, Arg^{127} and Tyr^{128} of human 14-3-3ζ being completely conserved in every 14-3-3 known. These residues are responsible for direct interaction with the phosphate moiety of the bound phosphoprotein, as illustrated in Figure 5.2.

The mapping of the phosphorylation sites on proteins that 14-3-3s recognize and bind to revealed a 14-3-3 docking motif. This was confirmed by the use of an oriented peptide library to define the preferred 14-3-3 binding peptides. This work produced two high affinity 14-3-3 binding peptides that required phosphorylation for interaction with an apparent preference for phospho-threonine over phospho-serine. These motifs, referred to as mode one and mode two respectively, are RSXpSXP and RXXXpSXP, where R is arginine, X is any amino acid, S is serine, pS is phosphoserine and P is proline. A binding constant (K_d) based on a mode two peptide showed the affinity to be very high ($K_d \sim 37$ nM). Although these are high affinity binding peptides, it turns out that many *in vivo* 14-3-3 binding proteins do not conform exactly to either of these sequences and several nonphosphorylated 14-3-3 binding proteins are known [77, 78]. One of the best-characterized examples of 14-3-3 interacting with a 'less than optimal' binding motif is the C-terminus of the plant plasma membrane H^+-ATPase [81]. *In vivo*, the C-terminus of the plant plasma membrane H^+-ATPase is autoinhibitory and inhibition is abolished when the second last residue of the enzyme is phosphorylated (QSYpTV-COOH) allowing 14-3-3 binding to this site. A phospho-peptide generated based on this motif interacts with 14-3-3 with a binding constant (K_d) 70-fold higher than the mode two peptide described above. It is thought that this is important *in vivo*, as it would likely yield a fast on–off rate.

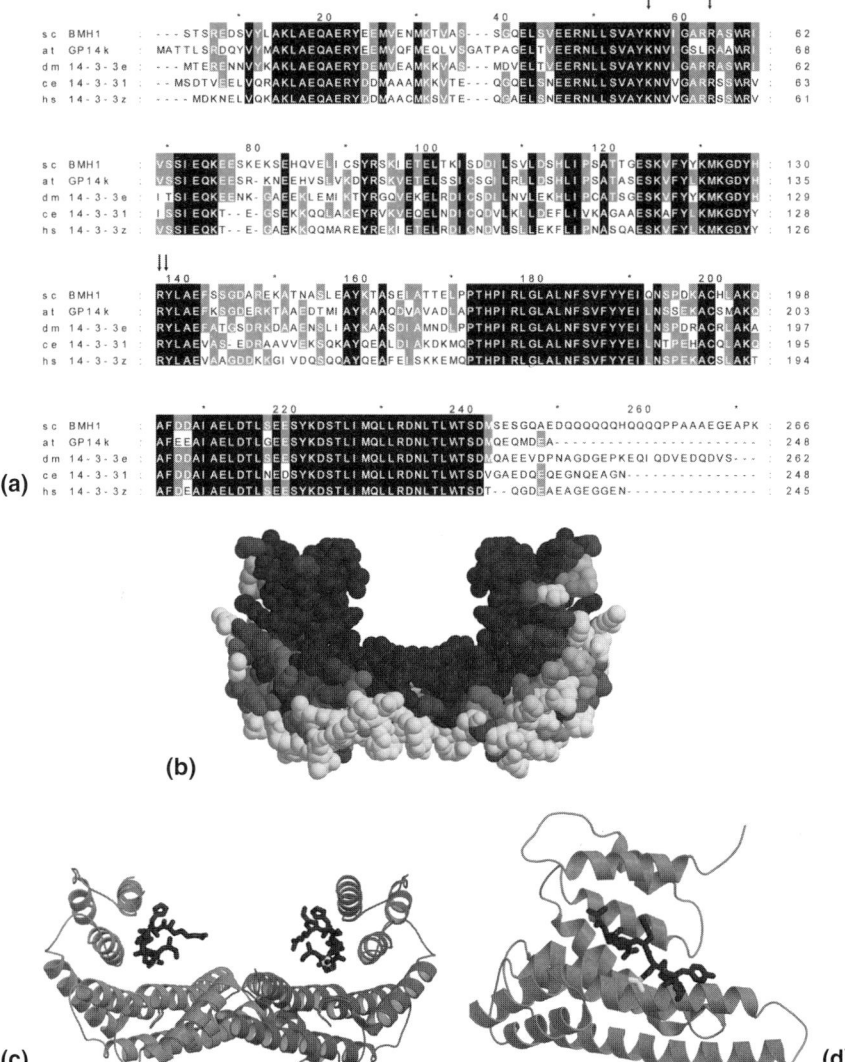

Figure 5.2 Conservation of peptide binding by 14-3-3 protein. (a) Sequence alignment of evolutionarily diverse 14-3-3 proteins. Residues which are 100% identical are shaded in black and residues which are >50% identical are shaded in gray. Included in the alignment are representative proteins from *S. cerevisiae* (Bmh1), *Arabidopsis thaliana* (GP14κ), *D. melanogaster* (14-3-3ε), *C. elegans* (14-3-3 isoform 1) and human (14-3-3ζ). Residues which directly contact the phosphate group of target proteins are indicated with dark arrows. (b) Structure of 14-3-3ζ shaded according to residue conservation with the darkest amino acids being most conserved. Structure analyzed using Consurf [77] with 14-3-3 isoforms from humans (7 isoforms), *A. thaliana* (12 isoforms), *S. cerevisiae* (2 isoforms), *D. melanogaster* (2 isoforms) and *C. elegans* (2 isoforms). Sequence alignment was performed using ClustalX [77] with residue conservation determined by a maximum likelihood method within Consurf. Structures (b, c and d) were based on 14-3-3ζ (PDB 1QJB). Legend denotes relative conservation levels in the protein. In (c) and (d) the peptide (ARSHpSYPA; shown as a dark stick representation with the phosphate group shown as light colored) binds in an extended conformation to 14-3-3 (shown as a ribbon diagram in different shades of gray for each monomer) within the conserved groove of each monomer. Figure 5.2d shows one half dimer from the perspective of the other monomer. Structure figures were generated using Molscript and Raster3D [31, 77] (figure from Bridges and Moorhead, *Sci STKE* 2004 RE10. Kindly provided by permission from *Sci STKE*).

Whether other factors are involved is not known, but nature has produced a toxin that exploits the interaction of 14-3-3 and the plasma membrane H$^+$-ATPase. The fungal toxin fusicoccin targets the plant plasma membrane H$^+$-ATPase resulting in constant activation of the pump and subsequent continuous opening of stomata and therefore wilting. The crystal structure of the 14-3-3, H$^+$-ATPase C-terminal phosphopeptide (QSYpTV-COOH) and fusicoccin beautifully illustrates the role of the toxin. Here fusicoccin fills a cavity at the 14-3-3-peptide binding site, which is empty due to the 'shortened' binding peptide. Biophysical analysis also demonstrated that in the presence of fusicoccin, the binding constant for 14-3-3 and QSYpTV dropped 90-fold. Although this 14-3-3 peptide binding is not 'tight', it has evolved to bind with precisely the correct affinity to regulate proton pumping. Interestingly, it has been demonstrated that 5'-AMP can promote the dissociation of 14-3-3 from this site on the H$^+$-ATPase implicating energy status as a regulator of at least some 14-3-3 functions [82].

5.4.2.2 *14-3-3 roles and control*

The diversity of binding partners and cellular processes that 14-3-3 participates in makes defining a single function for 14-3-3s very difficult. In fact, it now appears that 14-3-3s have several and in some cases overlapping functions. We recently attempted to classify their roles and have been able to ascribe three functions. Much of this understanding comes from insights derived from the crystal structures of 14-3-3 with and without bound ligands.

1. *14-3-3 directed conformational changes*. The solved structure of human 14-3-3 alone, with a phosphopeptide or with a large fragment of a target protein, has shown the 14-3-3 dimer to be a very rigid structure with little change in shape upon ligand binding. This is consistent with the highly α-helical nature of the protein. These structures have led to the molecular anvil or molecular clamp hypothesis for 14-3-3 function. It is thought that this rigid structure of the 14-3-3 imposes structural alterations in the target, leading to a change in some property of the target protein [78, 79]. The enzyme serotonin-*N*-acetyltransferase binds a 14-3-3 protein after phosphorylation at two separate sites. The binding of the 14-3-3 induces an alteration in the enzyme that increases catalytic activity and affinity for substrate. This is the only example where the 'molecular anvil' hypothesis has been clearly illustrated, but likely explains the effect of 14-3-3 on many other targets. A beautiful example of inactivation after 14-3-3 binding came from studies on the chloroplast and mitochondrial ATP synthase. The chloroplast ATP synthase β subunit was one of the original 14-3-3 binding proteins identified eluting from a 14-3-3-Sepharose affinity column [74]. Subsequent work showed that the binding of 14-3-3 to the ATP synthase β-subunit (chloroplast or mitochondrial) inactivated the enzyme [83]. This is thought to prevent the synthase from operating in reverse and consuming ATP when the chloroplast or mitochondrial membrane proton gradient is not maintained because of darkness or lack of oxygen, respectively.

2. *Masking of sequence specific or structural features*. The 'molecular anvil' hypothesis works on the concept that 14-3-3 binding could potentially induce

structural changes in a protein very distant from the site of binding. Yet it is also possible that the binding of 14-3-3 to a target protein could directly mask or block a structural or sequence specific feature on the protein. Again, this concept is supported by the co-crystal structure of serotonin-*N*-acetyltransferase and 14-3-3ζ. Here, the association of the 14-3-3 buries 2527 Å^2 of surface area on the target enzyme, thus demonstrating that 14-3-3s can mask a relatively large surface area. The classic example of this phenomenon is evidenced with the dual specificity phosphatase known as CDC25. CDC25 performs its function in the nucleus where it dephosphorylates a tyrosine and threonine in the N-terminus of CDK2, the first step in the activation of this mitotic kinase. The phosphatase CDC25 has a constitutively exposed nuclear export signal (NES) and a nuclear localization signal (NLS) that resides near a 14-3-3 binding site. Once CDC25 is phosphorylated at the appropriate site, 14-3-3 docking here is thought to mask the nearby NLS and the NES dictates shuttling out of the nucleus and no subsequent import back unless this NLS is exposed again. Once out of the nucleus, CDK2 is activated and mitosis is triggered. Other well-studied examples of 14-3-3 masking include several mammalian histone deacetylases and the potassium channel protein Kir6.2 [84].

3. *Co-localization of proteins*. The idea that 14-3-3s could co-localize two separate proteins, in addition to the 14-3-3 itself, arose from data showing the co-binding of Bcr and Raf-1 with 14-3-3 and the structural studies illustrating that each subunit of a 14-3-3 dimer could bind a phosphopeptide. This meant that a 14-3-3 molecule could act as a phospho-dependent scaffold. To date only a few studies have shown this principle [77].

5.5 The role of protein phosphorylation in the control of plant primary metabolism

Some of the best-understood roles for protein phosphorylation in plants come from studies in primary metabolism and key examples will be highlighted below. A number of conserved protein kinases in yeast and mammals play central roles in nutrient sensing and signalling and a greater understanding of their function has emerged recently. Many of these protein kinases, such as the TOR and general control nonderepressible (GCN2), appear to be conserved in plants. Recent work in yeast and mammalian cells has established that the protein kinase TOR is a coordinator between nutrient and energy charge sensing (amino acids and ATP) in the cytosol of cells and the control of protein synthesis, cell growth and proliferation. Glimpses of a related capacity for TOR have been uncovered in plant cells and will be discussed here.

5.5.1 *Nutrient sensing and signalling through conserved protein kinases*

The TOR pathway in mammalian and yeast cells has a variety of both upstream and downstream components that all work through the TOR protein to illicit a cellular response to a variety of conditions, such as amino acid or ATP depletion. TOR exerts its effects on the cell through its downstream targets, the most characterized of which are ribosomal protein S6 kinases (S6Ks) and eukaryotic initiation factor 4E binding proteins (4E-BPs) [85, 86]. The phosphorylation state of these two types of proteins, as well as other targets of TOR, is thought to be mediated by a combination of TOR kinase activity and PP2A phosphatase-like activity [87].

When amino acids are abundant, TOR phosphorylates S6K, which serves to activate the enzyme, leading to increased ribosomal S6 protein and eukaryotic initiation factor 4B phosphorylation (dephosphorylation occurs during amino acid starvation). Phosphorylation of these targets increases translation of ribosomal proteins and other translational regulators [88]. Despite several theories, it has not yet been shown conclusively how S6Ks mediate the effects of TOR signalling [89–91]. Mammalian TOR is also thought to function as a homeostatic ATP sensor due to its high K_m for ATP [92]. More recently, the energy charge sensing AMP-activated protein kinase (AMPK) has been demonstrated to function upstream in the TOR pathway and control signalling to mTOR in response to AMP/ATP levels [88].

In mammals, S6Ks and 4E-BPs (discussed below) are targeted to TOR through a conserved F(D/E)(hydrophobic)(D/E)(I/L) motif [93]. While plants do have two S6K proteins, neither has the TOR targeting motif, suggesting TOR functions through some alternate means. The mechanism of action of the other well-characterized targets of TOR, the 4E-BPs, has been largely determined in yeast and mammals. Eukaryotic initiation factor 4E (eIF4E) is responsible for binding to the 7-methyl guanosine cap on the 5' end of mRNA and then recruiting other proteins required for initiation through binding of eukaryotic initiation factor 4G (eIF4G). The binding of a 4E-BP protein prevents the binding of eIF4G, and thus prevents initiation of translation. The phosphorylation of 4E-BPs prevents them from binding to eIF4E and allows the initiation complex to form normally [86]. While plants do not have a close homolog of the 4E-BPs, another protein has been identified as capable of binding eIF4E in the same manner as 4E-BPs in yeast, with very little homology to them, outside a small docking motif [94]. This protein also does not have a homolog in plants. Despite the fact that the two best-known targets of TOR lack highly related homologs in plants, it has been demonstrated that TOR has a very similar function in plants compared to yeast and mammals. TOR knockouts in both Arabidopsis and *Drosophila* show similar phenotypes: an inability to progress past equivalent stages in development [95–97]. It is then very likely that the function of the TOR pathway will be similar in both organisms.

While the targets of TOR are relatively well defined, the means of activating TOR in response to nutrient availability and energy charge (both amino acids and ATP) is not. Clues to the mechanism have been found by identifying TOR interacting proteins, such as the regulatory associated protein of TOR (RAPTOR in mammals; KOG1 in yeast) and LST8 (sometimes called GβL in mammals) [95]. These proteins are essential for the nutrient sensing ability of TOR, and their complex with TOR may be the nutrient sensing complex. Both of these proteins are well conserved in plants, indicating a likely similar role in the regulation of TOR [92]. Even though TOR shares a common role in developmental pathways, it may be that the nutrient sensing roles may differ in plants and animals. While TOR is ubiquitously expressed in mammalian cells, it is only present in nondifferentiated tissues of plants, raising the question of how differentiated plant cells respond to nutrient stress [14].

The role of TOR in nutrient deprivation is only recently beginning to take shape, but the roles of other protein kinases, GCN2 and Snf1, have been studied for some time. The Snf1-like kinase has been studied in many organisms, including plants, while the focus of GCN2 research has been on yeast, the organism it was discovered in. Upon binding of uncharged tRNA (an indicator of amino acid starvation) to GCN2, the enzyme is activated and phosphorylates eukaryotic initiation factor 2α, which inhibits the entire eIF2 complex [98]. This leads to an inhibition of translation initiation, and thus protein synthesis as a whole. Snf1, known as AMPK in mammals, is responsible for the control of a wide variety of functions within the cell. AMP is a good indication of cellular energy status because all cells have an adenylate kinase which maintains the equilibrium $2ADP \leftrightarrow ATP + AMP$, which means that a very small decrease in ATP produces a comparatively large increase in AMP. AMPK, Snf1 and Snf1-like kinases have many known targets in yeast, mammals and plants, some being specific to certain kingdoms, such as nitrate reductase (NR) in plants, and others being more general, such as TOR [99, 100].

5.5.2 Nitrate reductase

One of the earliest and best-studied examples of protein phosphorylation as a regulatory mechanism in primary plant metabolism concerns NR. NR is a cytosolic enzyme that catalyzes the reduction of nitrate to nitrite. Several early observations indicated that NR was a highly regulated enzyme. During the artificial lowering of CO_2 levels or upon the transition from light to dark, the extractable activity of NR decreased dramatically and this was a phosphorylation-dependent event [101, 102]. Attempts to purify the 'inactive', phosphorylated form of the enzyme proved unfruitful, even in the presence of excess protein phosphatase inhibitors [102]. It was subsequently found that some inhibitory factor was dissociating from phospho-NR because of the dilution associated with the purification procedure. An enzyme assay set up to look for this inhibitory factor of phosphorylated NR revealed that this factor was a protein. Purification and structural analysis of this protein, originally called NIP (nitrate reductase inhibitory protein), disclosed its identity as 14-3-3. Interestingly, we were able to show that a phospho-peptide based on the 14-3-3 binding region of mammalian Raf-1 could abolish the inhibitory effect of the 14-3-3. Moreover, the inhibitory effect of the purified plant 14-3-3 on phosph-NR activity could be replicated with yeast or human 14-3-3 proteins [103]. Knowing that the inhibitory effect of the yeast and human 14-3-3 was as potent as the plant 14-3-3, we then used immobilized recombinant yeast 14-3-3 proteins as an affinity matrix in an attempt to purify and identify other plant 14-3-3 binding proteins. The strength of this approach relied on the affinity displacement of the proteins retained on the matrix with a phosphorylated peptide based on a 14-3-3 binding motif. This technique was enormously successful and to date more than 30 plant 14-3-3 interactors have been identified by this method [74].

5.5.3 Sucrose synthase

Sucrose synthase (SUS) catalyzes the synthesis of sucrose from UDP-glucose and fructose. This reaction occurs optimally at pH 8–8.8, whereas under hypoxic or anoxic conditions, the pH in the cell falls and the sucrose synthase enzyme degrades sucrose to UDP-glucose and fructose [104]. SUS is only one of two enzymes (the other being invertase) capable of sucrose degradation *in vivo*. The SUS enzyme has increased degradation activity under reduced oxygen conditions, as its pH optimum is between 6 and 6.5 [104]. It is this characteristic of the SUS enzyme that allows plants to acclimate to oxygen stresses, such as flooding. Changes in environmental conditions activate both translational and post-translational responses in SUS, as the SUS gene is regulated by the level of its own enzyme products, and the SUS enzyme is regulated by changes in subcellular localization, protein turnover and phosphorylation [104, 105]. The phosphorylation of SUS by a CDPK occurs on two conserved Ser residues. In *Zea mays*, the phosphorylation of the major site (Ser15) has been linked to subsequent phosphorylation on its second site (Ser170) [105]. Phosphorylation of the major site may activate the cleavage activity of the enzyme by altering the structure of the amino terminus [105]. In addition, Ser15 phosphorylation results in changes in the kinetic properties and subcellular localization of SUS, as the distribution of SUS between the soluble and membrane fractions is associated with the developmental stage of the organ examined and the phosphorylation status of Ser15. It is thought that membrane associated SUS provides UDP-glucose to the membrane localized cellulose synthase complex [104]. The minor phosphorylation site (Ser170) plays a role in regulating protein turnover, as it targets SUS for ubiquitin-mediated degradation by the proteosome [105].

5.5.4 Sucrose phosphate synthase and trehalose phosphate synthase

The synthesis of the nonreducing sugars sucrose (glucose and fructose) and trehalose (two glucose molecules) from hexose monophosphates proceed via the rate limiting steps catalyzed by sucrose phosphate synthase (SPS) and trehalose phosphate synthase (TPS), respectively. The phosphate is released by the respective 6-phosphatases to yield either sucrose or trehalose. The synthesis of sucrose and trehalose and the reactions catalyzed by SPS and TPS are highlighted below:

SPS
UDP-glucose + fructose-6-phosphate $\rightarrow$ sucrose-6-phosphate $\rightarrow$ sucrose + Pi

TPS
UDP-glucose + glucose-6-phosphate $\rightarrow$ trehalose-6-phosphate $\rightarrow$ trehalose + Pi

Both enzymes were retained and eluted from the 14-3-3 affinity matrix described above, placing phosphorylation and 14-3-3 binding at the forefront of regulation of the synthesis of these key disaccharides. Genomics has shown that there are four SPS genes in Arabidopsis and classic kinetics demonstrated that SPS is allosterically activated by glucose-6-phosphate and inhibited by inorganic phosphate. SPS was found to be phosphorylated many years ago [106] and the three sites in the spinach leaf enzyme are serines 158, 229 and 424. Although the role of each phosphorylation site is not completely resolved, serine 229 is the likely 14-3-3 docking site. Two different groups have reported different effects for 14-3-3 binding to SPS [74, 107]. Both activation and inhibition have been suggested, and as yet this controversy has not been resolved.

TPS was indicated to be a phospho-protein due to its retention on 14-3-3-Sepharose. Recent work with overexpressing plants suggest that trehalose-6-phosphate may be a key signalling molecule, as has been demonstrated in yeast [108]. The roles of phosphorylation and 14-3-3 binding of TPS are yet to be elucidated. Future research will be needed to define the regulation of TPS and the functions of trehalose-6-phosphate.

5.5.5 6-phosphofructo-2-kinase/fructose2,6-bisphosphatase

PFK2 is a bifunctional enzyme responsible for the reversible synthesis of the regulatory metabolite fructose-2,6-biphosphate ($F26P_2$) from fructose-6-phosphate. In mammals and yeast, $F26P_2$ stimulates glycolysis by activation of 6-phosphofructo-1-kinase (PFK1) and slows gluconeogenesis by inhibiting fructose-1,6-bisphosphatase (FBPase). Plant cytosolic FBPase is inhibited by $F26P_2$, while the ATP-dependent PFK1 is not $F26P_2$ stimulated. Interestingly, the PPi-dependent PFK (PFP) is activated by this metabolite. Although a clear role for PFP is yet to be defined in plants, various evidences suggest it is an adaptive enzyme that helps to contribute to the unique flexibility of primary plant metabolism thereby helping plants acclimate to unavoidable abiotic stresses (such as anoxia and Pi starvation) they encounter in their natural environment. Recently the plant PFK2 was demonstrated to be phosphorylated on serine and threonine residues [109] and a screen of an Arabidopsis 14-3-3 column eluate showed that this enzyme is retained on the matrix and eluted with a 14-3-3 competing phospho-peptide [110]. Recombinant PFK2, phosphorylated *in vitro* with an Arabidopsis extract, recombinant atCDPK3 or mammalian AMPK then has the ability to bind 14-3-3s. Mapping of the phosphorylation sites generated during conditions that result in 14-3-3 binding yielded three phosphorylation sites on PFK2, including one that resembles a mode one 14-3-3 binding site. The role of 14-3-3 binding to PFK2 is yet to be determined, as a thorough kinetic examination did not find any changes in kinetic parameters or ratio of $F26P_2$ synthesis to degradation activities. The precise function of 14-3-3 binding to PFK2 awaits discovery.

5.5.6 Starch synthase and starch branching enzyme

Starch is synthesized in chloroplasts during the light period from carbon assimilated through the reductive pentose phosphate pathway. Starch accumulates as an insoluble polymer or granule of amylose and amylopectin. Starch synthase adds glucose moieties from ADP-glucose to the nonreducing ends of amylose or amylopectin with specific isoforms of starch synthase bound to the starch granule, or found in the soluble fraction or stroma of the plastid. Branching, the generation of α1→6 bonds, is introduced in the growing chains by starch branching enzyme.

The first suggestion that protein phosphorylation plays a role in regulating starch metabolism came from 14-3-3 antisense plants which accumulate more starch compared to wild-type plants [111]. Western analysis of isolated starch and microscopy studies displayed the presence of 14-3-3s on the starch granule. An analysis of starch metabolic enzyme sequences showed that starch synthase III, a granule bound form, has a near 'perfect' mode one 14-3-3 binding motif and it was proposed to bind 14-3-3 on the starch granule [111]. More recently, labeling of intact wheat amyloplasts with γ^{32}P-ATP and purification of phosphoproteins on a phospho-affinity matrix detected one soluble and two granule associated starch branching enzymes, plus two granule bound starch synthases as phospho-proteins. Phosphorylation of starch branching enzyme II stimulated the activity of the enzyme [112].

5.5.7 Glutamine synthetase (GS_1 and GS_2)

The ATP-dependent glutamine synthetase (GS) reaction yields glutamine from glutamate and ammonium and is considered to be a pivotal interface of carbon and nitrogen metabolism as this catalyzes the reaction that incorporates inorganic nitrogen into organic form. Bacterial GS is highly regulated, being covalently modified through adenylylation, which allows feedback inhibition by no fewer than eight key downstream metabolites. Both cytosolic (GS_1) and plastidic (GS_2) forms of glutamine synthetase are present in plant cells. Because of the critical step catalyzed by this reaction, it is natural to predict that these are regulated enzymes. Perhaps it was no surprise that both GS_1 and GS_2 were purified on a 14-3-3 matrix, the first clue that they were likely phosphoproteins [74]. Additional work has revealed that both GS_1 and GS_2 are phosphoproteins; both bind 14-3-3s in their respective compartments and association with 14-3-3 activates the enzymes 1.5 to 2-fold [113, 114]. Chromatography of *Chlamydomonas reinhardtii* extracts on 14-3-3 Sepharose also identified one of the major 14-3-3- binding proteins as the cytosolic form of GS [115]. Here no effect on enzyme activity was observed.

5.5.8 Nonphosphorylating glycerladehyde-3-phosphate dehyrdrogenase

There are three different enzymes capable of oxidizing glyceraldehydes-3-phosphate (G3P) in plants. Two of them are phosphate dependent, with one being present in both photosynthetic and nonphotosynthetic tissues, whereas the other is present strictly in the chloroplast. The third enzyme converts G3P to 3-phosphoglycerate coupled to the production of NADPH. This enzyme, known as nonphosphorylating glyceraldehyde-3-phosphate dehydrogenase (GAPN), is unrelated to the other two enzymes with which it shares function, and is actually most similar to the superfamily of aldehyde dehydrogenases [116]. GAPN is proposed to be involved in the shuttling of NADPH generated through photosynthesis into the cytosol. However, GAPN is not specifically expressed in photosynthetic tissues, which suggests that it may have a nonphotosynthetic role as well. In support of a role outside photosynthesis, GAPN has been found to be phosphorylated in wheat (*Triticum aestivum*), but only in endosperm and shoots, but not leaves. Furthermore, the phosphorylation has been found to allow binding of the enzyme to 14-3-3 proteins. This binding changes the catalytic properties of the enzyme, producing an enzyme complex that inhibits GAPN activity in cells with high energy charge [117].

5.5.9 Phosphoenolpyruvate carboxylase and PEPC kinase

PEPC catalyzes the cytosolic carboxylation of phospho*enol*pyruvate to form oxaloacetate and phosphate (see Chapter 12). In C4 and CAM plants, the photosynthetic PEPC is responsible for the primary fixation of CO_2, while other isoforms are best known for their role as a replenisher of TCA cycle intermediates during nitrogen assimilation. All forms are allosterically activated by glucose-6-phosphate and inhibited by malate. Phosphorylation of a highly conserved serine residue near the N-terminus relieves malate inhibition of PEPC. The identification and purification of the protein kinase responsible for this phosphorylation was a daunting task. The protein kinase was partially purified from maize as a Ca^{2+}-independent enzyme of 30–32 kDa and was eventually cloned. The PEPC kinases are unusual in that they consist almost entirely of just a protein kinase catalytic domain with no N- or C-terminal extensions [8]. They are also unique in that the activity of this enzyme appears to be regulated solely by expression. As mentioned earlier, the residue of the activation loop that would be phosphorylated to activate the kinase is a glycine in the PEPC enzymes and genomics has shown that the PEPC kinases are unique to plants and a specific subgroup of protists.

5.6 Summary

Genomics, standing on the shoulders of the information derived from the long history of biochemistry and molecular biology of protein kinases, phosphatases and metabolism, has thrown light on many phospho-regulated processes in plants. By exploiting the recently developed techniques of microarray analysis, RNAi, peptide library analysis, proteomics and phospho-proteomics, we are now on the verge of witnessing an explosion of information on plant biology concerning the role protein phosphorylation plays in primary plant metabolism and other cellular processes. This chapter provides a framework of information on protein kinases and protein phosphatases across a broad spectrum of organisms and we hope it will function as a guide for plant biologists and stimulate additional interest in protein phosphorylation as a regulatory mechanism.

Acknowledgements

G.B.G. Moorhead is supported by the Natural Sciences and Engineering Research Council of Canada. The authors thank D. Bridges for assistance in modifying Figure 5.2.

References

1. P. Cohen (2002) The origins of protein phosphorylation. *Nature Cell Biology* **4**, E127–E130.
2. G.D. Plowman, S. Sudarsanam, J. Bingham, D. Whyte and T. Hunter (1999) The protein kinases of *Caenorhabditis elegans*: a model for signal transduction in multicellular organisms. *Proceedings of the National Academy of Sciences of the United States of America* **96**, 13603–13610.
3. A. Champion, M. Kreis, K. Mockaitis, A. Picaud and Y. Henry (2004) Arabidopsis kinome: after the casting. *Functional & Integrative Genomics* **4**, 163–187.
4. G. Manning, D.B. Whyte, R. Martinez, T. Hunter and S. Sudarsanam (2002) The protein kinase complement of the human genome. *Science* **298**, 1912–1934.
5. D. Wang, J.F. Harper, M. Gribskov (2003) Systematic trans-genomic comparison of protein kinases between Arabidopsis and *Saccharomyces cerevisiae*. *Plant Physiology* **132**, 2152–2165.
6. E.H. Fischer and E.G. Krebs (1955) Conversion of phosphorylase b to phosphorylase a in muscle extracts. *The Journal of Biological Chemistry* **216**, 121–132.
7. Z. Songyang, S. Blechner, N. Hoagland, M.F. Hoekstra, H. Piwnica-Worms and L.C. Cantley (1994) Use of an oriented peptide library to determine the optimal substrates of protein kinases. *Current Biology* **4**, 973–982.
8. E.M. Hrabak, C.W. Chan, M. Gribskov *et al.* (2003) The Arabidopsis CDPK–SnRK superfamily of protein kinases. *Plant Physiology* **132**, 666–680.
9. J.G. Williams and M. Zvelebil (2004) SH2 domains in plants imply new signalling scenarios. *Trends in Plant Science* **9**, 161–163.
10. C. Fankhauser, K.C. Yeh, J.C. Lagarias, H. Zhang, T.D. Elich and J. Chory (1999) PKS1, a substrate phosphorylated by phytochrome that modulates light signalling in Arabidopsis, *Science* **284**, 1539–1541.
11. T. Takahashi, K. Hara, H. Inoue *et al.* (2000) Carboxyl-terminal region conserved among phosphoinositide-kinase-related kinases is indispensable for mTOR function *in vivo* and *in vitro*. *Genes to Cells* **5**, 765–775.
12. M.A. Andrade and P. Bork (1995) HEAT repeats in the Huntington's disease protein. *Nature Genetics* **11**, 115–116.
13. M.R. Groves, N. Hanlon, P. Turowski, B.A. Hemmings and D. Barford (1999) The structure of the protein phosphatase 2A PR65/A subunit reveals the conformation of its 15 tandemly repeated HEAT motifs. *Cell* **96**, 99–110.
14. B. Menand, C. Meyer and C. Robaglia (2004) Plant growth and the TOR pathway. *Current Topics in Microbiology and Immunology* **279**, 97–113.
15. R.T. Abraham (2004) PI 3-kinase related kinases: 'big' players in stress-induced signalling pathways. *DNA Repair (Amsterdam)* **3**, 883–887.
16. P. Cohen (1989) The structure and regulation of protein phosphatases. *Annual Review of Biochemistry* **58**, 453–508.
17. C. MacKintosh, J. Coggins and P. Cohen (1991) Plant protein phosphatases. Subcellular distribution, detection of protein phosphatase 2C and identification of protein phosphatase 2A as the major quinate dehydrogenase phosphatase. *The Biochemical Journal* **273**, 733–738.
18. D. Kerk, J. Bulgrien, D.W. Smith, B. Barsam, S. Veretnik and M. Gribskov (2002) The complement of protein phosphatase catalytic subunits encoded in the genome of Arabidopsis. *Plant Physiology* **129**, 908–925.
19. P.T. Cohen (1997) Novel protein serine/threonine phosphatases: variety is the spice of life. *Trends in Biochemical Sciences* **22**, 245–251.
20. A. Alonso, J. Sasin, N. Bottini *et al.* (2004) Protein tyrosine phosphatases in the human genome. *Cell* **117**, 699–711.
21. T.L. Tootle, S.J. Silver, E.L. Davies *et al.* (2003) The transcription factor eyes absent is a protein tyrosine phosphatase. *Nature* **426**, 299–302.
22. J.A. Dominov, L. Stenzler, S. Lee, J.J. Schwarz, S. Leisner and S.H. Howell (1992) Cytokinins and auxins control the expression of a gene in *Nicotiana plumbaginifolia* cells by feedback regulation. *The Plant Cell* **4**, 451–461.

23. W. Li, S. Luan, S.L. Schreiber and S.M. Assmann (1994) Evidence for protein phosphatase 1 and 2A regulation of K+ channels in two types of leaf cells. *Plant Physiology* **106**, 963–970.
24. V. Raz and R. Fluhr (1993) Ethylene signal is transduced via protein phosphorylation events in plants. *The Plant Cell* **5**, 523–530.
25. S.J. Rundle, M.E. Nasrallah and J.B. Nasrallah (1993) Effects of inhibitors of protein serine/threonine phosphatases on pollination in Brassica. *Plant Physiology* **103**, 1165–1171.
26. J. Sheen (1993) Protein phosphatase activity is required for light-inducible gene expression in maize. *The EMBO Journal* **12**, 3497–3505.
27. S.M. Wolniak and P.M. Larsen (1992) Changes in the metaphase transit times and the pattern of sister chromatid separation in stamen hair cells of Tradescantia after treatment with protein phosphatase inhibitors. *Journal of Cell Science* **102**, 691–715.
28. S. Zhao and E.Y. Lee (1997) A protein phosphatase-1-binding motif identified by the panning of a random peptide display library. *The Journal of Biological Chemistry* **272**, 28368–28372.
29. P. Wakula, M. Beullens, H. Ceulemans, W. Stalmans and M. Bollen (2003) Degeneracy and function of the ubiquitous RVXF motif that mediates binding to protein phosphatase-1. *The Journal of Biological Chemistry* **278**, 18817–18823.
30. M.P. Egloff, D.F. Johnson, G. Moorhead, P.T. Cohen, P. Cohen and D. Barford (1997) Structural basis for the recognition of regulatory subunits by the catalytic subunit of protein phosphatase 1. *The EMBO Journal* **16**, 1876–1887.
31. R. Koradi, M. Billeter and K. Wuthrich (1996) MOLMOL: a program for display and analysis of macromolecular structures. *Journal of Molecular Graphics* **14**, 29–32, 51–55.
32. M. Terrak, F. Kerff, K. Langsetmo, T. Tao and R. Dominguez (2004) Structural basis of protein phosphatase 1 regulation. *Nature* **429**, 780–784.
33. C.G. Armstrong, M.J. Doherty and P.T. Cohen (1998) Identification of the separate domains in the hepatic glycogen-targeting subunit of protein phosphatase 1 that interact with phosphorylase a, glycogen and protein phosphatase 1. *The Biochemical Journal* **336**, 699–704.
34. M. Beullens, A. Van Eynde, V. Vulsteke et al. (1999) Molecular determinants of nuclear protein phosphatase-1 regulation by NIPP-1. *The Journal of Biological Chemistry* **274**, 14053–14061.
35. S. Wera and B.A. Hemmings (1995) Serine/threonine protein phosphatases. *The Biochemical Journal* **311**, 17–29.
36. G. Sun and J. Markwell (1992) Lack of type 1 and type 2A protein serine (P)/ threonine (P) phosphatase activities in chloroplasts. *Plant Physiology* **100**, 620–624.
37. D.H. Kim, J.G. Kang, S.S. Yang, K.S. Chung, P.S. Song and C.M. Park (2002) A phytochrome-associated protein phosphatase 2A modulates light signals in flowering time control in Arabidopsis. *The Plant Cell* **14**, 3043–3056.
38. V. Janssens and J. Goris (2001) Protein phosphatase 2A: a highly regulated family of serine/threonine phosphatases implicated in cell growth and signalling. *The Biochemical Journal* **353**, 417–439.
39. B. McCright and D.M. Virshup (1995) Identification of a new family of protein phosphatase 2A regulatory subunits. *The Journal of Biological Chemistry* **270**, 26123–26128.
40. C.S. Moreno, S. Park, K. Nelson et al. (2000) WD40 repeat proteins striatin and S/G(2) nuclear autoantigen are members of a novel family of calmodulin-binding proteins that associate with protein phosphatase 2A. *The Journal of Biological Chemistry* **275**, 5257–5263.
41. C. Camilleri, J. Azimzadeh, M. Pastuglia, C. Bellini, O. Grandjean and D. Bouchez (2002) The Arabidopsis TONNEAU2 gene encodes a putative novel protein phosphatase 2A regulatory subunit essential for the control of the cortical cytoskeleton. *The Plant Cell* **14**, 833–845.
42. A. Schweighofer, H. Hirt and I. Meskiene (2004) Plant PP2C phosphatases: emerging functions in stress signalling. *Trends in Plant Science* **9**, 236–243.
43. W.R. Scheible, R. Morcuende, T. Czechowski et al. (2004) Genome-wide reprogramming of primary and secondary metabolism, protein synthesis, cellular growth processes, and the regulatory infrastructure of Arabidopsis in response to nitrogen. *Plant Physiology* **136**, 2483–2499.
44. S. Mora-Garcia, G. Vert, Y. Yin, A. Cano-Delgado, H. Cheong and J. Chory (2004) Nuclear protein phosphatases with Kelch-repeat domains modulate the response to brassinosteroids in Arabidopsis. *Genes & Development* **18**, 448–460.

45. A. Ali, J. Zhang, S. Bao et al. (2004) Requirement of protein phosphatase 5 in DNA-damage-induced ATM activation. *Genes & Development* **18**, 249–254.
46. T. Wechsler, B.P. Chen, R. Harper et al. (2004) DNA-PKcs function regulated specifically by protein phosphatase 5. *Proceedings of the National Academy of Sciences of the United States of America* **101**, 1247–1252.
47. Z. Zuo, N.M. Dean and R.E. Honkanen (1998) Serine/threonine protein phosphatase type 5 acts upstream of p53 to regulate the induction of p21(WAF1/Cip1) and mediate growth arrest. *The Journal of Biological Chemistry* **273**, 12250–12258.
48. G.L. Blatch and M. Lassle (1999) The tetratricopeptide repeat: a structural motif mediating protein–protein interactions. *Bioessays* **21**, 932–939.
49. M.X. Chen and P.T. Cohen (1997) Activation of protein phosphatase 5 by limited proteolysis or the binding of polyunsaturated fatty acids to the TPR domain. *FEBS Letters* **400**, 136–140.
50. J. Skinner, C. Sinclair, C. Romeo, D. Armstrong, H. Charbonneau and S. Rossie (1997) Purification of a fatty acid-stimulated protein-serine/threonine phosphatase from bovine brain and its identification as a homolog of protein phosphatase 5. *The Journal of Biological Chemistry* **272**, 22464–22471.
51. J. Yang, S.M. Roe, M.J. Cliff et al. (2005) Molecular basis for TPR domain-mediated regulation of protein phosphatase 5. *The EMBO Journal* **24**, 1–10.
52. J. Adams, R. Kelso and L. Cooley (2000) The kelch repeat superfamily of proteins: propellers of cell function. *Trends in Cell Biology* **10**, 17–24.
53. H. Koiwa, S. Hausmann, W.Y. Bang et al. (2004) Arabidopsis C-terminal domain phosphatase-like 1 and 2 are essential Ser-5-specific C-terminal domain phosphatases. *Proceedings of the National Academy of Sciences of the United States of America* **101**, 14539–14544.
54. I. Landrieu, M. da Costa, L. De Veylder et al. (2004) A small CDC25 dual-specificity tyrosine-phosphatase isoform in *Arabidopsis thaliana*. *Proceedings of the National Academy of Sciences of the United States of America* **101**, 13380–13385.
55. S. Krishnamurthy, X. He, M. Reyes-Reyes, C. Moore and M. Hampsey (2004) Ssu72 is an RNA polymerase II CTD phosphatase. *Molecular Cell* **14**, 387–394.
56. R.S. Chambers and M.E. Dahmus (1994) Purification and characterization of a phosphatase from HeLa cells which dephosphorylates the C-terminal domain of RNA polymerase II. *The Journal of Biological Chemistry* **269**, 26243–26248.
57. J. Archambault, G. Pan, G.K. Dahmus et al. (1998) FCP1, the RAP74-interacting subunit of a human protein phosphatase that dephosphorylates the carboxyl-terminal domain of RNA polymerase IIO. *The Journal of Biological Chemistry* **273**, 27593–27601.
58. T. Kamenski, S. Heilmeier, A. Meinhart and P. Cramer (2004) Structure and mechanism of RNA polymerase II CTD phosphatases. *Molecular Cell* **15**, 399–407.
59. M. Yeo, P.S. Lin, M.E. Dahmus and G.N. Gill (2003) A novel RNA polymerase II C-terminal domain phosphatase that preferentially dephosphorylates serine 5. *The Journal of Biological Chemistry* **278**, 26078–26085.
60. S. Klumpp, J. Hermesmeier, D. Selke, R. Baumeister, R. Kellner and J. Krieglstein (2002) Protein histidine phosphatase: a novel enzyme with potency for neuronal signalling. *Journal of Cerebral Blood Flow and Metabolism* **22**, 1420–1424.
61. S. Klumpp, G. Bechmann, A. Maurer, D. Selke and J. Krieglstein (2003) ATP-citrate lyase as a substrate of protein histidine phosphatase in vertebrates. *Biochemical and Biophysical Research Communications* **306**, 110–115.
62. M.B. Yaffe and A.E. Elia (2001) Phosphoserine/threonine-binding domains. *Current Opinion in Cell Biology* **13**, 131–138.
63. M.B. Yaffe, K. Rittinger, S. Volinia et al. (1997) The structural basis for 14-3-3: phosphopeptide binding specificity. *Cell* **91**, 961–971.
64. I.A. Manke, D.M. Lowery, A. Nguyen and M.B. Yaffe (2003) BRCT repeats as phosphopeptide-binding modules involved in protein targeting. *Science* **302**, 636–639.
65. B.E. Moore and V.J. Perez (1967) In: *Physiological and Biochemical Aspects of Nervous Integration* (ed. F.D. Carlson), Prentice-Hall, Englewood Cliffs, NJ, pp. 343–359.

66. T. Ichimura, T. Isobe, T. Okuyama et al. (1988) Molecular cloning of cDNA coding for brain-specific 14-3-3 protein, a protein kinase-dependent activator of tyrosine and tryptophan hydroxylases. *Proceedings of the National Academy of Sciences of the United States of America* **85**, 7084–7088.
67. T. Ichimura, T. Isobe, T. Okuyama, T. Yamauchi and H. Fujisawa (1987) Brain 14-3-3 protein is an activator protein that activates tryptophan 5-monooxygenase and tyrosine 3-monooxygenase in the presence of Ca2+, calmodulin-dependent protein kinase II. *FEBS Letters* **219**, 79–82.
68. W.J. Fantl, A.J. Muslin, A. Kikuchi et al. (1994) Activation of Raf-1 by 14-3-3 proteins. *Nature* **371**, 612–614.
69. H. Fu, K. Xia, D.C. Pallas et al. (1994) Interaction of the protein kinase Raf-1 with 14-3-3 proteins. *Science* **266**, 126–129.
70. A. Toker, C.A. Ellis, L.A. Sellers and A. Aitken, (1990) Protein kinase C inhibitor proteins. Purification from sheep brain and sequence similarity to lipocortins and 14-3-3 protein. *European Journal of Biochemistry* **191**, 421–429.
71. A.J. Muslin, J.W. Tanner, P.M. Allen and A.S. Shaw (1996) Interaction of 14-3-3 with signalling proteins is mediated by the recognition of phosphoserine. *Cell* **84**, 889–897.
72. V. Cotelle, S.E. Meek, F. Provan, F.C. Milne, N. Morrice and C. MacKintosh (2000) 14-3-3s regulate global cleavage of their diverse binding partners in sugar-starved Arabidopsis cells. *The EMBO Journal* **19**, 2869–2876.
73. S.E. Meek, W.S. Lane and H. Piwnica-Worms (2004) Comprehensive proteomic analysis of interphase and mitotic 14-3-3-binding proteins. *The Journal of Biological Chemistry* **279**, 32046–32054.
74. G. Moorhead, P. Douglas, V. Cotelle et al. (1999) Phosphorylation-dependent interactions between enzymes of plant metabolism and 14-3-3 proteins. *The Plant Journal* **18**, 1–12.
75. J. Jin, F.D. Smith, C. Stark et al. (2004) Proteomic, functional, and domain-based analysis of *in vivo* 14-3-3 binding proteins involved in cytoskeletal regulation and cellular organization. *Current Biology* **14**, 1436–1450.
76. M. Pozuelo Rubio, K.M. Geraghty, B.H. Wong et al. (2004) 14-3-3-affinity purification of over 200 human phosphoproteins reveals new links to regulation of cellular metabolism, proliferation and trafficking. *The Biochemical Journal* **379**(Pt 2), 395–408.
77. D. Bridges and G.B. Moorhead (2004) 14-3-3 proteins: a number of functions for a numbered protein. *Science's STKE* **242**, re10.
78. C. Mackintosh (2004) Dynamic interactions between 14-3-3 proteins and phosphoproteins regulate diverse cellular processes. *The Biochemical Journal* **381**, 329–342.
79. M.B. Yaffe (2002) How do 14-3-3 proteins work? – Gatekeeper phosphorylation and the molecular anvil hypothesis. *FEBS Letters* **513**, 53–57.
80. T. Obsil, R. Ghirlando, D.C. Klein, S. Ganguly and F. Dyda (2001) Crystal structure of the 14-3-3 zeta:serotonin N-acetyltransferase complex. A role for scaffolding in enzyme regulation. *Cell* **105**, 257–267.
81. M. Wurtele, C. Jelich-Ottmann, A. Wittinghofer and C. Oecking (2003) Structural view of a fungal toxin acting on a 14-3-3 regulatory complex. *The EMBO Journal* **22**, 987–994.
82. L. Camoni, S. Visconti, M. Marra and P. Aducci (2001) Adenosine 5'-monophosphate inhibits the association of 14-3-3 proteins with the plant plasma membrane H(+)-ATPase. *The Journal of Biological Chemistry* **276**, 31709–31712.
83. T.D. Bunney, H.S. van Walraven and A.H. de Boer (2001) 14-3-3 protein is a regulator of the mitochondrial and chloroplast ATP synthase. *Proceedings of the National Academy of Sciences of the United States of America* **98**, 4249–4254.
84. H. Yuan, K. Michelsen and B. Schwappach (2003) 14-3-3 dimers probe the assembly status of multimeric membrane proteins. *Current Biology* **13**, 638–646.
85. P.E. Burnett, R.K. Barrow, N.A. Cohen, S.H. Snyder and D.M. Sabatini (1998) RAFT1 phosphorylation of the translational regulators p70 S6 kinase and 4E-BP1. *Proceedings of the National Academy of Sciences of the United States of America* **95**, 1432–1437.
86. G.J. Brunn, C.C. Hudson, A. Sekulic et al. (1997) Phosphorylation of the translational repressor PHAS-I by the mammalian target of rapamycin. *Science* **277**, 99–101.

87. C.J. Di Como and K.T. Arndt (1996) Nutrients, via the Tor proteins, stimulate the association of Tap42 with type 2A phosphatases. *Genes & Development* **10**, 1904–1916.
88. K. Inoki, H. Ouyang, Y. Li and K.L. Guan (2005) Signalling by target of rapamycin proteins in cell growth control. *Microbiology and Molecular Biology Reviews* **69**, 79–100.
89. H.B. Jefferies, S. Fumagalli, P.B. Dennis, C. Reinhard, R.B. Pearson and G. Thomas (1997) Rapamycin suppresses 5'TOP mRNA translation through inhibition of p70s6k. *The EMBO Journal* **16**, 3693–3704.
90. M. Stolovich, H. Tang, E. Hornstein *et al.* (2002) Transduction of growth or mitogenic signals into translational activation of TOP mRNAs is fully reliant on the phosphatidylinositol 3-kinase-mediated pathway but requires neither S6K1 nor rpS6 phosphorylation. *Molecular and Cellular Biology* **22**, 8101–8113.
91. H. Tang, E. Hornstein, M. Stolovich *et al.* (2001) Amino acid-induced translation of TOP mRNAs is fully dependent on phosphatidylinositol 3-kinase-mediated signalling, is partially inhibited by rapamycin, and is independent of S6K1 and rpS6 phosphorylation. *Molecular and Cellular Biology* **21**, 8671–8683.
92. G.W. Templeton and G.B. Moorhead (2004) A renaissance of metabolite sensing and signalling: from modular domains to riboswitches. *The Plant Cell* **16**, 2252–2257.
93. S.S. Schalm and J. Blenis (2002) Identification of a conserved motif required for mTOR signalling. *Current Biology* **12**, 632–639.
94. G.P. Cosentino, T. Schmelzle, A. Haghighat, S.B. Helliwell, M.N. Hall and N. Sonenberg (2000) Eap1p, a novel eukaryotic translation initiation factor 4E-associated protein in *Saccharomyces cerevisiae*. *Molecular and Cellular Biology* **20**, 4604–4613.
95. R. Loewith, E. Jacinto, S. Wullschleger *et al.* (2002) Two TOR complexes, only one of which is rapamycin sensitive, have distinct roles in cell growth control. *Molecular Cell* **10**, 457–468.
96. S. Oldham, J. Montagne, T. Radimerski, G. Thomas and E. Hafen (2000) Genetic and biochemical characterization of dTOR, the Drosophila homolog of the target of rapamycin. *Genes & Development* **14**, 2689–2694.
97. H. Zhang, J.P. Stallock, J.C. Ng, C. Reinhard and T.P. Neufeld (2000) Regulation of cellular growth by the Drosophila target of rapamycin dTOR. *Genes & Development* **14**, 2712–2724.
98. T.E. Dever, J.J. Chen, G.N. Barber *et al.* (1993) Mammalian eukaryotic initiation factor 2 alpha kinases functionally substitute for GCN2 protein kinase in the GCN4 translational control mechanism of yeast. *Proceedings of the National Academy of Sciences of the United States of America* **90**, 4616–4620.
99. N.G. Halford and D.G. Hardie (1998) SNF1-related protein kinases: global regulators of carbon metabolism in plants?. *Plant Molecular Biology* **37**, 735–748.
100. S.W. Cheng, L.G. Fryer, D. Carling and P.R. Shepherd (2004) Thr2446 is a novel mammalian target of rapamycin (mTOR) phosphorylation site regulated by nutrient status. *The Journal of Biological Chemistry* **279**, 15719–15722.
101. C. MacKintosh (1992) Regulation of spinach-leaf nitrate reductase by reversible phosphorylation. *Biochimica et Biophysica Acta* **1137**, 121–126.
102. C. Mackintosh, P. Douglas and C. Lillo (1995) Identification of a protein that inhibits the phosphorylated form of nitrate reductase from spinach (*Spinacia oleracea*) leaves. *Plant Physiology* **107**, 451–457.
103. G. Moorhead, P. Douglas, N. Morrice, M. Scarabel, A. Aitken and C. MacKintosh (1996) Phosphorylated nitrate reductase from spinach leaves is inhibited by 14-3-3 proteins and activated by fusicoccin. *Current Biology* **6**, 1104–1113.
104. K. Koch (2004) Sucrose metabolism: regulatory mechanisms and pivotal roles in sugar sensing and plant development. *Current Opinion in Plant Biology* **7**, 235–246.
105. S.C. Hardin, H. Winter and S.C. Huber (2004) Phosphorylation of the amino terminus of maize sucrose synthase in relation to membrane association and enzyme activity. *Plant Physiology* **134**, 1427–1438.
106. J.E. Lunn and E. MacRae (2003) New complexities in the synthesis of sucrose. *Current Opinion in Plant Biology* **6**, 208–214.

107. D. Toroser, G.S. Athwal and S.C. Huber (1998) Site-specific regulatory interaction between spinach leaf sucrose-phosphate synthase and 14-3-3 proteins. *FEBS Letters* **435**, 110–114.
108. A.J. van Dijken, H. Schluepmann and S.C. Smeekens (2004) Arabidopsis trehalose-6-phosphate synthase 1 is essential for normal vegetative growth and transition to flowering. *Plant Physiology* **135**, 969–977.
109. T. Furumoto, M. Teramoto, N. Inada, M. Ito, I. Nishida and A. Watanabe (2001) Phosphorylation of a bifunctional enzyme, 6-phosphofructo-2-kinase/fructose-2,6-bisphosphate 2-phosphatase, is regulated physiologically and developmentally in rosette leaves of *Arabidopsis thaliana*. *Plant & Cell Physiology* **42**, 1044–1048.
110. A. Kulma, D. Villadsen, D.G. Campbell *et al.* (2004) Phosphorylation and 14-3-3 binding of Arabidopsis 6-phosphofructo-2-kinase/fructose-2,6-bisphosphatase. *The Plant Journal* **37**, 654–667.
111. P.C. Sehnke, H.J. Chung, K. Wu and R.J. Ferl (2001) Regulation of starch accumulation by granule-associated plant 14-3-3 proteins. *Proceedings of the National Academy of Sciences of the United States of America* **98**, 765–770.
112. I.J. Tetlow, R. Wait, Z. Lu *et al.* (2004) Protein phosphorylation in amyloplasts regulates starch branching enzyme activity and protein–protein interactions. *The Plant Cell* **16**, 694–708.
113. J. Riedel, R. Tischner and G. Mack (2001) The chloroplastic glutamine synthetase (GS-2) of tobacco is phosphorylated and associated with 14-3-3 proteins inside the chloroplast. *Planta* **213**, 396–401.
114. J. Finnemann and J.K. Schjoerring (2000) Post-translational regulation of cytosolic glutamine synthetase by reversible phosphorylation and 14-3-3 protein interaction. *The Plant Journal* **24**, 171–181.
115. M. Pozuelo, C. MacKintosh, A. Galvan and E. Fernandez (2001) Cytosolic glutamine synthetase and not nitrate reductase from the green alga *Chlamydomonas reinhardtii* is phosphorylated and binds 14-3-3 proteins. *Planta* **212**, 264–269.
116. A. Habenicht (1997) The non-phosphorylating glyceraldehyde-3-phosphate dehydrogenase: biochemistry, structure, occurrence and evolution. *Biological Chemistry* **378**, 1413–1419.
117. D.M. Bustos and A.A. Iglesias (2003) Phosphorylated non-phosphorylating glyceraldehyde-3-phosphate dehydrogenase from heterotrophic cells of wheat interacts with 14-3-3 proteins. *Plant Physiology* **133**, 2081–2088.

6 Redox signal transduction in plant metabolism

Santiago Mora-Garcia, Fabiana G. Stolowicz and Ricardo A. Wolosiuk

6.1 Introduction

The advent of oxygenic photosynthesis was probably the second major event in the history of life on our planet, after the inception of life itself. The ability to use staples such as water and light as energy sources for the assimilation of carbon unleashed the potential of primary producers. At the same time, oxygen released as a by-product of the photosynthetic process became an effective electron acceptor for respiration. However, organisms that perform this particular type of reaction play, literally, with fire. Molecular oxygen is particularly prone to yield reduced and highly unstable reactive oxygen species (ROS), which avidly react with the electron-rich organic molecules. Oxygen builds up in the light, precisely at the same time as light-excited photosystems and transport chains handle a rich pool of free electrons. Whenever the abundance of suitable acceptors fails to match the rate of production, several mechanisms tend to transfer these electrons directly to oxygen. The most abundant protein in land plants, ribulose-1, 5-bisphosphate (RuBP) carboxylase (Rubisco), further complicates this scheme. Rubisco incorporates O_2 into RuBP almost as readily as CO_2, a process known as photorespiration. When the availability of CO_2 becomes restricted, for instance in response to water deficit, or when high temperatures alter the catalytic properties of Rubisco, photorespiration may take over, and 2-phosphoglycolate is produced at the expense of the building block for the Benson-Calvin cycle, 3-phosphoglycerate. Although this reaction is a starting point for several biosynthetic processes and a way to remove oxygen, the oxidation of glycolate in peroxisomes significantly contributes to the production of H_2O_2 during the light period. Photosynthetic organisms must, therefore, deal with elevated concentrations of a molecule that is both a sink of electrons and a source of wrecking intermediates (for a comprehensive review see [1]). In addition, as sessile organisms, plants must continually acclimate to changing conditions. The term acclimation involves both developmental plasticity in response to long-term environmental trends as well as the ability to tolerate a broad spectrum of transient changes.

The impact of these processes on central metabolic pathways is such that, in plants, departures from optimal conditions ultimately result in increases in the abundance of ROS and, thus, in unbalances of the redox homeostasis, the so-called oxidative stress. Given the crucial importance of this condition for the overall performance of the cells, ROS are powerful adaptive cues. Toxic oxygen derivatives are also effective means to fight pathogens. ROS contribute to an active defense strategy, strengthening cell walls and damaging the intruder, and eventually lead to the programmed death of the cells under attack [2]. In many cases, the ability of H_2O_2 to permeate lipid membranes allows it to diffuse throughout the cell and into neighboring cells, acting by itself or through intermediates as a systemic signal for disparate environmental injuries.

It has recently become clear that controlled and localized production of ROS fulfills much broader functions in plant cells. Local increases in ROS concentration precede stomatal closure in reaction to water shortage and abscisic acid, and are associated with polar root hair growth and differential cell expansion in tropic responses under the control of auxin. In all these processes, ROS activate hyperpolarization-dependent Ca^{2+}-permeable cation channels, which increase cytosolic Ca^{2+} concentration and prompt Ca^{2+}-dependent signal transduction [3]. Although the molecular mechanisms leading to the activation of these channels, whether direct or indirect, are still unknown, sulfur atoms in cysteines and methionines are especially attractive as targets of ROS. Alteration of the catalytic or structural features of proteins through the control of the oxidation number of cysteines offers a versatile framework for the control of physiological processes. Furthermore, the ability of a particular cysteine to undergo various oxidation states opens the way to a 'redox code', suited to trigger different cellular responses. In fact, an increasing number of signal transduction components and gene expression modulators are recognized to bear redox-sensitive residues.

An exogenous stimulus displaces the thermodynamic equilibrium with the consequent imbalance of associated processes. Although the midpoint redox potential gives an estimation of the driving force for transferring electrons from donor molecules to acceptors, catalysts that lower barriers of activation energy are key players. As a consequence, any attempt to establish the importance of redox cell signalling in metabolism requires the characterization of the interplay between stimuli, the abundance of reductants and oxidants (thermodynamic control) and the features of the catalysts involved (kinetic control). Here, we summarize our current knowledge about thiol redox signalling on a number of well-studied systems, with the final goal of clarifying, where possible, the complexity of the network. Experimental evidence on cysteine-based signals from organisms other than plants are mentioned when they give useful clues as to how the concept of cellular redox modulation can be furthered. Next, we describe in detail thiol/disulfide exchanges catalyzed by protein-disulfide oxido-reductases, because this mechanism plays paramount roles in the control of primary plant metabolism. We concentrate on important breakthroughs from the past few years and provide an overview of the components involved and the functional principles that govern these processes. Where appropriate, we remit the reader to well-documented reviews.

6.2 The reactivity of the sulfhydryl group

In proteins, covalent modifications of amino acids by small chemical groups usually affect the stability, intracellular location, catalytic function and/or their ability to interact with other partners [4]. Phosphorylation of serine, threonine, tyrosine or histidine, oxidation of methionine, methylation of lysine or hydroxylation of proline does not have the flexibility to adopt more than two states. At variance, sulfhydryl groups in cysteines are able to acquire a series of redox states ranging from thiol (-S-H; sulfur oxidation number: -2) to more oxidized forms such as disulfide (-S-SR; -1), sulfenic acid (-S-OH; 0) or S-nitrosothiol (-S-NO; 0). Since the pioneer review of Barron [5], the thiol-disulfide alternancy in cysteines has been implied in all sorts of biochemical events, from the modulation of catalysis to the stabilization of the tertiary structure in secretory proteins. More recently, the study of several case proteins has revealed the hidden subtleties of thiol chemistry. For example, the transformation of a single -SH group in OxyR, a redox sensing transcription factor in *Escherichia coli*, into either -S-S-glutathione, -S-OH or -S-NO yields transcriptionally active forms that differ in structure, binding affinity for DNA and promoter activity [6]. This is but a token of the flurry of novel mechanisms related to cysteine residues discovered in the last few years – a trend that shows no sign of slowing down. The emerging picture shows that, in a cellular context, redox transitions of sulfhydryls give rise to different products that in turn trigger specific responses.

As the predominant nonprotein thiol in cells, reduced glutathione (GSH) plays an important role as a reductant that helps to metabolize damaging oxidants in most aerobic cells, yielding the oxidized glutathione disulfide (GSSG) [7]. Shifts in the balance between reduced and oxidized forms of GSH in response to endogenous or environmental stimuli serve as sensors of stress and triggers for development [8–11]. Protein S-glutathionylation is a particular case of the well-known reversible thiol/disulfide exchange between reduced cysteines and GSSG:

$$\text{HS-Prot-SH} + \text{GSSG} \leftrightarrow \text{HS-Prot-S-SG} + \text{GSH}$$

or between protein cystines and GSH:

$$\text{Prot-(S)}_2 + \text{GSH} \leftrightarrow \text{HS-Prot-S-SG}$$

In some cases, however, a glutathione moiety can be bound to proteins through the intermediary of highly reactive species, i.e., glutathione S-oxide (also named glutathione thiosulfinate), a by-product of S-nitrosoglutathione transformations [12]:

$$\text{GSH} + \text{NOH} \rightarrow \text{GSNHOH} \rightarrow \text{GS(O)NH}_2$$

$$\text{GSH} + \text{GS(O)NH}_2 \rightarrow \text{GS(O)SG} + \text{NH}_3$$

$$\text{Prot-SH} + \text{GS(O)SG} \rightarrow \text{Prot-SSG} + \text{GSOH}$$

Ample experimental evidence has revealed that glutathionylation has considerable significance for the function of many proteins [9–11, 13]. A relatively large number of S-glutathionylated proteins have been detected in yeast and animal cells, including transcription factors such as the p50 subunit of NF-κB, critical enzymes such as glyceraldehyde 3-P dehydrogenase and the redox active thioredoxin (Trx) [14–16] Our current knowledge of such events in plant cells is still scarce, but is rapidly improving. Accordingly, a study identified about 20 proteins from cultured

Arabidopsis thaliana cells that incorporated biotinylated GSH. Two of these targets were characterized as triose-phosphate isomerase and a putative plastidic aldolase, key enzymes for sugar metabolism. The inactivation and reactivation of recombinant triose-phosphate isomerase by GSSG and GSH, respectively, further links the redox status of GSH to enzyme control [17]. Moreover, the recent finding that poplar thioredoxin-h2 (ptTrx-h2) (a mitochondrial Trx isoform) is glutathionylated, whereas two cytosolic counterparts (PtTrx-h1 and PtTrx-h3) are unreactive under similar experimental conditions, stresses the specificity of this type of modification [18].

Another remarkable mechanism of redox control involves the conversion of the sulfur atom to more oxidized states by the action of reactive oxygen and nitrogen species. Chemical and crystallographic studies have provided compelling evidence of the presence of functional sulfenates in many proteins upon mild oxidation [19]. The -Cys-SOH group of some proteins (i.e., 2-Cys peroxiredoxins (2-Cys Prx)) functions as a transient intermediate during the reduction of H_2O_2 and the peroxynitrite anion ($ONOO^-$) or coordinates metals in the active site (i.e., Fe(III) in nitrile hydratase) [20–22]. In other cases, the reversible oxidation of key Cys-SH to Cys-SOH may serve as a relay sensor of the intracellular redox status, modulating stress responses.

S-nitrosylation of proteins is another important post-translational modification that must be adequately poised to avoid the so-called nitrosative stress [23–25]. In plants, NADH:nitrate reductase or an inducible nitric oxide synthase produces the nitric oxide radical ($NO^{\bullet}$), which triggers the conversion of -Cys-SH into -Cys-SNO. The cell redox status and the proximity to the $NO^{\bullet}$ source control the reversal of this process [26–28]. A search of putative sites for the acid–base catalysis of nitrosylation using the degenerate motif (G/S/T/C/Y/N/Q)(K/R/H/D/E)C(D/E) [29] disclosed 103 matches in Arabidopsis, including proteins involved in the cell cycle, transport, signalling and metabolism [30]. Other timely findings uncovered that thiol groups of Trx in animal cells are targets for S-nitrosylation by N_2O_3-like species generated in a superoxide producing system containing xanthine and xanthine oxidase. In one set of experiments, this mechanism dissociates Trx from the apoptosis signal regulating kinase 1 which in turn becomes functional [31, 32]. This finding suggests the intriguing possibility that Trx, under some conditions, may rely on post-translational modifications other than the classical thiol/disulfide exchange. Another reactive nitrogen species that mediates the oxidation of sulfhydryl groups is the $ONOO^-$, the reaction product between the free radical species superoxide and $NO^{\bullet}$ ($O_2^- + NO^{\bullet} \rightarrow ONOO^- \leftrightarrow ONOOH \rightarrow OH^{\bullet} + NO_2$). Although the apparent second order rate constant for the reaction of protein thiols with the $ONOO^-$ (2700 $M^{-1}.s^{-1}$) is three orders of magnitude greater than the corresponding rate constant for the reaction with H_2O_2 (1.14 $M^{-1}.s^{-1}$) at pH 7.4, the regulatory role, if any, of $ONOO^-$ is yet to be established [33].

Data summarized above show that our understanding of thiol dynamics has evolved from the simple on/off switch associated with thiol/disulfide exchanges to encompass a range of redox-based modifications driven by reactive oxygen and nitrogen species (Figure 6.1). By virtue of this flexibility, sulfhydryls can process a wide spectrum of stimuli into different functional responses [34].

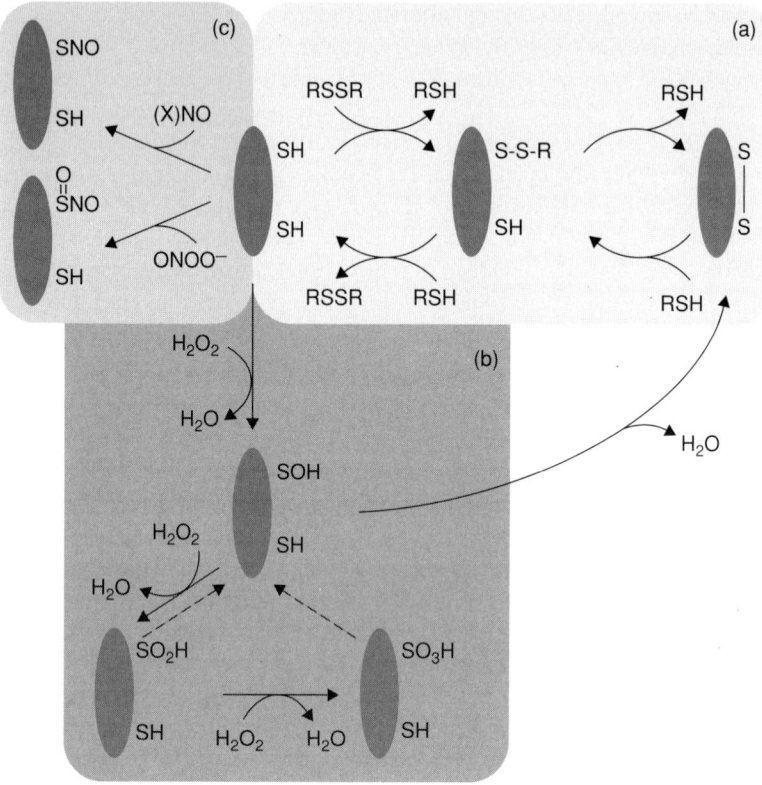

Figure 6.1 Redox-sensitive cysteine residues as multiple regulatory switches. (a) Thiol/disulfide exchange. Protein thiols react with disulfides located at other proteins or low molecular weight species (i.e., GSSG) yielding a heterodisulfide and subsequently a cystine. The couple GSH/GSSG is excluded from this scheme when glutathionylation proceeds via glutathione disulfide S-oxide (see the text). (b) ROS. The sequential oxidation of sulfhydryls with oxygen-bearing oxidants (i.e., hydroperoxides) yields sulfenic (-SOH), sulfinic (-SO$_2$H) and sulfonic (-SO$_3$H) acid derivatives. Dashed arrows indicate the reduction of overoxidized forms by sulfiredoxin [218]. (c) Reactive nitrogen species. The free radical nitric oxide (NO$^\bullet$) and the peroxynitrite anion (ONOO$^-$) can mediate the formation of S-nitrosothiol (-SNO) and S-nitrothiol (-S(O)NO), respectively.

Despite the ample room for variations at a single cysteine, thiol/disulfide exchanges are still the mechanisms studied in greater detail. These reactions are carried out by a specialized and diverse group of enzymes dubbed protein disulfide oxido-reductases (PDOR), seemingly one of the most ancient and widespread protein families among living organisms. In recent years, the availability of complete genome sequences and the burgeoning of high-throughput proteomic studies have boosted the interest in this group. The discovery of an unexpected variety of PDOR isoforms encoded in the genomes of oxygenic photosynthetic organisms goes hand in hand with a flood of structural and biochemical studies and with the identification of a plethora of previously unsuspected interacting partners. Therefore, we discuss this issue in detail below.

Table 6.1 Midpoint redox potentials of thiols and dithiol proteins

Protein/thiol	$E^{o'}$ (mV)	Reference
GSH	−252	[199]
PDI	−180	[200]
E. coli Grx	−230	[200]
E. coli Trx	−270	[200]
Spinach Trx-f	−290	[201]
Spinach Trx-m	−300	[201]

6.3 Protein-disulfide oxido-reductases

The relevant feature in the ubiquitous PDOR is the presence of the motif -CXXC-, whose cysteine residues undergo a cycle of intracatenary oxido-reduction. The midpoint redox potential of this reactive site grossly establishes the tendency of catalysis. Thus, a highly reducing motif (i.e., *E. coli* Trx, $Eo = -270$ mV; *E. coli* Grx = −233 mV) drives hydrogens for cleaving target disulfide bonds while a more oxidizing motif (i.e., *E. coli* DsbA, $Eo = -106$ mV) withdraws hydrogens from the sulfhydryls of target proteins, generating a cystine (Table 6.1). All members of this superfamily share a common tertiary fold composed of a central four-stranded β-sheet surrounded by three α-helices, named the Trx fold. The nucleophilic cysteine (located at the N-side of the active site) protrudes to the solvent while the remaining cysteine remains buried in the globular structure [35]. This basic unit seems to be extremely versatile, considering the growing number of proteins comprising one or more Trx modules linked to additional domains. Although these composite proteins play important roles in redox signalling – i.e., nucleoredoxin, a polypeptide containing three Trx-like modules that localizes preferentially in the nucleus of developing maize kernels [36] – space limitations make it impossible to cover all of them in this review. We thus circumscribe primarily our analysis to the most recent developments in functional aspects of plant Trx, glutaredoxins (Grx), protein disulfide isomerases (PDI) and some related PDOR; several excellent reviews provide an overview of the previous [37–40] and recent [41] literature in this field.

6.4 Thioredoxins

Trx are small (ca 12 kDa) single-domain proteins carrying a conserved -CGPC- motif that catalyzes the reduction of protein disulfides at rate orders of magnitude faster than those of free thiols such as dithiothreitol (DTT) or GSH. The unusual basic microenvironment around the Trx active cysteine allows the formation, under physiological pH, of a nucleophilic thiolate that attacks the disulfide bond of the target protein, forming a covalently linked heterodimer:

HS-Trx-SH → HS-Trx-S$^-$ + H$^+$

HS-Trx-S$^-$ + Prot-(S)$_2$ → [HS-Trx-S⋯S-Prot-S]$^-$ → $^-$S-Trx-S-S-Prot-SH

The heterodisulfide undergoes an intramolecular thiol/disulfide exchange by the action of the buried, resolving Cys, releasing the oxidized Trx and the reduced target protein:

$$^-S\text{-Trx-S-S-Prot-SH} + H^+ \rightarrow \text{Trx-}(S)_2 + \text{HS-Prot-SH}$$

Finally, for the functioning of this catalytic cycle, the Trx active site needs to retrieve the hydrogens provided to the target protein.

6.4.1 Thioredoxin isoforms

The gearing of CO_2 assimilation in the Benson-Calvin cycle to the light-driven electron transport in photosystems was one of the first examples of global metabolic redox control through thiol/disulfide exchanges [42]. Biochemical features and the location in leaf cells set the basis for originally grouping plant Trx into three subfamilies [43]. Chloroplast species that enhanced the activity of chloroplast fructose-1,6-bisphosphatase (CFBPase) and NADP-malate dehydrogenase (NADP-MDH) were named Trx-f and Trx-m, respectively, while that found in the cytosolic fraction was called Trx-c and afterward renamed Trx-h (for heterotrophic). Later, it was recognized that these forms belonged to different phylogenetic lineages, providing a first glimpse of the remarkable variety of Trx isoforms in plants [44]. Full genomic sequences have confirmed the considerable complexity of the Trx complement of oxygenic photosynthetic organisms, compared to nonphotosynthetic prokaryotes and eukaryotes [45]. Indeed, whereas *E. coli* encodes 2 divergent Trx genes, the cyanobacterium *Synechocystis* codes for 4 different forms, named -m, -x, -y and an unclassified one. On the eukaryotic side, brewer's yeast and humans have 3 and 2 isoforms, respectively, while the green algae *Chlamydomonas reinhardtii* and the land plant Arabidopsis have 8 and 19 isoforms, distributed in six different subfamilies (Trx-f, -m, -x, -y, -o and -h), respectively (Tables 6.2 and 6.3) [46–48]. Phylogenetic relationships indicate that subfamilies -h, -f and -o are akin to eukaryotic Trx, whereas subfamilies -m, -x and -y seem to be of prokaryotic origin; in fact, Trx-m in red algae is encoded by the plastidial genome [49]. Being all Trx nuclear encoded in Viridiplantae, a relatively large number of predicted primary structures that bear N-terminal extensions suggest a variety of final intracellular destinations. In line with this view, N-terminal sequences of purified mature Trx-f and Trx-m and fusions of Trx-m2 and Trx-y2 to green fluorescent protein (GFP) confirmed their localization in the stroma of chloroplasts [50]. The expression of these chloroplastic isoforms is stronger in leaves and is induced by light. By contrast, the expression of Trx-y1, highly similar to Trx-y2 and likely localized in plastids as well, is high in nonphotosynthetic tissues, especially in seed leucoplasts during the accumulation of storage lipids [50]. The genome of Arabidopsis also encodes Trx-o1 and Trx-o2, which do not keep homology to chloroplast counterparts [51, 52]. Recent import experiments showed that mature AtTrx-o1, devoid of transit peptide, is present in the mitochondrial matrix [53].

Table 6.2 Non Trx-h isoforms of *Arabidopsis thaliana*

Isoform	Polypeptide length	Active site sequence	Subcellular localization	MATDB entry
AtTrx-m1	179	-CGPC-	Chloroplast	At1g03680
AtTrx-m2	186	-CGPC-	Chloroplast	At4g03520
AtTrx-m3	173	-CGPC-	Chloroplast	At2g15570
AtTrx-m4	193	-CGPC-	Chloroplast	At3g15360
AtTrx-x	171	-CGPC-	Chloroplast	At1g50320
AtTrx-f1	178	-CGPC-	Chloroplast	At3g02730
AtTrx-f2	185	-CGPC-	Chloroplast	At5g16400
AtTrx-y1	172	-CGPC-	Chloroplast	At1g76760
AtTrx-y2	167	-CGPC-	Chloroplast	At1g43560
AtTrx-o1	194	-CGPC-	Mitochondria	At2g35010
AtTrx-o2	159	-CGPC-	Cytosol[a]	At1g31020

[a] Putative.

Arabidopsis encodes eight Trx-h isoforms; some of them conserve the classical site -CGPC- while others hold atypical -CPPC- and -CXXS- motifs (Table 6.3). Since these motifs are nonetheless followed by predicted α-helices, a feature conserved in redox enzymes but not in other proteins, they may impart alternative catalytic functions [54, 55]. Phylogenetic evaluations divided this subfamily into three main groups. In Arabidopsis, only one member of group I, AtTrx-h1, harbors the classic sequence at the active site whereas the other three hold the unusual motif -CPPC- without modification of the Trx fold [54, 56]. On the other hand, all Trx-h in group II host the typical -CGPC- active site, but the comparison with counterparts from many other plants led to further subdivision of this group into three sub-

Table 6.3 Trx-h isoforms of *Arabidopsis thaliana* [18, 58]

Group	Subgroup	Isoform	Polypeptide length	Active site sequence	Subcellular localization (putative)	MATDB entry
I		AtTrx-h1	114	CGPC	Cytosol	At3g51030
		AtTrx-h3	118	CPPC	Cytosol	At5g42980
		AtTrx-h4	119	CPPC	Cytosol	At1g19730
		AtTrx-h5	118	CPPC	Cytosol	At1g45145
II	II-A	AtTrx-h2	133	CGPC	Cytosol	At5g39950
	II-C	AtTrx-h7	129	CGPC	Cytosol	At1g59730
	II-C	AtTrx-h8	148	CGPC	Mitochondria	At1g69880
III		AtTrx-h9	140	CGPC	Cytosol	At3g08710
		AtCXXS1	118	CIPS	Cytosol	At1g11530
		AtCXXS2	154	CLPS	Cytosol	At2g40790

groups [18]: II-A and II-C comprise proteins related to AtTrx-h2 and AtTrx-h8, respectively, while II-B contains mainly Trx-h from cereals. The third group of Trx-h was initially identified in monocots but seems to be present in all land plants. The relevant feature of this group is that many members harbor the canonical bicysteinic -CGPC- active site while some hold a monocysteinic -CXXS- sequence [54, 57]. Interestingly, the reductive capacity of two isoforms from poplar belonging to this group and related to AtCXXS1 departs from that usually ascribed to Trx (cf. below) [58, 59]. Trx-h isoforms have been detected in different cell locations, except in chloroplasts. In addition to the cytosol [60], they appear in the nuclei of developing wheat seeds [61] and in the extracellular compartment, as a component of phloem sap [62, 63]. More recently, immunological detection and fusions to GFP showed that PtTrx-h2, grouped in II-C with AtTrx-h7 and AtTrx-h8, is targeted to plant mitochondria [18]. Not surprisingly, different Trx-h isoforms are expressed not only at different levels [64] but also in tissue- and developmental-stage specific manners. For example, in mature barley seeds, HvTrx-h1 is present in the endosperm, the aleurone layer and the embryo and HvTrx-h2 locates mainly at the embryo. At the onset of germination, the levels of HvTrx1 remain high in the embryo but fall in the other two compartments, while the abundance of HvTrx2 decreases [65].

The variety of proteins that contain a Trx domain but do not easily accommodate among the well-known PDOR is a clear indication of the multiple alternatives present in this superfamily for driving reducing power to specific physiological processes. Support for the idea that similar proteins will be likely found in the near future comes from a novel bipartite protein from Arabidopsis that bears a C-terminal Trx and an N-terminal tetratricopeptide repeat domain, similar to that observed in rat and human Hip (HSP70-interacting protein), a protein that stabilizes the ADP-bound form of the chaperone Hsp70 [66]. This 42-kDa protein, AtTDX (for Tetratricopeptide domain-containing Trx), exhibits disulfide reductase activity both *in vitro* and *in vivo* and interacts specifically with the yeast Hsp70 Ssb2 protein. This interaction is sensitive to the redox status, with the Trx domain acting as a redox switch that turns the complex with Ssb2 on and off.

6.4.2 Reductants of thioredoxins (sources of reducing power)

In cellular compartments that depend on reduced carbon skeletons as their main source of energy, NADPH ($E_m = -340$ mV) provides the reducing power to cleave the disulfide bond of Trx, assisted by NADP-Trx reductase (NTR) (Figure 6.2) [51, 67, 68]:

$$\text{NADPH} + \text{H}^+ + \text{Trx(S)}_2 \xrightarrow{\text{NTR}} \text{NADP}^+ + \text{HS-Trx-SH}$$

Along with dihydrolipoamide reductase, glutathione reductase, alkylhydroperoxide reductase and mercuric reductase [69, 70], NTR is a member of the superfamily of flavoprotein-disulfide oxido-reductases whose redox-active disulfide -CA(V/T)C- locates at a gaping hole between the FAD and NADP domains [71, 72].

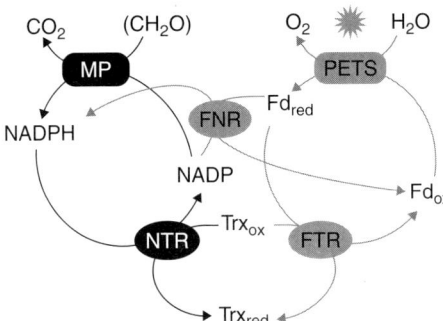

Figure 6.2 Reduction of Trx in plant cells. NTR mediates the reduction of Trx by NADPH, whereas FTR uses Fd as reductant. In the stroma of chloroplasts, reduced Fd is the initial source of electrons for the reduction of both NADP and Trx. It was generally believed that the NTR pathway was absent from chloroplasts; the finding of NTRC [81] suggests that both Trx-reducing systems may be operative. Abbreviations: PETS, photosynthetic electron transport system; FNR, Fd-NADP reductase; MP, metabolic pathways producing NADPH.

Contrasting with the bewildering diversity of Trxs, Arabidopsis seems to make do with a limited number of NTRs. Laloi et al. [51] found that most of the NTR function in Arabidopsis is served by two paralogous genes, *AtNTRA* and *AtNTRB*. In both *AtNTRA* and *AtNTRB*, two in-frame start codons produce two different polypeptides of ca 42 and 37 kDa. For *NTRB*, the former species yields a 37-kDa form after import into mitochondria, in a process that is sensitive to the electrochemical membrane potential. Although *AtNTRA* and *AtNTRB* seemingly have overlapping functions, given that individual knockout plants are viable, *AtNTRB* appears to produce the major form in mitochondria [73], whereas *AtNTRA* provides most of the cytosolic enzyme. The use of two ATGs for cytosolic and mitochondrial targeting has also been described for NTR in mammals, insects and parasites [74–76]. Hence, like plant aminoacyl tRNA synthetases [77, 78], alternative translation start seems to be a common feature of the *NTR* gene in eukaryotes.

Gelhaye et al. [59] have recently found evidence hinting at an alternative way for the reduction of vascular plants Trx-h. PtTrx-h4 and PtCXXS3 are members of the poplar Trx-h group III – harboring the typical -CGPC- and the unusual -CMPS- sequences, respectively. Surprisingly, both proteins are insensitive to Arabidopsis and *E. coli* NTR. Instead, PtTrx-h4 reduces several Trx targets, such as Prx or methionine sulfoxide reductases, accepting reducing equivalents from poplar or *E. coli* Grx, whereas PtCXXS3 drives the reaction commonly used for testing the activity of Grx, i.e. the GSH-dependent cleavage of hydroxyethyldisulfide. Hence, the transfer of reducing equivalents from GSH via members of Trx-h group III uncovers the presence (in plants) of Trx-like structures with Grx-like activities, linking Grx- and Trx-dependent systems (Figure 6.3). Apparently, a conserved cysteine residue located in the N-terminal extension in group III Trx-h helps to circumvent the unfavorable redox potential of Grx for the reduction of Trx in thiol/disulfide exchanges [58]. It will be extremely interesting to identify the Trx isoforms whose catalytic cycle is driven by GSH-dependent reductions.

In chloroplasts and cyanobacteria, a completely different system provides reducing equivalents to Trx (Figure 6.2). The product of the photosynthetic electron transport system, reduced ferredoxin (Fd) ($E_m = -420$ mV) and two protons

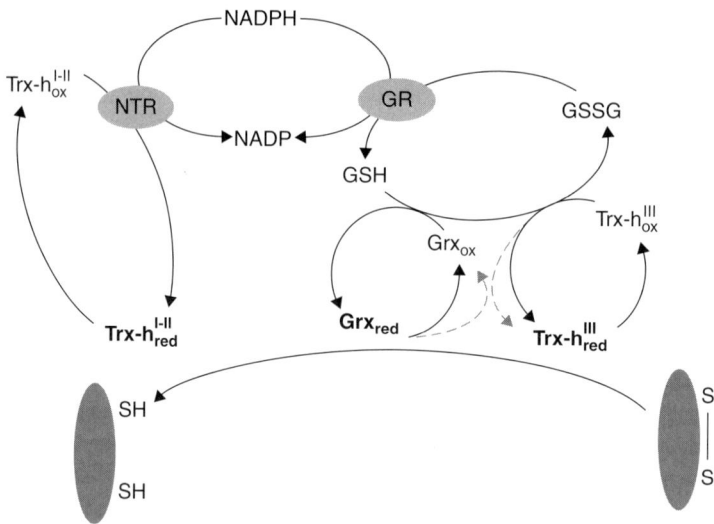

Figure 6.3 Different types of Trx-h accept reducing equivalents from different donors. On the left, the cleavage of the disulfide bond in groups I and II is catalyzed by NTR. On the right, the conversion of group III Trx-h to the reduced form occurs when Grx uses GSH, formed by the action of glutathione reductase (GR), as reductant. Reduced Grx *per se* has also the ability to cleave cystines in many proteins.

regenerate thiol groups in oxidized Trx, thus linking redox reactions to the operation of photosystems:

$$2\ Fd_{red} + 2\ H^+ + Trx(S)_2 \xrightarrow{FTR} 2\ Fd_{ox} + HS\text{-}Trx\text{-}SH$$

The iron–sulfur enzyme Fd-Trx reductase (FTR), a key element in this pathway, is formed by two different subunits [79]. The catalytic subunit (ca 13 kDa) contains seven conserved cysteine residues, six of which participate actively in the binding of the [4Fe–4S] iron–sulfur cluster and in the redox reactions with the target disulfide. The variable subunits (8–13 kDa), on the other hand, exhibit pronounced diversity in their primary structure; truncations in their N-terminal extensions significantly increase the stability of the complexes, without impairing the affinity for Fd and Trx-f [80]. The mechanism suggested for the two-electron process, the reduction of a disulfide, using a one-electron carrier, ferredoxin, is described elsewhere [79].

The Fd/Trx system was considered so far the only source of reducing power for Trx in oxygenic photosynthetic organisms. However, a novel gene with homology to the typical *NTR* was recently identified in both cyanobacteria and land plants. *NTRC* notably encodes an NTR with an N-terminal extension typical of a chloroplast transit peptide and a C-terminal region homologous to Trx [53, 81]. The mature form of this composite protein thus resembles the NTR of *Mycobacterium leprae*, which is also fused to a Trx domain [82]. Rice NTRC exhibits *in vitro* both

NTR and Trx activities but, surprisingly, lacks the capacity to function cooperatively as an NTR/Trx system [81]. Although the function of this NTR is still unknown, this notable finding brings a twist to the view that hitherto considered the plastid the exclusive residence of the FTR.

6.4.3 Targets of thioredoxins (oxidants of thioredoxins)

In plants, reduced Fd and NADPH poise the abundance of reduced Trx, which in turn serves as reductant for many cellular oxidants. However, the high number of isoforms located in almost every cell compartment and the broad range of midpoint redox potentials raise inquiries on target specificity and functional reductants, respectively. As an additional level of complexity, the expression levels of a given isoform may vary independently of other counterparts at the same cellular location during cell growth or in response to environmental cues. The understanding of redox signals, therefore, requires the coordinated analysis of structure and function of these PDOR in a spatial and temporal context. Despite considerable efforts and advances in the last three decades, the full extent of Trx physiological functions is still incomplete.

Until the early 1990s, biochemists were restricted to the study of specific metabolic steps; nowadays, genomic and proteomic analyses provide a much larger body of information. An exciting new approach is the isolation of Trx targets via affinity chromatography. As discussed in further detail by Lee Sweetlove (Chapter 2), Trx mutants holding a serine residue in place of the resolving cysteine were linked to a solid support. Thus, mixed heterodisulfides formed by the attack of the nucleophilic cysteine became stably linked. Subsequently, reduction with DTT released a population of Trx-bound polypeptides suitable for characterization by mass spectrometry [83, 84]:

$$-\boxed{\text{Matrix}}-\text{Trx-SH} + \mathbf{X}(S)_2$$
$$\Downarrow \qquad \text{Formation of the heterodisulfide}$$
$$-\boxed{\text{Matrix}}-\text{Trx-S-S}\mathbf{X}\text{SH}$$
$$\Downarrow \qquad \text{Elution with DTT}$$
$$\mathbf{X}(SH)_2 + -\boxed{\text{Matrix}}-\text{Trx-SH}$$

This experimental approach largely increased the number of proteins putatively targeted by Trx in chloroplasts, mitochondria [83–86] and in the endosperm and embryo of mature seeds (see below) [87–91]. Similar analyses in *Chlamydomonas* and *Synechocystis* also pinpointed numerous novel putative targets [92, 93]. Thus far, the relevance of the observed interactions has only been confirmed in few cases. For example, phosphoglucomutase, one of the novel Trx-binding enzymes isolated from cyanobacterial extracts, was found to be activated and inhibited *in vitro* by reducing and oxidizing conditions, respectively [92].

The following sections survey notorious plant biological processes that involve thiol/disulfide exchanges. We review at length chloroplast events, because connections between Trx and the metabolism in mitochondria and the cytosol are in general less explored. However, it must be kept in mind that this stage is about to change dramatically, as a wide variety of processes linked to Trx-h are being uncovered [94].

6.4.4 Control of chloroplast enzymes by thioredoxin

For more than 30 years, the function of Trx in chloroplasts has been associated with the control of enzyme activity (Table 6.4) and, as a consequence, this issue has been extensively reviewed [40, 95–97]. As the number of isoforms increases, a detailed account of functional aspects is necessary to establish the specificity for target proteins. For example, recombinant Trx-y1 and Trx-y2 only weakly enhanced the activity of the extensively used enzymes NADP-MDH and CFBPase *in vitro* [98]. Perhaps their midpoint redox potential, less negative than plastidial Trx-f and Trx-m but similar to ineffective Trx-x, is unlikely to be a significant source of energy for driving the reductive activation [50]. Beyond their crucial role in CO_2 fixation, Trx likely modulate many other stromal processes; in consequence, different metabolic pathways should be explored to determine the functions of the newly found isoforms.

Starch metabolism has emerged as an important target for Trx modulation. The activity of chloroplast ADP-glucose pyrophosphorylase (AGPase), which catalyzes a rate-determining step in starch biosynthesis, cf. [99], is activated by 3-phosphoglycerate, inhibited by inorganic phosphate and stimulated by the DTT-mediated reduction of the disulfide bond that links two catalytic small subunits [100]. *In vitro* studies established that reduced Trx-f and Trx-m activate, while the oxidized forms inactivate, potato tuber AGPase [101]. Similar mechanisms occur in Arabidopsis and pea chloroplasts. Moreover, the reduced (active) form of this enzyme builds up *in vivo* when the flow of carbon into starch is increased [102, 103]. Still, further studies are required to understand how conditions in the cytosol affect the redox control in plastids, because the activation of chloroplast AGPase by sucrose and glucose also depends on the activity of a cytosolic protein kinase (SNF1-related protein kinase-1, SnRK1) and hexokinase, respectively [104]. Trx has also been shown to control the activity of α-glucan water dikinase (GWD), an important enzyme that transfers the β-phosphate of ATP to the C-6 or the C-3 positions in amylopectin prior to the amylolytic degradation via β-amylases [105]:

$$(Glc)_n + A\text{-}P\text{-}\boxed{P}\text{-}P + H_2O \rightarrow \boxed{P}\text{-}(Glc)_n + AMP + Pi$$

When a disulfide bond links cysteine residues at the conserved site sequence -CFATC-, GWD is bound to the starch granule and is inactive [106]. The enzyme becomes both soluble and active after the *in vitro* reduction with Trx-f and, less efficiently, with Trx-m. Although these data collectively suggest that Trx controls both the synthesis and the degradation of starch in plastids, the spatial and temporal aspects of the modulation remain to be established.

3-deoxy-D-arabino-heptulosonate 7-phosphate (DAHP) synthase is also activated by reduced Trx [107]. This enzyme catalyzes the condensation of phosphoenolpyruvate and erythrose-4-phosphate yielding DAHP and inorganic phosphate,

Table 6.4 Stromal enzymes modulated by Trx

Target enzyme[a]	Regulatory site	Most efficient modulator	Species	References
Benson-Calvin cycle				
Fructose-1,6-bisphosphatase	-EC^{153}X$_{19}$C^{173}IVNVCQ-	Trx-f	Pea	[40, 202]
Sedoheptulose-1,7-bisphosphatase	-SC^{52}GGTAC^{57}V-	Trx-f	Wheat	[203]
Phosphoribulokinase	-GC^{16}X$_{38}$C^{55}L-	Trx-f	Spinach	[201, 204, 205]
Rubisco activase	-GC^{392}X$_{18}$C^{411}V-	Trx-f	Arabidopsis	[206]
Glyceraldehyde 3-P dehydrogenase	-FC^{364}X$_{10}$C^{375}K- (B subunit)	Trx-f	Pea	[207]
Photophosphorylation				
CF1-ATP synthase (γ-subunit)	-IC^{198}DINGNC^{204}V-	Trx-f	Pea	[40, 208]
C3 (redox shuttle) and C4 (carbon assimilation) metabolism				
NADP- malate dehydrogenase	-EC^{24}FGVFC^{29}T-KC^{365}X$_{11}$C^{377}D-	Trx-m	Sorghum	[209, 210]
Oxidative pentose phosphate cycle				
Glucose 6-phosphate dehydrogenase	-TC^{149}X$_7$C^{157}D-	Trx-m	Potato	[211]
Starch synthesis				
ADP-glucose pyrophosphorylase	C^{12} (small subunits) intercatenary disulfide	Trx-f	Potato	[100]
Starch degradation				
Glucan-water dikinase	N.D.	Trx-f	Potato	[106]
Fatty acid synthesis				
Acetyl-CoA carboxylase	C^{267} (α subunit) and C^{442} (β subunit) intercatenary disulfide	Trx-f	Pea	[212–214]

[a] Reduction by Trx stimulates the activity of most of the above mentioned enzymes. The only exception is glucose 6-phosphate dehydrogenase (italics), active in the oxidized state and inactivated by reduction [211].

N.D.: not determined

the first step in the shikimate pathway that leads to chorismate, the precursor of phenylalanine, tyrosine, tryptophan and numerous secondary metabolites derived from these aromatic amino acids. *In vitro*, reduced Trx-f is extremely efficient while Trx-m is orders of magnitude weaker in enhancing the catalytic capacity of the recombinant enzyme.

Proteomic studies have shown that Trx potentially affect almost every metabolic pathway in chloroplasts. Motohashi *et al.* [83] have assigned several chloroplast proteins as targets of Trx-m, some known previously from biochemical studies (Rubisco activase, 2-Cys Prx, glyceraldehyde-3-phosphate dehydrogenase, sedoheptulose-1,7-bisphosphatase) while others were novel targets (glutamine synthetase, cyclophilin, Prx-Q, Rubisco small subunit). In parallel, 26 stromal proteins were identified as targets of Trx-f and Trx-m in spinach, 11 in established Trx-regulated processes (Benson–Calvin cycle, nitrogen and sulfur metabolism, translation, pentose phosphate cycle and glycolysis), but 15 involved in processes not previously known to be linked to Trx (isoprenoid, porphyrin and vitamin biosynthesis, protein assembly/folding, protein and starch degradation, glycolysis, HCO_3^-/CO_2 equilibration, plastid division and DNA replication/transcription) [85]. Given that electrostatic interactions seem to play an important role in docking Trx to certain targets [108], adsorption of stromal proteins onto immobilized wild type Trx-f at low salt and elution at high salt concentrations [109] identified 18 possible targets of Trx, some of them missing in previous approaches. These proteins are involved in translation, protein assembly/folding, protein degradation, nitrogen metabolism, the C4/malate valve, the HCO_3^-/CO_2 equilibration and the biosynthesis of ATP, starch, fatty acids and tetrapyrroles. Out of 10 proteins not previously associated with Trx, nine are known members of chloroplast complexes that hold at least one component linked to Trx (the large subunit of Rubisco, phosphoglycerate kinase, RNA binding proteins (24 and 41 kDa), ribosomal proteins (S1, S5, L4 and L21) and the α-subunit of ATP synthase). Possibly, these proteins bind via one of the target enzymes rather than interacting directly with Trx-f itself. As usual in proteomic studies, further biochemical analyses are necessary to establish whether the observed interactions modulate the function of the identified proteins.

Ever since the discovery of the Fd-Trx system in chloroplasts, most target proteins for these reductants have been sought in the stroma; a number of recent developments have broadened the horizon. Immunophilins are a diverse group of proteins comprising parvulins, cyclophilins and proteins that bind the immunomodulator FK506 (FKBP); some of these proteins catalyze the *cis-trans* isomerization of the peptide bond located between proline and its preceding residue [110]. Notably, Arabidopsis encodes 19 putative chloroplastic immunophilins, with three of them presumably located in the stroma and the remaining in the thylakoid lumen [111]. The stromal cyclophilin AtCYP20-3 lacks peptidyl-prolyl *cis-trans* isomerase (PPIase) activity in the oxidized state, but regains it after reductive cleavage of two disulfide bridges by Trx-m [112]. On the other hand, Gopalan *et al.* [113] recently found that the thylakoid lumen resident AtFKBP13 exhibits a unique pair of disulfide bonds that are reduced *in vitro* by chloroplast Trx-m and *E. coli* Trx. Whereas in most cases Trx-dependent reduction increases the catalytic activity of stromal enzymes (Table 6.4), reduction inactivates the associated PPIase activity of AtFKBP13. It is speculated that AtFKBP13 is kept reduced by the Fd/Trx system while traveling across the stroma, but becomes oxidized upon its incorporation into the lumen of thylakoids. Again, the function of this protein is poorly understood, but it was recently shown that the precursor form of AtFKBP13 interacts with and modulates the level of the Rieske protein, a component of cytochrome b_6f on the luminal side of thylakoid membranes [114]. It is worth recalling that cytochrome b_6f is thought to be the main transducer of the other major redox signal that senses the function of photosystems, namely, the plastoquinol–plastoquinone (PQ) ratio in the thylakoid membranes (see below for other examples). If proteins in the thylakoid lumen are subject to redox modulation, a system for handling reducing equivalents must exist in this compartment. A recently identified thylakoid protein in Arabidopsis exhibits significant homology to CcdA and related plasma membrane bacterial proteins. These proteins are involved in the maturation of c-type cytochromes, to which cytochrome b_6f is related, and in the transfer of electrons from cytosolic Trx to disulfide exchange proteins resident in the periplasm using a thiol-disulfide cascade. Disruption of Arabidopsis *CCDA* impaired the accumulation of cytochrome b_6f [115]. Severe deficiency of cytochrome b_6f was also observed in a mutant for a membrane-bound Trx-like protein facing the thylakoid lumen, HCF164 [115]. Although the experimental evidence suggests that both proteins play a specific role in the formation of the cytochrome b_6f complex, it is tempting to speculate that this system may also constitute a more general mechanism for the transference of reducing equivalents from the stroma to the thylakoid lumen.

6.4.5 *Translation of chloroplast mRNA*

Light greatly enhances the translation of many mRNAs encoded by the chloroplast genome, i.e., *psbA* (coding for D1, a core protein of photosystem II (PSII)), *psbD* (coding for D2 protein), *rbcL* (coding for large subunit of Rubisco) and *psaA–psaB* (coding for 65- and 70-kDa chlorophyll *a* apoproteins of photosystem I (PSI)). Given that D1 and the products of *psaA* and *psaB* genes are the main targets of photoinhibition, the replacement of the oxidatively damaged forms results in one of the fastest turnover rates for plant proteins. Genetic and biochemical evidences indicate that, in *Chlamydomonas*, four nuclear-encoded proteins of ca 60, 55, 47 and 38 kDa must interact with the 5' UTR of the *psbA* mRNA for light to activate translation [116–118]. The 47-kDa species (RB47) is highly homologous to eukaryotic poly(A)-binding proteins whereas the 60-kDa protein (RB60) contains two Trx-like domains with putative catalytic sites -CGHC-, resembling a PDI. Surprisingly, this protein resides in chloroplasts of both vascular plants and *Chlamydomonas* even though it holds a C-terminal -KDEL- signal for retention in the endoplasmic reticulum (ER) [116, 119]. In fact, it has recently been found that, in *Chlamydomonas*, RB60 does co-localize in microsomal fractions along with ER markers, revealing a novel mechanism for dual targeting [120]. In addition to an ADP-dependent phosphorylation, the specific and reversible binding of RB47 to the 5' UTR of the *psbA* mRNA requires the participation of vicinal dithiols at the Trx domains of RB60

[121, 122]. The current set of data supports the idea that a light-driven priming signal oxidizes RB60 so that it becomes sensitive to the reductive signal. Apparently, the reduction of PQ by PSII triggers the priming signal while PSI transfers electrons via the Fd-Trx system to the PDI moiety of the 5' UTR binding complex. Hence, the relative activities of both PSII (signaled by the redox state of the PQ pool) and PSI (signaled by the reduction of Fd), acting in opposite directions on the redox state of RB60, modulate the synthesis of the chloroplast protein D1 [119]. The reader will find further information on different aspects of the light-actuated expression of plastid and nuclear genes in an authoritative review [123].

6.4.6 *Phosphorylation of chloroplast proteins*

In thylakoid membranes, phosphorylation of Lhcb1 and Lhcb2, chlorophyll a/b-binding proteins of the light-harvesting complex II (LHCII), redirects the excitation energy to PSI at the expense of PSII, thereby tuning the distribution of light energy between both photosystems in response to variations of light quality and intensity [124–126]. At high light intensities, binding of plastoquinol to the cytochrome-b_6f complex controls the activity of thylakoid-bound kinases responsible for the phosphorylation of LHCII [127–130]. Reduction by Trx-f and Trx-m counter the activity of these kinases, and conformational changes after phosphorylation shield further phosphorylation sites in LHCII [131–134]. In line with these observations, H_2O_2 serves as an oxidant that restores the activity of LHCII kinases in thylakoids isolated from illuminated leaves. The number and identity of thylakoid protein kinases in vascular plants are still uncertain, but both the thylakoid-associated Ser-Thr kinase Stt7 from *Chlamydomonas* and its recently identified orthologue STN7 from Arabidopsis participate in state transition modulation and hold two cysteines separated by four amino acids, a typical target for Trx [135, 136]. Other evidence suggests, in contrast, that LHCII phosphorylation–dephosphorylation as a function of irradiance is a thylakoid-sufficient phenomenon [137]. Experiments with isolated thylakoids of *Arachis hypogea* showed that the down-regulation of LHCII phosphorylation at high irradiance does not require assistance from stromal components, even though it becomes sensitive to modulation by the thiol redox state under specific experimental conditions. The apparent insensitivity to reduction may be the consequence of thiol-insensitive intrinsic thylakoid LHCII kinase(s), acting in parallel to thiol-sensitive peripheric thylakoid LHCII kinase(s). Thus, the way components of the redox signalling interact in response to environmental stimuli to modulate the amount of light harvested the individual photosystems is far from being completely understood.

6.4.7 *Control of mitochondrial proteins*

For more than a decade, it has been known that plant mitochondria possess a Trx system [60, 138], but its variety and functions are just starting to be unveiled

[51–53]. As outlined by Vanlerberghe and McDonald (Chapter 11) one potential target of mitochondrial Trx is the homodimeric alternative oxidase (AOX), a non-proton pumping bypass to cytochrome oxidase of classical respiration. AOX is inactive when the subunits are linked by a disulfide bond but becomes sensitive to activation by α-keto acids after reduction [139, 140]. The role of Trx in this reaction was confirmed by monitoring the pyruvate-dependent activation of soybean AOX after *in vitro* reduction with NTRA and PtTrx-h2 [18]. On this basis, some Trx-h isoforms may play a major role in lowering the levels of mitochondrial ROS through the activation of AOX, which can thereby effectively compete with the cytochrome oxidase for electrons donated by the ubiquinone pool.

6.4.8 Removal of reactive oxygen species

Reactive oxygen and nitrogen species modify a wide range of cellular components including lipids, proteins and nucleic acids. Although they may act in a random and destructive fashion, numerous studies suggest that their intracellular levels are tightly modulated and drive specific signalling cascades. As it is impossible to prevent the generation of these species *in vivo*, both enzymatic and nonenzymatic defenses have evolved in aerobic organisms. The most effective enzymes for removal of reactive species include superoxide dismutase, catalase, glutathione peroxidase and a large array of peroxidases, while the main nonenzymatic antioxidants are glutathione, ascorbic acid and tocopherol. Prx is a particular class of ubiquitous peroxidases that reduce reactive oxygen and nitrogen species, such as H_2O_2 and $ONOO^-$, using one or two conserved cysteines, thus defining the 1-Cys and 2-Cys Prx types, respectively [21, 141–145]. The reaction cycle of 2-Cys Prx starts with the formation of a sulfenic acid derivative at the nucleophilic thiolate:

$$\text{HS-[2-Cys Prx]-SH} + \text{ROOH} \rightarrow \text{HS-[2-Cys Prx]-SOH} + \text{ROH}$$

The unstable -SOH group reacts to form an internal disulfide bond with a nearby Cys:

$$\text{HS-[2-Cys Prx]-SOH} \rightarrow \text{[2-Cys Prx]-(S)}_2 + H_2O$$

Cleavage of the disulfide bond with reduced Trx or Grx closes the cycle [146]:

$$\text{[2-Cys Prx]-(S)}_2 + \text{[T/G]rx-(SH)}_2 \rightarrow \text{HS-[2-Cys Prx]-SH} + \text{[T/G]rx-(S)}_2$$

Hence, the biological role of Prx is linked to the capacity for poising the concentrations of reactive oxygen and nitrogen species, thereby preventing the damage of biomolecules or triggering the operation of signal transduction pathways [20, 147–149].

The Arabidopsis genome encodes 10 different Prx isoforms, many of which are predicted to localize in chloroplasts (Table 6.5). *In vitro*, chloroplast 2-Cys Prx [148] reduces H_2O_2 via NADPH → NTR → Trx → 2-Cys Prx → H_2O_2 [147]. On this basis, the *in vitro* action of several plastidial Trx isoforms on 2-Cys Prx was confronted with two well-known targets of chloroplast Trx, NADP-MDH and

Table 6.5 Prx isoforms of *Arabidopsis* thaliana [215]

Group	Polypeptide length	Subcellular localization (putative)	MATDB entry
1-Cys Prx	216	Nucleus	At1g48130
2-Cys PrxA	266	Chloroplast	At3g11630
2-Cys PrxB	271	Chloroplast	At5g06290
Type II PrxA	553	Peroxisome	At1g65990
Type II PrxB	162	Cytosol	At1g65980
Type II PrxC	162	Cytosol	At1g65970
Type II PrxD	174		At1g60740
Type II PrxE	234	Chloroplast	At3g52960
Type II PrxF	201	Mitochondria	At3g06050
Prx Q	215	Chloroplast	At3g26060

CFBPase. Notably, Trx-x provides reducing power to 2-Cys Prx but is unable to stimulate the other targets, suggesting a specific involvement in resistance against oxidative stress. Contrasting with all the Trx isoforms assayed, Trx-m3 is inactive with all three targets [50, 150]. In line with their possible role in the removal of reactive species, chloroplast Trx-x and most Trx-m isoforms, but not Trx-f or Trx-m3, restore the ability of yeasts devoid of endogenous Trx to thrive under oxidative conditions [151]. Similarly, cytosolic, mitochondrial and chloroplastic Trx efficiently reduce many alkyl hydroperoxides via PrxQ, a particular group of Prx, presumably chloroplastic, that carries a disulfide bridge formed by two cysteines separated by four amino acids [98, 152, 153]. The hypersensitive response of poplar to the causative agent of rust is accompanied by a marked increase in the mRNA levels of PrxQ, supporting previous evidence on the participation of peroxidases in poising the concentration of H_2O_2 during biotic stresses [154]. On the other hand, changes in the redox state affect the intraorganellar partition of chloroplast 2-Cys Prx. At redox potentials above the midpoint (–315 mV), the oxidized Prx remains in the stroma as a homodimer linked via two intercatenary disulfide bonds. Under reducing conditions, the excision of the cystine drives the oligomerization of dimers, with the subsequent attachment to thylakoid membranes [147]. Thus light intensity, as well as temperature and environmental oxidants (i.e., CO_2, NO_3^-), tunes 2-Cys Prx for optimum performance.

CDSP32 (chloroplastic drought-induced stress protein of 32 kDa), a stromal protein composed of two Trx domains whose expression is induced by oxidative stress [155], is closely linked to 2-Cys Prx. Potato plants with reduced levels of CDSP32 by cosuppression fared poorly under conditions that cause oxidative damage [156–158]. Affinity chromatography with a CDSP32 mutant form lacking the resolving Cys identified 2-Cys Prx as a major interactor. Five additional chloroplastic targets were also identified; three participate in photosynthesis-related processes (ATPase γ-subunit, Rubisco, aldolase) while the others are functional in responses to the oxidative damage (PrxQ and methionine sulfoxide reductase).

Co-immunoprecipitation experiments with extracts from plants overexpressing the wild type CDSP32 or the active site mutant counterpart confirmed the formation *in vivo* of heterodimeric complexes between CDSP32 and PrxQ for the reduction of H_2O_2. Despite tantalizing glimpses provided by these studies, significant gaps remain in our understanding of the mechanisms involved in the control of the oxidative stress, mainly the coordination of the redox events among the multiple isoforms of Trx and Prx. Even though some isoforms are undoubtedly addressed to a particular cellular location, validation of appropriate interactions is still a fertile ground for research.

6.4.9 Seed germination

During seed maturation and drying, the oxidation of thiol groups in reserve tissue proteins inactivates selected enzymes, activates inhibitors and increases the stability of storage proteins. Once adequate environmental conditions trigger germination, reduction of disulfide bonds reverses these processes. Earlier studies in wheat supported the idea that Trx-h, reduced with NADPH, takes part in the mobilization of nitrogen and carbon reserves in the starchy endosperm through the inactivation of amylolytic inhibitors and the activation of thiocalsin, a calcium-dependent protease [159, 160]. In fact, overexpression of Trx-h in barley endosperm not only enhanced the activity of the starch debranching enzyme but also speeded up the synthesis of α-amylase and the rate of germination [88, 91, 161, 162]. Trx-h also seems to participate in the transfer of compounds from maternal tissues to the developing seed and in the response to oxidative stress during dessiccation and germination [61].

Several proteomic approaches have been tried to identify the targets of Trx in cereal seeds during maturation or germination [91]. On the one hand, endosperm proteins from wheat and barley seeds were first reduced with the NADP/Trx system, subsequently labeled with the fluorescent, thiol-reactive probes monobromobimane or Cy5 maleimide and finally separated by 2-D electrophoresis [87, 89]. On the other hand, interacting proteins from young and mature wheat endosperm were covalently trapped on an affinity column prepared with a mutant Trx-h lacking the resolving cysteine [90]. Although results varied in each case, some proteins were consistently recovered. Among them were several members of the α-amylase/subtilisin inhibitor family, metabolic enzymes (alcohol dehydrogenase, fructose 1,6-bisphosphate aldolase, cytosolic glyceraldehyde 3-phosphate dehydrogenase, cytosolic malate dehydrogenase, glutamine synthase, vitamin B12-independent methionine synthase) and proteins involved in the oxidative stress response (ascorbate peroxidase, germin-like protein, monomeric type II Prx) and in protein biosynthesis (elongation factor 2, eukaryotic translation initiation factor 4A) and degradation (the regulatory subunit of the 26S proteasome). Some of these targets were also identified in another study that used mutant Trx-h isoforms to trap proteins from dark-grown Arabidopsis plants [163]. The biological relevance of these findings still needs to be thoroughly evaluated.

6.4.10 Modulation of receptor functions

Two recent discoveries suggest that cytosolic Trx or Trx-like proteins modulate the function of certain plasma membrane-bound receptors that recognize extracellular ligands. Self-incompatibility is a widespread mechanism by which flowering plants prevent inbreeding. Self-pollen rejection in the *Brassicaceae* family occurs when specific 'S-alleles' are expressed by the pollen and the pistil. The S-locus contains, among others, two polymorphic genes, one encoding the male determinant – the S-locus cysteine-rich (SCR) protein – and another the female counterpart – the membrane-spanning S-locus receptor kinase (SRK). The signalling cascade that leads to the rejection of the incompatible pollen starts with the autophosphorylation of SRK on the pistil stigma upon recognition of its cognate SCR on the surface of the pollen grain [164–166]. Two *Brassica* Trx-h, THL1 and THL2, have the capacity to interact with SRK in a phosphorylation-independent manner [167, 168]. In fact, it was shown that *Brassica oleracea* THL1 prevents the spontaneous autophosphorylation of SRK in the absence of the 'activating' component of the pollen coat, thus lowering the risk of constitutive pollen rejection [169]. Although it is not clear how THL1 affects the activity of SRK, there is evidence to suggest that the reduced form does so via conserved cysteines on the cytosolic side of the transmembrane domain of SRK. Since both THL-1 and THL-2 are expressed in a variety of organs, their action may not be restricted to reproductive tissues.

A Trx-like protein, CITRX, with an N-terminal variable extension and a typical C-terminal Trx domain, interacted in a yeast two-hybrid screen with the cytosolic domain of the receptor-like protein Cf-9 of *Solanaceae*. Tomato *Cf-9* codes for a transmembrane protein with extracellular leucine-rich repeat domain and a short cytosolic tail. This protein is involved in the specific resistance to *Cladosporium fulvum* strains carrying the avirulence determinant Avr9. CITRX was found to interact specifically with the cytosolic domain of Cf-9, but not with the same region of the related Cf-2. In addition, silencing of CITRX expression led to an exacerbated response to Avr9, suggesting that CITRX acts as a negative modulator of Cf-9 signalling. Since there are no cysteine residues in the Cf-9 fragment used as bait, it will be interesting to investigate whether these effects depend on the integrity of the active site of CITRX [170].

Briefly, it seems that some membrane receptors, in the absence of their specific ligands, are kept in a silent state through the interaction with Trx-like proteins. The physiological meaning of this phenomenon is still unclear. It is intriguing, however, that both pollen-stigma and pathogen-plant signalling pathways lead to rapid changes in Ca^{2+} fluxes and ultimately to processes of programmed cell death. In the light of these results, it would be worth revisiting earlier hypotheses on the links between both processes [171].

6.5 Glutaredoxins

In 1976, the presence of an alternative hydrogen donor for the reduction of ribonucleotides to deoxyribonucleotides in mutants of *E. coli* devoid of Trx led to the identification of Grx. These small proteins (ca 10 kDa) exhibit an overall three-dimensional structure similar to Trx and have been found in all living organisms [45]. An analysis of annotated genome sequences reveals that most organisms possess many isoforms with both classical (-CXXC-) or atypical (-CXXS-) sequences at the active site. Searches in the Arabidopsis genome revealed an extraordinary diversity comprising 14 bicysteinic and 17 monocysteinic isoforms (Table 6.6) [172, 173]. Based on the current level of genome annotation, rice, wheat, maize and barley also exhibit an apparent wealth of Grx. The Arabidopsis family can be classified into three groups, mainly on the basis of their active site sequences. Out of six members in the first group – five bicysteinic and one monocysteinic – four would

Table 6.6 Grx isoforms of *Arabidopsis thaliana* [173]

Isoform[a]	Polypeptide length	Active site sequence	Subcellular localization	MATDB entry
CxxC1	125	**CGYC**	Cytosol	At5g63030
CxxC2	111	**CPYC**	Secretory pathway	At5g40370
CxxC3	130	**CPYC**	Secretory pathway	At1g77370
CxxC4	135	**CPYC**	Secretory pathway	At5g20500
CxxC5	174	**CSYC**	Chloroplast	At4g28730
CxxC6	144	CCMC	Cytosol	At4g33040
CxxC7	136	CCMC	Cytosol	At3g02000
CxxC8	140	CCMC	Cytosol	At5g14070
CxxC9	137	CCMC	Cytosol	At1g28480
CxxC10	145	CCMC	Chloroplast	At5g11930
CxxC11	103	CCMC	Cytosol	At3g62950
CxxC12	103	CCMC	Cytosol	At2g47870
CxxC13	102	CCLC	Cytosol	At2g47880
CxxC14	102	CCLC	Cytosol	At3g62960
CxxS1	102	CCMS	Cytosol	At1g03020
CxxS2	102	CCMS	Cytosol	At5g18600
CxxS3	102	CCMS	Cytosol	At4g15700
CxxS4	102	CCMS	Cytosol	At4g15680
CxxS5	102	CCMS	Cytosol	At4g15690
CxxS6	102	CCMS	Cytosol	At3g62930
CxxS7	102	CCMS	Cytosol	At4g15670
CxxS8	102	CCMS	Cytosol	At4g15660
CxxS9	102	CCMS	Cytosol	At2g30540
CxxS10	102	CCMS	Mitochondria	At3g21460
CxxS11	99	CCLS	Cytosol	At1g06830
CxxS12	179	**CSYS**	Chloroplast	At2g20270
CxxS13	150	CCLG	Chloroplast	At1g03850
CxxS14	173	*CGFS*	Chloroplast	At3g54900
CxxS15	169	*CGFS*	Mitochondria	At3g15660
CxxS16	293	*CGFS*	Chloroplast	At2g38270
CxxS17	488	*CGFS*	Cytosol	At4g04950

[a]Isoforms in bold, italic and normal characters possess the active site sequences Cxx(C/S), CGFS and CCx(C/S/G), respectively.

be addressed to a cytosolic or perhaps extracellular localization; the other two, with active site sequences -CSYC- and -CSYS-, possess a long N-terminal extension characteristic of chloroplast transit peptides. Four different Grx containing the -CGFS- sequence cluster in the second group, three of which are possibly addressed to organelles. The fourth, surprisingly, codes for a putative cytosolic protein (488 amino acids) that hosts one Trx motif followed by three Grx modules, bearing striking resemblance to PICOT (for protein kinase C-interacting cousin of Trx), a human protein that plays a regulatory role in cellular stress responses associated with transcription factors AP-1 and NF-κB [174]. The third group, with 21 members, includes both bicysteinic and monocysteinic isoforms bearing unusual geminate cysteines in the active site -CC(M/L)(C/S/G)-. Remarkably, homologous sequences are present in all land plants analyzed, but are absent in other photosynthetic organisms such as *Synechocystis* or *Chlamydomonas*. It is important to bear in mind that intracellular localizations predicted in Table 6.6 are all tentative, since sound experimental evidence is lacking.

A more positive midpoint redox potential endows Grx with the capacity to use GSH to cleave disulfides or GSH-mixed disulfides, via a dithiol or a monothiol mechanism, respectively [45]. In the dithiol pathway, the nucleophilic cysteine attacks the protein disulfide forming a heterodisulfide resolved by the second cysteine, releasing oxidized Grx and the reduced protein:

$$\text{HS-Grx-SH} + \text{Prt-(S)}_2 \rightarrow \text{HS-Grx-S-S-Prt-SH} \rightarrow \text{Grx(S)}_2 + \text{Prt-(SH)}_2$$

Subsequently, GSH restores the reduced form of Grx:

$$\text{Grx(S)}_2 + 2\,\text{GSH} \rightarrow \text{Grx-(SH)}_2 + \text{GSSG}$$

The monothiol pathway accommodates the functioning of monocysteinic Grx because the single thiol of Grx releases the reduced protein and concurrently generates a mixed disulfide with GSH, which can be cleaved subsequently by GSH:

$$\text{Grx-SH} + \text{Prt-S-SG} \rightarrow \text{Grx-S-SG} + \text{Prt-SH}$$

$$\text{Grx-S-SG} + \text{GSH} \rightarrow \text{Grx-SH} + \text{GS-SG}$$

A crucial feature of the monothiol pathway is that the affinity of Grx for free GSH is higher than for the same moiety in glutathionylated proteins [175–177]. However, the finding that the nucleophilic cysteine of yeast Grx with a -CGFS- motif forms a disulfide with a second cysteine placed 50 residues upstream of the active site may change this view [178].

Few data are available on the distribution of Grx isoforms in plant tissues. Grx were initially identified in spinach leaves, developing rice seeds or *Ricinus communis* cotyledons [179–181]. The rice gene is expressed almost exclusively in the seed aleurone while the *R. communis* counterpart appears in cotyledons, hypocotyls and roots and to a lesser extent in leaves. On the other hand, the gene coding for AtCxxS14 is expressed in leaves, stems and roots, and is repressed in seedlings by ion treatment [182]. Functional information on this PDOR family in plants is also scant. Studying the activation of the Arabidopsis vacuolar H^+/Ca^{2+} antiporter (CAX1) in yeast hypersensitive to high levels of Ca^{2+}, Cheng and Hirschi [182] found that AtCxxS14, but not

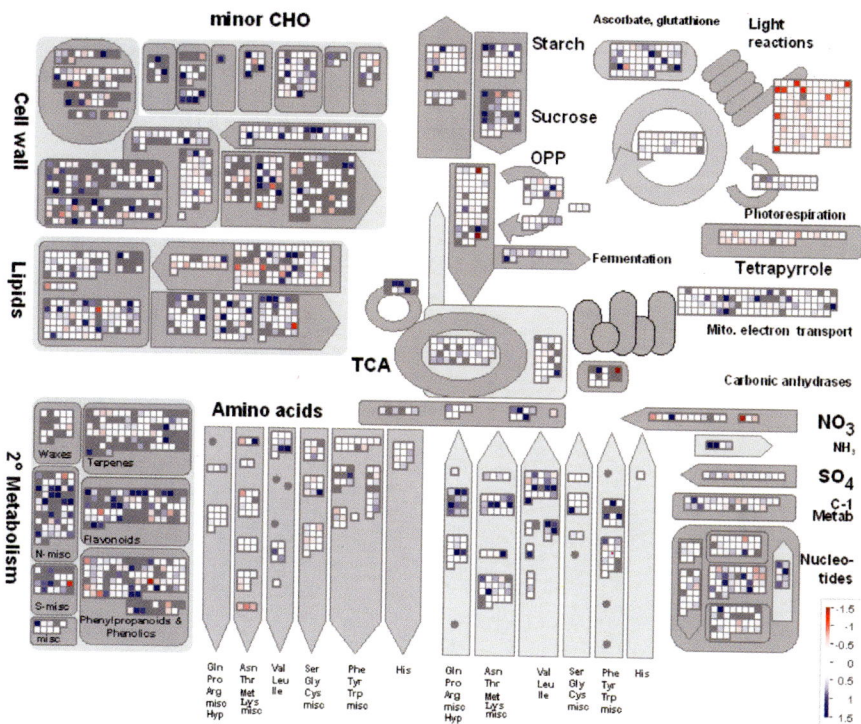

Plate 1 Comparison of the average levels of transcripts encoding enzymes from central metabolism, calculated throughout a day and night cycle, in *A. thaliana* Col0 wild type and its starchless mutant *pgm*. The results were calculated as the log 2 of the ratio between *pgm* and wild type. Rosette leaves of 5 weeks old plants grown in a 12-h photoperiod and at 20°C were harvested every 4 h, starting from the end of the night period. Abbreviations: CHO, carbohydrate metabolism; TCA, tricarboxylic acid cycle; OPP, oxidative pentose phosphate cycle; NO_3, nitrate assimilation; NH_3, nitrogen degradation; SO_4, sulfate assimilation. Arrows indicate synthesis or degradation processes pointing to or from associated pathways/processes.

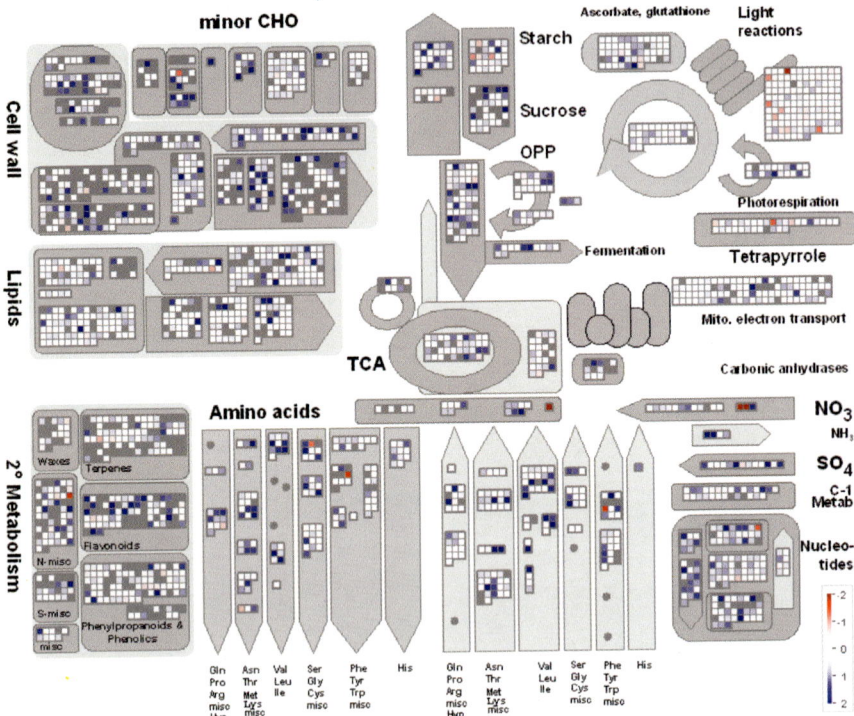

Plate 2 Comparison of the amplitudes of the changes in the levels of transcripts encoding enzymes from central metabolism, throughout a day and night cycle in Arabidopsis Col0 wild type and the starch-less mutant *pgm*. The results were calculated as the log 2 of the ratio between *pgm* and wild type. Rosette leaves of 5 weeks old plants grown in a 12-h photoperiod and at 20°C were harvested every 4 h, starting from the end of the night period. Abbreviations: CHO, carbohydrate metabolism; TCA, tricarboxylic acid cycle; OPP, oxidative pentose phosphate cycle; NO_3, nitrate assimilation; NH_3, nitrogen degradation; SO_4, sulfate assimilation. Arrows indicate synthesis or degradation processes pointing to or from associated pathways/processes.

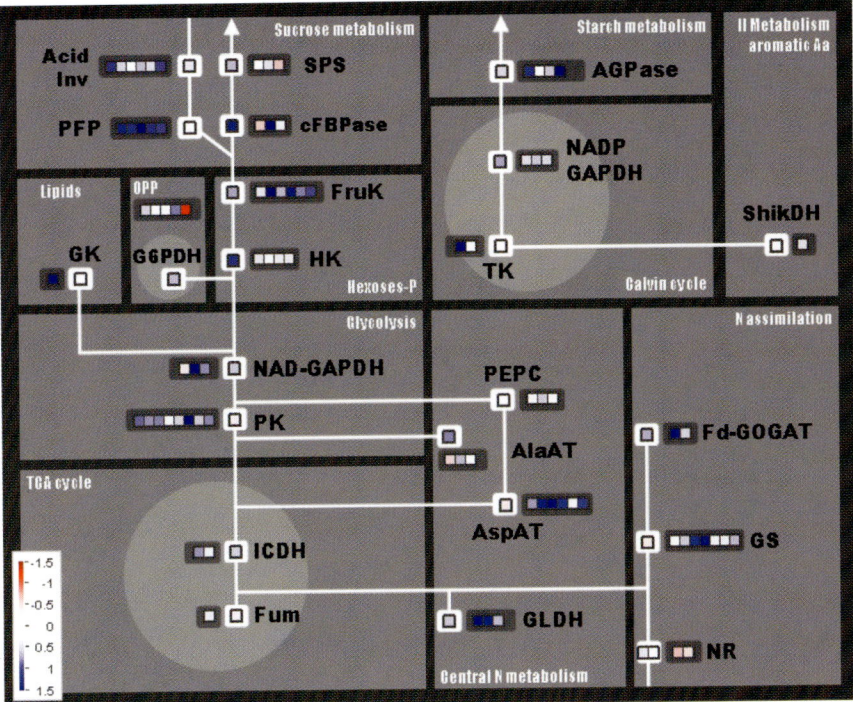

Plate 3 Comparison of the diurnal amplitudes of the changes in the levels of transcripts encoding enzymes from central metabolism and the corresponding activities in Arabidopsis Col0 wild type and the starchless mutant *pgm*. The results were calculated as the log 2 of the ratio between *pgm* and wild type. Rosette leaves of 5 weeks old plants grown in a 12-h photoperiod and at 20°C were harvested every 4 h, starting from the end of the night period.

Abbreviations: AcidInv, acidic invertase; SPS, sucrose phosphate synthase; PFP, pyrophosphate-fructose 6-phosphate 1-phosphotransferase; cFBPase, cytosolic fructose-1,6-bisphosphatase; FruK, fructokinase; HK, hexokinase; G6PDH, glucose-6-phosphate dehydrogenase; GK, glycerol kinase; NAD-GAPDH, NAD-dependent glyceraldehyde-3-phosphate dehydrogenase; PK, pyruvate kinase; ICDH, isocitrate dehydrogenase; Fum, fumarase; AGPase, ADPglucose pyrophosphorylase; NADP-GAPDH, NADP-glyceraldehyde 3-P dehydrogenase; TK, transketolase; ShikDH, shikimate dehydrogenase; PEPC, phosphoenolpyruvate carboxylase; AlaAT, alanine amino transferase; AspAT, aspartate aminotransferase; GLDH, glutamate dehydrogenase; Fd-GOGAT, ferredoxin-dependent glutamate synthase; GS, glutamine synthetase; NR, nitrate reductase.

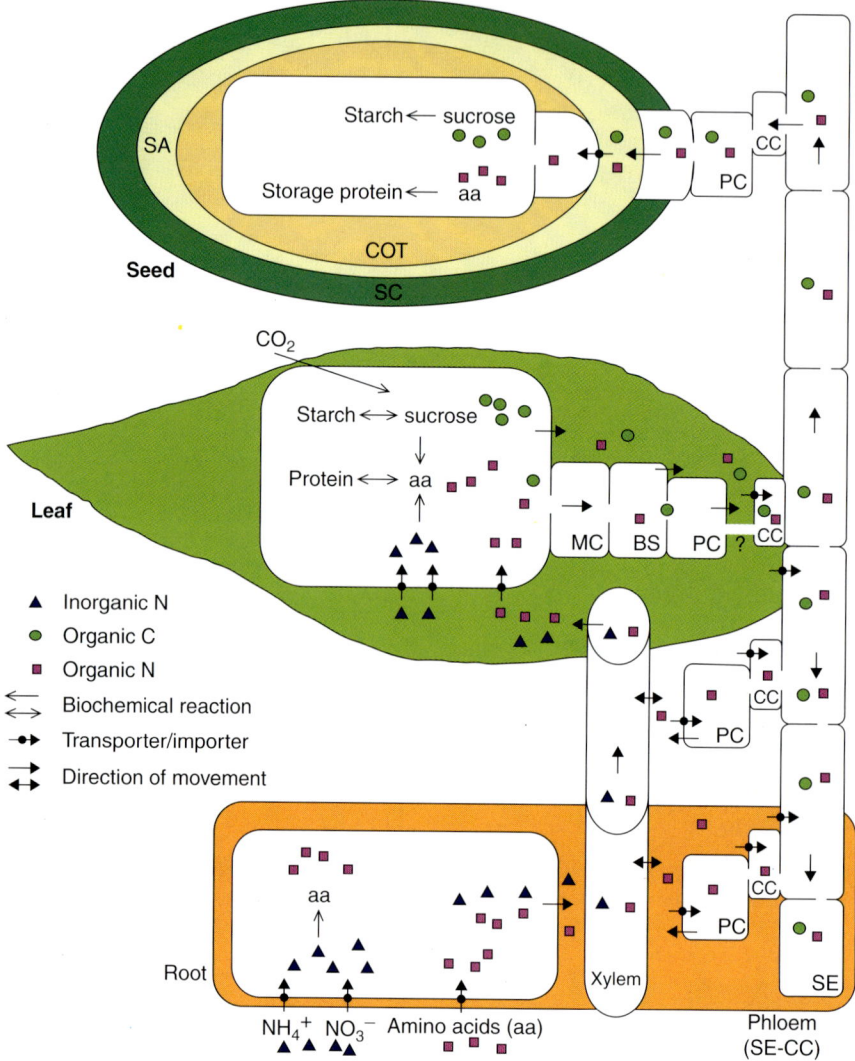

Plate 4 Model of long-distance transport of sucrose and amino acids relative to plant growth and synthesis of storage compounds. Depending on the plant species and/or the physiological stage, N assimilation might occur in roots or source leaves followed by long-distance transport of amino acids via the xylem and phloem, respectively. Exchange of amino acids between the xylem and phloem is also possible. CO_2 fixation and carbon assimilation take place in photosynthetically active source leaves, and reduced carbon and nitrogen in the form of sucrose (sugars) and amino acids are subsequently transported by the phloem pathway (bi-directional) to sink organs. The organic nitrogen and carbon can move symplasmically via plasmodesmata or apoplastically. The mechanism of release of sucrose and amino acids into the apoplast is unknown but might be by a passive (effluxer) or active (exporter) transport step, depending upon apoplastic concentrations. Apoplastic phloem loading or reloading of assimilates leaked into the apoplast back into the cell is by active transport mediated by proton symporters/importers located in the plasma membrane. Operation of plasma membrane H^+-ATPases provide the proton motive force and are coordinated with these transport activities. Abbreviations: BC, bundle sheath cell; CC, companion cell; COT, cotyledon; PC, parenchyma cell; MC mesophyll cell; SA, seed apoplast; SC, seed coat; SE, sieve element.

AtCxxS16, associates with the N-terminal region of CAX1 and suppresses the vacuolar transport defect. Interestingly, CAX1 contains a nine-amino-acid region required for Ca^{2+} transport in which resides a -CXXC- sequence [183]. There is consequently great interest in examining how each Grx interacts and modulates specific targets.

6.6 Protein-disulfide isomerases

Protein-disulfide isomerases are PDOR that, owing to a more positive redox potential at their active sites, tend to catalyze the formation and shuffling of disulfide bridges, rather than their reduction. Classical eukaryotic PDI are ER resident proteins with a modular structure, comprising domains a, b, b′, a′. The most conserved and best defined a and a' domains have Trx folds containing, at least one of them, the characteristic motif -CXXC- at the redox active site [184, 185]. b domains, which also have a Trx-like fold but lack sequence homology with a domains, play an important role in the specificity for target proteins. A set of 22, 19 and 22 proteins identified in Arabidopsis (Table 6.7), rice and maize, respec-

Table 6.7 PDI isoforms of *Arabidopsis thaliana* [186]

MATDB entry	Polypeptide length	C-terminal sequence	Trx domains	Structural class[a]	Phylogenetic group
At1g21750	501	KDEL	2	1	I
At1g77510	508	KDEL	2	1	I
At5g60640	597	KDEL	2	1	II
At3g54960	579	KDEL	2	1	II
At1g52260	537	KDEL	2	1	III
At3g16110	531	KDEL	2	1	III
At2g47470	361	VASS	2	2	IV
At2g32920	440	KDEL	2	2	V
At1g04980	443	KDDL	2	2	V
At1g07960	146	DKEL	1	5	VI
At1g35620	440	KKED	1	5	VII
At3g20560	483	GKNI	1	5	VIII
At4g27080	480	GKNF	1	5	VIII
At1g15020	528	EQER	1	5	IX
At2g01270	495	PRRR	1	5	IX
At4g04610	465	NLVR	1	5	X
At1g62180	454	NLLR	1	5	X
At4g21990	458	NLVR	1	5	X
At1g34780	310	SSSQ	1	5	X
At3g03860	300	SDQS	1	5	X
At4g08930	295	SASQ	1	5	X
At5g18120	289	SQSA	1	5	X

[a]Class 1 proteins have two Trx domains, one close to the N-terminus and another at the C-terminus. Class 2 (two Trx domains) and 4 (three Trx domains) proteins contain two Trx domains in tandem at the N- and C-terminal regions, respectively. Class 3 proteins are similar to Class 1 plus an additional Trx domain at the N-terminal region. Class 5 proteins have single Trx domains [188, 216, 217].

tively, led recently to the classification of plant PDI into 10 groups [186]. The structure of proteins from groups I to V, with two Trx domains, is similar to PDI from other eukaryotes. The remaining five groups bear a single Trx domain. Two of these exhibit nonisomerase enzymatic activities encoded by an additional domain, i.e. adenosine 5'-phosphosulfate reductase and quiescin-sulfhydryl oxidase. Further examination of predicted primary structures revealed two salient features [186]. First, single-domain PDI have neither KDEL-like ER retention signals characteristic of groups I–III and V, nor the conserved C-terminal ER retention domains found on multidomain PDI of group IV. Traditionally, PDI were characterized by the -CGHC- motif found in all of the multidomain PDI as well as in those from group VII, but members of groups VI and VIII surprisingly hold motifs -CKHC- and -CYW(C/S)-. Given that mining of maize expressed sequence tags and RNA-profiling databases indicates that members of the single-domain PDI are likely expressed in all plants, new insights into the enzymatic activities of single domain PDI are necessary for understanding their biological roles. Moreover, it is still unclear whether all these proteins, that exhibit distinct domain distribution, originated from the same ancestral protein containing either one [187] or two Trx domains [188].

Along with their capacity to isomerize cystine bonds, PDI may also act as thiol oxidases for the formation of disulfides in nascent proteins, mainly in the secretory pathway [189]. Congruent with this view, PDI enhances the oxidative refolding rate of the fully reduced horse gram protease inhibitor, a disulfide-rich Bowman–Birk class dual inhibitor (8–10 kDa) that binds and inhibits trypsin and chymotrypsin activities [190]. However, it is remarkable that PDI has evolved other 'moonlighting' functions devoid of redox activity, including a role as a chaperone that assists in the *in vitro* refolding of denatured proteins with no disulfide bonds [191] or as the β-subunit of prolyl hydroxylase – the enzyme that catalyzes the post-translational hydroxylation of proline [192]. Although it has long been known that ER-localized PDI catalyze the reversible formation and isomerization of disulfide bonds necessary for the proper folding, assembly, activity and secretion of numerous proteins, the identification of PDI-like proteins in mitochondria and chloroplasts clearly indicates that this family is not restricted to the ER [116, 193]. Proteins homologous to PDI are expressed in all tissues of durum wheat as well as peach shoots and leaves [194, 195]. Some evidence suggests that different PDI isoforms have overlapping functions. In fact, individual knock out of three PDI-like genes in the moss *Physcomitrella patens* had no effect on viability [196]. Nonetheless, PDI seems to play major roles during seed development. Three *esp2* mutants of rice, which lack PDI in the endosperm, contain larger amounts of the 57-kDa proglutelin precursor and levels of acidic and basic glutelin subunits lower than normal [197]. In line with a role of PDI in processing proglutelin within the ER lumen, PDI was expressed in wild type seeds only during early stages of seed development. Finally, PDI may also be involved in some aspects of the response to pathogens, as it is differentially expressed during the infection of wheat with the fungus *Mycosphaerella graminicola* [198].

6.7 Concluding remarks

The last decade has significantly extended our understanding of redox cell signalling, disclosing novel post-translational modifications of cysteines and introducing additional regulatory systems into a scene thus far dominated by thiol/disulfide exchanges. It is well established now that many oxidation states of thiol groups enable proteins to acquire multiple post-translational modifications on a single residue. Thiol-dependent reactions appear to affect almost every major metabolic pathway in plant cells, not only in chloroplasts where this process was first described, but also in mitochondria and in the cytosol. Although much progress has been made concerning the pivotal role of cysteine residues as exquisite sensors of cellular redox status, our knowledge on the participation of these mechanisms in plants is still fragmentary. However, recent progress with similarly complex systems in other organisms points to suitable experimental avenues and provides confidence that novel plant mechanisms will appear on time.

The ever growing collections of plant genes and the implementation of proteomic studies have dramatically increased the inventory of redox signalling players in plants and also raised numerous questions that remain to be answered. Photosynthetic cells turn out to be lush diversity hotspots for PDOR and related proteins. Indeed, over 100 genes in Arabidopsis code for polypeptides capable of catalyzing thiol-based reactions [186]. Alternative biochemical and cellular strategies are required to define how the redox status controls the large number of PDOR isoforms and their precise cellular location. Considering the close midpoint redox potential values of many members in each family of PDOR (cf. Table 6.1), noncovalent interactions during the approximation and the docking of interacting proteins should play important roles in the formation of complexes with their targets. These ideas have been revitalized by the identification of a plethora of PDOR-sensitive proteins through proteomic and complementary genomic and biochemical studies. Examination of these potential targets is experimentally challenging. In particular, the recent progress made in the analysis of transcription factors and the thylakoid lumen constitutes suitable experimental avenues off the well-trodden path of stromal metabolism that should provide novel alternatives about the generality of the redox signalling.

Acknowledgements

The authors gratefully acknowledge the funding from Agencia Nacional de la Promoción Científica y Técnica (ANPCyT), Consejo Nacional de Investigaciones Científicas y Técnicas (CONICET) and the Universidad de Buenos Aires (UBA), Argentina.

References

1. C.H. Foyer and I.G. Noctor (2000) Oxygen processing in photosynthesis: regulation and signalling. *New Phytologist* **146**, 359–388.
2. K.N. Apel and H. Hirt (2004) Reactive oxygen species: metabolism, oxidative stress, and signal transduction. *Annual Review of Plant Biology* **55**, 373–399.

3. I.C. Mori and J.I. Schroeder (2004) Reactive oxygen species activation of plant Ca^{2+} channels. A signalling mechanism in polar growth, hormone transduction, stress signalling, and hypothetically mechanotransduction. *Plant Physiology* **135**, 702–708.
4. O.N. Jensen (2004) Modification-specific proteomics: characterization of post-translational modifications by mass spectrometry. *Current Opinions in Chemical Biology* **8**, 33–41.
5. E.S.G. Barron (1951) Thiol groups of biological importance. *Advances in Enzymology* **11**, 201–266.
6. S.O. Kim, K. Merchant, R. Nudelman et al. (2002) Oxy R: a molecular code for redox-related signalling. *Cell* **109**, 383–396.
7. G. Noctor and C.G. Foyer (1998) Ascorbate and glutathione: keeping active oxygen under control. *Annual Review of Plant Physiology and Plant Molecular Biology* **49**, 249–279.
8. M.J. May, T. Vernoux, C. Leaver, M. Van Montagu and D. Inze (1998) Glutathione homeostasis in plants: implications for environmental sensing and plant development. *Journal of Experimental Botany* **49**, 649–667.
9. I.A. Cotgreave and R.G. Gerdes (1998) Recent trends in glutathione biochemistry – glutathione-protein interactions: a molecular link between oxidative stress and cell proliferation? *Biochemical and Biophysical Research Communications* **242**, 1–9.
10. P. Klatt and S. Lamas (2000) Regulation of protein function by S-glutathiolation in response to oxidative and nitrosative stress. *European Journal of Biochemistry* **267**, 4928–4944.
11. C.M. Grant (2001) Role of the glutathione/glutaredoxin and thioredoxin systems in yeast growth and response to stress conditions. *Molecular Microbiology* **39**, 533–541.
12. K.P. Huang and F.L. Huang (2002) Glutathionylation of proteins by glutathione disulfide S-oxide. *Biochemical Pharmacology* **64**, 1049–1056.
13. D.M. Sullivan, N.B. Wehr, M.M. Fergusson, R.L. Levine and T. Finkel (2000) Identification of oxidant-sensitive proteins: TNF-a induces protein glutathiolation. *Biochemistry* **39**, 11121–11128.
14. V. Ravichandran, T. Seres, T. Moriguchi, J.A. Thomas and R.B. Johnston (1994) S-thiolation of glyceraldehyde-3-phosphate dehydrogenase induced by the phagocytosis-associated respiratory burst in blood monocytes. *Journal of Biological Chemistry* **269**, 25010–25015.
15. E. Pineda-Molina, P. Katt, J. Vazquez et al. (2001) Glutathionylation of the p50 subunit of NF-κB: a mechanism for redox-induced inhibition of DNA binding. *Biochemistry* **40**, 14134–14142.
16. S. Casagrande, V. Bonetto, M. Fratelli et al. (2002) Glutathionylation of human thioredoxin: a possible crosstalk between the glutathione and thioredoxin systems. *Proceedings of the National Academy of Sciences of the United States of America* **99**, 9745–9749.
17. H. Ito, M. Iwabuchi and K. Ogawa (2003) The sugar-metabolic enzymes aldolase and triose-phosphate isomerase are targets of glutathionylation in *Arabidopsis thaliana*: detection using biotinylated glutathione. *Plant and Cell Physiology* **44**, 655–660.
18. E. Gelhaye, N. Rouhier, J. Gerard et al. (2004) A specific form of thioredoxin h occurs in plant mitochondria and regulates the alternative oxidase. *Proceedings of the National Academy of Sciences of the United States of America* **101**, 14545–14550.
19. A. Claiborne, J.I. Yeh, T.C. Mallett et al. (1999) Protein-sulfenic acids: diverse roles for an unlikely player in enzyme catalysis and redox regulation. *Biochemistry* **38**, 15407–15416.
20. K.J. Dietz (2003) Plant peroxiredoxins. *Annual Review of Plant Biology* **54**, 93–107.
21. R. Bryk, P. Griffin and C. Nathan (2000) Peroxynitrite reductase activity of bacterial peroxiredoxins. *Nature* **407**, 211–215.
22. S. Nagashima, M. Nakasako, N. Dohmae et al. (1998) Novel non-heme iron center of nitrile hydratase with a claw setting of oxygen atoms. *Nature Structural Biology* **5**, 347–351.
23. J.S. Stamler, S. Lamas and F.C. Fang (2001) Nitrosylation: the prototypic redox-based signalling mechanism. *Cell* **106**, 675–686.
24. A. Matsumoto, K.E. Comatas, L. Liu and J.S. Stamler (2003) Screening for nitric oxide-dependent protein-protein interactions. *Science* **301**, 657–661.
25. J.B. Mannick and C.M. Schonoff (2002) Nitrosylation: the next phosphorylation? *Archives of Biochemistry and Biophysics* **408**, 1–6.

26. K. Becker, S.N. Savvides, M. Keese, R.H. Schirmer and P.A. Karplus (1998) Enzyme inactivation through sulfhydryl oxidation by physiologic NO-carriers. *Nature Structural Biology* **5**, 267–271.
27. M.R. Chandok, A.J. Ytterberg, K.J. van Wijk and D.F. Klessig (2003) The pathogen-inducible nitric oxide synthase (iNOS) in plants is a variant of the P protein of the glycine decarboxylase complex. *Cell* **113**, 469–482.
28. P. Rockel, F. Strube, A. Rockel, J. Wildt and W.M. Kaiser (2002) Regulation of nitric oxide (NO) production by plant nitrate reductase *in vivo* and *in vitro*. *Journal of Experimental Botany* **53**, 103–110.
29. J.S. Stamler, E.J. Toone, S.A. Lipton and N.J. Sucher (1997) (S)NO signals: translocation, regulation, and a consensus motif. *Neuron* **18**, 691–696.
30. S.C. Huber and S.C. Hardin (2004) Numerous posttranslational modifications provide opportunities for the intricate regulation of metabolic enzymes at multiple levels. *Current Opinion in Plant Biology* **7**, 318–322.
31. V.V. Sumbayev (2003) S-nitrosylation of thioredoxin mediates activation of apoptosis signal-regulating kinase 1. *Archives of Biochemistry and Biophysics* **415**, 133–136.
32. I.M. Yasinska, A.V. Kozhukhar and V.V. Sumbayev (2004) S-nitrosation of thioredoxin in the nitrogen monoxide/superoxide system activates apoptosis signal-regulating kinase 1. *Archives of Biochemistry and Biophysics* **428**, 198–203.
33. R. Radil, J.S. Beckman, K.M. Bush and B.A. Freeman (1991) Peroxynitrite oxidation of sulfhydryls. The cytotoxic potential of superoxide and nitric oxide. *Journal of Biological Chemistry* **266**, 4244–4250.
34. C.E. Cooper, R. Patel, P.S. Brookes and V.M. Darley-Usmar (2002) Nanotransducers in cellular redox signalling: modification of thiols by reactive oxygen and nitrogen species. *Trends in Biochemical Sciences* **27**, 489–492.
35. J.L. Martin (1995) Thioredoxin – a fold for all reasons. *Structure* **3**(3), 245–250.
36. B.J. Laughner, P.C. Sehnke and R.J. Ferl (1998) A novel nuclear member of the thioredoxin superfamily. *Plant Physiology* **118**(3), 987–996.
37. P.J. Hogg (2003) Disulfide bonds as switches for protein function. *Trends in Biochemical Sciences* **28**, 210–214.
38. H. Kadokura, F. Katzen and J. Beckwith (2003) Protein disulfide bond formation in prokaryotes. *Annual Review of Biochemistry* **72**, 111–135.
39. D.B. Ritz and J. Beckwith (2001) Roles of thiol-redox pathways in bacteria. *Annual Review of Microbiology* **55**, 21–48.
40. P. Schurmann and J.P. Jacquot (2000) Plant thioredoxin systems revisited. *Annual Review of Plant Physiology and Plant Molecular Biology* **51**, 371–400.
41. B.B. Buchanan and Y. Balmer (2005) Redox regulation: a broadening horizon. *Annual Review of Plant Biology* **56**, 187–220.
42. R.A. Wolosiuk and B.B. Buchanan (1976) Thioredoxin and glutathione regulates photosynthesis in chloroplasts. *Nature* **266**, 565–567.
43. R.A. Wolosiuk, N.A. Crawford, B.C. Yee and B.B. Buchanan (1979) Isolation of three thioredoxins from spinach leaves. *Journal of Biological Chemistry* **254**, 1627–1632.
44. H. Hartman, M. Syvanen and B.B. Buchanan (1990) Contrasting evolutionary histories of chloroplast thioredoxins f and m. *Molecular Biology and Evolution* **7**, 247–254.
45. A. Vlamis-Gardikas and A. Holmgren (2002) Thioredoxin and glutaredoxin isoforms. *Methods in Enzymology* **347**, 286–296.
46. D. Mestres-Ortega and Y. Meyer (1999) The *Arabidopsis thaliana* genome encodes at least four thioredoxins m and a new prokaryotic-like thioredoxin. *Gene* **240**, 307–316.
47. Y. Meyer, F. Vignols and J.P. Reichheld (2002) Classification of plant thioredoxins by sequence similarity and intron position. *Methods in Enzymology* **347**, 394–402.
48. S.D. Lemaire, V. Collin, E. Keryer, A. Quesada and M. Miginiac-Maslow (2003) Characterization of thioredoxin y, a new type of thioredoxin identified in the genome of *Chlamydomonas reinhardtii*. *FEBS Letters* **543**, 87–92.

49. U. Baumann and J. Juttner (2002) Plant thioredoxins: the multiplicity conundrum. *Cellular and Molecular Life Sciences* **59**, 1042–1057.
50. V. Collin, E. Issakidis-Bourguet, C. Marchand *et al.* (2003) The Arabidopsis plastidial thioredoxins: new functions and new insights into specificity. *Journal of Biological Chemistry* **278**, 23747–23752.
51. C. Laloi, N. Rayapuram, Y. Chartier, J.M. Grienenberger, G. Bonnard and Y. Meyer (2001) Identification and characterization of a mitochondrial thioredoxin system in plants. *Proceedings of the National Academy of Sciences of the United States of America* **98**, 14144–14149.
52. Y. Balmer and B.B. Buchanan (2002) Yet another plant thioredoxin. *Trends in Plant Sciences* **7**, 191–193.
53. C. Brehelin, C. Laloi, A.T. Setterdahl, D.B. Knaff and Y. Meyer (2004) Cytosolic, mitochondrial thioredoxins and thioredoxin reductases in *Arabidopsis thaliana*. *Photosynthesis Research* **79**, 295–304.
54. C. Brehelin, N. Mouaheb, L. Verdoucq, J.M. Lancelin and Y. Meyer (2000) Characterization of determinants for the specificity of Arabidopsis thioredoxins h in yeast complementation. *Journal of Biological Chemistry* **275**, 31641–31647.
55. D.E. Fomenko and V.N. Gladyshev (2002) CxxS: fold-independent redox motif revealed by genome-wide searches for thiol/disulfide oxidoreductase function. *Protein Science* **11**, 2285–2296.
56. N. Coudevylle, A. Thureau, C. Hemmerlin, E. Gelhaye, J.P. Jacquot and M.T. Cung (2005) Solution structure of a natural CPPC active site variant, the reduced form of thioredoxin h1 from poplar. *Biochemistry* **44**, 2001–2008.
57. J. Juttner, D. Olde, P. Langridge and U. Baumann (2000) Cloning and expression of a distinct subclass of plant thioredoxins. *European Journal of Biochemistry* **267**, 7109–7117.
58. E. Gelhaye, N. Rouhier and J.P. Jacquot (2004) The thioredoxin h system of higher plants. *Plant Physiology and Biochemistry* **42**, 265–271.
59. E. Gelhaye, N. Rouhier and J.P. Jacquot (2003) Evidence for a subgroup of thioredoxin h that requires GSH/Grx for its reduction. *FEBS Letters* **555**, 443–448.
60. F. Marcus, S.H. Chamberlain, C. Chu *et al.* (1991) Plant thioredoxin h: an animal-like thioredoxin occurring in multiple cell compartments. *Archives of Biochemistry and Biophysics* **287**, 195–198.
61. A.J. Serrato and F.J. Cejudo (2003) Type-h thioredoxins accumulate in the nucleus of developing wheat seed tissues suffering oxidative stress. *Planta* **217**, 392–399.
62. Y. Ishiwatari, C. Honda, I. Kawashima *et al.* (1995) Thioredoxin h is one of the major proteins in rice phloem sap. *Planta* **195**, 456–463.
63. C. Schobert, L. Baker, J. Szederkenyi *et al.* (1998) Identification of immunologically related proteins in sieve-tube exudate collected from monocotyledonous and dicotyledonous plants. *Planta* **206**, 245–252.
64. J.P. Reichheld, D. Mestres-Ortega, C. Laloi and Y. Meyer (2002) The multigenic family of thioredoxin h in *Arabidopsis thaliana*: specific expression and stress response. *Plant Physiology and Biochemistry* **40**, 685–690.
65. K. Maeda, C. Finnie, O. Ostergaard and B. Svensson (2003) Identification, cloning and characterization of two thioredoxin h isoforms, HvTrxh1 and HvTrxh2, from the barley seed proteome. *European Journal of Biochemistry* **270**, 2633–2643.
66. F. Vignols, N. Mouaheb, D. Thomas and Y. Meyer (2003) Redox control of Hsp70-Co-chaperone interaction revealed by expression of a thioredoxin-like Arabidopsis protein. *Journal of Biological Chemistry* **278**, 4516–4523.
67. C.H. Williams, L.D. Arscott, S. Muller *et al.* (2000) Thioredoxin reductase two modes of catalysis have evolved. *European Journal of Biochemistry* **267**, 6110–6117.
68. D. Mustacich and G. Powis (2000) Thioredoxin reductase. *Biochemical Journal* **346**, 1–8.
69. M. Russel and P. Model (1988) Sequence of thioredoxin reductase from *Escherichia coli*. Relationship to other flavoprotein disulfide oxidoreductases. *Journal of Biological Chemistry* **263**, 9015–9019.
70. H.Z. Chae, S.J. Chung and S.G. Rhee (1994) Thioredoxin-dependent peroxide reductase from yeast. *Journal of Biological Chemistry* **269**, 27670–27678.

71. J. Kuriyan, T.S. Krishna, L. Wong et al. (1991) Convergent evolution of similar function in two structurally divergent enzymes. *Nature* **352**, 172–174.
72. S. Dai, M. Saarinen, S. Ramaswamy, Y. Meyer, J.P. Jacquot and H. Eklund (1996) Crystal structure of *Arabidopsis thaliana* NADPH dependent thioredoxin reductase at 2.5 Å resolution. *Journal of Molecular Biology* **264**, 1044–1057.
73. J.P. Reichheld, E. Meyer, M. Khafif, G. Bonnard and Y. Meyer (2005) AtNTRB is the major mitochondrial thioredoxin reductase in *Arabidopsis thaliana*. *FEBS Letters* **579**, 337–342.
74. Q.A. Sun, F. Zappacosta, V.M. Factor, P.J. Wirth, D.L. Hatfield and V.N. Gladyshev (2001) Heterogeneity within animal thioredoxin reductases. Evidence for alternative first exon splicing. *Journal of Biological Chemistry* **276**, 3106–3114.
75. F. Missirlis, J.K. Ulschmid, M. Hirosawa-Takamori et al. (2002) Mitochondrial and cytoplasmic thioredoxin reductase variants encoded by a single *Drosophila* gene are both essential for viability. *Journal of Biological Chemistry* **277**, 11521–11526.
76. A. Agorio, C. Chalar, S. Cardozo and G. Salinas (2003) Alternative mRNAs arising from trans-splicing code for mitochondrial and cytosolic variants of *Echinococcus granulosus* thioredoxin glutathione reductase. *Journal of Biological Chemistry* **278**, 12920–12928.
77. H. Mireau, D. Lancelin and I.D. Small (1996) The same Arabidopsis gene encodes both cytosolic and mitochondrial alanyl-tRNA synthetases. *Plant Cell* **8**, 1027–1039.
78. A.M. Duchene, N. Peeters, A. Dietrich, A. Cosset, I.D. Small and H. Wintz (2001) Overlapping destinations for two dual targeted glycyl-tRNA synthetases in *Arabidopsis thaliana* and *Phaseolus vulgaris*. *Journal of Biological Chemistry* **276**, 15275–15283.
79. S. Dai, C. Schwendtmayer, K. Johansson, S. Ramaswamy, P. Schurmann and H. Eklund (2000) How does light regulate chloroplast enzymes? Structure-function studies of the ferredoxin/thioredoxin system. *Quarterly Reviews of Biophysics* **33**, 67–108.
80. W. Manieri, L. Franchini, L. Raeber, S. Dai, A.L. Stritt-Etter and P. Schurmann (2003) N-terminal truncation of the variable subunit stabilizes spinach ferredoxin:thioredoxin reductase. *FEBS Letters* **549**, 167–170.
81. A.J. Serrato, J.M. Perez-Ruiz, M.C. Spinola and F.J. Cejudo (2004) A novel NADPH thioredoxin reductase, localized in the chloroplast, which deficiency causes hypersensitivity to abiotic stress in *Arabidopsis thaliana*. *Journal of Biological Chemistry* **279**, 43821–43827.
82. B. Wieles, J. van Noort, J.W. Drijfhout, R. Offringa, A. Holmgren and T.H. Ottenhoff (1995) Purification and functional analysis of the *Mycobacterium leprae* thioredoxin/thioredoxin reductase hybrid protein. *Journal of Biological Chemistry* **270**, 25604–25606.
83. K. Motohashi, A. Kondoh, M.T. Stumpp and T. Hisabori (2001) Comprehensive survey of proteins targeted by chloroplast thioredoxin. *Proceedings of the National Academy of Sciences of the United States of America* **98**, 11224–11229.
84. H. Yano, J.H. Wong, Y.M. Lee, M.J. Cho and B.B. Buchanan (2001) A strategy for the identification of proteins targeted by thioredoxin. *Proceedings of the National Academy of Sciences of the United States of America* **98**, 4794–4799.
85. Y. Balmer, A. Koller, G. del Val, W. Manieri, P. Schurmann and B.B. Buchanan (2003) Proteomics gives insight into the regulatory function of chloroplast thioredoxins. *Proceedings of the National Academy of Sciences of the United States of America* **100**, 370–375.
86. Y. Balmer, W.H. Vensel, C.K. Tanaka et al. (2004) Thioredoxin links redox to the regulation of fundamental processes of plant mitochondria. *Proceedings of the National Academy of Sciences of the United States of America* **101**, 2642–2647.
87. K. Maeda, C. Finnie and B. Svensson (2004) Cy5 maleimide labelling for sensitive detection of free thiols in native protein extracts: identification of seed proteins targeted by barley thioredoxin h isoforms. *Biochemical Journal* **378**(2), 497–507.
88. C. Marx, J.H. Wong and B.B. Buchanan (2003) Thioredoxin and germinating barley: targets and protein redox changes. *Planta* **216**, 454–460.
89. J.H. Wong, Y. Balmer, N. Cai et al. (2003) Unraveling thioredoxin-linked metabolic processes of cereal starchy endosperm using proteomics. *FEBS Letters* **547**, 151–156.
90. J.H. Wong, N. Cai, Y. Balmer et al. (2004) Thioredoxin targets of developing wheat seeds identified by complementary proteomic approaches. *Phytochemistry* **65**, 1629–1640.

91. K. Maeda, C. Finnie and B. Svensson (2005) Identification of thioredoxin h-reducible disulphides in proteomes by differential labelling of cysteines: insight into recognition and regulation of proteins in barley seeds by thioredoxin h. *Proteomics* **5**, 1634–1644.
92. M. Lindahl and F.J. Florencio (2003) Thioredoxin-linked processes in cyanobacteria are as numerous as in chloroplasts, but targets are different. *Proceedings of the National Academy of Sciences of the United States of America* **100**, 16107–16112.
93. S.D. Lemaire, B. Guillon, P. Le Marechal, E. Keryer, M. Miginiac-Maslow and P. Decottignies (2004) New thioredoxin targets in the unicellular photosynthetic eukaryote *Chlamydomonas reinhardtii*. *Proceedings of the National Academy of Sciences of the United States of America* **101**, 7475–7480.
94. C. Marchand, P. Le Marechal, Y. Meyer, M. Miginiac-Maslow, E. Issakidis-Bourguet and P. Decottignies (2004) New targets of Arabidopsis thioredoxins revealed by proteomic analysis. *Proteomics* **4**, 2696–2706.
95. B.B. Buchanan (1981) Role of light in the regulation of chloroplast enzymes. *Annual Review of Plant Physiology* **31**, 341–374.
96. R.A. Wolosiuk, M.A. Ballicora and K. Hagelin (1993) The reductive pentose phosphate cycle for photosynthetic CO_2 assimilation: enzyme modulation. *FASEB Journal* **7**, 622–637.
97. J.P. Jacquot, J.M. Lancelin and Y. Meyer (1997) Thioredoxins: structure and function in plant cells. *New Phytologist* **136**, 543–570.
98. V. Collin, P. Lamkemeyer, M. Miginiac-Maslow et al. (2004) Characterization of plastidial thioredoxins from Arabidopsis belonging to the new y-type. *Plant Physiology* **136**, 4088–4095.
99. E. Baroja-Fernandez, F.J. Munoz, A. Zandueta-Criado et al. (2004) Most of ADP x glucose linked to starch biosynthesis occurs outside the chloroplast in source leaves. *Proceedings of the National Academy of Sciences of the United States of America* **101**, 13080–13085.
100. Y. Fu, M.A. Ballicora, J.F. Leykam and J. Preiss (1998) Mechanism of reductive activation of potato tuber ADP-glucose pyrophosphorylase. *Journal of Biological Chemistry* **273**, 25045–25052.
101. M.A. Ballicora, J.B. Frueauf, Y. Fu, P. Schurmann and J. Preiss (2000) Activation of the potato tuber ADP-glucose pyrophosphorylase by thioredoxin. *Journal of Biological Chemistry* **275**, 1315–1320.
102. A. Tiessen, J.H. Hendriks, M. Stitt et al. (2002) Starch synthesis in potato tubers is regulated by post-translational redox modification of ADP-glucose pyrophosphorylase: a novel regulatory mechanism linking starch synthesis to the sucrose supply. *Plant Cell* **14**, 2191–2213.
103. J.H. Hendriks, A. Kolbe, Y. Gibon, M. Stitt and P. Geigenberger (2003) ADP-glucose pyrophosphorylase is activated by posttranslational redox-modification in response to light and to sugars in leaves of Arabidopsis and other plant species. *Plant Physiology* **133**, 838–849.
104. A. Tiessen, K. Prescha, A. Branscheid et al. (2003) Evidence that SNF1-related kinase and hexokinase are involved in separate sugar-signalling pathways modulating post-translational redox activation of ADP-glucose pyrophosphorylase in potato tubers. *The Plant Journal* **35**, 490–500.
105. T.S. Yu, H. Kofler, R.E. Hausler et al. (2001) The Arabidopsis sex1 mutant is defective in the R1 protein, a general regulator of starch degradation in plants, and not in the chloroplast hexose transporter. *Plant Cell* **13**, 1907–1918.
106. R. Mikkelsen, K.E. Mutenda, A. Mant, P. Schurmann and A. Blennow (2005) Alpha-glucan, water dikinase (GWD): a plastidic enzyme with redox-regulated and coordinated catalytic activity and binding affinity. *Proceedings of the National Academy of Sciences of the United States of America* **102**, 1785–1790.
107. R. Entus, M. Poling and K.M. Herrmann (2002) Redox regulation of Arabidopsis 3-deoxy-D-arabino-heptulosonate 7-phosphate synthase. *Plant Physiology* **129**, 1866–1871.
108. S. Mora-Garcia, R. Rodriguez-Suarez and R.A. Wolosiuk (1998) Role of electrostatic interactions on the affinity of thioredoxin for target proteins. Recognition of chloroplast fructose-1,6-bisphosphatase by mutant *Escherichia coli* thioredoxins. *Journal of Biological Chemistry* **273**, 16273–16280.
109. Y. Balmer, A. Koller, G. del Val, P. Schürmann and B.B. Buchanan (2004) Proteomics uncovers proteins interacting electrostatically with thioredoxin in chloroplasts. *Photosynthesis Research* **79**, 275–280.

110. P.G. Romano, P. Horton and J.E. Gray (2004) The Arabidopsis cyclophilin gene family. *Plant Physiology* **134**, 1268–1282.
111. Z. He, L. Li and S. Luan (2004) Immunophilins and parvulins. Superfamily of peptidyl prolyl isomerases in Arabidopsis. *Plant Physiology* **134**, 1248–1267.
112. K. Motohashi, F. Koyama, Y. Nakanishi, H. Ueoka-Nakanishi and T. Hisabori (2003) Chloroplast cyclophilin is a target protein of thioredoxin. Thiol modulation of the peptidyl-prolyl *cis-trans* isomerase activity. *Journal of Biological Chemistry* **278**, 31848–31852.
113. G. Gopalan, Z. He, Y. Balmer *et al.* (2004) Structural analysis uncovers a role for redox in regulating FKBP13, an immunophilin of the chloroplast thylakoid lumen. *Proceedings of the National Academy of Sciences of the United States of America* **101**, 13945–13950.
114. R. Gupta, R.M. Mould, Z. He and S. Luan (2002) A chloroplast FKBP interacts with and affects the accumulation of Rieske subunit of cytochrome bf complex. *Proceedings of the National Academy of Sciences of the United States of America* **99**, 15806–15811.
115. M.L. Page, P.P. Hamel, S.T. Gabilly *et al.* (2004) A homolog of prokaryotic thiol disulfide transporter CcdA is required for the assembly of the cytochrome b6f complex in Arabidopsis chloroplasts. *Journal of Biological Chemistry* **279**, 32474–32482.
116. J. Kim and S.P. Mayfield (1997) Protein disulfide isomerase as a regulator of chloroplast translational activation. *Science* **278**, 1954–1957.
117. C.B. Yohn, A. Cohen, A. Danon and S.P. Mayfield (1998) A poly(A) binding protein functions in the chloroplast as a message-specific translation factor. *Proceedings of the National Academy of Sciences of the United States of America* **95**, 2238–2243.
118. C.B. Yohn, A. Cohen, C. Rosch, M.R. Kuchka and S.P. Mayfield (1998) Translation of the chloroplast psbA mRNA requires the nuclear-encoded poly(A)-binding protein, RB47. *Journal of Cell Biology* **142**, 435–442.
119. T. Trebitsh and A. Danon (2001) Translation of chloroplast psbA mRNA is regulated by signals initiated by both photosystems II and I. *Proceedings of the National Academy of Sciences of the United States of America* **98**, 12289–12294.
120. A. Levitan, T. Trebitsh, V. Kiss, Y. Pereg, I. Dangoor and A. Danon (2005) Dual targeting of the protein disulfide isomerase RB60 to the chloroplast and the endoplasmic reticulum. *Proceedings of the National Academy of Sciences of the United States of America* **102**, 6225–6230.
121. T. Trebitsh, A. Levitan, A. Sofer and A. Danon (2000) Translation of chloroplast psbA mRNA is modulated in the light by counteracting oxidizing and reducing activities. *Molecular and Cellular Biology* **20**, 1116–1123.
122. J. Kim and S.P. Mayfield (2002) The active site of the thioredoxin-like domain of chloroplast protein disulfide isomerase, RB60, catalyzes the redox-regulated binding of chloroplast poly(A)-binding protein, RB47, to the 5' untranslated region of psbA mRNA. *Plant and Cell Physiology* **43**, 1238–1243.
123. T. Pfannschmidt (2003) Chloroplast redox signals: how photosynthesis controls its own genes. *Trends in Plant Sciences* **8**, 33–41.
124. A. Haldrup, P.E. Jensen, C. Lunde and H.V. Scheller (2001) Balance of power: a view of the mechanism of photosynthetic state transitions. *Trends in Plant Sciences* **6**(7), 301–305.
125. J.F. Allen and J. Forsberg (2001) Molecular recognition in thylakoid structure and function. *Trends in Plant Sciences* **6**, 317–326.
126. F.A. Wollman (2001) State transitions reveal the dynamics and flexibility of the photosynthetic apparatus. *EMBO Journal* **20**, 3623–3630.
127. E. Rintamaki, M. Salonen, U.M. Suoranta, I. Carlberg, B. Andersson and E.M. Aro (1997) Phosphorylation of light-harvesting complex II and photosystem II core proteins shows different irradiance-dependent regulation *in vivo*. Application of phosphothreonine antibodies to analysis of thylakoid phosphoproteins. *Journal of Biological Chemistry* **272**, 30476–30482.
128. A.V. Vener, I. Ohad and B. Andersson (1998) Protein phosphorylation and redox sensing in chloroplast thylakoids. *Current Opinion in Plant Biology* **1**, 217–223.
129. A.V. Vener, P.J. van Kan, P.R. Rich, I.I. Ohad and B. Andersson (1997) Plastoquinol at the quinol oxidation site of reduced cytochrome bf mediates signal transduction between light and protein

phosphorylation: thylakoid protein kinase deactivation by a single-turnover flash. *Proceedings of the National Academy of Sciences of the United States of America* **94**, 1585–1590.
130. F. Zito, G. Finazzi, R. Delosme, W. Nitschke, D. Picot and F.A. Wollman (1999) The Q_o site of cytochrome b6f complexes controls the activation of the LHCII kinase. *EMBO Journal* **18**, 2961–2969.
131. H. Zer, M. Vink, N. Keren *et al.* (1999) Regulation of thylakoid protein phosphorylation at the substrate level: reversible light-induced conformational changes expose the phosphorylation site of the light-harvesting complex II. *Proceedings of the National Academy of Sciences of the United States of America* **96**, 8277–8282.
132. H. Zer, M. Vink, S. Shochat, R.G. Herrmann, B. Andersson and I. Ohad (2003) Light affects the accessibility of the thylakoid light harvesting complex II (LHCII) phosphorylation site to the membrane protein kinase(s). *Biochemistry* **42**, 728–738.
133. M. Vink, H. Zer, N. Alumot *et al.* (2004) Light-modulated exposure of the light-harvesting complex II (LHCII) to protein kinase(s) and state transition in *Chlamydomonas reinhardtii* xanthophyll mutants. *Biochemistry* **43**, 7824–7833.
134. P. Martinsuo, S. Pursiheimo, E.M. Aro and E. Rintamaki (2003) Dithiol oxidant and disulfide reductant dynamically regulate the phosphorylation of light-harvesting complex II proteins in thylakoid membranes. *Plant Physiology* **133**, 37–46.
135. N. Depege, S. Bellafiore and J.D. Rochaix (2003) Role of chloroplast protein kinase Stt7 in LHCII phosphorylation and state transition in *Chlamydomonas*. *Science* **299**, 1572–1575.
136. S. Bellafiore, F. Barneche, G. Peltier and J.D. Rochaix (2005) State transitions and light adaptation require chloroplast thylakoid protein kinase STN7. *Nature* **433**, 892–895.
137. A. Hazra and M. DasGupta (2003) Phosphorylation-dephosphorylation of light-harvesting complex II as a response to variation in irradiance is thiol sensitive and thylakoid sufficient: modulation of the sensitivity of the phenomenon by a peripheral component. *Biochemistry* **42**, 14868–14876.
138. J. Bodenstein-Lang, A. Buch and H. Follmann (1989) Animal and plant mitochondria contain specific thioredoxins. *FEBS Letters* **258**, 22–26.
139. A.L. Umbach, J.T. Wiskich and J.N. Siedow (1994) Regulation of alternative oxidase kinetics by pyruvate and intermolecular disulfide bond redox status in soybean seedling mitochondria. *FEBS Letters* **348**, 181–184.
140. D. Pastore, D. Trono, M.N. Laus, N. Di Fonzo and S. Passarella (2001) Alternative oxidase in durum wheat mitochondria. Activation by pyruvate, hydroxypyruvate and glyoxylate and physiological role. *Plant and Cell Physiology*, **42**, 1373–1382.
141. L.H. Butterfield, A. Merino, S.H. Golub and H. Shau (1999) From cytoprotection to tumor suppression: the multifactorial role of peroxiredoxins. *Antioxidants and Redox Signalling* **1**, 385–402.
142. S.G. Rhee, S.W. Kang, T.S. Chang, W. Jeong and K. Kim (2001) Peroxiredoxin, a novel family of peroxidases. *IUBMB Life* **52**, 35–41.
143. T. Finkel (2003) Oxidant signals and oxidative stress. *Current Opinion in Cell Biology* **15**, 247–254.
144. Z.A. Wood, E. Schröder, J.R. Harris and L.B. Poole (2003) Structure, mechanism and regulation of peroxiredoxins. *Trends in Biochemical Sciences* **28**, 32–40.
145. J. Konig, K. Lotte, R. Plessow, A. Brockhinke, M. Baier and K.J. Dietz (2003) Reaction mechanism of plant 2-Cys peroxiredoxin. Role of the C terminus and the quaternary structure. *Journal of Biological Chemistry* **278**, 24409–24420.
146. N. Rouhier, E. Gelhaye, P.E. Sautiere *et al.* (2001) Isolation and characterization of a new peroxiredoxin from poplar sieve tubes that uses either glutaredoxin or thioredoxin as a proton donor. *Plant Physiology* **127**, 1299–1309.
147. J. Konig, M. Baier, F. Horling *et al.* (2002) The plant-specific function of 2-Cys peroxiredoxin-mediated detoxification of peroxides in the redox-hierarchy of photosynthetic electron flux. *Proceedings of the National Academy of Sciences of the United States of America* **99**, 5738–5743.
148. K.J. Dietz, F. Horling, J. Konig and M. Baier (2002) The function of the chloroplast 2-cysteine peroxiredoxin in peroxide detoxification and its regulation. *Journal of Experimental Botany* **53**, 1321–1329.

149. A. Sakamoto, S. Tsukamoto, H. Yamamoto et al. (2003) Functional complementation in yeast reveals a protective role of chloroplast 2-Cys peroxiredoxin against reactive nitrogen species. *The Plant Journal* **33**, 841–851.
150. E. Issakidis-Bourguet, N. Mouaheb, Y. Meyer and M. Miginiac-Maslow (2001) Heterologous complementation of yeast reveals a new putative function for chloroplast m-type thioredoxin. *The Plant Journal* **25**, 127–135.
151. N. Mouaheb, D. Thomas, L. Verdoucq, P. Monfort and Y. Meyer (1998) *In vivo* functional discrimination between plant thioredoxins by heterologous expression in the yeast *Saccharomyces cerevisiae*. *Proceedings of the National Academy of Sciences of the United States of America* **95**, 3312–3317.
152. W. Kong, S. Shiota, Y. Shi, H. Nakayama and K. Nakayama (2000) A novel peroxiredoxin of the plant Sedum lineare is a homologue of *Escherichia coli* bacterioferritin co-migratory protein (Bcp). *Biochemical Journal* **351**(1), 107–114.
153. N. Rouhier, E. Gelhaye, J.M. Gualberto et al. (2004) Poplar peroxiredoxin Q. A thioredoxin-linked chloroplast antioxidant functional in pathogen defense. *Plant Physiology* **134**, 1027–1038.
154. H.M. Do, J.K. Hong, H.W. Jung, S.H. Kim, J.H. Ham and B.K. Hwang (2003) Expression of peroxidase-like genes, H2O2 production, and peroxidase activity during the hypersensitive response to *Xanthomonas campestris* pv. vesicatoria in *Capsicum annuum*. *Molecular Plant-Microbe Interactions* **16**, 196–205.
155. M. Broin, S. Cuine, G. Peltier and P. Rey (2000) Involvement of CDSP 32, a drought-induced thioredoxin, in the response to oxidative stress in potato plants. *FEBS Letters* **467**, 245–248.
156. M. Broin, S. Cuine, F. Eymery and P. Rey (2002) The plastidic 2-cysteine peroxiredoxin is a target for a thioredoxin involved in the protection of the photosynthetic apparatus against oxidative damage. *Plant Cell* **14**, 1417–1432.
157. M. Broin and P. Rey (2003) Potato plants lacking the CDSP32 plastidic thioredoxin exhibit overoxidation of the BAS1 2-cysteine peroxiredoxin and increased lipid peroxidation in thylakoids under photooxidative stress. *Plant Physiology* **132**, 1335–1343.
158. P. Rey, S. Cuine, F. Eymery et al. (2005) Analysis of the proteins targeted by CDSP32, a plastidic thioredoxin participating in oxidative stress responses. *The Plant Journal* **41**, 31–42.
159. K. Kobrehel, J.H. Wong, A. Balogh, F. Kiss, B.C. Yee and B.B. Buchanan (1992) Specific reduction of wheat storage proteins by thioredoxin h. *Plant Physiology* **99**, 919–924.
160. R.M. Lozano, J.H. Wong, B.C. Yee, A. Peters, K. Kobrehel and B.B. Buchanan (1996) New evidence for a role for thioredoxin h in germination and seedling development. *Planta* **200**, 100–106.
161. M.J. Cho, J.H. Wong, C. Marx, W. Jiang, P.G. Lemaux and B.B. Buchanan (1999) Overexpression of thioredoxin h leads to enhanced activity of starch debranching enzyme (pullulanase) in barley grain. *Proceedings of the National Academy of Sciences of the United States of America* **96**, 14641–14646.
162. J.H. Wong, Y.B. Kim, P.H. Ren et al. (2002) Transgenic barley grain overexpressing thioredoxin shows evidence that the starchy endosperm communicates with the embryo and the aleurone. *Proceedings of the National Academy of Sciences of the United States of America* **99**, 16325–16330.
163. D. Yamazaki, K. Motohashi, T. Kasama, Y. Hara and T. Hisabori (2004) Target proteins of the cytosolic thioredoxins in *Arabidopsis thaliana*. *Plant and Cell Physiology* **45**, 18–27.
164. J.B. Nasrallah (2000) Cell-cell signalling in the self-incompatibility response. *Current Opinion in Plant Biology* **3**, 368–373.
165. D.R. Goring and J.C. Walker (2004) Plant sciences. Self-rejection – a new kinase connection. *Science* **303**, 1474–1475.
166. K. Murase, H. Shiba, M. Iwano et al. (2004) A membrane-anchored protein kinase involved in *Brassica* self-incompatibility signalling. *Science* **303**, 1516–1519.
167. M.S. Bower, D.D. Matias, E. Fernandes-Carvalho et al. (1996) Two members of the thioredoxin-h family interact with the kinase domain of a *Brassica* S locus receptor kinase. *Plant Cell* **8**, 1641–1650.

168. M. Mazzurco, W. Sulaman, H. Elina, J.M. Cock and D.R. Goring (2001) Further analysis of the interactions between the *Brassica* S receptor kinase and three interacting proteins (ARC1, THL1 and THL2) in the yeast two-hybrid system. *Plant Molecular Biology* **45**, 365–376.
169. D. Cabrillac, J.M. Cock, C. Dumas and T. Gaude (2001) The S-locus receptor kinase is inhibited by thioredoxins and activated by pollen coat proteins. *Nature* **410**, 220–223.
170. S. Rivas, A. Rougon-Cardoso, M. Smoker, L. Schauser, H. Yoshioka and J.D. Jones (2004) CITRX thioredoxin interacts with the tomato Cf-9 resistance protein and negatively regulates defence. *EMBO Journal* **23**, 2156–2165.
171. J.J. Rudd and V.E. Franklin-Tong (2003) Signals and targets of the self-incompatibility response in pollen of *Papaver rhoeas*. *Journal of Experimental Botany* **54**, 141–148.
172. S.D. Lemaire (2004) The glutaredoxin family in oxygenic photosynthetic organisms. *Photosynthesis Research* **79**, 305–318.
173. N. Rouhier, E. Gelhaye and J.P. Jacquot (2004) Plant glutaredoxins: still mysterious reducing systems. *Cellular and Molecular Life Sciences* **61**, 1266–1277.
174. N. Isakov, S. Witte and A. Altman (2000) PICOT-HD: a highly conserved protein domain that is often associated with thioredoxin and glutaredoxin modules. *Trends in Biochemical Sciences* **25**, 537–539.
175. K. Nordstrand, F. slund, A. Holmgren, G. Otting and K.D. Berndt (1999) NMR structure of *Escherichia coli* glutaredoxin 3-glutathione mixed disulfide complex: implications for the enzymatic mechanism. *Journal of Molecular Biology* **286**, 541–552.
176. N. Rouhier, E. Gelhaye and J.P. Jacquot (2002) Exploring the active site of plant. glutaredoxin by site-directed mutagenesis. *FEBS Letters* **511**, 145–149.
177. N. Rouhier, E. Gelhaye and J.P. Jacquot (2002) Glutaredoxin-dependent peroxiredoxin from poplar: protein-protein interaction and catalytic mechanism. *Journal of Biological Chemistry* **277**, 13609–13614.
178. J. Tamarit, G. Belli, E. Cabiscol, E. Herrero and J. Ros (2003) Biochemical characterization of yeast mitochondrial Grx5 monothiol glutaredoxin. *Journal of Biological Chemistry* **278**, 25745–25751.
179. S. Morell, H. Follmann and I. Haberlein (1995) Identification and localization of the first glutaredoxin in leaves of a higher plant. *FEBS Letters* **369**, 149–152.
180. K. Minakuchi, T. Yabushita, T. Masumura, K. Ichihara and K. Tanaka (1994) Cloning and sequence analysis of a cDNA encoding rice glutaredoxin. *FEBS Letters* **337**, 157–160.
181. J. Szederkenyi, E. Komor and C. Schobert (1997) Cloning of the cDNA for glutaredoxin, an abundant sieve-tube exudate protein from *Ricinus communis* L. and characterisation of the glutathione-dependent thiol-reduction system in sieve tubes. *Planta* **202**, 349–356.
182. N.H. Cheng and K.D. Hirschi (2003) Cloning and characterization of CXIP1, a novel PICOT domain-containing Arabidopsis protein that associates with CAX1. *Journal of Biological Chemistry* **278**, 6503–6509.
183. T. Shigaki, N.H. Cheng, J.K. Pittman and K. Hirschi (2001) Structural determinants of Ca2+ transport in the Arabidopsis H+/Ca2+ antiporter CAX1. *Journal of Biological Chemistry* **276**, 43152–43159.
184. D.M. Ferrari and H.D. Soling (1999) The protein disulphide-isomerase family: unravelling a string of folds. *Biochemical Journal* **339**(1), 1–10.
185. P.M. Clissold and R. Bicknell (2003) The thioredoxin-like fold: hidden domains in protein disulfide isomerases and other chaperone proteins. *Bioessays* **25**, 603–611.
186. N.L. Houston, C. Fan, J.Q. Xiang, J.M. Schulze, R. Jung and R.S. Boston (2005) Phylogenetic analyses identify 10 classes of the protein disulfide isomerase family in plants, including single-domain protein disulfide isomerase-related proteins. *Plant Physiology* **137**, 762–778.
187. A.G. McArthur, L.A. Knodler, J.D. Silberman, B.J. Davids, F.D. Gillin and M.L. Sogin (2001) The evolutionary origins of eukaryotic protein disulfide isomerase domains: new evidence from the amitochondriate protist *Giardia lamblia*. *Molecular Biology and Evolution* **18**, 1455–1463.
188. S. Kanai, H. Toh, T. Hayano and M. Kikuchi (1998) Molecular evolution of the domain structures of protein disulfide isomerases. *Journal of Molecular Evolution* **47**, 200–210.

189. A.R. Frand, J.W. Cuozzo and C.A. Kaiser (2000) Pathways for protein disulphide bond formation. *Trends in Cell Biology* **10**, 203–210.
190. R.R. Singh and A.G. Appu Rao (2002) Reductive unfolding and oxidative refolding of a Bowman-Birk inhibitor from horsegram seeds (*Dolichos biflorus*): evidence for 'hyperreactive' disulfide bonds and rate-limiting nature of disulfide isomerization in folding. *Biochimica et Biophysica Acta* **1597**, 280–291.
191. J.L. Song and C.C. Wang (1995) Chaperone-like activity of protein disulfide-isomerase in the refolding of rhodanese. *European Journal of Biochemistry* **231**, 312–316.
192. T. Pihlajaniemi, T. Helaakoski, K. Tasanen et al. (1987) Molecular cloning of the beta-subunit of human prolyl 4-hydroxylase. This subunit and protein disulphide isomerase are products of the same gene. *EMBO Journal* **6**, 643–649.
193. M.P. Rigobello, A. Donella-Deana, L. Cesaro and A. Bindoli (2001) Distribution of protein disulphide isomerase in rat liver mitochondria. *Biochemical Journal* **356**(2), 567–570.
194. M. Ciaffi, A.R. Paolacci, L. Dominici, O.A. Tanzarella and E. Porceddu (2001) Molecular characterization of gene sequences coding for protein disulfide isomerase (PDI) in durum wheat (*Triticum turgidum* ssp. durum). *Gene* **265**, 147–156.
195. S. Sugaya, A. Ohmiya, M. Kikuchi and T. Hayashi (2000) Isolation and characterization of a 60 kDa 2,4-D-binding protein from the shoot apices of peach trees (*Prunus persica* L.); it is a homologue of protein disulfide isomerase. *Plant and Cell Physiology* **41**, 503–508.
196. E. Meiri, A. Levitan, F. Guo et al. (2002) Characterization of three PDI-like genes in *Physcomitrella patens* and construction of knock-out mutants. *Molecular Genetics and Genomics* **267**, 231–240.
197. Y. Takemoto, S.J. Coughlan, T.W. Okita, H. Satoh, M. Ogawa and T. Kumamaru (2002) The rice mutant *esp2* greatly accumulates the glutelin precursor and deletes the protein disulfide isomerase. *Plant Physiology* **128**, 1212–1222.
198. S. Ray, J.M. Anderson, F.I. Urmeev and S.B. Goodwin (2003) Rapid induction of a protein disulfide isomerase and defense-related genes in wheat in response to the hemibiotrophic fungal pathogen *Mycosphaerella graminicola*. *Plant Molecular Biology* **53**, 701–714.
199. W.J. Lees and G.M. Whitesides (1993) Equilibrium constants for thiol-disulfide interchange reactions: a coherent view, corrected set. *Journal of Organic Chemistry* **58**, 642–647.
200. P.T. Chivers, K.E. Prehoda and R.T. Raines (1997) The CXXC motif: a rheostat in the active site. *Biochemistry* **36**, 4061–4066.
201. M. Hirasawa, P. Schurmann, J.P. Jacquot et al. (1999) Oxidation-reduction properties of chloroplast thioredoxins, ferredoxin:thioredoxin reductase, and thioredoxin f-regulated enzymes. *Biochemistry* **38**, 5200–5205.
202. M. Chiadmi, A. Navaza, M. Miginiac-Maslow, J.P. Jacquot and J. Cherfils (1999) Redox signalling in the chloroplast: structure of oxidized pea fructose-1,6-bisphosphate phosphatase. *EMBO Journal* **18**, 6809–6815.
203. R.P. Dunford, M.A. Catley, C.A. Raines, J.C. Lloyd and T.A. Dyer (1998) Purification of active chloroplast sedoheptulose-1,7-bisphosphatase expressed in *Escherichia coli*. *Protein Expression and Purification* **14**, 139–145.
204. M.A. Porter, C.D. Stringer and F.C. Hartman (1988) Characterization of the regulatory thioredoxin site of phosphoribulokinase. *Journal of Biological Chemistry* **263**, 123–129.
205. H.K. Brandes, F.C. Hartman, T.Y. Lu and F.W. Larimer (1996) Efficient expression of the gene for spinach phosphoribulokinase in *Pichia pastoris* and utilization of the recombinant enzyme to explore the role of regulatory cysteinyl residues by site-directed mutagenesis. *Journal of Biological Chemistry* **271**, 6490–6496.
206. N. Zhang and A.R. Portis, Jr. (1999) Mechanism of light regulation of Rubisco: a specific role for the larger Rubisco activase isoform involving reductive activation by thioredoxin-f. *Proceedings of the National Academy of Sciences of the United States of America* **96**, 9438–9443.
207. H. Brinkmann, R. Cerff, M. Salomon and J. Soll (1989) Cloning and sequence analysis of cDNAs encoding the cytosolic precursors of subunits GapA and GapB of chloroplast glyceraldehyde-3-phosphate dehydrogenase from pea and spinach. *Plant Molecular Biology* **13**, 81–94.

208. O. Schwarz, P. Schurmann and H. Strotmann (1997) Kinetics and thioredoxin specificity of thiol modulation of the chloroplast H$^+$-ATPase. *Journal of Biological Chemistry* **272**, 16924–16927.
209. K. Johansson, S. Ramaswamy, M. Saarinen *et al.* (1999) Structural basis for light activation of a chloroplast enzyme: the structure of sorghum NADP-malate dehydrogenase in its oxidized form. *Biochemistry* **38**, 4319–4326.
210. M. Hirasawa, E. Ruelland, I. Schepens, E. Issakidis-Bourguet, M. Miginiac-Maslow and D.B. Knaff (2000) Oxidation-reduction properties of the regulatory disulfides of sorghum chloroplast nicotinamide adenine dinucleotide phosphate-malate dehydrogenase. *Biochemistry* **39**, 3344–3450.
211. I. Wenderoth, R. Scheibe and A. von Schaewen (1997) Identification of the cysteine residues involved in redox modification of plant plastidic glucose-6-phosphate dehydrogenase. *Journal of Biological Chemistry* **272**, 26985–26990.
212. Y. Sasaki, A. Kozaki and M. Hatano (1997) Link between light and fatty acid synthesis: thioredoxin-linked reductive activation of plastidic acetyl-CoA carboxylase. *Proceedings of the National Academy of Sciences of the United States of America* **94**, 11096–11101.
213. A. Kozaki, K. Mayumi and Y. Sasaki (2001) Thiol-disulfide exchange between nuclear-encoded and chloroplast-encoded subunits of pea acetyl-CoA carboxylase. *Journal of Biological Chemistry* **276**, 39919–39925.
214. A. Kozaki, K. Kamada, Y. Nagano, H. Iguchi and Y. Sasaki (2000) Recombinant carboxyltransferase responsive to redox of pea plastidic acetyl-CoA carboxylase. *Journal of Biological Chemistry* **275**, 10702–10708.
215. F. Horling, P. Lamkemeyer, J. Konig *et al.* (2003) Divergent light-, ascorbate-, and oxidative stress-dependent regulation of expression of the peroxiredoxin gene family in Arabidopsis. *Plant Physiology* **131**, 317–325.
216. P. Norgaard, V. Westphal, C. Tachibana, L. Alsoe, B. Holst and J.R. Winther (2001) Functional differences in yeast protein disulfide isomerases. *Journal of Cell Biology* **152**, 553–562.
217. B. Wilkinson and H.F. Gilbert (2004) Protein disulfide isomerase. *Biochimica et Biophysica Acta* **1699**, 35–44.
218. B. Biteau, J. Labarre and M.B. Toledano (2003) ATP-dependent reduction of cysteine-sulphinic acid by *S. cerevisiae* sulphiredoxin. *Nature* **425**, 980–984.

7 Control of carbon fixation in chloroplasts

Brigitte Gontero, Luisana Avilan and Sandrine Lebreton

7.1 Introduction

The complex process of photosynthesis can be summarized with the following equation:

$$6CO_2 + 12H_2O + Light \rightarrow C_6H_{12}O_6 + 6O_2 + 6H_2O$$

In this process, water is split into hydrogen and oxygen, incorporating the electrons of hydrogen into the energy-rich bonds of sugar molecules. Photosynthesis is thus an endergonic redox process in which water is oxidized and carbon dioxide is reduced. The process comprises two, interrelated stages, the light reactions and the Calvin cycle. The light reactions produce ATP by photophosphorylation and split water, evolving oxygen and forming NADPH by transferring electrons from water to $NADP^+$. This occurs in the thylakoid membranes of chloroplasts. The ATP and NADPH generated by the light reactions of photosynthesis are then consumed in the chloroplasts to drive fixation of carbon dioxide into sugars (sucrose and starch) by a series of reactions first elucidated by Calvin and his colleagues in the 1950s. This series of reactions is hence often known as the 'Calvin cycle'. It occurs in the chloroplast stroma and is the only process of net photosynthetic carbon fixation in plants. In C_4 or CAM (Crassulacean acid metabolism) plants, the carbon dioxide is first fixed by other reactions but is then released and again incorporated into the Calvin cycle (see Chapter 8). As a consequence, this pathway is universal to all photosynthetic life forms.

The Calvin cycle is usually divided into three stages:

1. CO_2 fixation or carboxylation
2. Reduction of phosphoglycerate or formation of carbohydrate at the expense of ATP and NADPH
3. Regeneration of the CO_2 acceptor, ribulose-1,5-bisphosphate (RuBP), at the expense of ATP

The first step involves the enzyme, ribulose-1,5-bisphosphate carboxylase-oxygenase (RuBisCO, EC 4.1.1.39), the second step involves the enzymes phosphoglycerate kinase (EC 2.7.2.3) and NADP-dependent glyceraldehyde-3-phosphate dehydrogenase (GAPDH) (EC 1.2.1.13) and the third step involves a series of enzymes (from triose phosphate isomerase (EC 5.3.1.1) and fructose-1,6-bisphosphatase (FBPase) (EC 3.1.3.11) to phosphoribulokinase (PRK) (EC 2.7.1.19)) that together convert triose phosphate into RuBP.

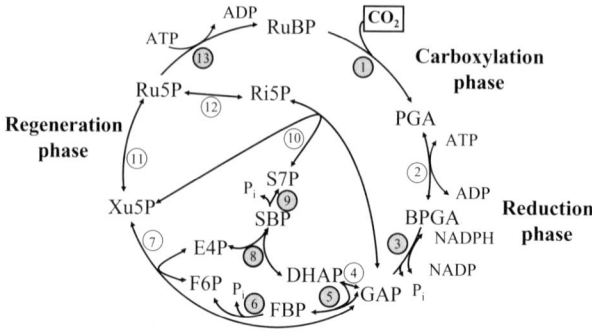

Figure 7.1 The Calvin cycle. 1: ribulose-1,5-bisphosphate carboxylase-oxygenase (RuBisCO); 2: phosphoglycerate kinase; 3: glyceraldehyde-3-phosphate dehydrogenase (GAPDH); 4: triose phosphate isomerase; 5: aldolase, 6: fructose-1,6-bisphosphatase (FBPase); 7: transketolase; 8: aldolase; 9: sedoheptulose-1,7-bisphosphatase (SBPase); 10: transketolase; 11: xylulose-5-phosphate epimerase; 12: phosphoribose isomerase (PRI); 13: phosphoribulokinase (PRK). The enzymes that have been shown to be sensitive to light regulation either at the gene or at the protein level are indicated by a number surrounded by a shaded circle. Abbreviations: PGA, 3-phosphoglycerate; BPGA, 1,3-bisphosphoglycerate; DHAP, dihydroxyacetone phosphate; E4P, erythrose-4-phosphate; F6P, fructose-6-phosphate; FBP, fructose-1,6-bisphosphate; GAP, glyceraldehyde-3-phosphate; Ri5P, ribose-5-phosphate; Ru5P, ribulose-5-phosphate; RuBP, ribulose-1,5-bisphosphate; S7P, sedoheptulose-7-phosphate; SBP, sedoheptulose-1,7-bisphosphate; Xu5P, xylulose-5-phosphate.

The sequence of the 13 reactions catalyzed by 11 different enzymes of this cycle is presented in Figure 7.1. This set of enzymes is also often referred to as the reductive pentose-phosphate pathway in contrast to the oxidative pentose-phosphate pathway (OPP). The stages in the 'oxidative' pathway are the conversion of glucose-6-phosphate into 6-phosphogluconate, followed by oxidation of the latter to ribose-5-phosphate.

There is an emerging consensus that the intact oxidative pentose-phosphate pathway occurs in plastids of spinach, pea, tobacco and sweet pepper fruit [1–3]. As the Calvin cycle and the oxidative pentose-phosphate pathway occur in the same compartment, a temporal separation is absolutely required, and so these pathways are influenced in an opposite fashion by the ferredoxin–thioredoxin system. This control prevents the Calvin cycle from being active in the dark, when it might function in a futile cycle either with the oxidative pentose-phosphate pathway or with glycolysis, wasting ATP and NADPH. The net effect of the ferredoxin–thioredoxin system is such that some of the Calvin cycle enzymes are up-regulated in the light and down-regulated in the dark (e.g. fructose-1,6-bisphosphatase, phosphoribulokinase) whereas some of the oxidative pentose-phosphate pathway and glycolysis enzymes are up-regulated in the dark and down-regulated in the presence of light (e.g. glucose-6-phosphate dehydrogenase).

Recently, the use of the technique of elementary mode analysis [4] has yielded new and useful insights into the Calvin cycle and the oxidative pentose-phosphate pathway [5]. Elementary mode analysis is a computer modeling technique concerned with identifying certain subsets of reactions, the so-called elementary modes, within a system. In the steady state, a mode has no net consumption or production of any internal substrate. An elementary mode can thus be thought of as a minimal independent pathway within a network of reactions. The term elementary refers to a mode that cannot be subdivided into further modes. By applying this technique to the Calvin cycle and the oxidative pentose-phosphate pathway, it has been shown that they are not separate elements but rather play complementary roles and should thus be possibly regarded as overlapping sets of components whose operation is governed by the thioredoxin system in response to ambient light intensity [5].

It was concluded from this analysis that it is possible for starch degradation to enhance photosynthetic triose phosphate export in the light. Indeed, the elementary modes degrading starch do not utilize fructose-1,6-bisphosphate (FBP) aldolase (EC 4.1.2.13) or FBPase. The flux that these reactions would otherwise have carried is supplied via the degradation of transitory starch and then becomes available for export via the triose phosphate/phosphate translocator that is able to exchange 3-phosphoglycerate, dihydroxyacetone phosphate or glyceraldehyde-3-phosphate for inorganic phosphate [6]. The same analysis also led to the conclusion that the reactions of the Calvin cycle alone are not capable of sustaining a triose phosphate output flux in the dark using transitory starch as the starting point [5].

These results clearly indicate that although the reactions involved in the Calvin cycle and the oxidative pentose-phosphate pathway were elucidated long ago (in the 1950s), their control and their functions still raise unsolved questions.

Long-known processes that control the enzymes of the Calvin cycle include light-dependent ion movements. As light-driven electron transport leads to the generation of a pH gradient across the thylakoid membrane, the pH of the stroma increases from around pH 7 (dark) to pH 8 (light). Regulated enzymes of the Calvin cycle work best at pH values close to 8 and are relatively inactive at pH 7. Moreover, as a result of the pH change, other ions move across the thylakoid membrane to compensate for charge differences. In particular, magnesium ions move out as protons move into the lumen of the thylakoid membranes.

The Calvin cycle was considered for many years to be controlled by light-dependent activation of at least five enzymes often referred to as 'key' enzymes. These are RuBisCO, NADP-glyceraldehyde-3-phosphate dehydrogenase, fructose-1,6-bisphosphatase, sedoheptulose-1,7-bisphosphatase (SBPase) and phosphoribulokinase. Of these five enzymes, all except RuBisCO are controlled by the ferredoxin–thioredoxin system and all are activated by light-induced increase in stromal pH.

This chapter will thus be mainly devoted to recent findings regarding the so-called key enzymes and to their control. However, transgenic plant analysis and large-scale gene expression data have recently revealed that other enzymes are also important and that their regulation at the gene level is sensitive to biological perturbations (e.g. alterations in light signals, mutations in light-signalling components, etc.) [7]. These data are most interesting as they overturn the preconception that the highly regulated ('key') enzymes exert the strongest control over carbon fixation. These findings will also be discussed in the present chapter.

Last but not least, there is no doubt now about the role of protein–protein interactions in many cellular processes. In plants for instance, it is well known that RuBisCO is regulated by a separate protein called RuBisCO activase. This protein binds to inactive RuBisCO and upon ATP hydrolysis, it changes the conformation of the enzyme, thus generating a highly active form of RuBisCO. Activase thus works along with pH and Mg^{2+} to adapt the function of RuBisCO to demand. This type of protein–protein interaction is not unique and many such interactions have been described in the Calvin cycle.

Protein–protein interactions allow communication at the level of a metabolic pathway. These communications rely on specific interactions in response to physical or chemical signals and will thus be reviewed here.

7.2 Ribulose-1,5-bisphosphate carboxylase-oxygenase

RuBisCO catalyzes CO_2 fixation and alternatively, an oxygenase reaction [8–10]. These two reactions can be summarized as follows:

ribulose-1,5-bisphosphate + CO_2 → 2(3-phosphoglycerate)

ribulose-1,5-bisphosphate + O_2 → (3-phosphoglycerate) + 2-phosphoglycolate

RuBisCO is the most abundant enzyme on earth and constitutes approximately 30% of the total protein in many leaves. For this reason, it is of considerable interest in relation to nitrogen nutrition of plants. In higher plants, RuBisCO is a complex made up of eight identical large subunits (L, 50–55 kDa) and eight identical small subunits (S, 12–16 kDa). The large subunits are chloroplast-encoded while the small subunits are nuclear-encoded [11]. The small subunits are targeted to the chloroplast via a transit peptide that is removed during import. In the chloroplasts, chaperones then assist in the assembly of the large and the small subunits into a hexadecameric form, via a complex process [9, 12, 13]. This hexadecameric form (designated type I or form I) is found not only in terrestrial plants but also in virtually all eukaryotic algae, cyanobacteria and phototrophic and chemoautotrophic proteobacteria. Another smaller form (designated type II or form II) that contains only the large subunit, the smallest functional unit being a homodimer L_2, is found in some bacteria (reviewed in [8]). Unusual RuBisCOs made up of five large subunit dimers $(L_2)_5$, and RuBisCO-like proteins grouped into forms III and IV, respectively, have been discovered recently [14–18].

Both the large and the small subunits of the RuBisCO are extensively modified during and/or after translation. Besides their phosphorylation, as part of a process regulating import into the chloroplast, the small subunits are methylated with still unknown effects. The N-terminus of the large subunits is acetylated in all vascular plant species examined and other modifications (e.g. formylation of methionine residues, trimethylation of lysine residues) have also been reported (reviewed in [19]). Some modifications on proline and cysteine residues have been found in RuBisCO of *Chlamydomonas reinhardtii* and may be unique to the algal forms [20, 21].

The crystal structures of RuBisCO from spinach [22–24], tobacco [25–27], cyanobacterium [28, 29], red alga [30], photosynthetic bacteria [31–33] and archaebacterium [34] and from *C. reinhardtii* [20, 21] have been reported.

Within RuBisCO, the large subunits are arranged as antiparallel dimers (head to tail) with the N-terminal domain of one monomer adjacent to the C-terminal domain of the other. Each active site is at the interface between monomers within a dimer, explaining the minimal requirement for a dimeric structure. In the holoenzyme, four L_2 dimers are assembled into a core of eight large subunits, $(L_2)_4$, displaying a fourfold axis with local 4-2-2 symmetry. The small subunits comprise two separate clusters of four subunits each $(S_4)_2$, which tightly interact with the large subunits. In the center of the L_8S_8 holoenzyme, along the fourfold axis, a channel extends throughout the molecule [22].

Besides solving the structure of this enzyme, searching for the molecular basis for its CO_2/O_2 specificity has been a challenge for many years, but has now greatly progressed as discussed below. The CO_2/O_2 specificity factor, Ω, is defined as the ratio between V_cK_o and V_oK_c, where V is the V_{max} of either carboxylation (c) or oxygenation (o) and K is the K_m for CO_2 or O_2 [35]. There is considerable variation in Ω values among RuBisCO enzymes of different species, with the enzyme from most land plants having Ω values in the range of 80–100 (reviewed in [9]).

Since the active site for which CO_2 and O_2 compete is borne by the large subunit, considerable data have been accumulated on this subunit. Just as many enzymes have an α/β-barrel-domain [36], the large subunit of RuBisCO has a loop (i.e. between β–strand 6 and α-helix 6 in the α/β-barrel) that folds over substrate during catalysis (reviewed in [10, 37]). RuBisCO may thus exist in two states; in the open state the active site is accessible to bulk solvent, whereas in the closed state the active site is completely shielded from the solvent. The transition between the two states is accompanied by significant structural changes that include the movement of the loop-6 region (residues 332–338), the movement of the C-terminal strand (residues 462 to the C-terminus) and a rigid-body rotation of the N-terminal domain of the adjacent L subunit. Numerous studies have indicated that loop 6 plays a major role in discriminating between CO_2 and O_2 in the competing RuBP carboxylation and oxygenation reactions of RuBisCO (reviewed in [10, 37]). Structural studies have demonstrated the importance of the C-terminus of the large subunit for the catalytic function [22, 28]. Moreover, removal of the C-terminal residues of the spinach or algal RuBisCO large subunit by carboxypeptidase-A reduces carboxylase activity by 60–70% [38]. Similarly, mutants in which the C-terminus of the large subunit is truncated lose catalytic activity and are no longer able to bind carboxy-arabinitol bisphosphate, a transition state analog [39]. The importance of the C-terminus and its folding over loop 6 was confirmed recently. It has been proposed that the aspartate residue in position 473 may serve as a latch, closing the RuBisCO structure that binds inorganic phosphate in one of the 'open' RuBisCO structures and thus supports the existence of a regulatory site [27]. Mutant enzymes in which the aspartate at position 473 has been replaced by an alanine or a glutamate residue have been produced in *C. reinhardtii*, and it has been shown that this aspartate residue is not essential for catalysis. However, both mutations (Asp 473/Ala/Glu) result in an 87% decrease in carboxylation catalytic efficiency and 16% decrease in CO_2/O_2 specificity. If the C terminus is required to stabilize loop 6 in the closed conformation, there must be additional residues at the carboxyl terminus/loop 6 interface that contribute to this mechanism.

Considering that the substitutions at position 473 can influence CO_2/O_2 specificity, further studies of the interactions between loop 6 and the C-terminus may define the precise role of this structural arrangement in catalysis and may provide clues to engineer RuBisCO to increase the carboxylation/oxygenation ratio that is expected to improve photosynthetic efficiency and consequently, crop yields [40].

In addition to the large subunit, attention has also been paid to the small subunits. Indeed, it has been shown that with expression of the large subunits of *Synechococcus* in *Escherichia coli* these proteins assemble into octameric cores [41–43]. These octameric forms have drastically reduced carboxylase activity. Thus the small subunits do more than just assemble or concentrate the large subunits of RuBisCO, but may also be partly responsible for the higher values characteristic of Ω obtained with plant form-I enzymes [11].

Besides CO_2/O_2 specificity, other interesting features of RuBisCO are: (i) changes in the activity of the enzyme are linked to the well-known 'activation' phenomenon by carbamylation [44, 45] of a lysine residue (lysine 201) by a non-substrate molecule of CO_2 and subsequent Mg^{2+} binding [9, 44] in the active site prior to binding RuBP as reviewed in [46] and (ii) its interaction with RuBisCO activase.

A reduction in the catalytic activity of RuBisCO has been observed *in vivo* as a consequence of either tight RuBP binding to the uncarbamylated enzyme [47, 48], or 2-carboxyarabinitol-1-phosphate (CA1P) binding to the carbamylated enzyme [49, 50]. CA1P is a naturally occurring, transition state analog of the carboxylase reaction of RuBisCO that binds tightly to the active site of this enzyme and thus inhibits catalytic activity. *In vitro*, a fall over (a characteristic time-dependent decrease in enzymatic activity) of RuBisCO has also been observed and attributed to the formation of a tight-binding inhibitor during catalysis [51, 52]. This process remains to be clarified since it was later shown that RuBP may be oxidized into a tight-binding inhibitor [53]; an observation that was further supported by an increase in fall over observed when catalysis is performed at higher temperatures, thereby enhancing RuBP oxidation [54].

The regulatory protein RuBisCO activase has been shown to facilitate the release of sugar phosphates such as RuBP or CA1P from the catalytic site of RuBisCO [55–57]. The activity of RuBisCO *in vivo* is modulated by the binding of these sugar phosphates, and the role of activase is thus crucial in the modulation of RuBisCO activity.

RuBisCO activase is a nuclear-encoded chloroplast protein that in most plants consists of two isoforms arising from alternative splicing of a pre-mRNA [58, 59]. It is an ATP hydrolyzing (ATPase) enzyme that causes a conformational change in RuBisCO from a closed to an open state. It is a member of the AAA^+ family (ATPases associated with diverse cellular activities), a family with chaperone-like functions. The activity of this protein is controlled by the ratio of ATP/ADP

[60]. The three-dimensional structure of the activase is still unknown although it possesses the conserved AAA^+ fold [61].

The function of each isoform has been examined by characterizing RuBisCO activation in transgenic plants of Arabidopsis that express only one or both isoforms of activase. In plants expressing only the shorter isoform (43 kDa), RuBisCO activity is as high as in the (nontransformed) wild type plants under saturating light, but activity in the transgenics is not down-regulated at limiting intensities for photosynthesis. In contrast, in plants expressing only the longer isoform (46 kDa), RuBisCO activity is down-regulated at limiting light intensities, but the activity is slightly lower and increases much more slowly at saturating light intensities when compared with the wild type [59]. The 46-kDa isoform contains C-terminal cysteine residues, and so when the capacity to redox-regulate the activity of the larger isoform is eliminated by replacement of these critical cysteine residues unique to this isoform, RuBisCO activity in saturating light is similar to the wild type. However, the ability of the larger isoform to down-regulate RuBisCO activity at limiting light intensities in the transgenic plants is almost abolished. These results indicate that the light modulation of RuBisCO under limiting light is mainly due to the ability of the RuBisCO activase to be regulated by redox changes in the stroma [59]. For those species that lack the larger isoform, the exact mode of activase/RuBisCO regulation remains unclear.

To understand and define the site of interaction between activase and RuBisCO, heterologous interactions between the two proteins have recently been analyzed.

RuBisCO activase from plants in the family Solanaceae (e.g. tobacco) fails to activate RuBisCO isolated from plants outside this family (e.g. spinach and the green alga *C. reinhardtii*). Conversely, activase from non-Solanaceous species cannot activate RuBisCO from Solanaceous species [62]. A comparison of the sequences and X-ray crystal structures reveals that there are seven residues clustered on the surface of the RuBisCO large subunit whose charges differ between the Solanaceae and non-Solanaceae enzymes (residues 86, 89, 94, 95, 356, 466 and 468) [63, 64]. Substitutions were then made to define the region of the large subunit that interacts with activase. A substitution was thus performed to change the specificity of the interaction by introducing a residue at position 89 that differs from either of the residues characteristic of Solanaceae (Arg 89) and non-Solanaceae (Pro 89) RuBisCO [65]. The results obtained seem to indicate that either of at least two residues (Arg 89 or Lys 94) on the surface of the algal RubisCO large subunit is sufficient for recognition by the tobacco activase, but that at least two residues (Pro 89 and Asp 94) are necessary for recognition by the spinach activase. However, a large subunit Pro89Ala substitution also blocks algal RuBisCO activation by spinach activase and allows activation by tobacco activase. It thus seems that it is more the absence of Pro than the presence of the Arg residue at position 89 that is required for activation by tobacco activase [65].

Considerable progress has been made in identifying natural variations in the catalytic properties of RuBisCO from different species, and in developing tools to introduce both novel and foreign RuBisCO genes into plants [10]. The manipulation of genes encoding RuBisCO activase has also provided a means to investigate the regulation of RuBisCO activity. For example, plants with reduced amounts of cytochrome b/f complex (through gene antisensing) have impaired electron transport and a low trans-thylakoid pH gradient that restrict ATP and NADPH synthesis. In these plants, CO_2 assimilation was further compromised by reduced carbamylation of RuBisCO. The authors thus propose that the low carbamylation of RuBisCO in the antisense b/f plants is due to reduced activity of RuBisCO activase. The possibilities that might explain this reduced activity include the trans-thylakoid pH gradient and the reduction state of the acceptor side of photosystem I and/or the degree of reduction of the thioredoxin pathway [66].

Extensive analyses have been carried out using antisense tobacco plants with reduced amounts of RuBisCO. The results obtained with the antisense plants showed that the flux control coefficient that may range from 0 to 1 [67, 68] is high in conditions where carboxylation efficiency is reduced; that is, when photorespiration is favored [69, 70]. In fact, the flux control coefficient exerted by RuBisCO may be below 0.2 [71–73] but can reach 0.8 or even higher when the antisense plants are grown in ambient CO_2 and moderate light, or in a greenhouse with high irradiance, high temperature and low humidity; hence the control exerted by RuBisCO can be quite significant [70, 74]. Recently, a regulatory potential of the genes in the Calvin cycle has been inferred from the expression patterns of these genes. In this study [7], the analysis of the expression patterns of genes from almost 100 Arabidopsis microarray experiments was reported. Interestingly, RuBisCO is one of the enzymes that is the most sensitive to biological perturbations, which is quite consistent with the important rate-limiting role of this enzyme in controlling the carbon assimilation flux of the Calvin cycle.

The complexity of assembling copies of the two distinct polypeptide subunits of RuBisCO into a functional holoenzyme *in vivo* (requiring sufficient expression and post-translational modifications), its interaction with RuBisCO activase and the very likely high control of this enzyme on the flux of carbon fixation constitute sufficient reasons to encourage further studies of this enzyme. Many of the structural and conformational changes have been shown to be crucial in the regulation and assembly of RuBisCO, but although data on RuBisCO have accumulated, the frequent reports of new information about this fascinating enzyme suggest that many important discoveries still remain to be made.

7.3 Glyceraldehyde-3-phosphate dehydrogenase

GAPDH exists in three main forms in higher plants and algae, two cytosolic forms and a chloroplast form involved in the Calvin cycle: (1) a cytosolic NADP-dependent nonphosphorylating GAPDH that occurs in the cytosol and catalyzes the irreversible oxidation of glyceraldehyde-3-phosphate into 3-phosphoglycerate coupled to the reduction of NADP into NADPH [75], (2) a classic phosphorylating NAD-specific GAPDH of glycolysis and (3) a phosphorylating NADP-dependent GAPDH involved in CO_2 fixation in the chloroplast that catalyzes the reversible reduction and dephosphorylation of 1,3-bisphosphoglycerate (BPGA) to produce glyceraldehyde-3-phosphate using NADPH.

GAPDH isolated from chloroplasts has a dual specificity and can use either NAD(H) or NADP(H). It has been suggested that in vascular plants GAPDH exists either as a heterotetramer of two A subunits (36 kDa) and two B subunits (39 kDa) (A_2B_2), or as a homotetramer of four A subunits (A_4) [76]. A 600-kDa aggregated form (A_8B_8) has also been isolated from vascular plants [77–80]. Besides these oligomers, heterooligomers have been found, involving association of GAPDH with PRK and CP12, a small nuclear encoded chloroplast protein [81–86]. It has been shown that in darkened chloroplasts, vascular plants possess both types of complexes (A_8B_8 and PRK/GAPDH/CP12) with a probable ratio of 1:1 based on total GAPDH activity in either type of association [87]. In algae, only the A subunit has been found. The A and B subunits are very similar (about 80% sequence identities), except that the B subunit has a highly negatively charged C-terminal extension of 30 amino acid residues that contains two additional cysteine residues. Interestingly, the C-terminus of CP12 is homologous to the C-terminal extension of the B subunit.

The C-terminal extension confers regulatory properties to B subunit-containing isoforms [88–91]. This extension is also responsible for the tendency of the A_2B_2 tetramer to aggregate into the A_8B_8 form [88, 92]. The polymerization state of the enzyme is linked to its control by dark-light transitions. The A_8B_8 form of GAPDH is considered to be a regulatory form, whose activity *in vitro* may be controlled by metabolites such as NADP(H) or BPGA in the presence of a reducing agent [92–94]. This control is mediated by the dissociation of the 'heavy' form of GAPDH in the presence of BPGA [94, 95] leading to the formation of a more active tetramer. In all cases, the modulation of GAPDH activity requires intact B subunits, as occurs in GAPDH having both A and B subunits, artificial isoforms composed only of B subunits [88, 89, 91] or alternatively, the presence of the regulatory peptide CP12 [84, 85, 87, 96]. The small CP12 protein can inhibit directly the 'non-regulatory' A_4-GAPDH isoform [96]. The resulting inhibitory effect is on the catalytic constant, k_{cat}, of the NADPH-dependent reaction. Indeed, the C-terminal extension and CP12 contain regulatory cysteine residues whose redox state is under the control of thioredoxin [84, 91] and these proteins exert their inhibitory effect following oxidation of the cysteine residues to form intrachain disulfide bridges.

Recent electron microscope data indicate that the A and the B subunits occur together *in situ* in pea (*Pisum sativum*), despite the apparent differences in distribution of these two subunits in the stroma (half of the A subunit, but only one-third of the B subunit seems to be associated with the thylakoid membranes) [97]. This difference suggests that the redox-insensitive A_4-GAPDH isoform and the redox-regulated tetramers might have different roles. It has been suggested that in the dark, the A_4-GAPDH more specific for NAD could participate in the metabolism when starch is broken down and phosphoglycerate is exported to cytosol. In the presence of light, A_4-GAPDH will convert spill-over BPGA from the Calvin cycle to glyceraldehyde-3-phosphate, which is then exported to the cytosol or returned to the Calvin cycle. In contrast the isozymes that contain the B subunit, whatever their composition, would be more adapted than those lacking the B subunit, to participate in photosynthetic CO_2 fixation [97].

Transcription is one of the primary steps by which light regulates gene expression in plants. In tobacco and Arabidopsis, expression of the nuclear genes that encode the A and B subunits of chloroplast GAPDH (*Gap*A and *Gap*B) is known to be regulated by light [98–100]. Several *cis*-acting elements and their cognate binding factors of both *Gap*A and *Gap*B have been identified [101], and it has been suggested that a single *cis*-acting element may respond to more than one photoreceptor [102]. It is known that blue and white light are much more efficient when compared with red light in inducing *Gap*A and *Gap*B expression. Further, continuous exposure of dark-treated mature plants to blue or white light is required for high-level expression of *Gap*A and *Gap*B [99, 103].

The crystal structure of spinach A_4-GAPDH has been solved [104, 105]. The coenzymes NADP and NAD show similar conformations and interactions with the apoprotein in NADP- and NAD-GAPDH complexes, respectively, except in the $2'$-phosphate region. The $2'$-phosphate group of NADP is stabilized by two hydrogen bonds with the hydroxylated side-chains of Thr 33 and Ser 188, the latter belonging to an S-shaped loop ('S loop') of a symmetry-related subunit [104]. On the other hand, NAD, in common with the NAD-specific glycolytic enzyme [106], is kept in place by hydrogen bonding between the $2'$-hydroxyl group of the NAD and the Asp32 carboxylate group of the same subunit [105, 106].

It was recently shown that a mutated spinach A_4-GAPDH in which the serine residue at position 188 was replaced by alanine, relaxes to a loosened, enlarged conformation [107]. This mutation results in a significant decrease of the maximum catalytic activity of the enzyme with NADPH alone [96]. Since Ser188 is surrounded by several positive charges (Arg residues 183, 191, 194, 195 and 197), and since both the C-terminal extension and the N-terminus of CP12 are negatively charged [84], an interaction between the S loop and the so-called 'inhibitory domain' was suggested [107]. This interaction might prevent Ser188 hydrogen bonding to the $2'$-phosphate group of NADP with a consequent dramatic change of enzyme conformation and a decrease in catalysis. The authors proposed that the oxidized C-terminal extension and (possibly) CP12 interacting with the S loop may inhibit the NADPH-dependent activity of the wild type enzyme [107]. Consistently, the lack of this hydrogen bond in Ser188 mutants of A_4-GAPDH produces kinetic effects similar to the inhibition linked to the C-terminal extension present in GAPDH having B and A subunits. The association of CP12 with algal A_4-GAPDH also results in a decrease of the maximum catalytic activity with NADPH [96]. Finally, the interaction between the S loop of algal GAPDH and CP12 protein has been clearly demonstrated. Indeed, a mutated algal GAPDH in which the arginine residue at position 195 that belongs to the S loop has been substituted by glutamate fails to reconstitute the complex GAPDH/CP12 [108]. This result is in very good agreement with the findings of Trost and colleagues [107].

The generation of antisense tobacco plants directed against GAPDH has shown that no effects on photosynthesis are observed until GAPDH activity is reduced to below 35% of the wild type plants even under saturating light [109]. Moreover, the flux control coefficient for this enzyme is quite low; in most conditions (high or low irradiance), it is less than 0.2. Nonetheless, even though understanding of the individual enzymes is necessary, it is not sufficient to understand the entire mechanisms that control metabolism. Consequently, the recent discovery of CP12 within a bienzyme complex that involves not only GAPDH but also phosphoribulokinase [82, 84, 85] indicates that further studies are required to understand the processes that operate *in vivo* to modulate these Calvin cycle activities, and the role they play in controlling carbon assimilation.

7.4 Fructose-1,6-bisphosphatase and sedoheptulose-1, 7-bisphosphatase

FBPase and SBPase (EC 3.1.3.37) catalyze irreversible reactions corresponding to dephosphorylation of FBP and sedoheptulose-1,7-bisphosphate (SBP) into fructose-6-phosphate and sedoheptulose-7-phosphate, respectively [110, 111]. While SBPase is a homodimer unique to the Calvin cycle, in vascular plants there are two distinct homotetrameric FBPases, a cytosolic FBPase that is involved in sucrose synthesis from triose phosphates exported from chloroplasts (see Chapter 9) and a chloroplastic FBPase involved in the Calvin cycle. Although there have long been doubts as to the existence of an SBPase distinct from FBPase, SBPases have been partially or fully purified. SBPases have been purified from a number of different C_3 plant species such as spinach [112–114], pea [115] and wheat [116]. An alternate phosphatase was also purified from spinach leaves but this enzyme is not a true SBPase as it hydrolyzes FBP more readily than SBP [117, 118]. This enzyme resembles more the one found in cyanobacterial cells that seem to contain a unique enzyme, fructose-1,6-/sedoheptulose-1,7-bisphosphatase (FBP/SBPase), although the cyanobacterial enzyme can hydrolyze both FBP and SBP with almost equal specific activities [119]. FBPase is widely distributed, and the plastid FBPase of red algae, green algae and plants are not of cyanobacterial origin but instead are related to cytosolic forms, suggesting that they originated through gene duplication [120]. SBPase is found in plastids of green algae and plants and in the cytosol of the kinetoplastid parasites such as *Trypanosoma brucei* [121]. The presence of a cytosolic SBPase, and the presence of an FBP aldolase in this kinetoplastid, was thus believed to be a good argument in favor of an ancestral plastid in kinetoplastid parasites. The discovery of a cytosolic SBPase in fungi, such as *Neurospora crassa* and *Magnaporthe grisea*, suggests that this enzyme may have been added later to plastid metabolism, having persisted in the cytosol of nonphotosynthetic eukaryotes prior to the initial acquisition of the chloroplast. Indeed, as mentioned above, cyanobacteria lack SBPase and use a dual-specificity FBPase, and therefore, the plastid SBPase probably originates from sources other than cyanobacteria [122]. Still, the function of SBPase in kinetoplastid parasites and fungi is puzzling; it may function in a modified pentose-phosphate pathway in concert with FBP aldolase [121].

The gluconeogenic FBPase is controlled in a similar way to the mammalian enzyme, that is, through fructose-2,6-bisphosphate and AMP [123]. The chloroplastic

FBPase is not controlled by AMP, but by light via the ferredoxin–thioredoxin system. SBPase is also light-regulated by this system [124]. Light-induced changes in stromal Mg^{2+} levels and pH also regulate the FBPase [125, 126] and SBPase [114, 116, 127–130] activities. The different regulatory pathways observed between the cytosolic enzyme and the chloroplastic enzyme rely on the insertion of a sequence within chloroplastic FBPases that bears three conserved cysteine residues [131]. It was shown by means of site-directed mutagenesis that this sequence contributes to redox regulation [132, 133]. While there have been extensive crystallographic studies on gluconeogenic FBPases from mammals, only a few reports deal with chloroplastic FBPase [123, 134]. These studies are valuable as they clearly show that the redox-sensitive cysteines are remote from the active site and accessible to thioredoxins at the surface of the enzyme [134]. As SBPase lacks this regulatory insertion [135], it has been suggested that the presence (in a loop) of redox active cysteine residues at positions 52 and 57 could fulfill the same role as that of the FBPase insertion. The interaction between FBPase and thioredoxin has also been investigated. It has been suggested that the FBPase binding site is a negatively charged cluster positioned in a protruding loop ('170 loop') in the middle of each FBPase subunit. This loop contains the two cysteine residues that seem to be involved in the thiol-disulfide exchange [136]. Also, it has been demonstrated that hydrophobic forces also play an essential role in FBPase-thioredoxin interaction [137]. Finally, the binding parameters for the interaction between chloroplast FBPase and the wild type pea thioredoxins f and m, as well as mutated thioredoxin m, determined by equilibrium dialysis in accordance with the induced fit model of saturation kinetics, provided additional support for the role of the two basic thioredoxin residues, Lys 70 and Arg 74, in the interaction with FBPase [138].

FBPase and SBPase catalyze irreversible reactions. The product of the reaction catalyzed by FBPase, fructose-6-phosphate, is the branch point for metabolites leaving the Calvin cycle and moving into starch biosynthesis. SBPase is at the branch point between regeneration of the CO_2 acceptor molecule and biosynthesis of starch and sucrose, and could thus influence the distribution of carbon between these three competing pathways. Finally, as briefly described above, these two phosphatases are thioredoxin targets. Thus transgenic approaches have been developed to assess the contributions of these two phosphatases to the control of carbon flux through the Calvin cycle.

Using transgenic potato plants, no effect on photosynthesis occurred until the FBPase activity was less than 34% of the wild type, and the flux control coefficient for this enzyme was equal to that of GAPDH (0.2) in most environmental conditions [139]. On the other hand, the SBPase control coefficient was higher than that of FBPase, and could be up to 0.75 in specific conditions such as high irradiance (reviewed in [70]). Furthermore, transgenic tobacco plants with levels of SBPase activity less than wild type presented decreased rates of photosynthetic carbon fixation and altered carbohydrate levels in mature leaves [140–142]. Interestingly, the introduction and overexpression of cyanobacterial FBP/SBPase in the chloroplasts of transgenic tobacco plants lead to increased photosynthetic capacity in leaves, carbohydrate accumulation and accelerated growth rate [119]. Quantitative flux control analysis of the photosynthetic data from the SBPase antisense plants shows that SBPase exerts considerable control on carbon assimilation, particularly under saturating light and CO_2 conditions [70].

These antisense plant approaches [140–142], the highly regulated catalytic activity of SBPase, together with modeling studies [143, 144] led to the hypothesis that SBPase, even more than FBPase, plays an important role in the control of carbon flux through the Calvin cycle [143, 145]. This enzyme, like RuBisCO, was the most sensitive to biological perturbations when analyzing large-scale gene expression profiling data of Arabidopsis [7]. These data clearly indicate that SBPase is a strategic enzyme in the Calvin cycle.

7.5 Phosphoribulokinase

PRK is an important regulatory enzyme in the process of carbon fixation. Unique to this metabolic pathway, PRK catalyzes the ATP-dependent phosphorylation of ribulose-5-phosphate, thus regenerating ribulose-1,5-bisphosphate, the CO_2 acceptor molecule and substrate for RuBisCO. A divalent cation is required for activity with, in order of preference, Mg^{2+} and then Mn^{2+} most efficiently supporting activity. PRK has been isolated from different sources and exhibits fundamental differences in structure as well as in catalytic and regulatory properties.

A comparison of prokaryotic and eukaryotic PRKs reveals many striking differences. While in prokaryotes, native PRKs appear as either hexameric or octameric oligomers composed of 32-kDa subunits, in eukaryotes native PRKs are dimers made up of 40-kDa subunits (reviewed in [146]). A high resolution (2.5 Å) three-dimensional structure of *Rhodobacter sphaeroides* PRK has been elucidated [147] but, to date, there are no reports of any such structures for eukaryotic PRK. At present, only modeling of a PRK structure of *C. reinhardtii* has been performed [148] although the low level of sequence identity shared between the *R. sphaeroides* and the *C. reinhardtii* PRK sequences (below 25%) have made this a difficult task. The authors thus took into consideration information about the secondary structure that is currently more conserved than the primary sequence. To do this, hydrophobic cluster analysis, a two-dimensional method of sequence analysis, was used to add to the one-dimensional lexical comparison, and so provide an analysis of the secondary structure [149].

As mentioned above, diverse regulatory mechanisms also distinguish prokaryotic from eukaryotic PRKs. The class of enzymes from prokaryotes are allosterically controlled, being activated by NADH and inhibited by AMP, while disulfide bridge reduction between cysteine 16 and cysteine 55 is a major determinant of activity for eukaryotic PRKs. Early studies had shown that eukaryotic PRKs lacking these two cysteine residues can be inhibited by chloroplast metabolites. The most effective inhibitor is 6-phosphogluconate, which is a competitive inhibitor of ribulose-5-phosphate [150, 151]. In later studies, light activation of PRK was shown to be linked to the ferredoxin–thioredoxin-mediated pathway [152], while more recent studies were mainly devoted to the relative activity of the enzyme in isolated or individual state versus multienzyme complexes. Here, PRK embedded in a five-enzyme complex in spinach leaves exhibited slow oxidation and consequently inactivation, in comparison with the individual or so-called 'isolated form' [153]. In contrast, more rapid thioredoxin-dependent activation of PRK was observed with PRK as part of either a five-enzyme complex from spinach [154] or a bienzyme complex purified from a green alga [155]. The bienzyme form, which is associated with GAPDH and CP12, but not the isolated PRK, is also activated by NADP(H) but not by NAD(H). These results, therefore, support previous findings of information transfer and imprinting effects between these two enzymes [156]. Information transfer corresponds to stabilization–destabilization energies. These arise from a conformational change in the enzyme within a complex, and so part of the energy stored during association between the enzymes serves to increase the rate of catalysis after disruption of the complex via imprinting or memory effects [157–160]. It has thus been shown that information transfer may also be modulated by binding of a ligand such as NADP or NADPH. Moreover, the signal triggered by these cofactors may have two targets within a single 'control unit' [156]. Thus protein–protein interactions may give rise to new and unexpected regulatory properties.

Expression of PRK is also regulated by light [161, 162] and is under the control of the circadian-clock. It has been shown in *C. reinhardtii* that regulation of the PRK messenger RNA is controlled directly or indirectly by photosynthetic electron transfer as it is DCMU (3-(3,4-dichlorophenyl)-1,1-dimethyl-urea)-sensitive. The authors suggest that the redox state of the plastoquinone pool might be a good candidate for this regulation [162].

Interestingly, PRK is also a good target to study direct effects between photosynthesis, growth and allocation of dry matter, using transgenic plant approaches. Indeed, a decrease in the amount of PRK does not result in a large redistribution of nitrogen because PRK, unlike RuBisCO, accounts for only a fraction of the nitrogen present within a leaf. As a consequence, decreased photosynthesis in PRK antisense plants occurs without any nitrogen redistribution [163]. In this case, a genetic alteration of photosynthesis will be independent of the effects on nitrogen or nitrate. The effects observed on growth and allocation will rely only on photosynthesis and not on any disruption of plant nitrogen balance. The analysis of transgenic tobacco plants grown in low light showed, however, that even when there is a large decrease in PRK activity (greater than 85%), only small effects on the rate of photosynthesis and on shoot growth were observed [164]. Moreover, the maximum flux control coefficient for this enzyme is 0.25 when transgenic plants are grown in low light and photosynthesis is measured under saturating light [70, 165]. Therefore, when placing the PRK gene to reflect the sensitivity of this gene to biological perturbations, PRK presents an intermediate sensitivity, in common with FBPase, GAPDH, the transketolase and the epimerase [7]. These results indicate that PRK, like GAPDH, seems to have little control over photosynthetic carbon fixation.

7.6 Other important enzymes in the Calvin cycle

7.6.1 Transketolase

Transketolase (EC 2.2. 1.1.) catalyzes reactions in the Calvin cycle and the oxidative pentose-phosphate pathway. It catalyzes the reversible transfer of a C2 ketol moiety and displays broad substrate specificity. In the Calvin cycle, it converts glyceraldehyde-3-phosphate and fructose-6-phosphate into xylulose-5-phosphate and erythrose-4-phosphate, and glyceraldehyde-3-phosphate and sedoheptulose-7-phosphate into xylulose-5-phosphate and ribose-5-phosphate. Plants contain one major isoform of transketolase [1]. The immediate substrates and products of the reaction catalyzed by the transketolase act as precursors for phenylpropanoid synthesis (erythrose-4-phosphate), nucleotide synthesis (pentose-phosphates), carbohydrate synthesis (fructose-6-phosphate) and of glycolysis (glyceraldehyde-3-phosphate) leading to respiratory metabolism, amino acid and lipid synthesis [166].

Only few reports describe the properties of this enzyme in vascular plants, and to our knowledge, only one deals with the purified enzyme. Transketolase purified from spinach leaves seems to be a homodimer composed of 74-kDa subunits. Although its optimum pH is close to 8, it is also active at least one pH value above and below the pH maximum. No influence of the redox status was observed [167]. It is thus not susceptible to the light-dependent regulation observed with the well-known 'key' enzymes, but the broad pH optima and its redox insensitivity may be advantageous to fulfill its function in both the Calvin cycle and the oxidative pentose-phosphate pathway.

Transketolase catalyzes a reversible reaction and is insensitive to 'fine' control. However, because of its strategic location between primary and secondary metabolism, the consequences of decreased activity of the transketolase have been studied using antisense tobacco plants. Surprisingly, the results showed that this enzyme is an important determinant of both photosynthetic carbon and phenylpropanoid metabolism [166]. Decreased transketolase activity alters photosynthate allocation to favor starch rather than sucrose synthesis. Sucrose levels decline in step with a reduction in transketolase activity, with a 25% decrease in activity corresponding to a 25% decline in sucrose level. In contrast, starch accumulation is unaffected until the transketolase activity reaches less than 60% of the wild type activity. A 50% decrease of plastid transketolase activity leads not only to a marked inhibition of ambient photosynthesis, but also to decreased growth and to a dramatic decrease in aromatic amino acids, phenylpropanoid intermediates, chlorogenic acid and lignin. Thus this enzyme is near-limiting for several important metabolic pathways.

By analyzing large-scale gene expression, the regulation of this enzyme was at the same level as PRK, GAPDH, FBPase and thus seems to have intermediate sensitivity toward biological perturbations [7].

A theory in which the main portion of the intermediates of the Calvin cycle remain part of the cycle during one turnover has been developed. It showed that not only the reactions of nonequilibrium enzymes such as RuBisCO, but also reactions that operate close to a thermodynamic equilibrium, especially the reduction of 3-phosphoglycerate and the transketolase reaction can significantly influence the total turnover period in the Calvin cycle [168].

Taken together, these observations show, quite unexpectedly, that this enzyme probably can play a significant role in regulating the flux through the Calvin cycle.

7.6.2. Aldolase

The FBP aldolases are divided into two classes (designated class I and class II) of phylogenetically and structurally unrelated enzymes characterized by several divergent properties. Both classes catalyze the cleavage of fructose-1,6-bisphosphate to glyceraldehyde-3-phosphate and dihydroxyacetone phosphate or the reverse aldol condensation in the Calvin cycle. In addition to the aldol condensation of triose sugars, plastid FBP aldolase I can condense dihydroxyacetone phosphate and erythrose-4-phosphate to form sedoheptulose-1,7-bisphosphate [169]. Class I aldolases are homotetramers (160 kDa) that form a Schiff base with their substrate and can be inhibited with borohydride reagents (e.g. $NaBH_4$). Class II aldolases occur as homodimers (80 kDa), require divalent cations as cofactors and are inhibited by EDTA. Class I FBP aldolases are found in eukaryotes and in some proteobacteria. Red algae, green algae and plants possess two distinct class I FBP aldolases, a cytosolic aldolase involved in gluconeogenesis and glycolysis, and a plastid aldolase involved in the Calvin cycle. Class II FBP aldolases are mainly found in eubacteria and are divided into two subgroups. Homology between class I aldolases is at least 50% [170]. Class II aldolases show no homology to class I aldolases and have probably developed independently during evolution [171]. The evolutionary history of this enzyme is quite complex, involving lateral gene transfer and retargeting between cellular compartments [122, 172].

Euglena gracilis is one of the few eukaryotic organisms that contains both a class I aldolase (mainly expressed in autotrophically grown cells) and a class II aldolase (predominantly expressed in heterotrophically grown cells). These enzymes have been purified and characterized [173]. In contrast, in *Xanthobacter flavus*, a gram-negative facultatively autotrophic bacterium, a class I FBP aldolase is employed during heterotrophic growth, whereas a class II FBP aldolase is synthetized during autotrophic growth [174]. Therefore, at present, no clear link between growth conditions and the relative activity of either class I or class II aldolases can be established and further investigations are required.

Surprisingly, of the nine enzymes of the Calvin cycle studied (gene expression of phosphoglycerate kinase and triose phosphate isomerase were not included), gene regulation of aldolase, in common with RuBisCO and SBPase, was one of the most, if not the most, sensitive to biological perturbations [7]. This high sensitivity of gene regulation may indicate a role of aldolase in controlling the activity of the Calvin cycle, and the results obtained with antisense aldolase plants grown in different conditions may support this possibility. These studies clearly show that the flux control value of aldolase may increase to 0.55 at saturating CO_2 concentration

[175]. In these plants, carbohydrate especially starch accumulation was reduced. Again, a nonregulated enzyme seems to have a significant share in the control over photosynthetic carbon flux.

7.7 Supramolecular complexes of the Calvin cycle

Because of the extremely high concentration of proteins in the stroma of chloroplasts, it is very likely that the enzymes involved in the Calvin cycle are not randomly distributed, but interact to give multienzyme complexes. This suggestion has been put forward because some of these proteins, for example PRK and GAPDH, are not easily isolated by conventional purification methods such as ion exchange, gel filtration and affinity chromatographies and copurify with other enzymes of the Calvin cycle. This organization into supramolecular edifices is a central feature of metabolism [176, 177]. If the enzymes inserted into these supramolecular structures do catalyze consecutive reactions, one of the major advantages of such structures is the transfer of reaction intermediates between catalytic sites without diffusion into the bulk of the cell. This so-called 'metabolic channeling' offers unique opportunities to enhance and regulate cell biochemistry. This metabolic channeling exists in primary and secondary metabolic systems in plants (reviewed in [177]). Although many multienzyme complexes catalyze consecutive reactions, others do not, and channeling thus cannot occur. In this case, a possible advantage of these complexes is alteration of the kinetic and regulatory intrinsic properties of the enzymes within the complex. New insights into both the function and mechanism of the formation of supramolecular complexes mainly come from studies on the green alga *C. reinhardtii*, and for the vascular plants mainly from spinach and pea. Different multi-enzyme complexes from pea and spinach with varying compositions have been isolated (reviewed in [176]). Smaller complexes that may be linked to the dissociation of higher order structures during the purification procedure may be considered as subcomplexes.

A five-enzyme complex composed of phosphoribose isomerase (PRI, EC 5.3.1.6), PRK, RuBisCO, phosphoglycerate kinase and GAPDH from spinach leaves has been studied in detail [153, 154, 178–180]. The same multienzyme complex with one additional enzyme, fructose-1,6-bisphosphatase has been identified in cotton (*Gossypium hirsutum*) [181]. The association of these enzymes results in alteration of the catalytic properties of RuBisCO and of PRK that results from an information transfer occurring in the complex [157, 179]. This complex may be bound to the thylakoid membranes [182]. Later, the use of cryomicroscopy has revealed that PRK and GAPDH from *C. reinhardtii* interact *in vivo* and are also associated with thylakoid membranes [183]. In addition to these two enzymes, PRI, FBP aldolase, SBPase and other enzymes that do not function as part of the Calvin cycle were also found bound to the thylakoid membrane [183]. A crucial role of thylakoid membranes in the supramolecular organization of the Calvin cycle has also been recently put forward by the use of differentially permeabilized cells of

Anacystis nidulans [184]. This study showed an association of Calvin cycle enzymes with the membranes near the sites of ATP and NADPH synthesis. Such an association would permit a link between carbon fixation and the light reactions of photosynthesis and may provide more efficient utilization of light-generated intermediates.

Clasper *et al.* [185] and Scheibe *et al.* [87] have isolated a PRK/GAPDH complex in spinach leaves, which is also found in green algae such as *C. reinhardtii* [83, 84], *Scenedesmus obliquus* [186, 187] and in the cyanobacterium *Synechocystis* PCC6803 [84]. In *Sinapis alba*, the presence of a binding fraction that was lost during purification [188, 189], and later proposed to be PRK [187], is required for the polymerization of GAPDH to form higher molecular mass structures. It has also been suggested that other chloroplastic components induce aggregation of PRKs that do not otherwise polymerize [190].

The PRK/GAPDH complex that may correspond to the core complex of a supercomplex involved in CO_2 assimilation is one of the best documented examples [85]. The small 8.5-kDa protein, designated CP12 [81, 82, 84], has also been identified in most of these complexes (spinach, pea, tobacco, some algae and some cyanobacteria). In *C. reinhardtii*, the complex can be dissociated by reduction and reversed by oxidizing conditions. Indeed, the oxidized partners can spontaneously re-form a complex *in vitro* that possesses the same kinetic properties as those of the native state [83]. Site-directed mutagenesis indicates that substitutions that eliminate basicity at residue 64 (Arg in wild type PRK) disrupt the bienzyme complex [191]. Quite recently, two residues of GAPDH involved in protein–protein interactions within this complex were also identified by changing residues 128 (lysine) and 197 (arginine) either to alanine or to glutamate residues [108]. A model based on surface plasmon resonance data for the assembly of PRK/CP12/GAPDH has been proposed, in which CP12 first binds to GAPDH [86]. As a consequence of this binding, there is a conformational change of the enzyme, which then allows association with PRK. To examine this further, interactions between mutant GAPDH and CP12 were first analyzed, followed by those between mutant GAPDH/CP12 sub-complexes and PRK. All the mutants, except the mutant in which the arginine at position 197 was replaced by glutamate, were able to reconstitute the GAPDH/CP12 sub-complex. All the mutant GAPDH/CP12 sub-complexes failed to interact with PRK to form the native complex. The absence of kinetic changes of all mutant GAPDH/CP12 sub-complexes compared to wild type GAPDH/CP12 suggests that the mutants do not undergo the conformation change required for PRK binding [96, 108].

The association of these two enzymes (PRK/GAPDH) also gives rise to new properties. To show this, further characterization of the activity of isolated PRK and PRK contained in the complex has been performed. An existing belief is that PRK in the 'oxidized' complex is active [159], but this result has been disputed [146] since the cysteinyl sulfhydryls of PRK within the complex had never been tested directly. Alkylation of the so-called oxidized complex coupled with mass spectrometry analyses revealed that there is one disulfide bridge, very likely the one involving cysteine 16 and cysteine 55 (using residues corresponding to the enzyme from spinach), in the PRK monomer within the complex [192]. The same study showed that GAPDH within the complex has four thiol groups whatever the redox state of this complex and so modulation of GAPDH activity observed upon reduction is not linked to disulfide reduction. It was proposed that the regulation of an enzyme such as PRK may modulate the regulation of another enzyme (GAPDH) via a 'domino-like' effect [192]. These data once more clearly demonstrate the role of protein–protein interactions in altering the properties of enzymes.

Gene expressions of GAPDH and PRK were investigated as mentioned above [98, 161, 162], but only recently was the coexpression of the *PRK*, *GapA*, *GapB* and *CP12* genes studied in Arabidopsis [193]. The expression of three Calvin cycle genes (*GapA*, *GapB*, *PRK*) and two *CP12* genes in the presence of light was inhibited by sucrose, although the response of *CP12-1* was somewhat delayed compared to the rapid disappearance of all other transcripts. All the transcripts were especially abundant in leaves and flower stalks, less prominent in flowers and virtually undetectable in roots and siliques; however, there was a higher expression of *CP12-1* in flowers and *CP12-2* in stalks. However, the coordinated regulation of *GapA-1, GapB, CP12-2* and *PRK* at the level of gene expression is fully consistent with the existence and likely physiological relevance of a supramolecular complex involving GAPDH, CP12 and PRK in chloroplasts [84, 85, 87].

Other interactions have been reported between PRI, PRK and RuBisCO in spinach [194] and PRK and RuBisCO in pea [195]. There is also both kinetic and physical evidence that GAPDH interacts with aldolase and with phosphoglycerate kinase in the chloroplasts of vascular plants [196–198]. Recently, a robust graphical method for identifying enzymes that are associated *in situ*, based on electron microscope analysis, has been developed [97]. This method also confirmed that GAPDH is close to phosphoglycerate kinase and aldolase *in situ*. The fact that interaction is detected with a variety of techniques such as fluorescence anisotropy, hybridization techniques, phase-partitioning and isoelectric focusing, and that these enzymes appear to be in close proximity *in situ*, increases the probability of functional aggregate of these enzymes within the chloroplast [97, 199].

The three-dimensional structure of the three-enzyme complex (PRI/PRK/RuBisCO) from spinach was obtained with a resolution of 3.5 Å and to our knowledge it is the only crystalline structure available of a complex involved in the Calvin cycle [200]. In this structure, certain conformational changes of RuBisCO are observed (the four axes of symmetry that are parallel in the isolated enzyme were at an angle of 70° in this structure). Using cryoelectron microscopy and image processing techniques for the *C. reinhardtii* PRK/GAPDH/CP12 complex, strong structural differences between the modeled PRK dimers (not embedded in the complex) and PRK in the three-dimensional reconstruction volume of the whole complex were suggested [148]. Most isolated particles had a rod-like shape, with overall dimensions (20×10 nm) that are in good agreement with the expected size of the complex (about 460 kDa). At that time, the authors had not detected CP12 in the bienzyme complex, but the data obtained clearly showed the presence of an additional material that was postulated to be CP12. At present, there is no doubt about the existence of CP12 in this structure.

It is thus clear that the Calvin cycle is characterized by different enzyme interactions, some of which are probably tighter than others. A tight association must be the case of PRK/GAPDH/CP12 found in different organisms. It should also be noted that as many enzymes from this cycle also participate in other pathways, the interactions may be transient to allow switches and interconnections between these pathways.

7.8 Conclusions

The Calvin cycle plays a pivotal role in most photosynthetic organisms and therefore, cellular functions and the so-called key enzymes in the cycle have been extensively studied. The model whereby the enzymes involved in this metabolic pathway are not randomly distributed in the chloroplast stroma but interact to give multienzyme complexes is widely accepted and provides new insights in the analysis of this pathway. Nonetheless, another organization scale now remains to be elucidated and the supramolecular organization around the thylakoid membranes remains to be investigated. It is also important to recall that in the Calvin cycle, there are only three enzymes (RuBisCO, PRK, SBPase) that are unique to this pathway, while the other enzymes are involved in additional pathways in chloroplasts. All these pathways are interconnected, and so many other interactions remain to be discovered. Of course, although easy to conceptualize, these 'interactomes' are not so easily mapped and analyzed, but probably progress made in both genomics and proteomics will make such studies more possible in the years to come.

Another powerful research tool that has emerged more recently is the use of transgenic plants. Over the past years, transgenic plant approaches and large-scale gene expression profiling data have revealed that enzymes to which only little attention had been paid to in the past are crucial; at least this seems to be the case for chloroplastic aldolase and transketolase. The findings that nonregulating enzymes catalyzing readily reversible reactions have a significant share in the control over the photosynthetic carbon flux are rather disturbing since 'regulatory' reactions are usually irreversible (thermodynamically nonequilibrium reactions, [201]). It is remarkable that the maximum rate of photosynthesis is almost completely limited by transketolase, an enzyme previously considered irrelevant to regulation. Plastid aldolase also exerts some control on the maximum rate of photosynthesis. Clearly, the investigations that were made on transgenic plants with decreased expression of specific enzymes of the Calvin cycle have led to a revision of ideas about the regulation of metabolism in this cycle. Probably, the development of new technologies will emphasize the fact that small changes in the activity of nonregulated enzymes can affect the Calvin cycle, while significant changes in the activity of so-called and long-known regulated enzymes do not. It will then be probably necessary to reassess these terms.

To conclude, photosynthesis and carbon fixation are co-limited by enzymes such as RuBisCO and SBPase that catalyze irreversible reactions and are subject to fine regulation. Moreover they are also co-limited by enzymes that are not regulated and freely catalyze reversible or equilibrium reactions such as the aldolase and the transketolase reactions. The interconnections with other metabolic pathways probably also require further investigation. As it is not possible to analyze every step in the photosynthetic and related processes, the predictive mathematical modeling approach will continue to be a desirable goal to identify likely targets. Utilization of clustering information with microarray data in the context of various pathways is also a challenging and promising approach to evaluate the effects and relationships of biological perturbations that may lead to new insights into the regulatory potentials of enzymes.

Acknowledgements

The authors are indebted to Dr A.L. Haenni for critical reading of the manuscript and helpful discussions.

References

1. C. Schnarrenberger, A. Flechner and W. Martin (1995) Enzymatic evidence for a complete oxidative pentose phosphate pathway in chloroplasts and an incomplete pathway in the cytosol of spinach leaves. *Plant Physiology* **108**, 609–614.
2. P.M. Debnam and M.J. Emes (1999) Subcellular distribution of enzymes of the oxidative pentose phosphate pathway in root and leaf tissues. *Journal of Experimental Botany* **340**, 1653–1661.
3. E. Thom, T. Mohlmann, W. Quick, B. Camara and H.E. Neuhaus. (1998) Sweet pepper plastids: enzymic equipment, characterisation of the plastidic oxidative pentose-phosphate pathway, and transport of phosphorylated intermediates across the envelope membrane. *Planta* **204**, 226–233.
4. S. Schuster, D.A. Fell and T. Dandekar (2000) A general definition of metabolic pathways useful for systematic organization and analysis of complex metabolic networks. *Nature Biotechnology* **18**, 326–332.
5. M.G. Poolman, D.A. Fell and C.A. Raines (2003) Elementary modes analysis of photosynthate metabolism in the chloroplast stroma. *European Journal of Biochemistry* **270**, 430–439.
6. U.I. Flügge (1999) Phosphate translocators in plastids. *Annual Review in Plant Physiology and Plant Molecular Biology* **50**, 27–45.
7. N. Sun, L. Ma, D. Pan, H. Zhao and X.W. Deng (2003) Evaluation of light regulatory potential of Calvin cycle steps based on large-scale gene expression profiling data. *Plant Molecular Biology* **53**, 467–478.
8. F.R. Tabita (1999) Microbial ribulose bisphosphate carboxylase/oxygenase: a different perspective. Invited review. *Photosynthesis Research* **60**, 1–28.
9. R.J. Spreitzer (1999) Questions about the complexity of chloroplast ribulose-1,5-bisphosphate carboxylase/oxygenase. *Photosynthesis Research* **60**, 29–42.
10. M.A.J. Parry, P.J. Andralojc, R.A.C. Mitchell, P.J. Madgwick and A.J. Keys (2003) Manipulation of Rubisco: the amount, activity, function and regulation. *Journal of Experimental Botany* **54**, 1321–1333.
11. R.J. Spreitzer (2003) Role of the small subunit in ribulose-1,5-bisphosphate carboxylase/oxygenase. *Archives of Biochemistry and Biophysics* **414**, 141–149.

12. F.C. Hartman and M.R. Harpel (1994) Structure, function, regulation, and assembly of D-ribulose-1,5-bisphosphate carboxylase/oxygenase. *Annual Review of Biochemistry* **63**, 197–234.
13. A.R. Portis (2001) The Rubisco activase-Rubisco system: an ATPase-dependent association that regulates photosynthesis. In: *Protein-Protein Interactions in Plant Biology*, Vol. 7, (eds M.T. McManus, W.A. Laing and A.C. Allan), pp. 30–52. Sheffield Academic, Sheffield.
14. G.M. Watson, J.P. Yu and F.R. Tabita (1999) Unusual ribulose 1,5-bisphosphate carboxylase/oxygenase of anoxic Archaea. *Journal of Bacteriology* **181**, 1569–1575.
15. S. Ezaki, N. Maeda, T. Kishimoto, H. Atomi and T. Imanaka (1999) Presence of a structurally novel type ribulose-bisphosphate carboxylase/oxygenase in the hyperthermophilic archaeon, *Pyrococcus kodakaraensis* KOD1. *Journal of Biological Chemistry* **274**, 5078–5082.
16. N. Maeda, K. Kitano, T. Fukui et al. (1999) Ribulose bisphosphate carboxylase/oxygenase from the hyperthermophilic archaeon *Pyrococcus kodakaraensis* KOD1 is composed solely of large subunits and forms a pentagonal structure. *Journal of Molecular Biology* **293**, 57–66.
17. F.J. Grundy and T.M. Henkin (1998) The S box regulon: a new global transcription termination control system for methionine and cysteine biosynthesis genes in gram-positive bacteria. *Molecular Microbiology* **30**, 737–749.
18. T.E. Hanson and F.R. Tabita (2001) A ribulose-1,5-bisphosphate carboxylase/oxygenase (RubisCO)-like protein from *Chlorobium tepidum* that is involved with sulfur metabolism and the response to oxidative stress. *Proceedings of the National Academy of Sciences of the United States of America* **98**, 4397–4402.
19. R.L. Houtz and A.R. Portis, Jr. (2003) The life of ribulose 1,5-bisphosphate carboxylase/oxygenase – posttranslational facts and mysteries. *Archives of Biochemistry and Biophysics* **414**, 150–158.
20. E. Mizohata, H. Matsumura, Y. Okano et al. (2002) Crystal structure of activated ribulose-1,5-bisphosphate carboxylase/oxygenase from green alga *Chlamydomonas reinhardtii* complexed with 2-carboxyarabinitol-1,5-bisphosphate. *Journal of Molecular Biology* **316**, 679–691.
21. T.C. Taylor, A. Backlund, K. Bjorhall, R.J. Spreitzer and I. Andersson (2001) First crystal structure of Rubisco from a green alga, *Chlamydomonas reinhardtii*. *Journal of Biological Chemistry* **276**, 48159–48164.
22. S. Knight, I. Andersson and C.I. Branden (1990) Crystallographic analysis of ribulose 1,5-bisphosphate carboxylase from spinach at 2.4 Å resolution. Subunit interactions and active site. *Journal of Molecular Biology* **215**, 113–160.
23. T.C. Taylor and I. Andersson (1996) Structural transitions during activation and ligand binding in hexadecameric Rubisco inferred from the crystal structure of the activated unliganded spinach enzyme. *Nature Structural Biology* **3**, 95–101.
24. N. Shibata, T. Inoue, K. Fukuhara et al. (1996) Orderly disposition of heterogeneous small subunits in D-ribulose-1,5-bisphosphate carboxylase/oxygenase from spinach. *Journal of Biological Chemistry* **271**, 26449–26452.
25. M.S. Chapman, S.W. Suh, D. Cascio, W.W. Smith and D. Eisenberg (1987) Sliding-layer conformational change limited by the quaternary structure of plant RuBisCO. *Nature* **329**, 354–356.
26. P.M. Curmi, D. Cascio, R.M. Sweet, D. Eisenberg and H. Schreuder (1992) Crystal structure of the unactivated form of ribulose-1,5-bisphosphate carboxylase/oxygenase from tobacco refined at 2.0-Å resolution. *Journal of Biological Chemistry* **267**, 16980–16989.
27. A.P. Duff, T.J. Andrews and P.M. Curmi (2000) The transition between the open and closed states of rubisco is triggered by the inter-phosphate distance of the bound bisphosphate. *Journal of Molecular Biology* **298**, 903–916.
28. J. Newman and S. Gutteridge (1993) The X-ray structure of *Synechococcus* ribulose-bisphosphate carboxylase/oxygenase-activated quaternary complex at 2.2-Å resolution. *Journal of Biological Chemistry* **268**, 25876–25886.
29. J. Newman (1993) Structure determination and refinement of ribulose 1,5-bisphosphate carboxylase/oxygenase from *Synechococcus* PCC6301. *Acta Crystallographica D* **49**, 548–560.
30. H. Sugawara, Y. Hiroki, N. Shibata, T. Inoue, S. Okada and C. Miyake (1999) Crystal structure of carboxylase reaction-oriented ribulose 1,5-bisphosphate carboxylase/oxygenase from

a thermophilic red alga, *Galdieria partita*. *Journal of Biological Chemistry* **274**, 15655–15661.
31. G. Schneider, Y. Lindqvist, I. Brandén and G. Lorimer (1986) Three-dimensional structure of ribulose-1,5-bisphosphate carboxylase/oxygenase from *Rhodospirillum rubrum* at 2.9 Å resolution. *EMBO Journal* **5**, 3409–3415.
32. G. Schneider, S. Knight, I. Andersson, C.I. Branden, Y. Lindqvist and T. Lundqvist (1990) Comparison of the crystal structures of L_2 and L_8S_8 Rubisco suggests a functional role for the small subunit. *EMBO Journal* **9**, 2045–2050.
33. S. Hansen, V.B. Vollan, E. Hough and K. Andersen (1999) The crystal structure of rubisco from *Alcaligenes eutrophus* reveals a novel central eight-stranded β-barrel formed by β-strands from four subunits. *Journal of Molecular Biology* **288**, 609–621.
34. K. Kitano, N. Maeda, T. Fukui, H. Atomi, T. Imanaka and K. Miki (2001) Crystal structure of a novel-type archaeal rubisco with pentagonal symmetry. *Structure* **9**, 473–481.
35. Z. Chen and R.J. Spreitzer (1992) How various factors influence the CO_2/O_2 specificity of ribulose-1,5-bisphosphate carboxylase/oxygenase. *Photosynthesis Research* **31**, 157–164.
36. R.K. Wierenga (2001) The TIM-barrel fold: a versatile framework for efficient enzymes. *FEBS Letters* **492**, 193–198.
37. R.J. Spreitzer and M.E. Salvucci (2002) Rubisco: structure, regulatory interactions, and possibilities for a better enzyme. *Annual Reviews in Plant Biology* **53**, 449–475.
38. A.R. Portis (1990) Partial reduction in ribulose 1,5-bisphosphate carboxylase/oxygenase activity by carboxypeptidase A. *Archives of Biochemistry and Biophysics* **283**, 397–400.
39. S. Gutteridge, D. Rhoades and C. Herrmann (1993) Site-specific mutations in a loop region of the C-terminal domain of the large subunit of ribulose bisphosphate carboxylase/oxygenase that influence substrate partitioning. *Journal of Biological Chemistry* **268**, 7818–7824.
40. S. Satagopan and R.J. Spreitzer (2004) Substitutions at the Asp-473 latch residue of *Chlamydomonas* ribulose bisphosphate carboxylase/oxygenase cause decreases in carboxylation efficiency and CO_2/O_2 specificity. *Journal of Biological Chemistry* **279**, 14240–14244.
41. B.G. Lee, B.A. Read and F.R. Tabita (1991) Catalytic properties of recombinant octameric, hexadecameric, and heterologous cyanobacterial/bacterial ribulose-1,5-bisphosphate carboxylase/oxygenase. *Archives of Biochemistry and Biophysics* **291**, 263–269.
42. T.J. Andrews (1988) Catalysis by cyanobacterial ribulose-bisphosphate carboxylase large subunits in the complete absence of small subunits. *Journal of Biological Chemistry* **263**, 12213–12219.
43. M.K. Morell, J.M. Wilkin, H.J. Kane and T.J. Andrews (1997) Side reactions catalyzed by ribulose-bisphosphate carboxylase in the presence and absence of small subunits. *Journal of Biological Chemistry* **272**, 5445–5451.
44. G.H. Lorimer, M.R. Badger and T.J. Andrews (1976) The activation of ribulose-1,5-bisphosphate carboxylase by carbon dioxide and magnesium ions. Equilibria, kinetics, a suggested mechanism, and physiological implications. *Biochemistry* **15**, 529–536.
45. W.W. Cleland, T.J. Andrews, S. Gutteridge, F.C. Hartman and G.H. Lorimer (1998) Mechanism of rubisco: the carbamate as general base. *Chemical Reviews* **98**, 549–562.
46. T.J. Andrews (1996) The bait in the Rubisco mousetrap. *Nature Structural Biology* **3**, 3–7.
47. A. Brooks and A.R. Portis (1988) Protein-bound ribulose-bisphosphate correlates with deactivation of ribulose bisphosphate carboxylase in leaves. *Plant Physiology* **87**, 244–249.
48. Z.G. Cardon and K.A. Mott (1989) Evidence that ribulose 1,5-bisphosphate (RuBP) binds to inactive sites of RuBP carboxylase *in vivo* and an estimate of the rate constant for dissociation. *Plant Physiology* **89**, 1253–1257.
49. J.R. Seemann, J.A. Berry, S.M. Freas and M.A. Krump (1985) Regulation of ribulose bisphosphate carboxylase activity *in vivo* by a light-modulated inhibitor of catalysis. *Proceedings of the National Academy of Sciences of the United States of America* **82**, 8024–8028.
50. E.T. Hammond, T.J. Andrews and I.E. Woodrow (1998) Regulation of ribulose-1,5-bisphosphate carboxylase/oxygenase by carbamylation and 2-carboxyarabinitol 1-phosphate in tobacco: insights from studies of antisense plants containing reduced amounts of rubisco activase. *Plant Physiology* **118**, 1463–1471.

51. D.L. Edmondson, M.R. Badger and T.J. Andrews (1990) Slow inactivation of ribulose bisphosphate carboxylase during catalysis is caused by accumulation of a slow, tight-binding inhibitor at the catalytic site. *Plant Physiology* **93**, 1390–1397.
52. D.L. Edmondson, M.R. Badger and T.J. Andrews (1990) A kinetic characterization of slow inactivation of ribulosebisphosphate carboxylase during catalysis. *Plant Physiology* **93**, 1376–1382.
53. H.J. Kane, J.M. Wilkin, A.R. Portis and T. John Andrews (1998) Potent inhibition of ribulose-bisphosphate carboxylase by an oxidized impurity in ribulose-1,5-bisphosphate. *Plant Physiology*, **117**, 1059–1069.
54. S.J. Crafts-Brandner and M.E. Salvucci (2000) Rubisco activase constrains the photosynthetic potential of leaves at high temperature and CO_2. *Proceedings of the National Academy of Sciences of the United States of America* **97**, 13430–13435.
55. Z.-Y. Wang and A.R. Portis (1992) Dissociation of ribulose-1,5-bisphosphate bound to ribulose-1,5-bisphosphate carboxylase/oxygenase and its enhancement by ribulose-1,5-bisphosphate carboxylase/oxygenase activase-mediated hydrolysis of ATP. *Plant Physiology* **99**, 1348–1353.
56. S.P. Robinson and A.R. Portis (1988) Release of the nocturnal inhibitor, carboxyarabinitol-1-phosphate, from ribulose bisphosphate carboxylase/oxygenase by Rubisco activase. *FEBS Letters* **233**, 413–416.
57. C.J. Mate, G.S. Hudson, S. von Caemmerer, J.R. Evans and T.J. Andrews (1993) Reduction of ribulose biphosphate carboxylase activase levels in tobacco (*Nicotiana tabacum*) by antisense RNA reduces ribulose biphosphate carboxylase carbamylation and impairs photosynthesis. *Plant Physiology* **102**, 1119–1128.
58. N. Zhang and A.R. Portis, Jr. (1999) Mechanism of light regulation of Rubisco: a specific role for the larger Rubisco activase isoform involving reductive activation by thioredoxin-f. *Proceedings of the National Academy of Sciences of the United States of America* **96**, 9438–9443.
59. N. Zhang, R.P. Kallis, R.G. Ewy and A.R. Portis (2002) Light modulation of Rubisco in Arabidopsis requires a capacity for redox regulation of the larger Rubisco activase isoform. *Proceedings of the National Academy of Sciences of the United States of America* **99**, 3330–3334.
60. V.J. Streusand and A.R. Portis (1987) Rubisco activase mediates ATP-dependent activation of ribulose bisphosphate carboxylase. *Plant Physiology* **85**, 152–154.
61. A.R. Portis (2003) Rubisco activase: Rubisco's catalytic chaperone. *Photosynthesis Research* **75**, 11–27.
62. Z.-Y. Wang, G.W. Snyder, B.D. Esau, A.R. Portis and W.L. Ogren (1992) Species-dependent variation in the interaction of substrate-bound rubisco and rubisco activase. *Plant Physiology* **100**, 1858–1862.
63. E.M. Larson, C.M. O'Brien, G. Zhu, R.J. Spreitzer and A.R. Portis Jr. (1997) Specificity for activase is changed by a Pro-89 to Arg substitution in the large subunit of ribulose-1,5-bisphosphate carboxylase/oxygenase. *Journal of Biological Chemistry* **272**, 17033–17037.
64. A.R. Portis (1995) The regulation of Rubisco by Rubisco activase. *Journal of Experimental Botany* **46**, 1285–1291.
65. C.M. Ott, B.D. Smith, A.R. Portis, Jr. and R.J. Spreitzer (2000) Activase region on chloroplast ribulose-1,5-bisphosphate carboxylase/oxygenase. Non conservative substitution in the large subunit alters species specificity of protein interaction. *Journal of Biological Chemistry* **275**, 26241–26244.
66. S.A. Ruuska, T.J. Andrews, M.R. Badger, G.D. Price and S. von Caemmerer (2000) The role of chloroplast electron transport and metabolites in modulating rubisco activity in tobacco. Insights from transgenic plants with reduced amounts of cytochrome b/f complex or glyceraldehyde 3-phosphate dehydrogenase. *Plant Physiology* **122**, 491–504.
67. D.A. Fell (1992) Metabolic control analysis: a survey of its theoretical and experimental development. *Biochemical Journal* **286**, 313–330.
68. H. Kacser and J.A. Burns (1973) The control of flux. *Symposium of the Society for Experimental Biology* **27**, 65–104.
69. M. Stitt and D. Schulze (1994) Does Rubisco control the rate of photosynthesis and plant growth? An exercise in molecular ecophysiology. *Plant Cell and Environment* **17**, 465–487.

70. C.A. Raines (2003) The Calvin cycle revisited. *Photosynthesis Research* **75**, 1–10.
71. M. Stitt, W. Quick, U. Schurr, E.D. Schulze, S.R. Rodermel and L. Bogorad (1991) Decreased ribulose-1,5-bisphosphate carboxylase-oxygenase in transgenic tobacco transformed with 'antisense' rbcS. II. Flux-control coefficients for photosynthesis in varying light, CO_2, and air humidity. *Planta* **183**, 555–566.
72. W. Quick, U. Schurr, R. Scheibe *et al.* (1991) Decreased ribulose-1,5-bisphosphate carboxylase-oxygenase in transgenic tobacco transformed with 'antisense' rbcS. I. Impact on photosynthesis in ambient growth conditions. *Planta* **183**, 542–554.
73. G.S. Hudson, M.K. Morell, Y.B.C. Arvidsson and T.J. Andrews. (1992) Synthesis of spinach phosphoribulokinase and ribulose 1,5-bisphosphate in *Escherichia coli*. *Australian Journal of Plant Physiology* **19**, 213–221.
74. A. Krapp, M.M. Chaves, M.M. David, M.L. Rodriguez, J.S. Pereira and M. Stitt. (1994) Decreased ribulose-1,5-bisphosphate carboxylase-oxygenase in transgenic tobacco transformed with 'antisense' rbs S. VII. Impact on photosynthesis and growth in tobacco growing under extreme high irradiance and temperature. *Plant Cell and Environment* **17**, 945–953.
75. A. Habenicht, U. Hellman and R. Cerff (1994) Non-phosphorylating GAPDH of higher plants is a member of the aldehyde dehydrogenase superfamily with no sequence homology to phosphorylating GAPDH. *Journal of Molecular Biology* **237**, 165–171.
76. R. Cerff (1979) Quaternary structure of higher plant glyceraldehyde-3-phosphate dehydrogenases. *European Journal of Biochemistry* **94**, 243–247.
77. P. Pupillo and G. G. Piccari (1973) The effect of NADP on the subunit structure and activity of spinach chloroplast glyceraldehyde-3-phosphate dehydrogenase. *Archives of Biochemistry and Biophysics* **154**, 324–331.
78. G. Ferri, G. Comerio, P. Iadarola, M.C. Zapponi and M.L. Speranza (1978) Subunit structure and activity of glyceraldehyde-3-phosphate dehydrogenase from spinach chloroplasts. *Biochimica and Biophysica Acta* **522**, 19–31.
79. R. Cerff and S.E. Chambers (1979) Subunit structure of higher plant glyceraldehyde-3-phosphate dehydrogenases (EC 1.2.1.12 and EC 1.2.1.13). *Journal of Biological Chemistry* **254**, 6094–6098.
80. G.R. Yonuschot, B.J. Ortwerth and O.J. Koeppe (1970) Purification and properties of a nicotinamide adenine dinucleotide phosphate-requiring glyceraldehyde 3-phosphate dehydrogenase from spinach leaves. *Journal of Biological Chemistry* **245**, 4193–4198.
81. K. Pohlmeyer, B.K. Paap, J. Soll and N. Wedel (1996) CP12: a small nuclear-encoded chloroplast protein provides novel insights into higher-plant GAPDH evolution. *Plant Molecular Biology* **32**, 969–978.
82. N. Wedel, J. Soll and B.K. Paap (1997) CP12 provides a new mode of light regulation of Calvin cycle activity in higher plants. *Proceedings of the National Academy of Sciences of the United States of America* **94**, 10479–10484.
83. L. Avilan, B. Gontero, S. Lebreton and J. Ricard (1997) Memory and imprinting effects in multienzyme complexes-I. Isolation, dissociation, and reassociation of a phosphoribulokinase-glyceraldehyde-3-phosphate dehydrogenase complex from *Chlamydomonas reinhardtii* chloroplasts. *European Journal of Biochemistry* **246**, 78–84.
84. N. Wedel and J. Soll (1998) Evolutionary conserved light regulation of Calvin cycle activity by NADPH-mediated reversible phosphoribulokinase/CP12/glyceraldehyde-3- phosphate dehydrogenase complex dissociation. *Proceedings of the National Academy of Sciences of the United States of America* **95**, 9699–9704.
85. E. Graciet, S. Lebreton and B. Gontero (2004) Emergence of new regulatory mechanisms in the Benson-Calvin pathway via protein-protein interactions: a glyceraldehyde-3-phosphate dehydrogenase/CP12/phosphoribulokinase complex. *Journal of Experimental Botany* **55**, 1245–1254.
86. E. Graciet, P. Gans, N. Wedel, S. Lebreton, J.M. Camadro and B. Gontero (2003) The small protein CP12: a protein linker for supramolecular assembly. *Biochemistry* **42**, 8163–8170.
87. R. Scheibe, N. Wedel, S. Vetter, V. Emmerlich and S.M. Sauermann (2002) Co-existence of two regulatory NADP-glyceraldehyde 3-P dehydrogenase complexes in higher plant chloroplasts. *European Journal of Biochemistry* **269**, 5617–5624.

88. E. Baalmann, R. Scheibe, R. Cerff and W. Martin (1996) Functional studies of chloroplast glyceraldehyde-3-phosphate dehydrogenase subunits A and B expressed in *Escherichia coli*: formation of highly active A_4 and B_4 homotetramers and evidence that aggregation of the B_4 complex is mediated by the B subunit carboxy terminus. *Plant Molecular Biology* **32**, 505–513.
89. A.D. Li and L.E. Anderson (1997) Expression and characterization of pea chloroplastic glyceraldehyde-3-phosphate dehydrogenase composed of only the B-subunit. *Plant Physiology* **115**, 1201–1209.
90. S. Scagliarini, P. Trost and P. Pupillo (1998) The non-regulatory isoform of NAD(P)-glyceraldehyde-3-phosphate dehydrogenase from spinach chloroplasts. *Journal of Experimental Botany* **49**, 1307–1315.
91. F. Sparla, P. Pupillo and P. Trost (2002) The C-terminal extension of glyceraldehyde-3-phosphate dehydrogenase subunit B acts as an autoinhibitory domain regulated by thioredoxins and nicotinamide adenine dinucleotide. *Journal of Biological Chemistry* **277**, 44946–44952.
92. R. Scheibe, E. Baalmann, J.E. Backhausen, C. Rak and S. Vetter (1996) C-terminal truncation of spinach chloroplast NAD(P)-dependent glyceraldehyde-3-phosphate dehydrogenase prevents inactivation and reaggregation. *Biochimica and Biophysica Acta* **1296**, 228–234.
93. P. Pupillo and G.G. Piccari. (1975) The reversible depolymerization of spinach chloroplast glyceraldehyde-phosphate dehydrogenase. *European Journal of Biochemistry* **51**, 475–482.
94. E. Baalmann, J.E. Backhausen, C. Kitzmann and R. Scheibe (1994) Regulation of NADP-dependent glyceraldehyde 3-phosphate dehydrogenase activity in spinach chloroplast. *Botanica Acta* **107**, 313–320.
95. E. Baalmann, J.E. Backhausen, C. Rak, S. Vetter and R. Scheibe (1995) Reductive modification and nonreductive activation of purified spinach chloroplast NADP-dependent glyceraldehyde-3-phosphate dehydrogenase. *Archives of Biochemistry and Biophysics* **324**, 201–208.
96. E. Graciet, S. Lebreton, J.M. Camadro and B. Gontero (2003) Characterization of native and recombinant A_4 glyceraldehyde 3-phosphate dehydrogenase. *European Journal of Biochemistry* **270**, 129–136.
97. J.B. Anderson, A.A. Carol, V.K. Brown and L.E. Anderson (2003) A quantitative method for assessing co-localization in immunolabeled thin section electron micrographs. *Journal of Structural Biology* **143**, 95–106.
98. M.C. Shih and H.M. Goodman (1988) Differential light regulated expression of nuclear genes encoding chloroplast and cytosolic glyceraldehyde-3-phosphate dehydrogenase in *Nicotiana tabacum*. *EMBO Journal* **7**, 893–898.
99. J. Dewdney, T.R. Conley, M.C. Shih and H.M. Goodman (1993) Effects of blue and red light on expression of nuclear genes encoding chloroplast glyceraldehyde-3-phosphate dehydrogenase of *Arabidopsis thaliana*. *Plant Physiology* **103**, 1115–1121.
100. C.S. Chan, H.P. Peng and M.C. Shih (2002) Mutations affecting light regulation of nuclear genes encoding chloroplast glyceraldehyde-3-phosphate dehydrogenase in Arabidosis. *Plant Physiology* **130**, 1476–1486.
101. T.R. Conley, S.C. Park, H.B. Kwon, H.P. Peng and M.C. Shih (1994) Characterization of cis-acting elements in light regulation of the nuclear gene encoding the A subunit of chloroplast isozymes of glyceraldehyde-3-phosphate dehydrogenase from *Arabidopsis thaliana*. *Molecular and Cell Biology* **14**, 2525–2533.
102. C.S. Chan, L. Guo and M.C. Shih (2001) Promoter analysis of the nuclear gene encoding the chloroplast glyceraldehyde-3-phosphate dehydrogenase B subunit of *Arabidopsis thaliana*. *Plant Molecular Biology* **46**, 131–141.
103. T.R. Conley and M.C. Shih (1995) Effects of light and chloroplast functional state on expression of nuclear genes encoding chloroplast glyceraldehyde-3-phosphate dehydrogenase in long hypocotyl (hy) mutants and wild-type *Arabidopsis thaliana*. *Plant Physiology* **108**, 1013–1022.
104. S. Fermani, A. Ripamonti, P. Sabatino, *et al.* (2001) Crystal structure of the non-regulatory A(4) isoform of spinach chloroplast glyceraldehyde-3-phosphate dehydrogenase complexed with NADP. *Journal of Molecular Biology* **314**, 527–542.

105. G. Falini, S. Fermani, A. Ripamonti et al. (2003) Dual coenzyme specificity of photosynthetic glyceraldehyde 3-phosphate dehydrogenase interpreted by the crystal structure of A_4 isoform complexed with NAD. *Biochemistry* **42**, 4631–4639.
106. C. Didierjean, S. Rahuel-Clermont, B. Vitoux, O. Dideberg, G. Branlant and A. Aubry (1997) A crystallographic comparison between mutated glyceraldehyde-3- phosphate dehydrogenases from *Bacillus stearothermophilus* complexed with either NAD^+ or $NADP^+$. *Journal of Molecular Biology* **268**, 739–759.
107. F. Sparla, S. Fermani, G. Falini et al. (2004) Coenzyme site-directed mutants of photosynthetic A(4)-GAPDH show selectively reduced NADPH-dependent catalysis, similar to regulatory AB-GAPDH inhibited by oxidized thioredoxin. *Journal of Molecular Biology* **340**, 1025–1037.
108. E. Graciet, G. Mulliert, S. Lebreton and B. Gontero (2004) Involvement of two positively charged residues of *Chlamydomonas reinhardtii* glyceraldehyde-3-phosphate dehydrogenase in the assembly process of a bi-enzyme complex involved in CO_2 assimilation. *European Journal of Biochemistry* **271**, 4737–4744.
109. G.D. Price, J.R. Evans, S. von Caemmerer, J.W. Yu and M.R. Badger (1995) Specific reduction of chloroplast glyceraldehyde-3-phosphate dehydrogenase activity by antisense RNA reduces CO_2 assimilation via a reduction in ribulose bisphosphate regeneration in transgenic tobacco plants. *Planta* **195**, 369–378.
110. G. Zimmermann, G.J. Kelly and E. Latzko (1976) Efficient purification and molecular properties of spinach chloroplast fructose 1,6-bisphosphatase. *European Journal of Biochemistry* **70**, 361–367.
111. V.D. Breazale, B.B. Buchanan and R.A. Wolosiuk. (1978) Chloroplast sedoheptulose-1,7-bisphosphatase: evidence for regulation by the ferredoxin/thioredoxin system. *Zeitschrift fur Naturforschung* **33c**, 521–528.
112. F. Cadet, J.C. Meunier and F. Ferté (1987) Isolation and purification of chloroplastic spinach (*Spinacia oleracea*) sedoheptulose-1,7-bisphosphatase. *Biochemical Journal* **241**, 71–74.
113. F. Cadet and J.C. Meunier (1988) Spinach (*Spinacia oleracea*) chloroplast sedoheptulose-1,7-bisphosphatase. Activation and deactivation, and immunological relationship to fructose-1,6-bisphosphatase. *Biochemical Journal* **253**, 243–248.
114. F. Cadet and J.C. Meunier (1988) pH and kinetic studies of chloroplast sedoheptulose-1,7-bisphosphatase from spinach (*Spinacia oleracea*). *Biochemical Journal* **253**, 249–254.
115. L.E. Anderson (1974) Activation of pea chloroplast sedoheptulose 1,7-diphosphate phosphatase by light and dithiothreitol. *Biochemical and Biophysical Research Communications* **59**, 907–913.
116. I.E. Woodrow and D.A. Walker (1982) Activation of wheat chloroplast sedoheptulose bisphosphatase: a continuous spectrophotometric assay. *Archives of Biochemistry and Biophysics* **216**, 416–422.
117. B. Gontero, J.C. Meunier and J. Ricard (1984) Purification and properties of a chloroplastic phosphatase distinct from fructose bisphosphatase. *Plant Science Letters* **36**, 137–142.
118. B. Gontero, J.C. Meunier and J. Ricard (1984) On the activation of two chloroplastic phosphatases by fructose bisphosphate, sedoheptulose bisphosphate and magnesium. *Plant Science Letters* **36**, 195–199.
119. M. Tamoi, T. Ishikawa, T. Takeda and S. Shigeoka (1996) Molecular characterization and resistance to hydrogen peroxide of two fructose-1,6-bisphosphatases from *Synechococcus* PCC 7942. *Archives of Biochemistry and Biophysics* **334**, 27–36.
120. W. Martin, A.Z. Mustafa, K. Henze and C. Schnarrenberger (1996) Higher-plant chloroplast and cytosolic fructose-1,6-bisphosphatase isoenzymes: origins via duplication rather than prokaryote-eukaryote divergence. *Plant Molecular Biology* **32**, 485–491.
121. V. Hannaert, E. Saavedra, F. Duffieux et al. (2003) Plant-like traits associated with metabolism of *Trypanosoma* parasites. *Proceedings of the National Academy of Sciences of the United States of America* **100**, 1067–1071.
122. M. Rogers and P.J. Keeling (2004) Lateral transfer and recompartmentalization of Calvin cycle enzymes of plants and algae. *Journal of Molecular Evolution* **58**, 367–375.
123. V. Villeret, S. Huang, Y. Zhang and W.N. Lipscomb (1995) Structural aspects of the allosteric inhibition of fructose-1,6-bisphosphatase by AMP: the binding of both the substrate analogue

2,5-anhydro-D-glucitol 1,6-bisphosphate and catalytic metal ions monitored by X-ray crystallography. *Biochemistry* **34**, 4307–4315.
124. P. Schurmann and J.P. Jacquot (2000) Plant thioredoxin systems revisited. *Annual Reviews in Plant Physiology and Plant Molecular Biology* **51**, 371–400.
125. S.A. Charles and B. Halliwell (1980) Properties of freshly purified and thiol treated spinach chloroplast fructose bisphosphatase. *Biochemical Journal* **185**, 689–693.
126. R. Minot, J.C. Meunier, J. Buc and J. Ricard (1982) The role of pH and magnesium concentration in the light activation of chloroplastic fructose bisphosphatase. *FEBS Letters* **142**, 118–120.
127. A.R. Portis, C.J.A. Chon, A. Mosbac and H.W. Heldt (1977) Fructose- and sedoheptulose-bisphosphatase: the sites of a possible control of CO2 fixation by light dependent changes of the stromal Mg^{2+} concentration. *Biochimica and Biophysica Acta* **461**, 313–325.
128. P. Purczeld, C.J. Chon, A.R. Portis, Jr., H.W. Heldt and U. Heber (1978) The mechanism of the control of carbon fixation by the pH in the chloroplast stroma. Studies with nitrite-mediated proton transfer across the envelope. *Biochimica and Biophysica Acta* **501**, 488–498.
129. A.N. Nishizawa and B.B. Buchanan (1981) Enzyme regulation in C_4 photosynthesis: purification and properties of thioredoxin-linked fructose bisphosphatase and sedoheptulose bisphosphatase from corn leaves. *Journal of Biological Chemistry* **256**, 6119–6126.
130. I.E. Woodrow, D.J. Murphy and E. Latzko (1984) Regulation of stromal sedoheptulose-1,7-bisphosphatase activity by pH and Mg^{2+} concentration. *Journal of Biological Chemistry* **259**, 3791–3795.
131. F. Marcus, L. Moberly and S.P. Latshaw (1988) Comparative amino acid sequence of fructose-1,6-bisphosphatases: identification of a region unique to the light-regulated chloroplast enzyme. *Proceedings of the National Academy of Sciences of the United States of America* **85**, 5379–5383.
132. J.P. Jacquot, J. Lopez-Jaramillo, M. Miginiac-Maslow *et al.* (1997) Cysteine-153 is required for redox regulation of pea chloroplast fructose-1,6-bisphosphatase. *FEBS Letters* **401**, 143–147.
133. R.J. Rodriguez-Suarez, S. Mora-Garcia and R.A. Wolosiuk (1997) Characterization of cysteine residues involved in the reductive activation and the structural stability of rapeseed (*Brassica napus*) chloroplast fructose-1,6-bisphosphatase. *Biochemical and Biophysical Research Communications* **232**, 388–393.
134. M. Chiadmi, A. Navaza, M. Miginiac-Maslow, J.P. Jacquot and J. Cherfils (1999) Redox signalling in the chloroplast: structure of oxidized pea fructose-1,6-bisphosphate phosphatase. *EMBO Journal* **18**, 6809–6815.
135. R.P. Dunford, M.C. Durrant, M.A. Catley and T.A. Dyer (1998) Location of the redox-active cysteines in chloroplast sedoheptulose-1,7-bisphosphatase indicates that its allosteric regulation is similar but not identical to that of fructose-1,6-bisphosphatase. *Photosynthesis Research* **58**, 221–230.
136. R. Hermoso, M. Castillo, A. Chueca, J.J. Lazaro, M. Sahrawy and J.L. Gorge (1996) Binding site on pea chloroplast fructose-1,6-bisphosphatase involved in the interaction with thioredoxin. *Plant Molecular Biology* **30**, 455–465.
137. A. Pla and J. Lopez Gorge (1981) Thioredoxin/fructose-1,6-bisphosphatase affinity in the enzyme activation by the ferredoxin-thioredoxin system. *Biochimica and Biophysica Acta* **636**, 113–118.
138. O.S. Wangensteen, A. Chueca, M. Hirasawa, M. Sahrawy, D.B. Knaff and J. Lopez Gorge (2001) Binding features of chloroplast fructose-1,6-bisphosphatase-thioredoxin interaction. *Biochimica and Biophysica Acta* **1547**, 156–166.
139. J. Kossmann, U. Sonnewald and L. Willmitzer (1994) Reduction of the chloroplastic fructose 1,6-bisphosphatase in transgenic potato plants impairs photosynthesis and plant growth. *Plant Journal* **6**, 637–650.
140. E.P. Harrison, N.M. Willingham, J.C. Lloyd and C.A. Raines (1998) Reduced sedoheptulose-1,7-bisphosphatase levels in transgenic tobacco lead to decreased photosynthetic capacity and altered carbohydrate accumulation. *Planta* **204**, 27–36.
141. E.P. Harrison, H. Olcer, J.C. Lloyd, S.P. Long and C.A. Raines (2001) Small decreases in SBPase cause a linear decline in the apparent RuBP regeneration rate, but do not affect Rubisco carboxylation capacity. *Journal of Experimental Botany* **52**, 1779–1784.

142. H. Olcer, J.C. Lloyd and C.A. Raines (2001) Photosynthetic capacity is differentially affected by reductions in sedoheptulose-1,7-bisphosphatase activity during leaf development in transgenic tobacco plants. *Plant Physiology* **125**, 982–989.
143. G. Petterson and U. Ryde-Petterson (1989) Dependence of the Calvin cycle activity on kinetic parameters for the interaction of non-equilibrium cycle enzymes with their substrates. *European Journal of Biochemistry* **186**, 683–687.
144. M.G. Poolman, H. Olcer, J.C. Lloyd, C.A. Raines and D.A. Fell (2001) Computer modelling and experimental evidence for two steady states in the photosynthetic Calvin cycle. *European Journal of Biochemistry* **268**, 2810–2816.
145. M.G. Poolman, D.A. Fell and S. Thomas (2000) Modeling photosynthesis and its control. *Journal of Experimental Botany* **51**, 319–328.
146. H.M. Miziorko (2000) Phosphoribulokinase: current perspectives on the structure/function basis for regulation and catalysis. *Advances in Enzymology and Related Areas of Molecular Biology* **74**, 95–127.
147. D.H.T. Harrison, J.A. Runquist, A. Holub and H.N. Miziorko (1998) The crystal structure of phosphoribulokinase from *Rhodobacter sphaeroides* reveals a fold similar to that of adenylate kinase. *Biochemistry* **37**, 5074–5085.
148. F. Mouche, B. Gontero, I. Callebaut, J.P. Mornon and N. Boisset (2002) Striking conformational change suspected within the phosphoribulokinase dimer induced by interaction with GAPDH. *Journal of Biological Chemistry* **277**, 6743–6749.
149. I. Callebaut, G. Labesse, P. Durand *et al.* (1997) Deciphering protein sequence information through hydrophobic cluster analysis (HCA): current status and perspectives. *Cellular and Molecular Life Sciences* **53**, 621–645.
150. L.E. Anderson (1973) Regulation of pea leaf ribulose 5-phosphate kinase activity. *Biochimica and Biophysica Acta* **321**, 484–488.
151. A. Gardemann, M. Stitt and H.W. Heldt (1983) Regulation of spinach ribulose-5-phosphate kinase by stromal metabolite levels. *Biochimica and Biophysica Acta* **722**, 51–60.
152. B.B. Buchanan (1980) Role of light in the regulation of chloroplast enzymes. *Annual Review in Plant Physiology* **31**, 341–374.
153. M. Rault, B. Gontero and J. Ricard (1991) Thioredoxin activation of phosphoribulokinase in a chloroplast multi-enzyme complex. *European Journal of Biochemistry* **197**, 791–797.
154. B. Gontero, G. Mulliert, M. Rault, M.T. Giudici-Orticoni and J. Ricard (1993) Structural and functional properties of a multi-enzyme complex from spinach chloroplasts. 2. Modulation of the kinetic properties of enzymes in the aggregated state. *European Journal of Biochemistry* **217**, 1075–1082.
155. L. Avilan, S. Lebreton and B. Gontero (2000) Thioredoxin activation of phosphoribulokinase in a bi-enzyme complex from *Chlamydomonas reinhardtii* chloroplasts. *Journal of Biological Chemistry* **275**, 9447–9451.
156. E. Graciet, S. Lebreton, J.M. Camadro and B. Gontero (2002) Thermodynamic analysis of the emergence of new regulatory properties in a phosphoribulokinase-glyceraldehyde 3-phosphate dehydrogenase complex. *Journal of Biological Chemistry* **277**, 12697–12702.
157. J. Ricard, M.T. Giudici-Orticoni and B. Gontero (1994) The modulation of enzyme reaction rates within multi-enzyme complexes. 1. Statistical thermodynamics of information transfer through multi-enzyme complexes. *European Journal of Biochemistry* **226**, 993–998.
158. S. Lebreton, B. Gontero, L. Avilan and J. Ricard (1997) Information transfer in multienzyme complexes-1. Thermodynamics of conformational constraints and memory effects in the bienzyme glyceraldehyde-3-phosphate-dehydrogenase-phosphoribulokinase complex of *Chlamydomonas reinhardtii* chloroplasts. *European Journal of Biochemistry* **250**, 286–295.
159. S. Lebreton, B. Gontero, L. Avilan and J. Ricard (1997) Memory and imprinting effects in multienzyme complexes-II. Kinetics of the bienzyme complex from *Chlamydomonas reinhardtii* and hysteretic activation of chloroplast oxidized phosphoribulokinase. *European Journal of Biochemistry* **246**, 85–91.
160. S. Lebreton and B. Gontero (1999) Memory and imprinting in multienzyme complexes. Evidence for information transfer from glyceraldehyde-3-phosphate dehydrogenase to phospho-

ribulokinase under reduced state in *Chlamydomonas reinhardtii. Journal of Biological Chemistry* **274**, 20879–20884.

161. C.A. Raines, M. Longstaff, J.C. Lloyd and T.A. Dyer (1989) Complete coding sequence of wheat phosphoribulokinase: developmental and light-dependent expression of the mRNA. *Molecular and General Genetics* **220**, 43–48.

162. S.D. Lemaire, M. Stein, E. Issakidis-Bourguet *et al.* (1999) The complex regulation of ferredoxin/thioredoxin-related genes by light and the circadian clock. *Planta* **209**, 221–229.

163. F.M. Banks, S.P. Driscoll, M.A.J. Parry *et al.* (1999) Decrease in phosphoribulokinase activity by antisense RNA in transgenic tobacco. Relationship between photosynthesis, growth, and allocation at different nitrogen levels. *Plant Physiology* **119**, 1125–1136.

164. M.J. Paul, J.S. Knight, D. Habash *et al.* (1995) Reduction in phosphoribulokinase activity by antisense RNA in transgenic tobacco: effect on CO_2 assimilation and growth in low irradiance. *Plant Journal* **7**, 535–542.

165. M.J. Paul, S.P. Driscoll, P.J. Andralojc, J.S. Knight, J.C. Gray and D.W. Lawlor (2000) Decrease of phosphoribulokinase activity by antisense RNA in transgenic tobacco: definition of the light environment under which phosphoribulokinase is not in large excess. *Planta* **211**, 112–119.

166. S. Henkes, U. Sonnewald, R. Badur, R. Flachmann and M. Stitt (2001) A small decrease of plastid transketolase activity in antisense tobacco transformants has dramatic effects on photosynthesis and phenylpropanoid metabolism. *Plant Cell* **13**, 535–551.

167. M. Teige, M. Melzer and K.H. Suss (1998) Purification, properties and *in situ* localization of the amphibolic enzymes D-ribulose 5-phosphate 3-epimerase and transketolase from spinach chloroplasts. *European Journal of Biochemistry* **252,** 237–244.

168. L.E. Fridlyand and R. Scheibe (1999) Regulation of the Calvin cycle for CO_2 fixation as an example for general control mechanisms in metabolic cycles. *Biosystems* **51**, 79–93.

169. A. Flechner, W. Gross, W.F. Martin and C. Schnarrenberger (1999) Chloroplast class I and class II aldolases are bifunctional for fructose-1,6-bisphosphate and sedoheptulose-1,7-bisphosphate cleavage in the Calvin cycle. *FEBS Letters* **447**, 200–202.

170. C. Schnarrenberger, B. Pelzer-Reith, H. Yatsuki, S. Freund, S. Jacobshagen and K. Hori (1994) Expression and sequence of the only detectable aldolase in *Chlamydomonas reinhardtii. Archives of Biochemistry and Biophysics* **313**, 173–178.

171. J.J. March and H.G. Lebherz (1992) Fructose bisphosphate aldolases: an evolutionary history. *Trends in Biochemistry Sciences* **17**, 110–113.

172. N.J. Patron, M.B. Rogers and P.J. Keeling (2004) Gene replacement of fructose-1,6-bisphosphate aldolase supports the hypothesis of a single photosynthetic ancestor of chromalveolates. *Eukaryotic Cell* **3**, 1169–1175.

173. B. Pelzer-Reith, S. Wiegand and C. Schnarrenberger (1994) Plastid class I and cytosol class II aldolase of *Euglena gracilis* (purification and characterization). *Plant Physiology* **106**, 1137–1144.

174. E.R. van der Bergh, S.C. Baker, R.J. Raggers *et al.* (1996) Primary structure and phylogeny of the Calvin cycle enzymes transketolase and fructosebisphosphate aldolase of *Xanthobacter flavus. Journal of Bacteriology* **178**, 888–893.

175. V. Haake, M. Geiger, P. Walch-Liu, C. Engels, R. Zrenner and M. Stitt (1999) Changes in aldolase activity in wild-type potato plants are important for acclimation to growth irradiance and carbon dioxide concentration, because plastid aldolase exerts control over the ambient rate of photosynthesis across a range of growth conditions. *Plant Journal* **17**, 479–489.

176. B. Gontero, S. Lebreton and E. Graciet. (2001) Multienzyme complexes involved in the Benson–Calvin cycle and in fatty acid metabolism. In: *Protein-Protein Interactions in Plant Biology*, Vol. 7**,** (eds M.T. McManus, W.A. Laing and A.C. Allan), Sheffield Academic, Sheffield, pp. 120–150.

177. B.S.J. Winkel (2004) Metabolic channeling in plants. *Annual Reviews in Plant Biology* **55**, 85–107.

178. B. Gontero, M.L. Cardenas and J. Ricard. (1988) A functional five-enzyme complex of chloroplasts involved in the Calvin cycle. *European Journal of Biochemistry* **173**, 437–443.

179. B. Gontero, M.T. Giudici-Orticoni and J. Ricard (1994) The modulation of enzyme reaction rates within multi-enzyme complexes. 2. Information transfer within a chloroplast multi-enzyme complex containing ribulose bisphosphate carboxylase-oxygenase. *European Journal of Biochemistry* **226**, 999–1006.

180. M. Rault, M.T. Giudici-Orticoni, B. Gontero and J. Ricard (1993) Structural and functional properties of a multi-enzyme complex from spinach chloroplasts. 1. Stoichiometry of the polypeptide chains. *European Journal of Biochemistry* **217**, 1065–1073.
181. M.A. Babadzhanova, N.P. Bakaeva and M.P. Babadzhanova (2000) Functional properties of the multienzyme complex of Calvin cycle key enzymes. *Russian Journal of Plant Physiology* **47**, 23–31.
182. M.T. Giudici-Orticoni, B. Gontero, M. Rault and J. Ricard (1992) Organisation structurale et fonctionnelle d'enzymes du cycle de Benson-Calvin à la surface des thylakoïdes des chloroplastes d'Epinard. *Comptes Rendus de l' Académie des Sciences. Paris* **314**(série III), 477–483.
183. K.H. Süss, I. Prokhorenko and K. Adler (1995) *In situ* association of Calvin cycle enzymes, ribulose-1,5-bisphosphate carboxylase/oxygenase activase, ferredoxine-NADP reductase, and nitrite reductase with thylakoid and pyrenoid membranes of *Chlamydomonas reinhardtii* chloroplasts as revealed by immunoelectron microscopy. *Plant Physiology* **107**, 1387–1397.
184. J.K. Sainis, D.N. Dani and G.K. Dey (2003) Involvement of thylakoid membranes in supramolecular organisation of Calvin cycle enzymes in *Anacystis nidulans. Journal of Plant Physiology* **160**, 23–32.
185. S. Clasper, J.S. Easterby and R. Powls (1991) Properties of two high-molecular-mass forms of glyceraldehyde-3- phosphate dehydrogenase from spinach leaf, one of which also possesses latent phosphoribulokinase activity. *European Journal of Biochemistry* **202**, 1239–1246.
186. M.J. O'Brien, J.S. Easterby and R. Powls (1976) Algal glyceraldehyde-3-phosphate dehydrogenases. Conversion of the NADH-linked enzyme of *Scenedesmus obliquus* into a form which preferentially uses NADPH as coenzyme. *Biochimica and Biophysica Acta* **449**, 209–223.
187. S. Nicholson, J.S. Easterby and R. Powls (1987) Properties of a multimeric protein complex from chloroplasts possessing potential activities of NADPH-dependent glyceraldehyde-3-phosphate dehydrogenase and phosphoribulokinase. *European Journal of Biochemistry* **162**, 423–431.
188. R. Cerff (1978) Glyceraldehyde-3-phosphate dehydrogenase (NADP) from *Sinapis alba* L. NAD(P)-induced conformation changes of the enzyme. *European Journal of Biochemistry* **82**, 45–53.
189. R. Cerff and S.E. Chambers (1978) Glyceraldehyde-3-phosphate dehydrogenase (NADP) from *Sinapis alba* L. Isolation and electrophoretic characterization of isoenzymes. *Hoppe Seylers Zeitschrift fur Physiologische Chemie* **359**, 769–772.
190. M.A. Porter (1990) The aggregation states of spinach phosphoribulokinase. *Planta* **181**, 349–357.
191. L. Avilan, B. Gontero, S. Lebreton and J. Ricard (1997) Information transfer in multienzyme complexes-2. The role of Arg64 of *Chlamydomonas reinhardtii* phosphoribulokinase in the information transfer between glyceraldehyde-3-phosphate dehydrogenase and phosphoribulokinase. *European Journal of Biochemistry* **250**, 296–302.
192. S. Lebreton, E. Graciet and B. Gontero (2003) Modulation, via protein-protein interactions, of glyceraldehyde-3-phosphate dehydrogenase activity through redox phosphoribulokinase regulation. *Journal of Biological Chemistry* **278**, 12078–12084.
193. L. Marri, F. Sparla, P. Pupillo and P. Trost (2005) Co-ordinated gene expression of photosynthetic glyceraldehyde-3-phosphate dehydrogenase, phosphoribulokinase, and CP12 in *Arabidopsis thaliana. Journal of Experimental Botany* **56**, 73–80.
194. J.K. Sainis and G.C. Harris (1986) The association of ribulose-1,5-bisphosphate carboxylase with phosphoriboisomerase and phosphoribulokinase. *Biochemical and Biophysical Research Communications* **139**, 947–954.
195. J.K. Sainis, K. Merriam and G.C. Harris (1989) The association of D-ribulose bisphosphate carboxylase-oxygenase with phosphoribulokinase. *Plant Physiology* **89**, 368–374.
196. J. Maciozek, J.B. Anderson and L.E. Anderson (1990) Isolation of chloroplastic phosphoglycerate kinase. Kinetics of the two enzyme phosphoglycerate kinase/glyceraldehyde-3-phosphate dehydrogenase couple. *Plant Physiology* **94**, 291–296.

197. L.E. Anderson, I.M. Goldhaber-Gordon, D. Li, X.Y. Tang, M. Xiang and N. Prakash (1995) Enzyme–enzyme interaction in the chloroplast: glyceraldehyde-3-phosphate dehydrogenase, triose phosphate isomerase and aldolase. *Planta* **196**, 245–255.
198. X. Wang, X.Y. Tang and L.E. Arderson (1996) Enzyme–enzyme interaction in the chloroplast: physical evidence for association between phosphoglycerate kinase and glyceraldehyde-3-phosphate dehydrogenase *in vitro*. *Plant Science* **117**, 45–53.
199. L.O. Persson and G. Johansson (1989) Studies of protein–protein interaction using countercurrent distribution in aqueous two-phase systems. Partition behaviour of six Calvin-cycle enzymes from a crude spinach (*Spinacia oleracea*) chloroplast extract. *Biochemical Journal* **259**, 863–870.
200. J.K. Sainis and N. Jawali (1994) Channeling of the intermediates and catalytic facilitation to Rubisco in a multienzyme complex of Calvin cycle enzymes. *Indian Journal of Biochemistry and Biophysics* **31**, 215–220.
201. E.A. Newsholme and C. Start (1973) *Regulation in Metabolism*. Wiley, Toronto.

8 Control of phosphoenolpyruvate carboxylase in plants

Hugh G. Nimmo

8.1 Introduction

In both C_4 and Crassulacean acid metabolism (CAM) plants, photosynthetic CO_2 fixation displays metabolic adaptations that improve water use efficiency and negate the oxygenase activity of Rubisco. In the various permutations of C_4 metabolism (NADP-malic enzyme type, NAD-malic enzyme type or phosphoenolpyruvate carboxykinase type), the C_4 cycle brings about the pumping of CO_2 into the vicinity of ribulose 1,5-bisphosphate carboxylase/oxygenase (Rubisco) in the bundle sheath cells. In CAM plants, the nocturnal fixation of atmospheric CO_2 and accumulation of malic acid permit the decarboxylation of malate and re-fixation of CO_2 by Rubisco behind closed stomata during the following day. Central to both processes is the primary fixation of CO_2 catalysed by phosphoenolpyruvate carboxylase (PEPC, EC 4.1.1.31). Photosynthetic CO_2 assimilation in C_4 and CAM plants is controlled at many different levels. The focus in this chapter is on control exerted through the reversible phosphorylation of PEPC and particularly on the properties and control of PEPC kinase, which is encoded by the *PPCK* gene family. PEPC plays many different roles in the metabolism of vascular plants apart from its involvement in photosynthesis in C_4 and CAM species [1–3], and several recent developments in our understanding of PEPC in green algae and C_3 plants are also included.

The basic features of the PEPC system have been known for many years. Most PEPCs are homotetramers, comprising identical subunits of ~110 kDa (but see an exception in Section 8.4). Crystal structures have been established for the maize and *Escherichia coli* enzymes [3–6]. All PEPCs show allosteric properties: vascular plant PEPCs are inhibited by L-malate and activated by glucose 6-phosphate [1–3]. In addition, (i) aspartate and glutamate are potent inhibitors of PEPCs in non-green plant tissues active in N-assimilation and/or transamination reactions, thus providing a link between C- and N-metabolism [7–10] and (ii) the activity of vascular plant PEPC is controlled by phosphorylation of a single, highly conserved serine residue close to the amino-terminal end of the polypeptide [2, 3, 11, 12] (but see in Section 8.4). This results in reduced sensitivity of the enzyme to inhibition by malate and increased sensitivity to activation by glucose 6-phosphate, and hence to greater flux through the enzyme *in vivo*. For prolonged CO_2 fixation in C_4 leaves, PEPC must function in the face of the high concentration of malate in mesophyll cells that is required to sustain its diffusion to bundle sheath cells. Similarly, PEPC must function during the accumulation of large amounts of malic acid in CAM leaves. There is now extensive evidence that phosphorylation of PEPC is required [2, 11]. For example, in elegant work on the C_4 species *Sorghum vulgare*, Bakrim *et al.* [13] prevented light-induced phosphorylation of PEPC by pre-treatment of leaves with cycloheximide. This blocked the primary fixation of CO_2 by PEPC but allowed a low rate of CO_2 assimilation directly via Rubisco. In several studies of CAM species, good correlations have been observed between nocturnal CO_2 fixation and the phosphorylation state of PEPC [11]. Hence much attention has been focused on the protein kinase that phosphorylates PEPC (PEPC kinase) [12] and, to a lesser extent, on the protein phosphatase 2A that dephosphorylates PEPC [14, 15].

This area of research has been reviewed recently [12]. In outline, the phosphorylation state of PEPC is controlled mainly by the activity of PEPC kinase. This enzyme is the smallest known protein kinase, comprising a catalytic domain with no extensions that could contribute regulatory properties. It is present at low abundance (less than 1 in 10^6 of soluble protein [16]), turns over rapidly and is controlled largely at the level of transcript abundance [17–19]. PEPC kinase activity is increased in response to a range of different signals, for example light in C_4 leaves, a circadian oscillator in CAM leaves, light and N supply in C_3 leaves and photosynthate in legume root nodules [2, 3, 11, 12]. In some cases the *PPCK* genes that are expressed in response to these signals have been identified [12]. The potential role of a protein inhibitor of PEPC kinase [20] has not yet been clarified. Malate inhibits PEPC kinase activity [17, 21], possibly by interacting with PEPC [21] but it is not clear if this effect is physiological. Izui's group has suggested that C_4 PEPC kinase may be controlled by thiol-disulfide interchange [16, 19], but evidence for the operation of this mechanism *in vivo* is still lacking (see below). There are no other known ways of controlling PEPC kinase activity *in vivo* apart from synthesis/ degradation. A protein phosphatase 2A holoenzyme active against PEPC has recently been purified [15]. This appears to be a heterotrimer, similar to protein phosphatase 2A holoenzymes from other eukaryotes. However it is not specific for PEPC and no physiologically relevant control mechanisms have been identified [15].

8.2 *PPCK* genes and their roles

Following the first cloning of a *PPCK* cDNA from the CAM species *Kalanchoë fedtschenkoi* and the resulting identification of Arabidopsis *PPCK1* [17], *PPCK* genes and cDNAs have now been cloned and sequenced from many plant species. The data available in early 2003 were summarised in [12]. Some additional sequences are now available and these are included in the phylogenetic tree shown in Figure 8.1. This work has shown quite clearly that many (perhaps all) plants contain a small *PPCK* gene family. However it has proved very difficult to ascribe particular functions to individual gene family members.

Arabidopsis contain two *PPCK* genes, rice three, and both soybean and maize contain at least four genes (Figure 8.1) [12, 17, 22, 23]. The different genes in any one species can show marked differences in expression pattern. However there is no

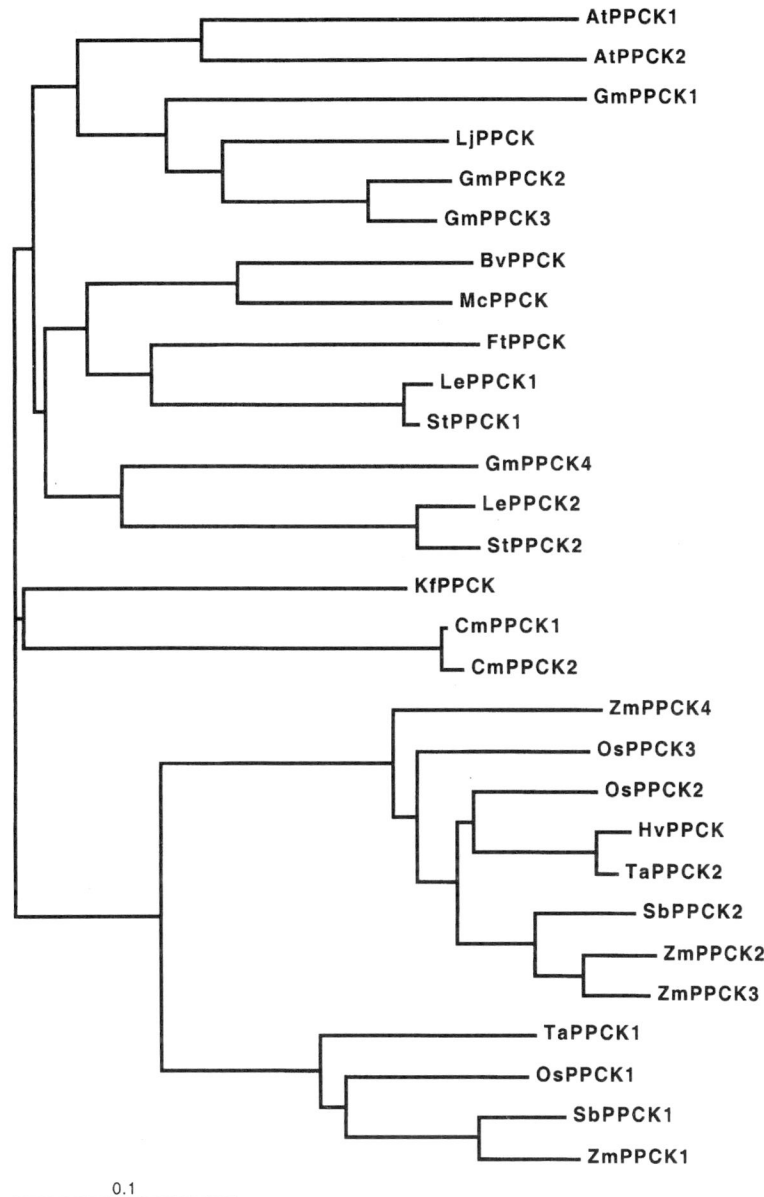

Figure 8.1 Phylogenetic analysis of *PPCK* genes. The tree is based on full-length deduced amino acid sequences. At, *Arabidopsis thaliana*, Bv, *Beta vulgaris*, Cm, *Clusia minor*, Ft, *Flaveria trinervia*, Gm, *Glycine max*, Hv, *Hordeum vulgare*, Kf, *Kalanchoë fedtschenkoi*, Le, *Lycopersicon esculentum*, Lj, *Lotus japonicus*, Mc, *Mesembryanthemum crystallinum*, Os, *Oryza sativa*, Pt, *Pinus taeda*, Sb, *Sorghum bicolor*, St, *Solanum tuberosum*, Ta, *Triticum aestivum* and Zm, *Zea mays*. Accession numbers are as given in [8, 19, 27]; *Z. mays PPCK1-4* are AY911413-AY911416; *C. minor PPCK1-2* are AY478419 and AY478420, *Oryza sativa PPCK1-3* are AK101080, XM466833 and AK066885 respectively. *S. bicolor PPCK1* (TC95314), *T. aestivum PPCK1-2* (TC241735 and TC255931) and *H. vulgare PPCK* (TC148750) are full length, unambiguous tentative consensus sequences in the TIGR database (see http://www.tigr.org/tdb/tgi/). The scale bar represents 0.1 substitution per site.

evidence that expression of any one gene is restricted to a single organ or cell type. For example, soybean contains four *PPCK* genes [23] (more than originally reported [24]). Three are expressed in nodules, of which two (*GmPPCK2* and *GmPPCK3*) show expression controlled by photosynthate; these are thought to contribute the PEPC kinase activity that is up-regulated by photosynthate. However these genes are also expressed in leaves, roots and flowers [23]. The *GmPPCK2:GmPPCK3* transcript ratio is much lower in flowers than in other tissues [23]. Xu *et al.* [24], who only studied *GmPPCK1* and *GmPPCK2* (which they termed *NE-PpcK*), suggested that *GmPPCK2* and the soybean nodule-enhanced PEPC gene (*GmPEPC7*) might comprise expression partners, with the implication that the PEPC GmPEPC7 might be a dedicated target of the kinase GmPPCK2. However *GmPEPC7* is not expressed in leaves, whereas *GmPPCK2* and *GmPPCK3* are [23]. Thus the targets of these kinases in leaves cannot be the same as their target in nodules (GmPEPC7). Although no fully systematic studies of specificity have yet been reported, it seems clear that PEPC kinases are able to phosphorylate plant PEPCs in general, even those from other species, albeit not necessarily at the same rates [2,20,25]. Hence at this stage one can suggest that any given PEPC kinase will probably phosphorylate the PEPCs present in the same cell without much distinction between them.

The difficulty in ascribing functions to *PPCK* genes is further illustrated by our recent work on maize (M. Shenton *et al.*, unpublished; see Figure 8.1). Of the four genes that we have identified, *ZmPPCK1* clearly encodes the 'C_4-type PEPC kinase'. This gene is expressed in leaf mesophyll cells but not bundle sheath cells and its transcript abundance is strongly and rapidly increased by high-intensity light. However it is also expressed in other organs such as ears, tassels and roots where it must clearly play a different role. Surprisingly, since the role of PEPC in C_4 photosynthesis is mesophyll cell-specific, another gene (*ZmPPCK2*) is expressed in the bundle sheath cells. The expression of this gene is higher in the dark than the light, and its role is not clear. The focus on PEPC in C_4 leaves has been on the mesophyll cell enzyme whose role is the primary fixation of CO_2. There is little information about the identity and possible roles of PEPC in the bundle sheath. One possibility is that the *ZmPPCK2* gene product may control an anaplerotic PEPC. This gene also is expressed in a range of organs.

The phylogenetic analysis shown in Figure 8.1 provides few guides to the function of *PPCK* genes. In general, the closest relative to any particular *PPCK* gene is another gene from the same species or a gene from a closely related species. However there are some interesting exceptions. *ZmPPCK1* is clearly the 'C_4-type' gene, and is very similar to a sorghum *PPCK* (*SbPPCK1*, for whose function there is as yet no evidence). However these genes are more closely related to *PPCK* genes from the C_3 grasses rice and wheat than to the other *PPCK* genes in maize or sorghum. Light does increase the expression and activity of PEPC kinase in C_3 leaves [2,12,22,23] but it is not yet known whether the *ZmPPCK1*-like genes of rice and wheat are expressed in leaves or are light-induced. In any event, their physiological functions clearly could not be the same as that of *ZmPPCK1* in maize leaves.

It is noteworthy that, while the soybean genes *GmPPCK1-3* group with the *PPCK* genes from other legumes, *GmPPCK4* is more closely related to the *PPCK2* genes of tomato and potato than to other soybean *PPCKs*. Similarly, the *PPCK1* genes of tomato and potato are more closely related to the well-authenticated 'C$_4$-type' *PPCK* of *Flaveria trinervia* [19] than to the tomato and potato *PPCK2* genes. These observations suggest that a duplication of an ancestral *PPCK* gene must have occurred relatively early in the evolution of flowering plants. Possibly the evolution of Arabidopsis (which has only two, closely related *PPCK* genes) involved the loss of any descendent from one of the two ancestral genes.

GmPPCK4 exhibits another property that seems to confuse the issue of the functions of *PPCK* genes. It is well known that CAM species show persistent circadian rhythms of CO$_2$ metabolism [11,26]. One factor that contributes to these rhythms is the circadian phosphorylation and dephosphorylation of PEPC, driven by circadian expression of *PPCK* [17,27,28]. The role of the phosphorylation is probably in controlling the amplitude of the rhythms rather than as the sole driving force [11,27]. However, until recently, the available evidence suggested that *PPCK* expression is not controlled circadianly in C$_3$ species [22]. While such species can show circadian rhythms of photosynthetic CO$_2$ assimilation [29,30] there is no reason to implicate PEPC in this.

Sullivan *et al.* [23] showed that the transcript abundance of *GmPPCK4* cycles robustly in both constant light and contrast darkness, peaking in subjective afternoon. This is the first example of circadian control of *PPCK* expression other than in CAM species. Cycling was detectable in leaves but not in roots. This observation is itself interesting, because it represents the first example of organ specificity in the circadian control of gene expression in plants. Of course, it is yet to be shown that the circadian control of *GmPPCK4* expression in leaves is physiologically important. Assuming this to be the case, the function of *GmPPCK4* presumably pertains to some circadian feature of metabolism in soybean leaves that is not relevant in roots. However, it is difficult to envisage why a *PPCK* gene should be under circadian control in the leaves of some C$_3$ species but not others such as Arabidopsis [22]. The promoter of *GmPPCK4*, but not those of the other soybean *PPCK* genes, contains an 'evening element' thought to be responsible for this temporal expression pattern [23]. It will be intriguing to see whether any other *PPCK* genes contain similar elements that might also give rise to circadian expression.

Another surprising feature of some *PPCK* genes is the existence and potential role of alternative splicing. Most *PPCK* genes contain a single intron close to the 3' end of the coding sequence, interrupting the codon for a conserved Arg residue in the sequence Arg.X.aromatic.hydrophobic at the end of the protein kinase catalytic domain [12]. However all solanaceous species examined to date (tomato, potato, aubergine and tobacco) contain a *PPCK* gene (termed *PPCK2* in tomato and potato) that contains a second intron [31]. This is in the middle of the coding sequence, shortly beyond the end of the N-terminal ATP-binding domain of the protein kinase. Alternative splicing gives rise to three transcripts, all of which can be detected *in vivo*, representing unspliced, incorrectly spliced and correctly spliced transcripts. The intron contains an in-frame stop codon such that both the unspliced and the incorrectly spliced transcripts code for a non-functional, truncated protein. Two lines of evidence suggest that this alternative splicing may have some functional significance. First, the nucleotide sequence of the intron is as highly conserved between the different solanaceous species as is the coding sequence, indicating conservation pressure. Secondly, the ratio of the different transcripts in tomato depends on tissue, conditions and metabolic context [31]. However, while several possibilities have been suggested [31], there is no firm evidence in favour of a function and this phenomenon is still a curiosity. As shown in Figure 8.1, the closest known relative of the tomato and potato *PPCK2* genes is *GmPPCK4*, which contains only one intron, as for most *PPCK* genes.

Overall, no clear correlation between sequence, expression and function has so far emerged from studies of *PPCK* genes in different species. They contain different numbers of genes controlled in different ways, likely reflecting the many different roles of PEPC in plant metabolism. Even where clear functions can be ascribed to specific genes, such as the role of *ZmPPCK1* in the light activation of C_4 PEPC, these genes probably play other roles in other cell types, and very similar genes in other species also play different roles. In general, the question of the number and function of *PPCK* genes in any plant species must be treated on its own merits and may give rise to unique answers.

8.3 Signalling pathways that control *PPCK* expression in CAM and C_4 plants

Following the first demonstration that light stimulates the phosphorylation of PEPC in maize leaves [32], the control of PEPC kinase expression and activity has been studied in range of C_4 species. Izui's group cloned a *PPCK* gene from the C_4 dust *F. trinervia* and showed that light increased the transcript abundance of this gene in leaves [19]. As noted above, maize contains four *PPCK* genes, one of which (*ZmPPCK1*) shows that properties expected of a 'C_4-type *PPCK*', namely mesophyll-specific expression and light induction. However it is not yet clear whether light affects the rate of transcription or the turnover of *PPCK* mRNA or both.

In analysis of the signalling pathways by which light affects *PPCK* transcript abundance and activity, and the phosphorylation of PEPC, use of mesophyll-cell protoplasts from *Digitaria sanguinalis* has proved particularly important. First it was shown that both light and increases in cytoplasmic pH are required for the induction of PEPC kinase activity, and that induction was blocked by pharmacological agents that perturb Ca^{2+} signalling [33]. Further work [34] showed that the signalling pathway includes transient activation of a phosphoinositide-dependent phospholipase C and production of inositol 1, 4, 5-trisphosphate which is thought to open tonoplast calcium channels. This places changes in $[Ca^{2+}]$ as a signalling component upstream of the induction of PEPC kinase activity, which is itself Ca^{2+}-independent. The role of Ca^{2+} might be to stimulate one of the calcium-dependent protein kinases [33,34], which are very numerous in plants [35]. One of the best understood components of Ca^{2+}-dependent signalling systems is animal protein kinase C (PKC), whose activity depends on both Ca^{2+} and phospholipids. No clear PKC gene has been found in plants, although there are several reports of PKC-like activities. Recently Osuna et al. [36] showed the existence of a PKC-like activity and protein in *D. sanguinalis* mesophyll protoplasts. The properties of this activity are consistent with the view that it may be involved in the *PPCK* signalling pathway: for example, inhibitors of the PKC activity also block the light-induced phosphorylation of PEPC in mesophyll-cell protoplasts [36]. However these inhibitors are non-specific and they probably inhibit many, if not all, the conventional plant calcium-dependent protein kinases. Hence the precise nature and target of the Ca^{2+}-dependent step in the signal transduction pathway in C_4 plants remains unclear. Further progress is likely to depend on identification of the transcription factors involved in controlling the expression of the C_4 *PPCK* genes and hence the mechanisms through which transcription is controlled. It will also be important to assess whether the light induction of *PPCK* expression in C_3 species involves a similar pathway.

Izui's group has suggested another mechanism through which the activity (but not expression) of PEPC kinase may be controlled in C_4 plants [16]. Maize PEPC kinase was inactivated rapidly by incubation with oxidised glutathione. This effect was reversed by subsequent incubation with reducing agents. Dithiothreitol-dependent reactivation was greatly enhanced by the presence of thioredoxin [16]. Similar findings have been made with the product of the *F. trinervia PPCK* gene expressed heterologously [19]. Thus it is possible that C_4 PEPC kinase is light-activated via thiol-disulfide exchange. However there is as yet no evidence that this mechanism is physiologically significant. It will be important to prepare tissue extracts and then measure PEPC kinase activity using conditions (such as extracts from darkened or illuminated plants, assayed in the presence or absence of dithiothreitol) that could reveal control by this mechanism.

Several other factors may impact on *PPCK* expression in C_4 species. Ueno et al. [37] noted that the phosphorylation state of PEPC in maize leaves was increased in N-limited seedlings, apparently compensating for a decrease in the abundance of PEPC itself. Another interesting finding of this study was that the phosphorylation state of PEPC did not correlate with illumination. In field-grown plants (cultivar H84) phosphorylation started to increase before dawn and declined well before sunset. Similar observations were made using glasshouse-grown seedlings (cultivar Golden Cross Bantam) but not with H84. Given the nocturnal increase in *ZmP-PCK2* transcripts, which are located in bundle sheath cells (see Section 8.2), it will be important to test whether any of the dark phosphorylation of PEPC occurs in bundle sheath tissue. In *Sorghum vulgare* phosphorylated PEPC and PEPC kinase activity can be detected in the dark in salt-stressed plants but again the cells involved have not been identified [38,39]. Little is known about the signalling through which N status and salt stress can affect PEPC kinase activity.

It is clear that CO_2 fixation in CAM species can be under circadian control [11,26,40]. Associated with this, the phosphorylation state of PEPC exhibits circadian oscillations [27] driven by circadian changes in the activity and transcript abundance of PEPC kinase [17,28]. Recent work has focused on the possible involvement of metabolites in controlling *PPCK* expression and on the nature of the circadian oscillator that controls *PPCK* and the overall process of CAM.

Early studies showed that the timing of the activation/inactivation of PEPC in CAM depends on the storage capacity of the vacuole for malate [41–43]. Borland *et al.* [44] examined the effects of treatments that disturb malate accumulation in the expression and activity of PEPC kinase. The data were consistent with the view that cytoplasmic malate (or a related metabolite) causes feedback inhibition of *PPCK* expression. Assuming that a 'central oscillator' controls gene expression as is the case in Arabidopsis [45, 46] (see below) the hypothesis was developed that circadian control of *PPCK* in CAM is actually secondary to a primary effect on malate transport across the tonoplast [11]. Consistent with this, the effects on *PPCK* expression of temperature changes (which affect permeability of the tonoplast) seem to be 'gated' by the circadian clock [47]. While the precise nature of the metabolite(s) that control *PPCK* expression remain elusive, other work has also provided evidence for such control. For example, Taybi *et al.* [48] studied *PPCK* expression in four species of *Clusia* with different photosynthetic patterns. The level of *PPCK* transcripts showed pronounced diurnal variations in species performing CAM but not in species performing C_3. However this appeared to represent a down-regulation of *PPCK* expression during the day in CAM rather than an up-regulation at night. The diurnal control of *PPCK* transcript abundance seemed to reflect the diurnal cycling of metabolites rather than to act as the primary controlling factor that drives the temporal separation of the two carboxylation processes in CAM.

Apart from CO_2 fixation in CAM, many other processes in vascular plants are under circadian control, including hypocotyl elongation, leaf movements and the expression of numerous genes such as *CAB* and *CHS*. Many of the components of a central 'clock' that drives these rhythms have been identified in Arabidopsis. In particular, the two related single Myb transcription factors 'circadian clock associated1' (*CCA1*) and 'late elongated hypocotyl' (*LHY*) and the pseudo response regulator 'timing of CAB expression1' (*TOC1*) form at least part of the central clock [45,46]. These form an autoregulatory negative feedback loop in which CCA1/LHY are synthesised around dawn and repress expression of *TOC1*. As dusk approaches, the levels of CCA1 and LHY decrease, TOC1 is synthesised and acts as a positive factor in the control of *CCA1* and *LHY* expression. Clearly a similar clock could control expression of genes involved in CAM, including *PPCK*.

An alternative view is that the machinery of CAM could itself be a self-sustaining oscillator. Transport of malate into and out of the vacuole, with the tonoplast functioning as a discrete hysteresis switch, could form the basis of a biophysical oscillator [40]. Indeed computer modelling of this oscillator showed that it could sustain robust rhythmicity of CAM compatible with experimental data [49]. However this model predicts that prevention of nocturnal malate accumulation should cause phase delays in the rhythm of CO_2 uptake [50]. In a careful study of *Kalanchoë daigremontiana*, such delays were not observed [50]. This seems to rule out the possibility that the accumulation of malic acid in the vacuole is the *central* pacemaker in CAM.

In contrast, recent evidence supports the view that CAM is under the control of a central oscillator similar to that in Arabidopsis. Boxall *et al.* [51] reported the cloning of seven cDNAs including *CCA1/LHY* and *TOC1* from the inducible CAM species *Mesembryanthemum crystallinum* that are orthologous to clock-associated genes in Arabidopsis. The transcript abundance of all seven *M. crystallinum* genes oscillated both in normal diurnal conditions and in constant conditions. Furthermore these oscillations were little affected by development or by salt stress, as would be expected of central clock components. It is also clear that numerous genes that encode components involved in CAM are under circadian control at the level of expression. Dodd *et al.* [52] showed that starch phosphorylase and β-amylase transcripts are under circadian control in CAM-induced leaves of *M. crystallinum*. In both cases, the oscillation was 180° out of phase with that of *PPCK* transcripts [52]. Hartwell [53] has now reported that many CAM genes fall under circadian control following induction of CAM in *M. crystallinum* by salt stress. Moreover some genes that are under circadian control in both C_3 and CAM-induced leaves show a pronounced phase shift in expression after induction of CAM [53], even though the phase of the central clock is unaffected [53]. This work suggests that the induction of CAM leads to the appearance of some components that bring about or affect the coupling of gene expression to the central oscillator.

The machinery involved in this coupling remains obscure. Pharmacological studies showed that the nocturnal appearance of PEPC kinase activity in detached CAM leaves could be blocked by inhibitors of Ca^{2+}/calmodulin-like interactions, phosphoinositide-dependent phospholipase C, tonoplast calcium channels and protein phosphatases [47,54,55]. Bakrim *et al.* [54] suggested that increases in cytoplasmic pH might initiate the signalling pathway leading to *PPCK* expression in CAM, rather as in C_4 leaves. However the treatment of *Kalanchoë fedtschenkoi* leaf discs with NH_4Cl to increase cytoplasmic pH had no effect on the nocturnal phosphorylation of PEPC [56]. There have been several reports of circadian oscillators of free Ca^{2+} in plants [i.e., 57], but at this stage the view that Ca^{2+} might be involved in the signalling pathway between the central oscillator and the expression of *PPCK* in CAM remains merely one attractive hypothesis.

Metabolite sensing may well be involved in the control of *PPCK* genes other than those in CAM leaves. For example, in soybean nodules expression of *GmPPCK2* and *GmPPCK3* seems to be strongly affected by the availability of photosynthate [23,24]. In leaves, both genes are more highly expressed in the light than the dark, though not as strikingly as in nodules [23]. Metabolite sensing may operate in both organs. Several other systems which may involve metabolite sensing of *PPCK* expression, or in which the phosphorylation of PEPC is at least contingent upon metabolism, have been reviewed in [12].

Most of the attention on the control of PEPC kinase in both C_4 and CAM has been at the level of expression. Of course, turnover of the PEPC kinase protein may also be involved. One important question concerning PEPC kinase has remained unresolved for several years. Analysis of PEPC kinase polypeptides by renaturation and activity staining after SDS gel electrophoresis has consistently resolved two polypeptides of 30–33 kDa and 37–39 kDa [2] in several systems. However the *PPCK* genes cloned to date encode 30–33 kDa polypeptides [12]. This led to the suggestion that PEPC kinase may be turned over by the ubiquitin/proteasome system and that the higher M_r form of PEPC kinase may be a mono-ubiquinated species [12]. Agetsuma *et al.* [58] have now obtained strong evidence for this hypothesis from a study in which tagged *F. trinervia* PEPC kinase was expressed in maize mesophyll protoplasts. Degradation of PEPC kinase polypeptide was blocked by the proteasomal inhibitor MG132, which enhanced the accumulation of a 'ladder' of PEPC kinase species with sizes separated by about 8 kDa owing to ubiquitination. It will now be important to ascertain whether the turnover of PEPC kinase is under separate control, for example by light in C_4 leaves or by the circadian clock in CAM leaves.

8.4 The 'bacterial-type' PEPC

As noted in Section 8.1, it had been assumed for many years that all vascular plant PEPCs can be controlled by phosphorylation at a serine residue close to the N-terminus of the PEPC polypeptide. However Sanchez and Cejudo [59] made the groundbreaking discovery that Arabidopsis and rice also contain a gene apparently encoding a non-phosphorylatable PEPC that is more closely related to bacterial PEPCs. A similar gene has been observed in soybean and several ESTs from various plant species corresponding to a 'bacterial-type' PEPC have been observed [23]. There are as yet no reports showing that the protein has been expressed successfully, but the possible function of this putative PEPC is a topic of great interest.

The 'bacterial-type' PEPCs are most closely related to the *Ppc2* sequence of the green alga *Chlamydomonas reinhardtii* [23,60]. Analysis of the PEPCs from both *C. reinhardtii* and (in more detail) another green alga, *Selenastrum minutum*, has shown the existence of two PEPC classes. In *S. minutum*, the Class 1 enzyme is a homotetramer of 102 kDa subunits while there are three Class 2 enzymes, which are larger complexes containing not only the 102 kDa subunit but also immunologically unrelated 130, 73 and 65 kDa subunits. The 130 kDa polypeptide seems to be another PEPC catalytic subunit that is very sensitive to an endogenous protease [61]. Similarly, in *C. reinhardtii* [62] the Class 1 PEPC is a homotetramer of 100 kDa subunits while the Class 2 PEPC is a larger complex containing additional subunits. Chollet's group [60] has now cloned, sequenced and expressed two *Ppc* genes from *C. reinhardtii*. *Ppc1* clearly encodes the 102 kDa subunits common to the Class 1 and Class 2 enzymes. While biochemical evidence is still lacking, largely due to high protease sensitivity of the product, *Ppc2* seems to encode the 130 kDa subunit of the Class 2 forms. Neither of the *C. reinhardtii* PEPC catalytic subunits contains the phosphorylation site typical of the vascular plant forms of the enzyme. Expression studies suggested that both play a non-photosynthetic, anaplerotic role [60]. In earlier work on *S. minutum*, two forms of PEPC were partially purified [63]. Both forms were inhibited by Glu, Asp and 2-oxoglutarate and activated by Gln, with the smaller form being appreciably more sensitive to these effectors. Analysis of the flux through PEPC in the dark following re-supply of NH_4^+ to N-limited cultures showed that the function of PEPC was anaplerotic and suggested that it was controlled by the Glu/Gln ratio [63,64].

Recent work has shown that vascular plants can also contain heteromeric PEPCs. Blonde and Plaxton [10] reported that developing castor oilseed endosperm contains high and low molecular mass forms of PEPC that bear remarkable physical and kinetic similarities to previously characterized Class 1 and Class 2 PEPCs from unicellular green algae. The smaller castor oilseed PEPC1 is a homotetramer of 107 kDa subunits, whereas the larger PEPC2 (native size 681 kDa) contains the same 107 kDa subunit found in PEPC1 associated with an immunologically unrelated 64 kDa polypeptide. It was hypothesised that PEPC1 (allosterically inhibited by malate, Asp and Glu) and PEPC2 (much less sensitive to effectors) support respectively storage protein synthesis and storage lipid synthesis in developing castor oilseed endosperm [10]. Sequence data from the 64 kDa polypetide of PEPC2 obtained by mass spectrometry showed marked similarity to the C-terminal half of the deduced sequences of the putative Arabidopsis and rice 'bacterial-type' PEPCs. However the authors did not determine whether the 64 kDa species might be a degradation product of a larger polypeptide. Nevertheless, one attractive possibility is that, as is the case with high molecular mass Class 2 PEPCs from green algae, the vascular plant 'bacterial-type' PEPC polypeptide usually exists as a heteromeric complex with the typical plant-type subunit. However, this area of research presents fascinating challenges. Isolation of antibodies specific to the 'bacterial-type' PEPC polypeptide will be crucial.

The expression pattern of the 'bacterial-type' PEPC gene in plants does not give strong clues as to its function. In Arabidopsis this gene is expressed exclusively in siliques, flowers and roots [59]. However the corresponding soybean gene shows a much broader expression pattern [23]. Although neither the Class 1 nor the Class 2 PEPCs of the green algae contain the phosphorylation site of the 'vascular plant-type' PEPCs, there is some evidence that both the 102 kDa and 130 kDa PEPC subunits of *S. minutum* can be phosphorylated [65]. The possibility that plants can phosphorylate the 'non-phosphorylatable, bacterial-type' PEPC should not be discounted at this stage.

8.5 Conclusions

The cloning of *PPCK* genes has allowed huge advances in our understanding of the control of PEPC but at the same time has opened up many further, often unexpected, fields for investigation. One example is the relation between the function of particular *PPCK* genes and cell specificity of their expression. Another is the recent evidence for turnover of PEPC kinase via ubiquitination which will lead to important questions about whether (and, if so, how) this process is controlled. In addition, the identification of the 'bacterial-type' PEPC gene in unicellular green algae and vascular plants and studies of the properties of the encoded polypeptide seem likely to show that the views about the control of PEPC which have been held for many years are in fact a great oversimplification. There have been several recent reports on the expression of heterologous PEPCs in transgenic plants. Altering the amount of PEPC activity and its properties can have profound effects on metabolism [66–69]. Hence studies of the functions and control of PEPC kinases and of the expression and roles of the 'bacterial-type' PEPC are not only of great intellectual interest but also of potential biotechnological importance.

Acknowledgements

I would like to thank several colleagues who have provided information prior to publication.

References

1. C.S. Andreo, D.H. Gonzales and A.A. Iglesias (1987) Vascular plant phosphoenolpyruvate carboxylase. Structure and regulation. *FEBS Letters* **213**, 1–8.
2. R. Chollet, J. Vidal and M.H. O'Leary (1996) Phosphoenolpyruvate carboxylase: a ubiquitous, highly regulated enzyme in plants. *Annual Reviews of Plant Physiology and Plant Molecular Biology* **47**, 273–296.
3. K. Izui, H. Matsumura, T. Furumoto and Y. Kai (2004) Phosphoenolpyruvate carboxylase: a new era of structural biology. *Annual Reviews of Plant Biology* **55**, 69–84.
4. Y. Kai, H. Matsumura, T. Inoue *et al.* (1999) Three-dimensional structure of phospho*enol*pyruvate carboxylase: a proposed mechanism for allosteric inhibition. *Proceedings of the National Academy of Sciences of the United States of America* **96**, 823–828.
5. H. Matsumura, M. Terada, S. Shirakata *et al.* (1999) Plausible phosphoenolpyruvate binding site revealed by 2.6 Å structure of Mn^{2+}-bound phosphoenolpyruvate carboxylase from *Escherichia coli*. *FEBS Letters* **458**, 1–4.
6. H. Matsumura, Y. Xie, S. Shirakata *et al.* (2002) Crystal structures of C_4 form maize and quaternary complex of *E. coli* phosphoenolpyruvate carboxylases. *Structure* **10**, 1721–1730.
7. R.D. Law and W.C. Plaxton (1995) Purification and characterization of a novel phosphoenolpyruvate carboxylase from banana fruit. *Biochemical Journal* **307**, 807–816.
8. S. Golombek, U. Heim, C. Horstmann, U. Wobus and H. Weber (1999) Phosphoenolpyruvate carboxylase in developing seeds of *Vicia faba* L.: gene expression and metabolic regulation. *Planta* **208**, 66–72.
9. T.F. Moraes and W.C. Plaxton (2000) Purification and characterization of phosphoenolpyruvate carboxylase from *Brassica napus* (rapeseed) suspension cell cultures. Implications for phosphoenolpyruvate carboxylase regulation during phosphate starvation, and the integration of glycolysis with nitrogen assimilation. *European Journal of Biochemistry* **267**, 4465–4476.
10. J.D. Blonde and W.C. Plaxton (2003) Structural and kinetic properties of high and low molecular mass phosphoenolpyruvate carboxylase isoforms from the endosperm of developing castor oilseeds. *Journal of Biological Chemistry* **278**, 11867–11873.
11. H.G. Nimmo (2000) The regulation of phosphoenolpyruvate carboxylase in CAM plants. *Trends in Plant Sciences* **5**, 75–80.
12. H.G. Nimmo (2003) Control of the phosphorylation of phosphoenolpyruvate carboxylase in vascular plants. *Archives of Biochemistry and Biophysics* **414**, 189–196.
13. N. Bakrim, J.L. Prioul, E. Dellens *et al.* (1993) Regulatory phosphorylation of C_4 phosphoenolpyruvate carboxylase. *Plant Physiology* **101**, 891–897.
14. P.J. Carter, H.G. Nimmo, C.A. Fewson and M.B. Wilkins (1990) *Bryophyllum fedtschenkoi* protein phosphatase type 2A can dephosphorylate phosphoenolpyruvate carboxylase. *FEBS Letters* **263**, 233–236.

15. L. Dong, N.V. Ermolova and R. Chollet (2001) Partial purification and biochemical characterization of a heteromeric protein phosphatase 2A holoenzyme from maize (*Zea mays* L.) leaves that dephosphorylates C_4 phosphoenolpyruvate carboxylase. *Planta* **213**, 379–389.
16. H. Saze, Y. Ueno, T. Hisabori, H. Hayashi and K. Izui (2001) Thioredoxin-mediated reductive activation of a protein kinase for the regulatory phosphorylation of C_4-form phosphoenolpyruvate carboxylase from maize. *Plant and Cell Physiology* **42**, 1295–1302.
17. J. Hartwell, A. Gill, G.A. Nimmo, M.B. Wilkins, G.I. Jenkins and H.G. Nimmo (1999) Phosphoenolpyruvate carboxylase kinase is a novel protein kinase regulated at the level of expression. *Plant Journal* **20**, 333–342.
18. T. Taybi, S. Patil, R. Chollet and J.C. Cushman (2000) A minimal serine/threonine protein kinase circadianly regulates phosphoenolpyruvate carboxylase activity in Crassulacean acid metabolism-induced leaves of the common ice plant. *Plant Physiology* **123**, 1471–1481.
19. Y. Tsuchida, T. Furumoto, A. Izumida, S. Hata and K. Izui (2001) Phosphoenolpyruvate carboxylase kinase involved in C_4 photosynthesis in *Flaveria trinervia*: cDNA cloning and characterization. *FEBS Letters* **507**, 318–322.
20. G.A. Nimmo, M.B. Wilkins and H.G. Nimmo (2001) Partial purification and characterization of a protein inhibitor of phosphoenolpyruvate carboxylase kinase. *Planta* **213**, 250–257.
21. Y.H. Wang and R. Chollet (1993) Partial purification and characterization of phospho*enol*pyruvate carboxylase protein-serine kinase from illuminated maize leaves. *Archives of Biochemistry and Biophysics* **304**, 496–502.
22. V. Fontaine, J. Hartwell, G.I. Jenkins and H.G. Nimmo (2002) *Arabidopsis thaliana* contains two phosphoenolpyruvate carboxylase kinase genes with different expression patterns. *Plant Cell and Environment* **25**, 115–122.
23. S. Sullivan, G.I. Jenkins and H.G. Nimmo (2004) Roots, cycles and leaves. Expression of the phosphoenolpyruvate carboxylase kinase gene family in soybean. *Plant Physiology* **135**, 2078–2087.
24. W. Xu, Y. Zhou and R. Chollet (2003) Identification and expression of a soybean nodule-enhanced PEP-carboxylase kinase gene *(NE-PpcK)* that shows striking up-/down-regulation *in vivo*. *Plant Journal* **34**, 441–452.
25. N.V. Ermolova, M.A. Cushman, T. Taybi, S.A. Condon, J.C. Cushman and R. Chollet (2003) Expression, purification, and initial characterization of a recombinant form of plant PEP-carboxylase kinase from CAM-induced *Mesembryanthemum crystallinum* with enhanced solubility in *Escherichia coli*. *Protein Expression and Purification* **29**, 123–131.
26. M.B. Wilkins (1992) Circadian rhythms – their origin and control. *New Phytologist* **121**, 347–375.
27. G.A. Nimmo, H.G. Nimmo, C.A. Fewson and M.B. Wilkins (1987) Persistent circadian rhythms in the phosphorylation state of phosphoenolpyruvate carboxylase from *Bryophyllum fedtschenkoi* leaves and in its sensitivity to inhibition by malate. *Planta* **170**, 408–415.
28. P.J. Carter, H.G. Nimmo, C.A. Fewson and M.B. Wilkins (1991) Circadian rhythms in the activity of a plant protein kinase. *EMBO Journal* **10**, 2063–2068.
29. T.L. Hennessey and C.B. Field (1987) Environmental effects of circadian rhythms in photosynthesis and stomatal opening. *Planta* **189**, 369–376.
30. A.N. Dodd, K. Parkinson and A.A.R. Webb (2004) Independent circadian regulation of assimilation and stomatal conductance in the *ztl-1* mutant of Arabidopsis. *New Phytologist* **162**, 63–70.
31. J.T. Marsh, S. Sullivan, J. Hartwell and H.G. Nimmo (2003) Structure and expression of phosphoenolpyruvate carboxylase kinase genes in Solanaceae. A novel gene exhibits alternative splicing. *Plant Physiology* **133**, 2021–2028.
32. G.A. Nimmo, G.A.L. McNaughton, C.A. Fewson, M.B. Wilkins and H.G. Nimmo (1987) Changes in the kinetic properties and phosphorylation state of phosphoenolpyruvate carboxylase in *Zea mays* leaves in response to light and dark. *FEBS Letters* **213**, 18–22.
33. N. Giglioli-Guivarc'h, J-N. Pierre, S. Brown, R. Chollet, J. Vidal and P. Gadal (1996) The light-dependent transduction pathway controlling the regulatory phosphorylation of C_4 phosphoenolpyruvate carboxylase in protoplasts from *Digitaria sanguinalis*. *Plant Cell* **8**, 573–586.
34. S. Coursol, N. Giglioli-Guivarc'h, J. Vidal and J-N. Pierre (2000) An increase in phosphoinositide-specific phospholipase C activity precedes induction of C_4 phosphoenolpyruvate carboxylase

phosphorylation in illuminated and NH$_4$Cl-treated protoplasts from *Digitaria sanguinalis*. *Plant Journal* **23**, 497–506.

35. E.M. Hrabak, C.W.M. Chan, M. Gribskov *et al.* (2003) The Arabidopsis CDPK-SnRK superfamily of protein kinases. *Plant Physiology* **132**, 666–680.
36. L. Osuna, S. Coursol, J.N. Pierre and J. Vidal (2004) A Ca^{2+}-dependent protein kinase with characteristics of protein kinase C in leaves and mesophyll cell protoplasts from *Digitaria sanguinalis*: possible involvement in the C$_4$-phosphoenolpyruvate carboxylase phosphorylation cascade. *Biochemical and Biophysical Research Communications* **314**, 428–433.
37. Y. Ueno, E. Imanari, J. Emura *et al.* (2000) Immunological analysis of the phosphorylation state of maize C$_4$-form phosphoenolpyruvate carboxylase with specific antibodies raised against a synthetic phosphorylated peptide. *Plant Journal* **21**, 17–26.
38. C. Echevarria, S. Garcia-Mauriño, R. Alvarez, A. Soler and J. Vidal (2001) Salt stress increases the Ca^{2+}-independent phosphoenolpyruvate carboxylase kinase activity in Sorghum leaves. *Planta* **214**, 283–287.
39. S. Garcia-Mauriño, J.A. Monreal, R. Alvarez, J. Vidal and C. Echeverria (2003) Characterization of salt stress-enhanced phosphoenolpyruvate carboxylase kinase activity in leaves of Sorghum vulgare: independence from osmotic stress, involvement of ion toxicity and significance of dark phosphorylation. *Planta* **216**, 648–655.
40. U. Lüttge (2000) The tonoplast functioning as the master switch for circadian regulation of crassulacean acid metabolism. *Planta* **211**, 761–769.
41. K. Winter and J.D. Tenhunen (1982) Light-stimuated burst of carbon dioxide uptake following nocturnal acidification in the Crassulacean acid metabolism plant *Kalanchoë daigremontiana*. *Plant Physiology* **70**, 1718–1722.
42. A. Fischer and M. Kluge (1984) Studies on carbon flow in Crassulacean acid metabolism during the initial light period. *Planta* **160**, 121–128.
43. A.M. Borland and H. Griffiths (1997) A comparative study on the regulation of C$_3$ and C$_4$ carboxylation processes in the constitutive crassulacean acid metabolism (CAM) plant Kalanchoë daigremontiana and the C$_3$-CAM intermediate Clusia minor. *Planta*, **201**, 368–378.
44. A.M. Borland, J. Hartwell, G.I. Jenkins, M.B. Wilkins and H.G. Nimmo (1999) Metabolite control overrides circadian regulation of phosphoenolpyruvate carboxylase kinase and CO$_2$ fixation in Crassulacean acid metabolism. *Plant Physiology* **121**, 889–896.
45. C.R. McClung (2001) Circadian rhythms in plants. *Annual Reviews of Plant Physiology and Plant Molecular Biology* **52**, 139–162.
46. M.E. Eriksson and A.J. Millar (2003) The circadian clock. A plant's best friend in a spinning world. *Plant Physiology* **132**, 732–738.
47. J. Hartwell, G.A. Nimmo, M.B. Wilkins, G.I. Jenkins and H.G. Nimmo (2002) Probing the circadian control of phospho*enol*pyruvate carboxylase kinase expression in *Kalanchoë fedtschenkoi*. *Functional Plant Biology* **29**, 663–668.
48. T. Taybi, H.G. Nimmo and A.M. Borland (2004) Expression of phosphoenolpyruvate carboxylase and phosphoenolpyruvate carboxylase kinase genes. Implications for genotypic capacity and phenotypic plasticity in the expression of Crassulacean acid metabolism. *Plant Physiology* **135**, 587–598.
49. B. Blasius, R. Neff, F. Beck and U. Lüttge (1999) Oscillatory model of Crassulacean acid metabolism with a dynamic hysteresis switch. *Philosophical Transactions of the Royal Society of London, Series B* **266**, 93–101.
50. T.P. Wyka, A. Bohn, H.M. Duarte, F. Kaiser and U.E. Lüttge (2004) Perturbations of malate accumulation and the endogenous rhythms of gas exchange in the Crassulacean acid metabolism plant *Kalanchoë daigremontiana*: testing the tonoplast-as-oscillator model. *Planta* **219**, 705–713.
51. S.F. Boxall, J.M. Foster, H.J. Bohnert, J.C. Cushman, H.G. Nimmo and J. Hartwell (2005) Conservation and divergence of circadian clock operation in a stress-inducible crassulacean acid metabolism species reveals clock compensation against stress. *Plant Physiology* **137**, 969–982.
52. A.N. Dodd, H. Griffiths, T. Taybi, J.C. Cushman and A.M. Borland (2003) Integrating diel starch metabolism with the circadian and environmental regulation of Crassulacean acid metabolism in *Mesembryanthemum crystallinum*. *Planta* **216**, 789–797.

53. J. Hartwell (2005) The circadian clock in CAM plants. In: *Annual Plant Reviews: Endogenous plant rhythms* (eds A. Hall and H.G. McWatters), pp. 211–236. Blackwell Publishing, Oxford, UK.
54. N. Bakrim, J. Brulfert, J. Vidal and R. Chollet (2001) Phosphoenolpyruvate carboxylase kinase is controlled by a similar signalling cascade in CAM and C_4 plants. *Biochemical and Biophysical Research Communications* **286**, 1158–1162.
55. H.G. Nimmo, V. Fontaine, J. Hartwell, G.I. Jenkins, G.A. Nimmo and M.B. Wilkins (2001) PEP carboxylase kinase is a novel protein kinase controlled at the level of expression. *New Phytologist* **151**, 91–97.
56. K.M. Paterson and H.G. Nimmo (2000) Effects of pH on the induction of phosphoenolpyruvate carboxylase kinase in *Kalanchoe fedtschenkoi*. *Plant Science* **154**, 135–141.
57. A.A.R. Webb (2003) The physiology of circadian rhythms in plants. *New Phytologist* **160**, 281–303.
58. M. Agetsuma, T. Furumoto, S. Yanagisawa and K. Izui (2005) The ubiquitin-proteasome pathway is involved in rapid degradation of phosphoenolpyruvate carboxylase kinase for C_4 photosynthesis. *Plant and Cell Physiology* **46**, 389–398.
59. R. Sanchez and F.J. Cejudo (2003) Identification and expression analysis of a gene encoding a bacterial-type phosphoenolpyruvate carboxylase from Arabidopsis and rice. *Plant Physiology* **132**, 949–957.
60. T.G. Mamedov, E.R. Moellering and R. Chollet (2005) Identification and expression analysis of two inorganic C- and N-responsive genes encoding novel and distinct molecular forms of eukaryotic phosphoenolpyruvate carboxylase in the green microalga *Chlamydomonas reinhardtii*. *Plant Journal* **42**, 832–843.
61. J. Rivoal, S. Trzos, D.A. Gage, W.C. Plaxton and D.H. Turpin (2001) Two unrelated phosphoenolpyruvate carboxylase polypeptides physically interact in the high molecular mass isoforms of this enzyme in the unicellular green alga *Selenastrum minutum*. *Journal of Biological Chemistry* **276**, 12588–12597.
62. J. Rivoal, W.C. Plaxton and D.H. Turpin (1998) Purification and characterization of high- and low-molecular-mass isoforms of phosphoenolpyruvate carboxylase from *Chlamydomonas reinhardtii*. *Biochemical Journal* **331**, 201–209.
63. K.A. Schuller, W.C. Plaxton and D.H. Turpin (1990) Regulation of phosphoenolpyruvate carboxylase from the green alga *Selenastrum minutum*. *Plant Physiology* **93**, 1301–1311.
64. G.C. Vanlerberghe, K.A. Schuller, R.G. Smith, R.Feil, W.C. Plaxton and D.H. Turpin (1990) Relationship between NH_4^+ assimilation rate and *in vivo* phosphoenolpyruvate carboxylase activity. *Plant Physiology* **94**, 284–290.
65. J. Rivoal, D.H. Turpin and W.C. Plaxton (2002) *In vitro* phosphorylation of phosphoenolpyruvate carboxylase from the green alga *Selenastrum minutum*. *Plant and Cell Physiology* **43**, 785–792.
66. M.S.B. Ku, S. Agarie, M. Nomura *et al.* (1999) High-level expression of maize phosphoenolpyruvate carboxylase in transgenic rice plants. *Nature Biotechnology* **17**, 76–80.
67. M. Jeanneau, J. Vidal, A. Gousset-Dupont *et al.* (2002) Manipulating PEPC levels in plants. *Journal of Experimental Botany* **53**, 1837–1845.
68. L.M. Chen, K.Z. Li, T. Miwa and K. Izui (2004) Over-expression of a cyanobacterial phospho*enol*pyruvate carboxylase with diminished sensitivity to feedback inhibition in Arabidopsis changes amino acid metabolism. *Planta* **219**, 440–449.
69. H. Rolletschek, L. Borisjuk, R. Radchuk *et al.* (2004) Seed-specific expression of a bacterial phosphoenolpyruvate carboxylase in *Vicia narbonensis* increases protein content and improves carbon economy. *Plant Biotechnology Journal* **2**, 214–219.

9 Control of sucrose biosynthesis

Elspeth MacRae and John Lunn

9.1 Introduction

Sucrose lies at the heart of plant metabolism. This disaccharide sugar is one of the main products of photosynthesis and the most common carbohydrate transported by vascular plants as a source of carbon and energy for non-photosynthetic tissues. Sucrose also acts as a signal molecule in plants, influencing both metabolism and morphological development by regulation of gene expression at multiple levels [1, 2]. It is a storage reserve in many plant species and serves as a compatible solute when they are faced with osmotic or low temperature stress. Sucrose-derived oligosaccharides, e.g. raffinose, are also used for transport and storage and some plants use sucrose esters as insecticides [3]. Few metabolites can rival the dominance of sucrose in plant metabolism, and none can replace it entirely.

In plant physiology textbooks the pathway of sucrose synthesis is usually described in the context of photosynthesis, and it is in leaves that we have the best understanding of how flux through the pathway is controlled and where most recent advances have been made. Therefore, in this chapter we focus mainly on sucrose synthesis in leaves. We begin by outlining the main pathways of sucrose synthesis in leaves, both in the light and in the dark, and then describe what is known about regulation of each step in the pathway. We then re-examine current models of how regulation of individual steps is integrated to control flux through the pathway as a whole, and how these might need to be updated in light of recent discoveries.

9.2 Pathways of sucrose biosynthesis in leaves

The committed pathway of sucrose biosynthesis contains just two steps and is essentially the same in all tissues. The first reaction, catalysed by sucrose-phosphate synthase (SPS; EC 2.4.1.14), is the synthesis of sucrose-6^F-phosphate (Suc6P) and UDP from UDPglucose (UDPGlc) and fructose 6-phosphate (Fru6P) (Figure 9.1) [4]. In the second reaction, Suc6P is hydrolysed by sucrose–phosphatase (SPP; EC 3.1.3.24), producing free sucrose and inorganic phosphate (Pi) (Figure 9.1) [5]. This reaction is irreversible and pulls the reversible SPS reaction in the direction of sucrose synthesis *in vivo* [6, 7]. Although some early work appeared to indicate that sucrose could be synthesised within chloroplasts, later studies established that SPS and SPP, and consequently sucrose synthesis, are restricted to the cytosol in both photosynthetic and non-photosynthetic tissues [8–11]. Another

CONTROL OF SUCROSE BIOSYNTHESIS 235

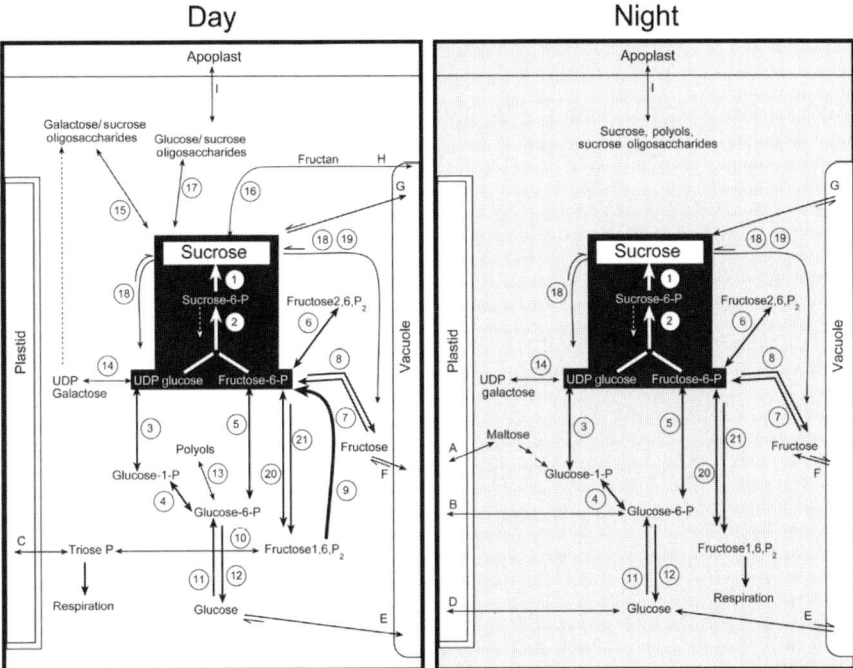

Figure 9.1 Sucrose biosynthetic pathways in the leaves during the day or night. The committed steps of sucrose synthesis are given in the black boxes. The other steps can contribute to other biosynthetic pathways as well as to sucrose synthesis. Precursor compounds enter the cytosol from the chloroplast or the vacuole, and sucrose is transported from the leaf via the apoplast and phloem to other parts of the plant. During the day, sucrose biosynthesis is primarily controlled by availability of precursor from the plastid and activity of cytFBPase and SPS. $F2,6P_2$, the product of F2KP, acts primarily as a regulator of cytFBPase, and Pi produced and consumed during sucrose synthesis are important for communication between cytosolic and plastidial metabolism. During the night, the precursors exported from the plastid change and cytFBPase is no longer an important control point. Instead SPS, transport of available precursor across both apoplast and plastid (and or vacuole) membranes and Pi regulate sucrose synthesis. In sink tissues, it is assumed (but not yet clearly demonstrated) that sucrose synthesis is regulated in a manner more analogous to the night pathway in leaves. The enzymes are numbered as follows (arrows indicate direction): 1. SPP, 2. SPS (dotted reverse arrow indicates a theoretical reverse reaction, not yet proven *in vivo*), 3. UGPase, 4. PGI, 5. PGM, 6. F2KP, 7. fructokinase or hexokinase, 8. sugar phosphate phosphatase, 9. cytFBPase, 10. aldolase and triose-phosphate isomerase, 11. glucokinase or hexokinase, 12. sugar phosphate phosphatase, 13. reductases, isomerases and polyol phosphate phosphatases (the precursor can be either Glu6P e.g. sorbitol or Fru6P e.g. mannitol), 14. UDPgalactose/glucose epimerase, 15. galactinol synthase/α-galactosyltransferase, 16. fructosyltransferase, 17. glucosyltransferase, 18. sucrose synthase, 19. invertase, 20. PFP, 21. ATP dependent phosphofructokinase. Compartmentation and membranes for transport across or diffusion through are denoted by arrows and letters as follows: A. maltose transporter, B. hexose phosphate transporter, C. TPT, D. glucose transporter, E. glucose transfer, F. fructose transfer, G. sucrose transfer, H. fructan and sucrose-oligosaccharide transfer and I. sugar transport across the plasmamembrane. (Abbreviations are contained in the text.)

cytosolic enzyme, sucrose synthase (SuSy; EC2.4.1.13), catalyses the reversible synthesis of sucrose and UDP from UDPGlc and fructose. This enzyme can produce sucrose under some circumstances in sink tissues *in vivo* [12, 13], but it usually operates in the direction of sucrose cleavage [14], and labelling studies in leaves showed that SuSy makes little or no contribution to photosynthetic sucrose synthesis [15].

Photosynthetic CO_2 fixation directly supplies the necessary precursors for sucrose synthesis during the day, but at night UDPGlc and Fru6P are supplied by breakdown of transitory starch in the chloroplasts or metabolism of other carbohydrates, such as hexoses or fructans stored in the vacuole. These pathways overlap to some extent, but there are also important differences, and in the following two sections we chart the main pathways of sucrose synthesis in illuminated and darkened leaves.

9.2.1 Sucrose synthesis in leaves during the day

In photosynthesising leaves, Fru6P and UDPGlc come from the carbon fixed in the chloroplasts by photosynthesis. The initial product of CO_2 fixation, 3-phosphoglycerate (3PGA), is converted to triose-phosphates – glyceraldehyde 3-phosphate (G3P) and dihydroxyacetone-phosphate (DHAP) – in the chloroplast stroma [16]. Triose-phosphates are exported to the cytosol in exchange for Pi via the triose-phosphate/phosphate translocator (TPT) in the inner membrane of the chloroplast envelope (Figure 9.1). In addition to triose-phosphates, this translocator can equally well transport the divalent anionic form of 3PGA. However, the high stromal pH in the light and the equilibrium position of the triose-phosphate isomerase reaction in the stroma favour the export of DHAP [16].

In the cytosol, DHAP is equilibrated with G3P by the cytosolic triose-phosphate isomerase, and these two triose-phosphates are condensed by aldolase to form fructose-1,6-bisphosphate (Fru1,6P$_2$) (Figure 9.1). Both of these reactions are freely reversible *in vivo*, unlike the next step in the pathway, the dephosphorylation of Fru1,6P$_2$ to Fru6P by the cytosolic fructose-1,6-bisphosphatase (cytFBPase). Although the reaction catalysed by the cytFBPase is irreversible, another enzyme, pyrophosphate: fructose-6-phosphate 1-phosphotransferase (PFP), reversibly catalyses the conversion of Fru1,6P to Fru6P, using Pi as co-substrate and producing PPi.

Fru6P enters the hexose-phosphate pool, where it is converted to glucose 6-phosphate (Glc6P) by phosphoglucose isomerase (PGI), and then to glucose 1-phosphate (Glc1P) by phosphoglucomutase (PGM) (Figure 9.1). Glc1P is used to synthesise UDPGlc by UDPGlc pyrophosphorylase (UGPase), with the consumption of UTP and production of inorganic pyrophosphate (PPi). All three reactions interconverting hexose-phosphates and UDPGlc are close to equilibrium and freely reversible *in vivo* [6, 16].

Up to this point, carbon is not committed to sucrose synthesis. For example, triose-phosphates can be respired via glycolysis, hexose-phosphates can enter the oxidative pentose phosphate pathway or be used to form polyols and UDPGlc has multiple fates, including production of cell wall precursors and protein glycosylation. Nevertheless, much of the flux through these reactions is directed towards sucrose synthesis. Several reactions in the pathway release Pi, which is returned to the chloroplast in exchange for further triose-phosphate. As will be discussed in more detail below (Section 9.5), the transport and levels of Pi in the cytosol and stroma play a major role in the control of sucrose synthesis and photoassimilate partitioning in the light.

9.2.2 Sucrose synthesis in leaves at night

Leaves continue to export sucrose even at night when photosynthetic carbon fixation can no longer supply the precursors for sucrose synthesis *de novo*. Some of this comes from pools of sucrose stored in the leaves during the day, but these are small in many species, and sucrose for export is synthesised *de novo* from other carbohydrate accumulated during the day. Transitory starch, which is made in the chloroplasts, is one of the most common and quantitatively important sources of precursors for sucrose synthesis in leaves at night. There have been some major advances in our understanding of this process over the last few years, especially from studies of *Arabidopsis thaliana* mutants with impaired starch catabolism, and transgenic approaches with potato. These have shown that, at least in Arabidopsis, starch is broken down predominantly by β-amylases to form maltose. Some glucose is also produced by the stromal disproportionating enzyme acting in concert with debranching enzymes (Figure 9.1) [17, 18]. Other enzymes, including α-amylase, starch phosphorylase, glucan:water dikinase and phosphoglucan:water dikinase play less well defined roles [19–21]. Maltose is exported to the cytosol via a specific maltose transporter on the inner membrane of the chloroplast envelope [17]. The conversion of maltose to sucrose in the cytosol is poorly understood. It appears to involve a cytosolic disproportionating enzyme and synthesis of a transitory glycan, which is presumably degraded to supply hexoses or hexose-phosphates for sucrose synthesis, although details remain sketchy (Figure 9.1) [22, 23]. Glucose is also exported from the chloroplasts to the cytosol by another transporter, where it is phosphorylated by hexokinase or glucokinase to produce Glc6P, which then enters the hexose-phosphate pool to become available for sucrose synthesis (Figure 9.1).

Some plants store appreciable amounts of glucose and fructose rather than sucrose in their vacuoles [24, 25]. Fructans are polymers of fructose attached to a sucrose molecule and are also stored in the vacuoles in some species. The turnover of fructans by fructan exohydrolase (FEH) releases free fructose and a small amount of sucrose (Figure 9.1) [26]. Stored hexoses or fructose from fructan breakdown are transported to the cytosol, where they can be used to synthesise sucrose after phosphorylation by hexo-, gluco- or fructokinases (Figure 9.1). It is worth noting here that the flow of carbon into sucrose from starch, hexose or fructan in leaves at night bypasses the cytFBPase, which, as we shall see in a later section (Section 9.5), is a key regulatory point in photosynthetic sucrose synthesis. Similarly, in non-photosynthetic tissues, such as germinating seeds, sprouting tubers or ripening fruits, sucrose synthesis from starch, fructan, oligosaccharides (e.g. planteose and raffinose), polyols or hexoses does not need cytFBPase activity either. In contrast, gluconeogenesis from oil reserves in germinating oilseeds does require this enzyme [27, 28].

9.3 Control of sucrose biosynthesis – the precursors

Sucrose biosynthesis in plants is controlled in many ways. Expression of genes encoding enzymes in the pathway is often developmentally regulated at the level of transcription, including during the sink–source transition in developing leaves, or the mobilisation of storage reserves in germinating seeds and ripening fruits. Post-transcriptional regulation of mRNA stability and translation also influence gene expression, and sucrose itself can affect expression of some genes [29]. However, at present, we know very little about the contribution such regulation might make to control of sucrose synthesis, nor what proteins might be involved. At the protein level, several of the enzymes in the pathway are activated or inhibited by metabolite effectors. Post-translational protein modification, especially by phosphorylation, plays a particularly important role in the control of sucrose synthesis and can influence interactions between different proteins and protein turnover. In the following sections we examine how each of the individual steps in the cytosolic pathway of sucrose synthesis is regulated and then discuss how these regulatory mechanisms are integrated to control flux through the pathway as a whole.

9.3.1 The conversion of triose-phosphate to hexose-phosphate

The TPT is not known to be regulated, but its total activity can be an important factor affecting photoassimilate partitioning between sucrose and starch. It is generally expressed only in photosynthetic tissues [30, 31]. Partial reduction in activity through antisense approaches led to a reduction in the maximal rate of photosynthesis and greater partitioning of photoassimilate into starch in potato and tobacco [32, 33]. TPT knock-out mutants in Arabidopsis compensated for complete loss of translocator activity by up-regulating the synthesis and turnover of starch and other high molecular weight polysaccharides in the light. This allowed carbon to be exported from the chloroplasts as maltose or glucose for sucrose synthesis in the cytosol, thus bypassing the missing translocator and also cytFBPase [34, 35]. The use of these 'night time' options indicates considerable flexibility in metabolic response in the leaf.

The reactions catalysed by the cytosolic triose-phosphate isomerase and aldolase are freely reversible *in vivo*. These enzymes have few known regulatory properties and are usually considered to be unimportant in control of flux through the pathway. There is also little information on the control of expression of the genes encoding these enzymes. However, a recent report that the cytosolic triose-phosphate isomerase is a target of glutathionylation, and is inhibited by oxidised glutathione, raises the possibility that its activity is sensitive to some form of redox regulation [36]. A cytosolic aldolase-like protein was also reported to be a binding target for the cytosolic thioredoxin h, suggesting that this enzyme might also be under redox control [37].

In contrast to the last two enzymes, the reaction catalysed by the cytFBPase is known to be irreversible *in vivo* and highly regulated. There are no reports on gene promoter studies, and only limited information on the regulation of expression of the genes encoding the protein, or on the levels of protein accumulation. In pea leaves, the enzyme and mRNA accumulated slowly in the presence of light, and transcripts and enzyme activity but not protein level declined with a dark treatment [38], suggesting some control by reversible enzyme modifications. Minimal variation in transcript and activity was also found during a diurnal cycle in Arabidopsis [39]. Antisense repression of the enzyme in potato and Arabidopsis plants significantly reduced the rate of sucrose synthesis, redirecting photoassimilates into starch [40, 41]. The enzyme is a homotetrameric protein and is sensitive to a wide range of metabolites including AMP and the signal metabolite fructose-2,6-bisphosphate ($Fru2,6P_2$). This has been shown both with purified enzyme [16] and recombinant protein [42]. $Fru2,6P_2$ is a potent competitive inhibitor of the cytosolic FBPase, which, at submicromolar concentrations, decreases the enzyme's affinity for $Fru1,6P_2$ by up to 100-fold, and induces strongly sigmoidal substrate saturation kinetics [16]. $Fru2,6P_2$ is synthesised and degraded by a bifunctional enzyme that has both fructose-6-phosphate,2-kinase and fructose-2,6-bisphosphatase activities (F2KP) [43].

The F2KP enzyme has been purified and genes encoding the protein from potato, Arabidopsis and spinach have been cloned and expressed *in vitro* [44–46]. Although there is no information yet on transcriptional regulation, the protein is highly regulated. The enzyme is encoded by a single gene in Arabidopsis, and the same appears to be true for other plants [43]. The individual catalytic activities are sensitive to a wide range of allosteric effectors, effectively driving the direction of enzyme activity. The kinase activity is activated by Pi, Fru6P and pyruvate and inhibited by PEP, 3PGA, DHAP, 2PGA and PPi, whereas the phosphatase activity is inhibited by Pi, Fru6P, 6-phosphogluconate and $Fru1,6P_2$ [43]. It is worth noting that both Pi and Fru6P have reciprocal effects on the two activities, amplifying the impact of changes in their concentrations on the level of $Fru2,6P_2$. In addition to the two catalytic domains, the F2KP protein has a long N-terminal domain, which modifies the kinetic properties of the catalytic domains, and is essential for assembly of the native homotetrameric protein [47]. In Arabidopsis, this domain has two phosphorylation sites, Ser220 and Ser303, of which the latter is conserved in all known plant F2KPs [43, 48]. Both of these sites can be phosphorylated by the Arabidopsis calcium-dependent protein kinase isoform 3 (CPK3), and Ser303 can also be phosphorylated by a mammalian homologue of the SNF1-related protein kinases. When phosphorylated at Ser303, F2KP can bind 14-3-3 proteins, although this appears to have no effect on the enzyme's activities [48]. Nevertheless, the phosphorylation status of the enzyme does change diurnally [49], and a comparison with analogous enzymes in animals suggests that this could affect the ratio of kinase:phosphatase activity, and therefore the level of $Fru2,6P_2$ in the cell. Transgenic plants with heterologous expression of mammalian F2K and thus higher levels of $F2,6P_2$ showed less carbon flux into sucrose in leaves. In contrast, down-regulation of F2KP in leaves of several species led to increased flux into sucrose (and/or hexose) [43]. These results indicate that the enzyme is an important regulator in directing photosynthate into sucrose.

As noted previously (Section 9.2.1), the enzyme PFP can also catalyse the conversion of $Fru1,6P_2$ to Fru6P but, in contrast to cytFBPase, the reaction is freely reversible, with PPi serving as the phosphate donor for phosphorylation of Fru6P in the reverse reaction. Interestingly, both the forward and reverse reactions of PFP are strongly activated by $Fru2,6P_2$. It seems almost perverse that this enzyme occurs in plants as it seems to override the complex regulatory mechanisms in place for controlling the cytosolic FBPase, and this apparent paradox has not yet been resolved satisfactorily. The maximum extractable activity of PFP is usually lower than that of the FBPase, and perhaps its activity *in vivo* is even lower due to inhibition by inorganic anions (e.g. Pi, sulfate and nitrate) to which the enzyme is very sensitive. Thus the function of PFP remains enigmatic, particularly as antisense plants with greatly reduced PFP activity showed no morphological phenotype and little alteration in metabolic fluxes [50, 51].

9.3.2 The hexose-phosphate pool and UDP-glucose pyrophosphorylase

The interconversions of Fru6P, Glc6P and Glc1P catalysed by PGI and PGM are freely reversible *in vivo*. There have been no reports on control of transcription or translation for these enzymes, and neither appears to have significant regulatory properties, or make an appreciable contribution to control of flux through the pathway of sucrose synthesis under normal conditions [52]. Nevertheless, the activity of PGM is not present in huge excess and antisense repression of a cytosolic PGM in potato did have a severe effect on the photosynthetic capacity and growth of the plants, but sucrose synthesis was relatively unaffected [53, 54]. Labelling studies and metabolite profiling showed that photoassimilate was diverted away from other pathways to maintain the rate of sucrose synthesis, indicating that sucrose synthesis can take priority when available resources are scarce.

UGPase does not appear to make a major contribution to control of sucrose synthesis in potato leaves. In antisense transgenic plants, a loss of up to 96% of total extractable enzyme activity had no appreciable effect on flux into sucrose, or on growth and development [55]. However, a smaller reduction (30%) in activity was reported to lead to lower sugar levels in Arabidopsis leaves [56]. Ectopic expression of an *Escherichia coli* pyrophosphatase in the cytosol of tobacco and potato led to a reduction in the amount of PPi and an increase in the UDPGlc/hexose-phosphate ratio [57], consistent with a higher flux through the UGPase reaction. The growth of the plants was retarded, and it is unclear whether the higher levels of sucrose, hexoses and starch observed in the leaves [57] were mainly attributable to the increased provision of UDPGlc for sucrose synthesis or reduced sucrose export, due to impaired sucrose metabolism in sink tissues.

There are two *Ugp* genes encoding UGPase in Arabidopsis, and two genes have also been isolated from several other species [56]. Seven *Ugp* genes have been isolated from potato, although some of these probably represent homologues from the tetraploid genome and/or allelic variants [58]. Expression of one of the *Ugp* genes in Arabidopsis is upregulated by sucrose, and this response is mediated via changes in the phosphorylation status of an unknown phosphoprotein [59]. When the phosphoprotein is dephosphorylated, the *Ugp* gene is unresponsive to sucrose. Low temperature or Pi deficiency upregulates expression of at least one of the *Ugp* genes expressed in Arabidopsis leaves, and in potato tubers one *Ugp* gene is upregulated by low temperature [59–63]. In addition to transcriptional control, the activity of the barley UGPase is affected by the degree of oligomerisation of the enzyme, with only the monomeric form having catalytic activity [64].

Potato and barley UGPases have been successfully expressed in bacteria, and both enzymes have been subjected to site-directed mutagenesis to identify residues involved in substrate binding and catalysis [64, 65]. The structure of the barley UGPase has been modelled on the experimentally determined crystal structure of the related human AGX-1 protein, and the predicted structure of the enzyme indicated that residues Gly91, Gly99, Trp191, Lys260 and Trp302 are likely to be involved in substrate binding and catalysis, in agreement with mutational analysis of the barley and potato enzymes [56].

9.4 The committed enzymes of sucrose biosynthesis

9.4.1 Sucrose-phosphate synthase

As the first enzyme in the committed pathway of sucrose synthesis, SPS lies at a strategic point for controlling flux of photoassimilate into sucrose, and the enzyme is one of the most highly regulated in plant metabolism. Numerous studies on transgenic plants with reduced or increased SPS activity have clearly shown that this enzyme makes a major contribution to control of flux through the pathway of sucrose synthesis, and changes in the activity of the enzyme often have far reaching effects on photosynthesis, growth and stress tolerance [66–78]. The flux control coefficient of SPS for sucrose synthesis was calculated to be 0.5 in photosynthesising potato leaves of antisense SPS plants with different levels of reduced enzyme activity [79], showing that SPS is an important regulator of the rate of sucrose synthesis. Leaf SPS activity has often been found to be highly correlated with growth rates and/or yield in crop plants, suggesting that the enzyme has been under strong selective pressure during domestication and breeding for higher yielding varieties ([80] and references therein).

For several years after the first genes encoding SPS were cloned from maize and spinach (*Spinacia oleracea*) [66, 81, 82], it was thought that plants contained a single *SPS* gene. However, multiple *SPS* genes were later found in several species, including Satsuma orange (*Citrus unshiu*), sugarcane and the resurrection plant *Craterostigma plantagineum* [83–85]. Full genome sequencing of Arabidopsis revealed that this species has four *SPS* genes: one each on chromosomes 1 (*AtSPS1*; At1g04920) and 4 (*AtSPS4*; At4g10120) and two on chromosome 5 (*AtSPS5a*;

At5g11110 and *AtSPS5b*; At5g20280) [86]. Many more SPS sequences are now available from the flurry of genome and expressed sequence tag (EST) sequencing projects in recent years, and detailed phylogenetic analysis has shown that these fall into four distinct families, designated A, B, C and D, with the first three having distinct mono- and dicotyledonous branches [1, 80, 86]. Of the four *SPS* genes in Arabidopsis, *AtSPS5a* and *AtSPS5b* belong to the A family, *AtSPS1* to the B family and *AtSPS4* to the C family. The fully sequenced genome of rice contains five *SPS* genes: *OsSPS1* (family B), *OsSPS2* (family D), *OsSPS6* (family D), *OsSPS8* (family A) and *OsSPS11* (family C), and the whole genome shotgun sequence of the black cottonwood tree (*Populus trichocarpa*) contains representatives from family A (three genes), family B (two genes) and family C (one gene) ([80] and unpublished data). In addition to these three species, EST or cDNA sequences from each of the A, B and C families have now been identified from a wide range of mono- and dicotyledonous species, including tomato, Satsuma orange, maize, wheat, barley, sugarcane and sorghum (*Sorghum bicolor*) [1, 80]. This leads us to predict that all Angiosperms have at least one gene from each of these gene families. Maize, wheat, barley, sugarcane and sorghum, like rice, each possess at least two *SPS* genes belonging to the D family [80]. So far, D-family genes have only been found in species from the family Poaceae (grasses), and the proteins they encode have some unique structural properties that are likely to affect their regulation. In at least three polyploid species: kiwifruit (*Actinidia* spp.), wheat and sugarcane, one or more *SPS* gene families are represented by homologous genes from the different sub-genomes and several homologues are expressed [80, 87].

The functional significance of this multiplicity of *SPS* genes is as yet only partially understood. Microarray experiments on Arabidopsis indicate that each of the four genes has different spatial and temporal expression patterns [39, 88, 89]. Similarly, northern hybridisation analysis in wheat, and analysis *in silico* of ESTs from several species, pointed to organ- and developmental stage-specific expression of the different *SPS* genes [80, 86]. Although some overlap in gene expression has often been observed at the tissue level [80, 83, 86, 90], it is still unclear whether there is co-expression of different *SPS* genes in the same cell, as most tissue samples would contain several different cell types. Native SPS is known to be a dimeric or tetrameric enzyme, but whether the holoenzyme contains different types of subunits is unknown. So far it has not been possible to assign specific functions to any one *SPS* gene family, as no consistent patterns of expression have been observed across species. For example, B-family genes appear to encode the major leaf isoforms in maize and rice, but expression of the wheat and barley B-family genes is restricted to anthers and germinating seedlings [80]. Although current evidence suggests that no broad generalisations about the function of each *SPS* gene family can be made, it has been speculated that selective pressure might have greatly altered the expression patterns of some *SPS* genes [80]. Therefore, as most of the available expression data are from domesticated species this might obscure any conserved functionality.

Analysis of publicly available microarray data from Arabidopsis indicated expression of each gene can respond to different signals (http://www.weigelworld.org;

http://web.uni-frankfurt.de/fb15/botanik/). The C-family gene, *AtSPS4*, shows a diurnal pattern of expression in leaves, peaking during the night, whereas *AtSPS5b* (A-family) transcript abundance increased during the day and decreased at night [39, 88]. Transcript levels of the second A-family gene, *AtSPS5a*, did not change during the diurnal cycle, but were greatly increased by osmotic or cold stress treatments. This suggests that the *AtSPS5a* gene encodes a stress-induced isoform, which might contribute to the increased rates of sucrose synthesis and accumulation observed under osmotic or cold stress conditions ([91–94] and references therein). This gene has also been shown to be upregulated by glucose in a manner requiring new protein synthesis [95]. Transcripts of the B-family gene, *AtSPS1*, were below the level of detection in leaf samples using microarrays. Interestingly, transcripts from this gene are relatively abundant in stamens and siliques, resembling to some extent the expression pattern of the B-family genes observed in wheat and barley.

The rice B-family gene, *OsSPS1*, is the only *SPS* gene whose promoter region has been investigated in any detail. A short region just upstream of the ATG translation start codon was found to direct gene expression in photosynthetically active cells and not in the vascular tissue of leaves [96, 97]. This matched localisation *in situ* of the corresponding mRNA and was not overridden by elements contained within a further 1700 bp upstream. The full promoter directed the gene to express in parallel with the source–sink transition in leaves. Expression was regulated by light and dependent on plastid development. The *OsSPS1* promoter also directed GUS expression in pollen and the scutellum of germinating seeds. In the latter, expression was independent of light or plastid development, and was co-regulated with secretory α-amylase expression. It was suggested that gibberellin responsive elements common to both promoters were responsible for this co-regulation. In rice, there appears to be an endogenous antisense gene that has been suggested to repress expression of the homologous *OsSPS11* gene (C-family) by post-transcriptional gene silencing [80].

SPS proteins have been purified to near homogeneity from only a few tissues, including spinach leaves (A-type), maize leaves (B-type) and germinating pea seeds [6, 81, 98], and even the most highly purified preparations have usually contained proteolytic cleavage products of the enzyme. Most of our knowledge of the leaf enzyme's properties comes from spinach and maize, which historically have been the main model species for photosynthesis research. The spinach enzyme was estimated to have a V_{max} of 150 U mg^{-1} protein [81, 99]. The spinach (A) and maize (B) leaf enzymes, and the C-family isoform from Arabidopsis, AtSPS4, have all been expressed in *E. coli* and shown to have catalytic activity and lack Pi sensitivity [66, 81, D. Gong and E.A. MacRae, unpublished data], but we still do not know whether the D-type SPSs from grasses are catalytically active.

Leaf SPSs are allosterically activated by Glc6P and inhibited by Pi [100]. The allosteric site of the protein has not yet been identified. Although the enzyme does not appear to be subject to redox control, thiol-inactivating agents were reported to affect the enzyme's allosteric properties [101], suggesting that the allosteric site contains a reduced Cys residue. The smaller SPS enzyme found in the cyanobacterium

Synechocystis sp. PCC6803 is not allosterically regulated [102], and when its sequence is aligned with plant SPSs, it can be seen to lack several regions containing highly conserved Cys residues. Although these regions are obvious candidates for the allosteric site, to our knowledge they have not yet been investigated experimentally.

In some species, especially those that store a large amount of starch in their leaves, SPS is reported to be inhibited by sucrose, but most reports have not been verified using preparations free from SuSy or SPP activities, which can interfere with studies of sucrose inhibition of SPS. The purified pea seed enzyme, which was shown to be free of contamination by SuSy and SPP, was not inhibited even by 0.5 M sucrose [6]. Neither the pea seed nor spinach leaf enzymes are inhibited by Suc6P at physiological concentrations [6, 7].

In many plants, the activation state of the leaf SPS increases in the light and decreases in the dark. Spinach leaf SPS (A-family) is deactivated in the dark by phosphorylation of Ser158, which can be catalysed by several protein kinases, including an SNF1-related protein kinase (SnRK1) and a calcium-dependent protein kinase (CDPK) ([94] and references therein). Activation in the light is brought about by dephosphorylation of the phospho-Ser158 by a type 2A protein phosphatase [103, 104]. At least one SPS kinase from spinach is inhibited by Glc6P, and the protein phosphatase is inhibited by Pi [105]. The sensitivity of the phosphatase to Pi inhibition was found to change during the diurnal cycle, possibly due to synthesis and binding of different regulatory subunits [106, 107]. In the B-family maize leaf SPS, Ser162 is in a comparable position to Ser158 in the spinach enzyme and has been shown to be the light-dark regulated phosphorylation site [100, 108]. Light activation alters the kinetic properties of these two enzymes in a broadly similar manner; the affinity for the substrates and Glc6P is increased, and sensitivity to inhibition by Pi is decreased. Contrary to some early reports, the V_{max} of the enzyme does not change in either species [109]. A consensus motif: B·H·X·B·X·X·S*·X·X·X·H, where B is a basic residue, H is a hydrophobic residue, X is any amino acid and S* is the phosphorylation site, has been identified [110], and this is present in almost all known A, B and C family SPS sequences from higher plants [1, 80]. The corresponding motif in the Arabidopsis AtSPS1 lacks a basic residue at the –6 position, but an oligopeptide containing the AtSPS1 motif was still phosphorylated by plant kinases [111]. Interestingly, the unusual D-family SPSs from grasses contain a homologous Ser residue, but lack most other elements of the consensus phosphorylation site motif [80]. A Pro residue in the –4 position, as found in most A- and D-family SPSs, is reported to block phosphorylation by CDPKs [112], and so determine the specificity of interaction with protein kinases. Other residues outside the phosphorylation site motifs themselves, in what have been called docking domains, may also influence the interaction between SPS and protein kinases [94].

The spinach leaf SPS contains a second phosphorylation site on Ser424, which is involved in osmotic stress-induced activation of the enzyme, and occurs within a similar motif to Ser158 [113]. However, in this instance, phosphorylation of Ser424 activates the enzyme. All known A- and C-family SPSs, except the A-type from mistletoe [114], contain this motif, as do the B-family SPSs but not the D-family

SPSs from grasses. The motif is also absent from the B-family SPSs from dicotyledonous plants. A third phosphorylation site, Ser229 in spinach, within the following motif R·X·X·S*·X·P, is involved in the binding of 14-3-3 proteins [115, 116]. The Ser residue, or a conservative substitution by Thr, is found in all known A-, B- and C-family SPSs, except an A-type from alfalfa and the B-type from Arabidopsis (AtSPS1), but is universally absent from the D-family enzymes [1, 80]. There is one report that binding of 14-3-3 proteins inhibits V_{max} activity [115]. However, another study found the opposite effect of 14-3-3 protein binding on 'sugar-phosphate synthase' activity, which included both SPS and trehalose-phosphate synthase activities, although the contribution of the latter was almost certainly insignificant [116]. This discrepancy remains unresolved. Plants contain several isoforms of 14-3-3 proteins. For example, there are 13 genes in Arabidopsis, but in tobacco only two isoforms appear to bind to SPS [117]. The possible significance of 14-3-3 protein binding to SPS is discussed further in Section 9.4.3. (See also Moorhead et al., Chapter 5, this volume) for further information on the role of 14-3-3 proteins in the control of primary metabolic enzymes.)

9.4.2 Sucrose-phosphatase

The characterisation of the last enzyme in the pathway of sucrose synthesis, SPP, has been something of a 'Cinderella' story. SPP was for many years dismissed as unimportant because its activity was thought to be in large excess of SPS and so unlikely to exert any control on flux through the pathway. However, re-evaluation of older data suggested that SPS activity had often been underestimated, and more recent measurements have shown that the maximal catalytic activity of SPP is not much more than that of SPS in several tissues, [118, 119]. Nevertheless, results from antisense tobacco plants indicate that, even though the enzyme is not present in large excess over SPS, it has a low flux control coefficient [119]. This suggests that SPP makes little contribution to control of flux through the pathway in tobacco leaves.

The maize *ZmSPP1* gene was the first *SPP* gene to be cloned [118], and homologous genes from many other plants have now been identified. Like SPS, plants contain multiple *SPP* genes, with at least two in maize, three in rice and the black cottonwood tree (*P. trichocarpa*) and four in Arabidopsis. Phylogenetic analysis indicates that these cluster into two broad families ([120] and unpublished data). Northern analysis in wheat and analysis *in silico* of microarray and EST data from other species suggest that one family predominates and is expressed in most tissues ([80] and unpublished data). As yet, little is known about regulation of transcription or post-transcriptional modifications of SPP.

Crystal structures of the cyanobacterial SPP from *Synechocystis* sp. PCC6803, alone and as a complex with various ligands i.e. crystallised as protein only and also crystallised with SPP or sucrose or glucose, etc., have recently been solved and suggest a two-step reaction mechanism involving transfer of the phosphate group of Suc6P to a conserved Asp residue (Asp9) in the active site, followed by hydrolysis of the phospho-Asp-enzyme intermediate to release free sucrose and Pi [121]. The crystal structures also revealed the importance of a glucose-binding site for substrate recognition and showed that the weak competitive inhibition of the enzyme by sucrose (typical $Ki > 200$ mM) is due to binding in the active site.

9.4.3 Evidence for a metabolon in sucrose biosynthesis

Studies on partially purified rice leaf SPS and SPP indicated that the two enzymes can form a stable complex [122]. Intriguingly, cloning of the *ZmSPP1* gene revealed that SPS contains an SPP-like domain at its C-terminus (Figure 9.2), and it was speculated that this might be involved in binding SPP to form a complex [118]. Modelling of the SPP-like domain of SPS on the *Synechocystis* SPP structure suggested that all the residues involved in Suc6P-binding, except the critical Asp residue needed for catalysis, were conserved in the SPS [121], indicating that the function of this domain might be to channel Suc6P within an SPS-SPP metabolon. However, at present, the existence of an SPS-SPP metabolon must be regarded as speculative. Additional binding studies and detailed kinetic analyses using purified enzymes are urgently required to test this hypothesis.

Although its existence remains unproven, the presence of an SPS-SPP metabolon might offer further levels of regulation of sucrose synthesis. When Arabidopsis cells were starved of sucrose, the SPS protein was no longer bound by 14-3-3 proteins and was proteolytically cleaved [123]. The truncated enzyme retained catalytic activity and was about the size expected (~90 kDa) if the SPP-like domain had been removed. If the model for an SPS-SPP metabolon is correct, the truncated SPS would presumably be unable to form a complex with SPP and metabolite channelling between the enzymes would be abolished. Thus there is circumstantial

SPS and SPP domain structure

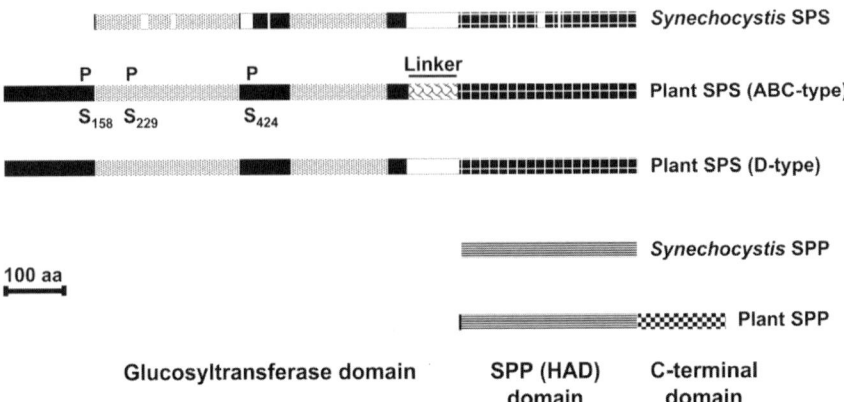

Figure 9.2 Domain structure comparison for SPS and SPP from vascular plants and cyanobacteria. The position of each of the known regulatory phosphoserines is indicated; Ser158 is light regulated, Ser229 interacts with 14-3-3 proteins and Ser424 is osmotically activated. HAD – haloacid dehalogenase.

evidence that proteolytic processing of SPS might play some role in controlling the activity of the two enzymes in the committed pathway of sucrose synthesis [1]. Interestingly, purified SPS preparations often contain protein fragments of about 30 kDa and 90 kDa in size, suggesting that the enzyme is particularly prone to proteolytic cleavage in a linker region, of about 80–90 residues, between the larger N-terminal catalytic domain and the SPP-like domain. Curiously, the D-family SPSs from grasses, which appear to lack most of the protein phosphorylation site motifs, also lack this linker region with its putative proteolytic cleavage site. Consequently, these unique isoforms appear to have much less regulatory potential than other types of SPS.

9.5 Integrated pathway control

The pathway of sucrose synthesis in the cytosol must be coordinated with the rate of photosynthetic CO_2 fixation in illuminated leaves. If flux is too high the release of excessive Pi in the cytosol and exchange via the triose-phosphate/phosphate translocator could drain too much triose-phosphate from the chloroplast stroma, limiting the regeneration of ribulose-1,5-bisphosphate. Conversely, if the flux is too low, the build-up of phosphorylated intermediates in the stroma would deplete the stromal Pi pool and so inhibit photophosphorylation. In some species, especially those that store relatively little sucrose in their leaves, the rate of sucrose synthesis decreases as sucrose accumulates in the leaf and photoassimilates are diverted into other pathways, e.g. starch synthesis. A model, based originally on data from spinach leaves, was proposed to explain how changes in stromal and cytosolic concentrations of metabolites, e.g. triose-phosphates, $Fru2,6P_2$, hexose-phosphates and Pi, could link sucrose synthesis in the cytosol to photosynthetic CO_2 fixation and starch synthesis in the chloroplast in the light [16]. The original model described two basic processes: (i) feed-forward activation and (ii) feedback inhibition. In brief, triose-phosphate export from the plastid following photosynthesis in exchange for Pi increases triose-phosphate and decreases Pi levels in the cytosol. These changes allosterically regulate F2KP lowering the concentration of $Fru2,6P_2$. This change and the synthesis of $Fru1,6P_2$ from triose-phosphates increase flux through the cytFBPase into hexose-phosphates, and higher hexose-phosphate and UDPGlc levels feed-forward to activate SPS. In contrast, feedback inhibition of SPS and/or SPP by sucrose was proposed to lead to accumulation of hexose-phosphates, including Fru6P, which would override the allosteric regulation of F2KP causing an increase in $Fru2,6P_2$ and inhibition of the cytFBPase. Together these changes would restrict the release of Pi in the cytosol and so limit export of triose-phosphate from the chloroplast. The resulting build-up of 3PGA and fall in stromal Pi levels would activate ADPglucose pyrophosphorylase (AGPase), diverting surplus photoassimilate into starch synthesis.

This basic model has evolved considerably over time as new discoveries have been incorporated [1, 94, 100, 109, 124], with a greater emphasis on the importance of changes in amount and activation state of key enzymes, such as SPS by protein phosphorylation, and AGPase by redox-modulation [125]. Studies with transgenic plants with constitutively higher or lower levels of Fru2,6P$_2$ have confirmed the importance of this metabolite in coordination of chloroplast and cytosolic metabolism, but downplay its role in control of photoassimilate partitioning between sucrose and starch [126–132]. The recent discovery that F2KP too might be subject to regulation by protein phosphorylation [43] adds yet another dimension to the control of the pathway, as does the potential for redox-modulation of several of the enzymes (triose-phosphate isomerase, aldolase and UGPase) in the pathway.

Our knowledge of short-term feedback regulation of sucrose synthesis is still fragmentary, and there appear to be important differences between species. In fact, there may not even be short-term feedback regulation in plants that store a lot of sucrose e.g. maize and wheat [133, 134]. Sucrose-induced inactivation of SPS has been reported for spinach leaves [135], and attributed to changes in phosphorylation status, but this has not yet been verified experimentally. Although sucrose does inhibit SPP, under most circumstances this is probably unimportant as the K_i is so high. There is also no obvious mechanism for subsequent feedback inhibition of SPS, as the enzyme is not inhibited by Suc6P [6, 7]. Direct inhibition of SPS by sucrose may occur in some starch-storing species. In pea, potato and Arabidopsis leaves (all starch storers), accumulation of sucrose induces redox activation of AGPase [125], and competition from increased rates of starch synthesis for precursors would also be expected to slow the rate of sucrose synthesis. In fructan-storing plants, accumulation of sucrose induces transcription and functional expression of the fructosyltransferases that synthesise fructan, and this response is mediated by protein kinases and phosphatases [136]. Long-term source–sink imbalances that lead to accumulation of sucrose in the leaves are compensated for by down-regulation of photosynthesis-related genes [137].

As already noted, sucrose synthesis in leaves at night derived from other stored carbohydrates, be it starch, hexoses or fructans, effectively bypasses regulation by the cytFBPase, and presumably any involvement of Fru2,6P$_2$. Therefore, it seems likely that much of the regulation of the cytosolic pathway falls on SPS. Microarray data from Arabidopsis suggest that different SPS isoforms, perhaps with different substrate affinities and allosteric properties, are expressed at different times of day, possibly indicating that there is a 'day shift'' and a 'night shift'. Analysis of diurnal changes in the various SPS proteins and their kinetic properties will be needed to test this hypothesis.

9.6 Future perspectives

It is clear that we know a great deal about the control of sucrose synthesis in leaves, at least in the light, and arguably, this is one of the best understood of all pathways in plant primary metabolism. Nevertheless, there seems little doubt that our understanding of pathway control will continue to evolve as new information becomes available, and we conclude this chapter with some crystal ball gazing into the future.

The wealth of publicly available microarray data from Arabidopsis, and increasingly other species, is already leading to identification of co-regulated genes and discovery of the promoter regulatory elements they have in common. This in turn should help us to identify the transcription factors that control their expression. Although microarray data can offer us great insight into control of gene expression at the level of transcription, we must not lose sight of the many other possible ways in which gene expression, in its broadest sense, can be controlled, ranging from regulation of mRNA stability and translation to post-translational modification and proteolytic turnover. Perhaps a new version of the old proverb – 'There's many a slip betwixt the cup and the lip' – might read: 'There's many a slip betwixt the chip and the clip'. This caution is especially justified in the case of sucrose synthesis, where we know that transcript abundance does not always mirror enzyme activity or vice versa [39]. Nor should we forget that several glycolytic enzymes can also act as transcription factors [138], and therefore bear in mind that enzymes in the pathway of sucrose synthesis might also have dual functions.

Quantitative trait loci (QTL) affecting leaf sucrose accumulation, and activities of SPS and sucrose-metabolising enzymes have been identified in maize and sugarcane [139, 140]. Evidence suggesting coordinated regulation of several sugar metabolising genes has been demonstrated in Arabidopsis through QTL analysis, with some genes mapping very close to their QTL for enzyme activity [141]. Hence the promoter regions may have some common element and use a common regulatory molecule. Genome sequencing, EST projects and generation of markers based on microsatellites or single nucleotide polymorphisms have opened up a rich seam of genetic resources for narrowing down the search for QTL and the genes that contribute to multi-genic traits.

Large scale sequencing efforts have also revealed an often unexpected multiplicity and diversity of genes encoding enzymes and other proteins of interest, where previously only single genes were known – SPS being a good example. This presents us with the challenge of discovering the functional significance of the different isoforms. Analysis of spatial and temporal expression patterns, and heterologous expression and characterisation of proteins will be required for a fuller understanding of gene diversity. Resolution of crystal structures is starting to reveal the intimate secrets of some of the enzymes of sucrose synthesis, including their catalytic mechanisms and how they are regulated. SPS, with its wealth of regulatory properties, must surely be a high priority for such studies. Three-D structures are needed to understand options for oligomeric arrangements and how the catalytic site and regulatory sites function. For example, do all subunits become phosphorylated, or does only one? Do different subunits have different serines phosphorylated? Do different subunits come from different genes? How might interactions with other proteins take place?

Redox regulation in plants is one area that is receiving increasing attention, and there is clear evidence that it extends far beyond the classical redox-sensitive enzymes, such as the plastidial FBPase, NADP-malate dehydrogenase and phosphoribulokinase, in the chloroplasts [142]. There are already tantalising hints that several of the cytosolic enzymes involved in sucrose synthesis (e.g. triose-phosphate isomerase, aldolase and UGPase) might be regulated in this way. Another type of post-translational modification, protein phosphorylation, is well established as a major player in control of sucrose synthesis, but we still know too little about the signal cascades that lead to modulation of the target enzymes, such as SPS, or the ramifications of having promiscuous protein kinases and phosphatases. Other signalling pathways and mechanisms might also be involved. For example, it has been suggested that hexokinase signalling could coordinate starch degradation and sucrose synthesis during the night in leaves linked to the transfer of non-phosphorylated intermediates between the chloroplasts and the cytosol [78]. Our knowledge of protein turnover is even more rudimentary than our understanding of these various signalling pathways.

In this chapter we have only been able to touch on sucrose synthesis in non-photosynthetic tissues. Essential functions of non-photosynthetic sucrose synthesis include remobilisation of storage reserves in germinating seeds, ripening fruit, sprouting tubers or woody stems of perennial plants, generation of concentration gradients to drive phloem unloading and protection against drought, salinity or cold stress. Its function in other tissues is more obscure. For example, there appear to be isoforms of SPS and SPP that are almost specific to pollen in Arabidopsis and some grasses ([80] and unpublished data), but the need for sucrose synthesis in pollen is unclear. Futile cycles of sucrose cleavage by SuSy or invertases, and resynthesis via SPS and SPP have been proposed to regulate net rates of sucrose utilisation in sink tissues [143], with possible implications for sink strength and plant yield. The little information that is available suggests that regulatory mechanisms similar to those in leaves at night also operate in non-photosynthetic tissues, although they may be adapted to meet the special needs of different tissues.

The flexibility of plant metabolism, demonstrated by several examples of mutants or transgenic plants with unexpected phenotypes, indicates that there are many regulatory networks still to be discovered. A good example of this is the unexpected regulation of secondary metabolism through carbohydrate accumulation caused by a mutation in a sucrose transporter [144]. Ironically, increasing amounts of information and better knowledge of these networks will probably make it more difficult for any one individual to conceptualise control of metabolic pathways. It seems likely that only computer modelling and simulation will allow us to handle such complexity, and the development of such tools will be essential to understand control of sucrose synthesis in plants.

References

1. J.E. Lunn and E.A. MacRae (2003) New complexities in the synthesis of sucrose. *Current Opinion in Plant Biology* **6**, 208–214.
2. S.I. Gibson (2005) Control of plant development and gene expression by sugar signalling. *Current Opinion in Plant Biology* **8**, 93–102.
3. G. Avigad and P.M. Dey (1997) Carbohydrate metabolism: storage carbohydrates. In: *Plant Biochemistry* (eds P.M. Dey and J.B. Harborne), pp. 143–204. Academic, San Diego, London, Boston, New York, Sydney, Tokyo, Toronto.

4. L.F. Leloir and C.E. Cardini (1955) The biosynthesis of sucrose phosphate. *Journal of Biological Chemistry* **214**, 157–165.
5. J.S. Hawker and M.D. Hatch (1966) A specific sucrose phosphatase from plant tissues. *The Biochemical Journal* **99**, 102–107.
6. J.E. Lunn and T.ap Rees (1990) Apparent equilibrium constant and mass-action ratio for sucrose–phosphate synthase from seeds of *Pisum sativum*. *The Biochemical Journal* **267**, 739–743.
7. K.P. Krause and M. Stitt (1992) Sucrose-6-phosphate levels in spinach leaves and their effects on sucrose–phosphate synthase. *Phytochemistry* **31**, 1143–1146.
8. S.P. Robinson and D.A. Walker (1979) The site of sucrose synthesis in isolated leaf protoplasts. *FEBS Letters* **107**, 295–299.
9. M. Nishimura and H. Beevers (1979) Subcellular-distribution of gluconeogenetic enzymes in germinating castor bean endosperm. *Plant Physiology* **64**, 31–37.
10. E. Echeverria and G. Salerno (1993) Intracellular localization of sucrosephosphate phosphatase in photosynthetic cells of lettuce (*Lactuca sativa*). *Physiologia Plantarum* **88**, 434–438.
11. E. Echeverria (1995) Intracellular localization of sucrose–phosphate phosphatase in storage cells. *Physiologia Plantarum* **95**, 559–562.
12. P. Geigenberger and M. Stitt (1993) Sucrose synthase catalyses a readily reversible reaction *in vivo* in developing potato tuber and other plant tissues. *Planta* **189**, 329–339.
13. U. Römer, H. Schrader, N. Günther, N. Nettelstroth, W.B. Frommer and L. Elling (2004) Expression, purification and characterization of recombinant sucrose synthase 1 from *Solanum tuberosum* L. for carbohydrate engineering. *Journal of Biotechnology* **107**, 135.
14. R. Wendler, R. Veith, J. Dancer, M. Sitt and E. Komor (1990) Sucrose storage in cell suspension cultures of *Saccharum* sp. (sugarcane) is regulated by a cycle of synthesis and degradation. *Planta* **183**, 31–39.
15. H.G. Pontis (1977) Riddle of sucrose. In: *International Review of Biochemistry*, Vol. 13 (ed. D.H. Northcote), University Park Press, Baltimore, pp. 80–117. Plant Biochemistry II.
16. M. Stitt, S. Huber and P. Kerr (1987) Control of photosynthetic sucrose formation. In: *The Biochemistry of Plants*, Vol. 10, Photosynthesis (eds M.D. Hatch and N.K. Broadman), Academic, New York, pp. 327–409.
17. T. Niittylä, G. Messerli, M. Trevisan, J. Chen, A.M. Smith and S.C. Zeeman (2004) A previously unknown maltose transporter essential for starch degradation. *Science* **303**, 87–89.
18. S.E. Weise, A.P.M. Weber and T.D. Sharkey (2004) Maltose is the major form of carbon exported from the chloroplast at night. *Planta* **218**, 474–482.
19. D. Stanley, A.M. Fitzgerald, K.J.F. Farnden and E.A. MacRae (2002) Characterisation of putative alpha-amylases from apple (*Malus domestica*) and *Arabidopsis thaliana*. *Biologia* **57**, 137–148.
20. S.C. Zeeman, S.M. Smith and A.M. Smith (2004) The breakdown of starch in leaves. *New Phytologist* **163**, 247–261.
21. O. Kotting, K. Pusch, A. Tiessen, P. Geigenberger, M. Steup and G. Ritte (2005) Identification of a novel enzyme required for starch metabolism in Arabidopsis leaves. The phosphoglucan, water dikinase. *Plant Physiology* **137**, 242–252.
22. T. Chia, D. Thorneycroft, A. Chapple *et al.* (2004) A cytosolic glucosyltransferase is required for conversion of starch to sucrose in Arabidopsis leaves at night. *The Plant Journal* **37**, 853–863.
23. Y. Lu and T.D. Sharkey (2004) The role of amylomaltase in maltose metabolism in the cytosol of photosynthetic cells. *Planta* **218**, 466–473.
24. A.H. Kingston-Smith, N. Galtier, C.J. Pollock and C.H. Foyer (1998) Soluble acid invertase in leaves is independent of species differences in leaf carbohydrates, diurnal sugar profiles and paths of phloem loading. *New Phytologist* **139**, 283–292.
25. J. Müller, R.A. Aeschbacher, N. Sprenger, T. Boller and A. Wiemken (2000) Disaccharide-mediated regulation of sucrose:fructan-6-fructosyltransferase, a key enzyme of fructan synthesis in barley leaves. *Plant Physiology* **123**, 265–273.
26. A. Morvan-Bertrand, J. Boucaud, J. LeSaos and M.P. Prud'homme (2001) Roles of the fructans from leaf sheaths and from the elongating leaf bases in the regrowth following defoliation of *Lolium perenne* L. *Planta* **213**, 109–120.

27. T. apRees (1987) Compartmentation of plant metabolism. In: *The Biochemistry Of Plants*, Vol 12, Physiology of metabolism (ed. D.D. Davies), Academic, San Diego, London, pp. 87–115.
28. H.W. Heldt (1997) *Plant Biochemistry and Molecular Biology*. Oxford University Press, Oxford.
29. A. Wiese, N. Elzinga, B. Wobbes and S. Smeekens (2004) A conserved upstream open reading frame mediates sucrose-induced repression of translation. *Plant Cell* **16**, 1717–1729.
30. B. Schulz, W.B. Frommer, U.I. Flügge, S. Hummel, K. Fischer and L. Willmitzer (1993) Expression of the triose phosphate translocator gene from potato is light dependent and restricted to green tissues. *Molecular General Genetics* **238**, 357–361.
31. B. Kammerer, K. Fischer, B. Hilpert, S. Schubert, M. Gutensohn, A. Weber and U.I. Flügge (1998) Molecular characterization of a carbon transporter in plastids from heterotrophic tissues: the glucose 6-phosphate/phosphate antiporter. *Plant Cell* **10**, 105–117.
32. J.W. Riesmeier, U.I. Flügge, B. Schulz *et al.* (1993) Antisense repression of the chloroplast triose phosphate translocator affects carbon partitioning in transgenic potato plants. *Proceedings of the National Academy of Sciences of the United States of America* **90**, 6160–6164.
33. S.A. Barnes, J.S. Knight and J.C. Gray (1994) Alteration of the amount of the chloroplast phosphate translocator in transgenic tobacco affects the distribution of assimilate between starch and sugar. *Plant Physiology* **106**, 1123–1129.
34. A. Schneider, R.E. Häusler, Ü. Kolukisaoglu *et al.* (2002) An *Arabidopsis thaliana* knock-out mutant of the chloroplast triose phosphate/phosphate translocator is severely compromised only when starch synthesis, but not starch mobilisation is abolished. *The Plant Journal* **32**, 685–699.
35. R.G. Walters, D.G. Ibrahim, P. Horton and N.J. Kruger (2004) A mutant of Arabidopsis lacking the triose-phosphate/phosphate translocator reveals metabolic regulation of starch breakdown in the light. *Plant Physiology* **135**, 891–906.
36. H. Ito, M. Iwabuchi and K. Ogawa (2003) The sugar-metabolic enzymes aldolase and triose-phosphate isomerase are targets of glutathionylation in *Arabidopsis thaliana*: detection using biotinylated glutathione. *Plant Cell Physiology* **44**, 655–660.
37. D. Yamazaki, K. Motohashi, T. Kasama, Y. Hara and T. Hisabori (2004) Target proteins of the cytosolic thioredoxins in *Arabidopsis thaliana*. *Plant Cell Physiology* **45**, 18–27.
38. S.W. Lee and T.R. Hahn (2003) Light-regulated differential expression of pea chloroplast and cytosolic fructose-1,6-bisphosphatases. *Plant Cell Reports* **21**, 611–618.
39. Y. Gibon, O.E. Bläsing, N. Palacios-Rojas *et al.* (2004) Adjustment of diurnal starch turnover to short days: depletion of sugar during the night leads to a temporary inhibition of carbohydrate utilization, accumulation of sugars and post-translational activation of ADP-glucose pyrophosphorylase in the following light period. *The Plant Journal* **39**, 847–862.
40. R. Zrenner, K-P. Krause, P. Apel and U. Sonnewald (1996) Reduction of the cytosolic fructose-1,6-bisphosphatase in transgenic potato plants limits photosynthetic sucrose biosynthesis with no impact on plant growth and tuber yield. *The Plant Journal* **9**, 671–681.
41. A. Strand, R. Zrenner, S. Trevanion, M. Stitt, P. Gustaffson and P. Gardeström (2000) Decreased expression of two enzymes in the sucrose biosynthesis pathway, cytosolic fructose-1,6-bisphosphatase and sucrose phosphate synthase, has remarkably different consequences for photosynthetic carbon metabolism in transgenic *Arabidopsis thaliana*. *Plant Journal* **23**, 759–770.
42. H.K. Jang, S.W. Lee, Y.H. Lee and T.R. Hahn (2003) Purification and characterization of a recombinant pea cytoplasmic fructose-1,6-bisphosphatase. *Protein Expression and Purification* **28**, 42–48.
43. T.H. Nielsen, J.H. Rung and D. Villadsen (2004) Fructose-2,6-bisphosphate: a traffic signal in plant metabolism. *Trends in Plant Science* **9**, 556–563.
44. H. Draborg, D. Villadsen and T.H. Nielsen (1999) Cloning, characterization and expression of a bifunctional fructose-6-phosphate, 2-kinase/fructose-2,6-bisphosphatase from potato. *Plant Molecular Biology* **39**, 709.
45. D. Villadsen, J.H. Rung, H. Draborg and T.H. Nielsen (2000) Structure and heterologous expression of a gene encoding fructose-6-phosphate 2-kinase/fructose-2,6-bisphosphatase from *Arabidopsis thaliana*. *Biochimica Biophysica Acta* **1492**, 406–413.
46. J.E. Markham and N.J. Kruger (2002) Kinetic properties of bifunctional 6-phosphofructo-2-kinase/fructose-2,6-bisphosphatase from spinach leaves. *European Journal of Biochemistry* **269**, 1267–1277.

47. D. Villadsen and T.H. Nielsen (2001) N-terminal truncation affects the kinetics and structure of fructose-6-phosphate 2-kinase/fructose-2,6-bisphosphatase from *Arabidopsis thaliana*. *The Biochemical Journal* **359**, 591–597.
48. A. Kulma, D. Villadsen, D.G. Campbell *et al.* (2004) Phosphorylation and 14-3-3 binding of Arabidopsis 6-phosphofructo-2-kinase/fructose-2,6-bisphosphatase. *The Plant Journal* **37**, 654–667.
49. T. Furomoto, M. Teramoto, N. Imada, M. Ito, I. Nishoda and A. Watanabe (2001) Phosphorylation of a bifunctional enzyme, 6-phosphofructo-2-kinase/fructose-2,6-bisphosphate 2-phosphatase, is regulated physiologically and developmentally in rosette leaves of *Arabidopsis thaliana*. *Plant Cell Physiology* **42**, 1044–1048.
50. M. Hajirezaei, U. Sonnewald, R. Viola, S. Carlisle, D. Dennis and M. Stitt (1994) Transgenic potato plants with strongly decreased expression of pyrophosphate: fructose-6-phosphate phosphotransferase show no visible phenotype and only minor changes in metabolic fluxes in their tubers. *Planta* **192**, 277–283.
51. M. Paul, U. Sonnewald, M. Hajirezaei, D. Dennis and M. Stitt (1995) Transgenic tobacco plants with strongly decreased expression of pyrophosphate: fructose-6-phosphate 1 phosphotransferase do not differ significantly from the wildtype in photosynthate partitioning, plant growth or ability to cope with limiting phosphate, limiting nitrogen and suboptimal temperatures. *Planta* **196**, 277–283.
52. M. Sitt and U. Sonnewald (1995) Regulation of metabolism in transgenic plants. *Annual Review of Plant Physiology and Plant Molecular Biology* **46**, 341–368.
53. A.R. Fernie, E. Tauberger, A. Lytovchenko, U. Roessner, L. Willmitzer and R.N. Trethewey (2002) Antisense repression of cytosolic phosphoglucomutase in potato (*Solanum tuberosum*) results in severe growth retardation, reduction in tuber number and altered carbon metabolism. *Planta* **214**, 510–520.
54. A. Lytovchenko, L. Sweetlove, M. Pauly and A.R. Fernie (2002) The influence of cytosolic phosphoglucomutase on photosynthetic carbohydrate metabolism. *Planta* **215**, 1013–1021.
55. R. Zrenner, L. Willmitzer and U. Sonnewald (1993) Analysis of the expression of potato uridinediphosphate-glucose pyrophosphorylase and its inhibition by antisense RNA. *Planta* **190**, 247–252.
56. L.A. Kleczkowski, M. Giesler, I. Ciereszko and H. Johansson (2004) UDP-glucose pyrophosphorylase. An old protein with new tricks. *Plant Physiology* **134**, 912–918.
57. T. Jelitto, U. Sonnewald, L. Willmitzer, M. Hajirezaei and M. Stitt (1992) Inorganic pyrophosphate content and metabolites in potato and tobacco plants expressing *E. coli* pyrophosphatase in their cytosol. *Planta* **188**, 238–244.
58. J.R. Sowokinos, V. Vigdorovich and M. Abrahamsen (2004) Molecular cloning and sequence variation of UDP-glucose pyrophosphorylase cDNAs from potatoes sensitive and resistant to cold sweetening. *Journal of Plant Physiology* **161**, 947–955.
59. I. Ciereszko, H. Johansson and L.A. Kleczkowski (2001) Sucrose and light regulation of a cold-inducible UDP-glucose pyrophosphorylase gene via a hexokinase-independent and abscisic acid-insensitive pathway in Arabidopsis. *The Biochemical Journal* **354**, 67–72.
60. J.P. Spychalla, B.E. Scheffler, J.R. Sowokinos and M.W. Bevan (1994) Cloning, antisense RNA inhibition, and the coordinated expression of UDP-glucose pyrophosphorylase with starch biosynthetic genes in potato-tubers. *Journal of Plant Physiology* **144**, 444–453.
61. A.Y. Borovkov, P.E. McClean and G.A. Secor (1997) Organization and transcription of the gene encoding UDP-glucose pyrophosphorylase. *Gene* **186**, 293–297.
62. I. Ciereszko, H. Johansson, V. Hurry and L.A. Kleczkowski (2001) Phosphate status affects the gene expression, protein content and enzymatic activity of UDP-glucose pyrophosphorylase in wild-type and *pho* mutants of Arabidopsis. *Planta* **212**, 598–605.
63. O. Repetto, G. Bestel-Corre, E. Dumas-Gaudot, G. Berta, V. Gianinazzi-Pearson and S. Gianinazzi (2003) Targeted proteomics to identify cadmium-induced protein modifications in Glomus mosseae-inoculated pea roots. *New Phytologist* **157**, 555–567.
64. F. Martz, M. Wilczynska and L.A. Kleczkowski (2002) Oligomerization status, with the monomer as active species, defines catalytic efficiency of UDP-glucose pyrophosphorylase. *The Biochemical Journal* **367**, 295–300.

65. Y. Kazuta, Y. Omura, M. Tagaya, K. Nakano and T. Fukui (1991) Identification of lysyl residues located at the substrate-binding site in UDP-glucose pyrophosphorylase from potato tuber: affinity labeling with uridine di- and triphosphopyridoxals. *Biochemistry* **30**, 8541–8545.
66. A.C. Worrell, J.M. Brunneau, K. Summerfelt, M. Boersig and T.A. Voelker (1991) Expression of maize sucrose phosphate synthase in tomato alters leaf carbohydrate partitioning. *Plant Cell* **3**, 1121–1130.
67. N. Galtier, C.H. Foyer, J. Huber, T.A. Voelker and S.C. Huber (1993) Effects of elevated sucrose-phosphate synthase activity on photosynthesis, assimilate partitioning, and growth in tomato (*Lycopersicon esculentum* var UC82B). *Plant Physiology* **101**, 535–543.
68. N. Galtier, C.H. Foyer, E. Murchie *et al.* (1995) Effects of light and atmospheric carbon dioxide enrichment on photosynthesis and carbon partitioning in the leaves of tomato (*Lycopersicon esculentum* L.) plants over-expressing sucrose-phosphate synthase. *Journal of Experimental Botany* **46**, 1335–1344.
69. B.J. Micallef, K.A. Haskins, P.J. Vanderveer, K.S. Roh, C.K. Shewmaker and T.D. Sharkey (1995) Altered photosynthesis, flowering, and fruiting in transgenic tomato plants that have an increased capacity for sucrose synthesis. *Plant* **196**, 327–334.
70. M.M. Laporte, J.A. Galagan, J.A. Shapiro, M.R. Boersig, C.K. Shewmaker and T.D. Sharkey (1997) Sucrose-phosphate synthase activity and yield analysis of tomato plants transformed with maize sucrose-phosphate synthase. *Planta* **203**, 253–259.
71. K. Ono, K. Ishimaru, N. Aoki and R. Ohsugi (1999) Transgenic rice with low sucrose-phosphate synthase activities retain more soluble protein and chlorophyll during leaf senescence. *Plant Physiology and Biochemistry* **37**, 949–953.
72. K. Ono, K. Ishimaru, N. Aoki *et al.* (1999) Characterization of a maize sucrose-phosphate synthase protein and its effect on carbon partitioning in transgenic rice plants. *Plant Production Science* **2**, 172–177.
73. E.H. Murchie, C. Sarrobert, P. Contard, T. Betsche, C.H. Foyer and N. Galtier (1999) Overexpression of sucrose-phosphate synthase in tomato plants grown with CO_2 enrichment leads to decreased foliar carbohydrate accumulation relative to untransformed controls. *Plant Physiology and Biochemistry* **3**, 251–260.
74. B. Nguyen-Quoc, H. N'Tchobo, C.H. Foyer and S. Yelle (1999) Overexpression of sucrose phosphate synthase increases sucrose unloading in transformed tomato fruit. *Journal of Experimental Botany* **50**, 785–791.
75. A. Strand, C.H. Foyer, P. Gustaffson, P. Gardeström and V. Hurry (2003) Altering flux through the sucrose biosynthesis pathway in transgenic *Arabidopsis thaliana* modifies photosynthetic acclimation at low temperatures and the development of freezing tolerance. *Plant, Cell and Environment* **26**, 523–535.
76. J.M. Mouillon and V. Hurry (2001) Effects of elevated sucrose-phosphate synthase activity on photosynthesis, carbohydrate partitioning and growth rate in hybrid aspen (*Populus tremula* × *P. tremuloides*). PS2001 *Proceedings of the 12th Inernational Congress on Photosynthesis*, S28-007.
77. J.E. Lunn, V.J. Gillespie and R.T. Furbank (2003) Expression of a cyanobacterial sucrose-phosphate synthase from *Synechocystis* sp. PCC 6803 in transgenic plants. *Journal of Experimental Botany* **54**, 1–15.
78. T.D. Sharkey, M. Laporte, Y. Lu, S. Weise and A.P.M. Weber (2004) Engineering plants for elevated CO_2: a relationship between starch degradation and sugar sensing. *Plant Biology* **6**, 280–288.
79. P. Geigenberger, K.-P. Krause, L.M. Hill *et al.* (1995) The regulation of sucrose synthesis in leaves and tubers of potato plants. In: *International Symposium on Sucrose Metabolism* (eds H.G. Pontis, G.L. Salerno and E.J. Echeverria), pp. 14–24. American Society of Plant Physiologists.
80. C.K. Castleden, N. Aoki, V.J. Gillespie *et al.* (2004) Evolution and function of the sucrose-phosphate synthase gene families in wheat and other grasses. *Plant Physiology* **135**, 1753–1764.
81. U. Sonnewald, W.P. Quick, E. MacRae, K.-P. Krause and M. Stitt (1993) Purification, cloning and expression of spinach leaf sucrose-phosphate synthase in *Escherichia coli*. *Planta* **189**, 174–181.

82. R.R. Klein, S.J. Crafts-Brandner and M.E. Salvucci (1993) Cloning and developmental expression of the sucrose-phosphate synthase gene from spinach. *Planta* **190**, 498–510.
83. A. Komatsu, Y. Takanokura, M. Omura and T. Akihima (1996) Cloning and molecular analysis of cDNAs encoding three sucrose-phosphate synthase isoforms from citrus fruit (*Citrus unshiu* Marc.). *Molecular and General Genetics* **252**, 346–351.
84. B. Sugiharto, H. Sakakibara, Samaudi and T. Sugiyama (1997) Differential expression of two genes for sucrose-phosphate synthase in sugarcane: molecular cloning of the cDNAs and comparative analysis of gene expression. *Plant Cell Physiology* **38**, 961–965.
85. J. Ingram, J.W. Chandler, L. Gallagher, F. Salamani and D. Bartels (1997) Analysis of cDNA clones encoding sucrose-phosphate synthase in relation to sugar interconversions associated with dehydration in the resurrection plant *Craterostigma plantagineum* Hochst. *Plant Physiology* **115**, 113–121.
86. G. Langenkämper, R.W.M. Fung, R.D. Newcomb, R.G. Atkinson, R.G. Gardner and E.A. MacRae (2002) Sucrose phosphate synthase genes in plants belong to three different families. *Journal of Molecular Evolution* **54**, 322–332.
87. R.W.M. Fung, G. Langenkämper, R.C. Gardner and E.A. MacRae (2003) Differential expression within an SPS gene family. *Plant Science* **164**, 459–470.
88. S.L. Harmer, J.B. Hogenesch, M. Straume *et al.* (2000) Orchestrated transcription of key pathways in Arabidopsis by the circadian clock. *Science* **290**, 2110–2113.
89. L. Hennig, W. Gruissem, U. Grossniklaus and C. Köhler (2004) Transcriptional programs of early stages of plant reproduction. *Plant Physiology* **135**, 1765–1775.
90. A. Komatsu, Y. Takanokura, T. Moriguchi, M. Omura and T. Akihima (1999) Differential expression of three sucrose-phosphate synthase isoforms during sucrose accumulation in citrus fruits (*Citrus unshiu* Marc.). *Plant Science* **140**, 169–178.
91. P. Quick, G. Siegl, E. Neuhaus, R. Feil and M. Stitt (1989) Short-term water stress leads to a stimulation of sucrose synthesis by activating sucrose-phosphate synthase. *Planta* **177**, 535–546.
92. V.M. Hurry, Å. Strand, M. Tobiaeson, P. Gardeström and G. Öquist (1995) Cold hardening of spring and winter wheat and rape results in differential effects on growth, carbon metabolism, and carbohydrate content. *Plant Physiology* **109**, 697–706.
93. V. Hurry, Å. Strand, R. Furbank and M. Stitt (2000) The role of phosphate in the development of freezing tolerance and the acclimatization of photosynthesis to low temperature is revealed by the *pho* mutants of *Arabidopsis thaliana*. *The Plant Journal* **24**, 383–396.
94. H. Winter and S.C. Huber (2000) Regulation of sucrose metabolism in higher plants: localization and regulation of activity of key enzymes. *Critical Reviews in Biochemistry and Molecular Biology* **35**, 253–289.
95. J. Price, A. Laxmi, S.K. St.Martin and J.C. Jang (2004) Global transcription profiling reveals multiple sugar signal transduction mechanisms in Arabidopsis. *Plant Cell* **16**, 2128–2150.
96. A.T. Chávez-Bárcenas, J.J. Valdez-Alarcón, M. Martinez-Trujillo *et al.* (2000) Tissue-specific and developmental pattern of expression of the rice *sps1* gene. *Plant Physiology* **124**, 641–653.
97. M. Martinez-Trujillo, T. Chávez-Bárcenas, V. Limones-Briones, J. Simpson and L. Herrera-Estrella (2004) Functional analysis of the promoter of the rice sucrose phosphate synthase gene (*sps1*). *Plant Science* **166**, 131–140.
98. J.M. Bruneau, A.C. Worrell, B. Cambou, D. Lando and T.A. Voelker (1991) Sucrose phosphate synthase, a key enzyme for sucrose biosynthesis in plants. *Plant Physiology* **96**, 473–478.
99. J.L. Walker and S.C. Huber (1989) Purification and preliminary characterisation of sucrose-phosphate synthase using monoclonal antibodies. *Plant Physiology* **89**, 518–525.
100. S.C. Huber and J.L. Huber (1996) Role and regulation of sucrose-phosphate synthase in higher plants. *Annual Review of Plant Physiology and Plant Molecular Biology* **47**, 431–444.
101. D.C. Doehlert and S.C. Huber (1985) The role of sulfhydryl groups in the regulation of spinach leaf sucrose phosphate synthase. *Biochimica et Biophysica Acta* **830**, 267–273.
102. J.E. Lunn, G.E. Price and R.T. Furbank (1999) Cloning and expression of a prokaryotic sucrose-phosphate synthase gene from the cyanobacterium *Synechocystis* sp. PCC 6803. *Plant Molecular Biology* **40**, 297–305.

103. G. Siegl and M. Stitt (1990) Partial purification of two forms of spinach leaf sucrose-phosphate synthase which differ in their kinetic properties. *Plant Science* **66**, 205–210.
104. D. Toroser, R. McMichael Jr., K.-P. Krause (1999) Site-directed mutagenesis of serine 158 demonstrates its role in spinach leaf sucrose-phosphate synthase modulation. *The Plant Journal* **17**, 407–413.
105. H. Weiner, R.W. McMichael and S.C. Huber (1992) Identification of factors regulating the phosphorylation status of sucrose-phosphate synthase *in vivo*. *Plant Physiology* **99**, 1435–1442.
106. H. Weiner, H. Weiner and M. Stitt (1993) Sucrose-phosphate synthase phosphatase, a type 2A protein phosphatase, changes its sensitivity towards inhibition by inorganic phosphate in spinach leaves. *FEBS Letters* **333**, 159–164.
107. D. Toroser, Z. Plaut and S.C. Huber (2000) Regulation of a plant SNF1-related protein kinase by glucose-6-phosphate. *Plant Physiology* **123**, 403–411.
108. S. Takahashi, K. Ono, M. Ugaki, K. Ishimura, N. Aoki and R. Ohsugi (2000) Ser162-dependent inactivation of overproduced sucrose-phosphate synthase protein of maize in transgenic rice plants. *Plant Cell Physiology* **41**, 977–981.
109. J.E. Lunn and R.T. Furbank (1999) Sucrose biosynthesis in C_4 plants. *New Phytologist* **143**, 221–237.
110. R.W. McMichael, J. Kochansky, R.R. Klein and S.C. Huber (1995) Characterization of the substrate specificity of sucrose-phosphate synthase. *Archives of Biochemistry and Biophysics* **321**, 71–75.
111. M. Glinski, T. Romeis, C.-P. Witte, S. Weinkoop and W. Weckwerth (2003) Stable isotope labeling of phosphopeptides for multiparallel kinase target analysis and identification of phosphorylation sites. *Rapid Communications in Mass Spectrometry* **17**, 1579–1584.
112. J.Z. Huang and S.C. Huber (2001) Phosphorylation of synthetic peptides by a CDPK and plant SNF1-related protein kinase. Influence of proline and basic amino acid residues at selected positions. *Plant Cell Physiology* **42**, 1079–1087.
113. D. Toroser and S.C. Huber (1997) Protein phosphorylation as a mechanism for osmotic-stress activation of sucrose-phosphate synthase in spinach leaves. *Plant Physiology* **114**, 947–955.
114. X. Li, M. Pfiz, M. Küppers, W. Einig, H. Rennenberg and R. Hampp (2003) Sucrose phosphate synthase in leaves of mistletoe: its regulation in relation to host (*Abies alba*) and season. *Trees* **17**, 221–227.
115. D. Toroser, G.S. Athwal and S.C. Huber (1998) Site-specific regulatory interaction between spinach leaf sucrose-phosphate synthase and 14-3-3 proteins. *FEBS Letters* **435**, 110–114.
116. G. Moorhead, P. Douglas, V. Cotelle *et al.* (1999) Phosphorylation-dependent interactions between enzymes of plant metabolism and 14-3-3 proteins. *The Plant Journal* **18**, 1–12.
117. F. Börnke (2005) The variable C-terminus of 14-3-3 proteins mediates isoform-specific interaction with sucrose-phosphate synthase in the yeast two-hybrid system. *Journal of Plant Physiology* **162**, 161–168.
118. J.E. Lunn, A.R. Ashton, M.D. Hatch and H.W. Heldt (2000) Purification, molecular cloning, and sequence analysis of sucrose-6^F-phosphate phosphohydrolase from plants. *Proceedings of the National Academy of Sciences of the United States of America* **97**, 12914–12919.
119. S. Chen, M. Hajirezaei, M. Peisker, H. Tschiersch, U. Sonnewald and F. Börnke (2005) Decreased sucrose-6-phosphate phosphatase level in transgenic tobacco inhibits photosynthesis, alters carbohydrate partitioning, and reduces growth. *Planta* **221**, 179–492.
120. J.E. Lunn (2003) Sucrose-phosphatase gene families in plants. *Gene* **303**, 187–196.
121. S. Fieulaine, J.E. Lunn, F. Borel and J.-L. Ferrer (2005) The structure of a cyanobacterial sucrose-phosphatase reveals the 'sugar tongs' that release free sucrose in the cell. *The Plant Cell* **17**, 2049–2058.
122. E. Echeverria, M.E. Salvucci, P. Gonzalez, P. Paris and G. Salerno (1997) Physical and kinetic evidence for an association between sucrose-phosphate synthase and sucrose-phosphate phosphatase. *Plant Physiology* **115**, 223–227.
123. V. Cotelle, S.E.M. Meek, F. Provan, F.C. Milne, N. Morrice and C. MacKintosh (2000) 14-3-3s regulate global cleavage of their diverse binding partners in sugar-starved Arabidopsis cells. *EMBO Journal* **19**, 2869–2876.

124. W.P. Quick and A.A. Schaffer (1996) Sucrose metabolism in sources and sinks. In *Photoassimilate Distribution in Plants and Crops: Source–Sink Relationships* (eds E. Zamski and A.A. Schaffer), Marcel Dekker, New York, pp. 115–156.
125. J.H. Hendriks, A. Kolbe, Y. Gibon, M. Stitt and P. Geigenberger (2003) ADP-glucose pyrophosphorylase is activated by posttranslational redox-modification in response to light and to sugars in leaves of Arabidopsis and other plant species. *Plant Physiology* **133**, 838–849.
126. P. Scott and N.J. Kruger (1995) Influence of elevated fructose-2,6-bisphosphate levels on starch mobilization in transgenic tobacco leaves in the dark. *Plant Physiology* **108**, 1569–1577.
127. P. Scott, A.J. Lange, S.J. Pilkis and N.J. Kruger (1995) Carbon metabolism in leaves of transgenic tobacco (*Nicotiana tabacum* L.) containing elevated fructose 2,6-bisphosphate levels. *The Plant Journal* **7**, 461–469.
128. N.J. Kruger and P. Scott (1995) Integration of cytosolic and plastidic carbon metabolism by fructose 2,6-bisphosphate. *Journal of Experimental Botany* **46**, 1325–1333.
129. P. Scott, A.J. Lange and N.J. Kruger (2000) Photosynthetic carbon metabolism in leaves of transgenic tobacco (*Nicotiana tabacum* L.) containing decreased amounts of fructose 2,6-bisphosphate. *Planta* **211**, 864–873.
130. M.R. Truesdale, O. Toldi and P. Scott (1999) The effect of elevated concentrations of fructose 2,6-bisphosphate on carbon metabolism during deacidification in the crassulacean acid metabolism plant *Kalanchöe diagremontiana*. *Plant Physiology* **121**, 957–964.
131. H. Draborg, D. Villadsen and T.H. Nielsen (2001) Transgenic Arabidopsis plants with decreased activity of fructose-6-phosphate, 2-kinase/fructose-2,6-bisphosphatase have altered carbon partitioning. *Plant Physiology* **126**, 750–758.
132. S.J. Trevanion (2002) Regulation of sucrose and starch in wheat (*Triticum aestivum* L.) leaves: role of fructose 2,6-bisphosphate. *Planta* **215**, 653–665.
133. J.E. Lunn and M.D. Hatch (1887) The role of sucrose-phosphate synthase in the control of photosynthate partitioning in *Zea mays* leaves. *Australian Journal of Plant Physiology* **24**, 1–8.
134. S.J. Trevanion, C.K. Castleden, C.H. Foyer, R.T. Furbank, W.P. Quick and J.E. Lunn (2004) Regulation of sucrose-phosphate synthase in wheat (*Triticum aestivum*) leaves. *Functional Plant Biology* **31**, 685–695.
135. M. Sitt, I. Wilke, R. Feil and H.W. Heldt (1988) Coarse control of sucrose-phosphate synthase in leaves: alterations of the kinetic properties in response to the rate of photosynthesis and the accumulation of sucrose. *Planta* **174**, 217–230.
136. G.M. Noël, J.A. Tognetti and H.G. Pontis (2001) Protein kinase and phosphatase activities are involved in fructan synthesis initiation mediated by sugars. *Planta* **213**, 640–646.
137. K. Koch (2004) Sucrose metabolism: regulatory mechanisms and pivotal roles in sugar sensing and plant development. *Current Opinion in Plant Biology* **7**, 235–246.
138. J.-W. Kim and C.V. Dang (2005) Multifaceted roles of glycolytic enzymes. *Trends in Biochemical Sciences* **30**, 142–150.
139. J.-L. Prioul, S. Pelleschi, M. Séne *et al.* (1999) From QTLs for enzyme activity to candidate genes in maize. *Journal of Experimental Botany*, **50**, 1281–1288.
140. R. Ming, Y.-W. Wang, X. Draye, P.H. Moore, J.E. Irvine and A.H. Paterson (2002) Molecular dissection of complex traits in autopolyploids: mapping QTLs affecting sugar yield and related traits in sugarcane. *Theoretical and Applied Genetics* **105**, 332–345.
141. T. Mitchell-Olds and D. Pedersen (1998) The molecular basis of quantitative genetic variation in central and secondary metabolism in Arabidopsis. *Genetics* **149**, 739–747.
142. B.B. Buchanan and Y. Balmer (2005) Redox regulation: a broadening horizon. *Annual Review of Plant Biology* **56**, 187–220.
143. D. Rontein, M. Dieuaide-Noubhani, E.J. Dufourc, P. Raymond and D. Rolin (2002) The metabolic architecture of plant cells. Stability of central metabolism and flexibility of anabolic pathways during the growth cycle of tomato cells. *Journal of Biological Chemistry* **277**, 43948–43960.
144. J.C. Lloyd and O.V. Zakhleniuk (2004) Responses of primary and secondary metabolism to sugar accumulation revealed by microarray expression analysis of the Arabidopsis mutant, pho3. *Journal of Experimental Botany* **55**, 1221.

10 Control of starch biosynthesis in vascular plants and algae

Matthew K. Morell, Zhongyi Li, Ahmed Regina, Sadiq Rahman, Christophe d'Hulst and Steven G. Ball

10.1 Introduction

Vascular plants store reserves of carbohydrate on both short- and long-term time basis in order to be able to meet changing metabolic requirements. Short-term storage is required in leaves to provide a store of carbohydrate that can be accessed to maintain metabolic functions when photosynthesis is inactive. Longer term storage of carbohydrate is necessary in tubers and seeds to support reproductive tissue development based on tuber stores and germination of seeds respectively. Given that water is limiting for terrestrial plants, that pathogen attack be resisted and that long-term survival of seeds is often required, it is necessary that the storage compound be dehydrated. Plants have found four general solutions to this problem: starch, nonstarch polysaccharides such as the mannans, glucans and xylans, lipids and seed storage proteins. Because of the ubiquitous nature of starch and its importance in economically important crops that provide staple foods, it is important to understand the intricacies of the starch biosynthetic pathway and its control. There are significant opportunities to manipulate starch synthesis in plants for improved food and industrial functionality.

Starch in vascular plants is synthesized within the plastid by a complex metabolic pathway. There are four basic steps in the biosynthetic process: substrate activation (catalyzed by ADPglucose pyrophosphorylase), polymer elongation (starch synthases), polymer branching (branching enzyme) and debranching (debranching enzymes). For recent reviews on starch biosynthesis see [1–3]. For each of the four steps in the pathway, multiple isoforms of each enzyme are found, encoded by complex families of genes. Mutational analysis demonstrates that while these genes have overlapping functions, complete functional redundancy is rare, indicating that each class of genes has a defined role in the pathway. Understanding how each of these genes contributes to the synthesis of starch in the various tissues of the plant and in different orders of plants remains a key objective of the field. In working toward this understanding, studies of nonvascular plant systems such as algae, as well as bacterial glycogen synthesis, have proved highly informative [2] and will be discussed in this review. While the major steps in the synthesis of the starch polymer are now well understood, the control of the synthesis of the starch granule is much less understood and remains an open field of research.

Starch is a deceptively simple material, being composed of just one monomeric unit, glucose, polymerized through two types of linkages, $\alpha,1$-4 and $\alpha,1$-6. However, differences in chain length and the frequency and clustering of $\alpha,1$-6 linkages mean that very complex polydisperse molecules result from this simple building block. The polydisperse molecules making up starch are usually classified into two groups based on the frequency of $\alpha,1$-6 linkages and molecular size. Amylopectin contains high molecular weight species (degree of polymerization, typically ranging from 5000 to 50000) and frequent $\alpha,1$-6 linkages (4–5% of $\alpha,1$-6 linkages). In contrast, the amylose fraction contains molecules of lower molecular weight (DP 500–3000) and infrequent $\alpha,1$-6 linkages (less than 1%). The starch granule shows complex organization across a wide range of dimensions, leading to description of the granule as a 'hierarchical material' in which structure at one scale defines the properties of the material at the next scale [4]. Such materials are widely known in nature and frequently have capacity for self-assembly.

Starch synthesis in higher plants is subject to complex control. The first level of control is transcriptional. Multiple genes for each of the classes of starch biosynthetic genes have been identified in the genomes of *Arabidopsis thaliana* and rice. However, differing subsets of these genes are involved in synthesis in differing tissues, indicating precise control over transcription. Secondly, there is evidence that the expression of some of these genes is under circadian control. Thirdly, there is redox control of some enzymes in the pathway, and finally, there is evidence for the formation of complexes of starch biosynthetic enzymes and their potential regulation by phosphorylation.

In this chapter, starch synthesis will be described from a range of species and tissues, with a view to illustrating the different physiological requirement for starch synthesis and the ways in which control mechanisms may act in shaping the synthesis of starch.

10.2 Synthesis of bacterial glycogen

The blueprint for higher plant starch synthesis is found in bacterial glycogen synthesis, where synthesis proceeds via an ADPglucose dependent pathway (Figure 10.1),

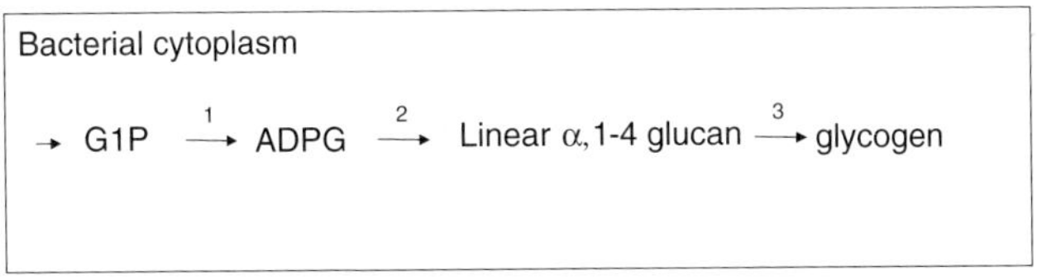

```
Bacterial cytoplasm

   → G1P  →¹→  ADPG  →²→  Linear α,1-4 glucan  →³→  glycogen
```

Figure 10.1 Bacterial glycogen synthesis. 1. ADPglucose pyrophosphorylase (catalyzes the synthesis of ADPglucose and PPi from glucose-1-P and ATP). 2. Glycogen synthase (transfers the glucosyl moiety of ADPglucose to the nonreducing end of an $\alpha,1$-4 glucan polymer, forming an additional $\alpha,1$-4 linkage) 3. Glycogen branching enzyme (cleaves a region of $\alpha,1$-4 glucan polymer, transferring the cleaved chain to a second $\alpha,1$-4 glucan polymer via an $\alpha,1$-6 linkage).

involving single ADPglucose pyrophosphorylase, glycogen synthase and glycogen branching enzyme genes [5]. However, the end result of this pathway is a relatively unstructured glycogen rather than starch, indicating that either these enzymes do not possess the specificities required to synthesize starch, or other activities are required. A glycogen debranching enzyme, *glg*X, is present in bacteria but elimination of its activity leads to overaccumulation of glycogen, suggesting its role is primarily in glycogen degradation rather than synthesis [6]. Interestingly, no genes have been identified in *E. coli* saturation mutagenesis programs that are required for the initiation of the glycogen molecule, suggesting that glycogen synthase is able to synthesize glycogen from ADPglucose without a separate priming activity or precursor molecule. Given that this core set of enzymes alone is insufficient to drive starch synthesis, the differences between bacterial glycogen synthesis and vascular plant starch synthesis warrant further analysis.

10.3 Synthesis of starch in vascular plants

As in bacterial glycogen synthesis, starch synthesis in vascular plants involves substrate activation, polymer elongation, polymer branching and an involvement from debranching enzymes. However, vascular plant starch synthesis involves a number of key differences. Firstly, multiple copies of each of the biosynthetic enzymes is typically present, and secondly, there is spatial separation of elements of the pathway. These enzymatic steps provide an explanation for the synthesis of the polymers present within the starch granule, but they do not provide a facile explanation for the synthesis of the starch granular structure [7]. The absence of a clear class of mutants that disrupts granular structure suggests that the granule is largely the result of 'self-assembly' driven by the physicochemical properties of the nascent starch polymers. Schematic representations of starch synthesis in the cereal endosperm and leaf are shown in Figures 10.2 and 10.3 respectively.

10.3.1 Substrate supply and activation

The first step in starch biosynthesis can be defined as the first reaction that irreversibly draws carbon from general metabolic pools into starch synthesis. Given this definition, the synthesis of ADPglucose from glucose-1-phosphate, catalyzed by ADPglucose pyrophosphorylase, can be considered the first step in the pathway [8]. However, it is clear that mutations affecting the supply of glucose-1-phosphase, such as the sucrose synthase mutant (*sh1*) in maize [9], reduce starch content and can influence starch structure by influencing the concentration of substrate pools available to the biosynthetic enzymes.

As the first step in the pathway, ADPglucose pyrophosphorylase is a key site for regulation of starch synthesis. However, this regulation is complex and involves a range of mechanisms including transcriptional control, differential compartmentation, redox control and allosteric regulation [3]. Initial studies

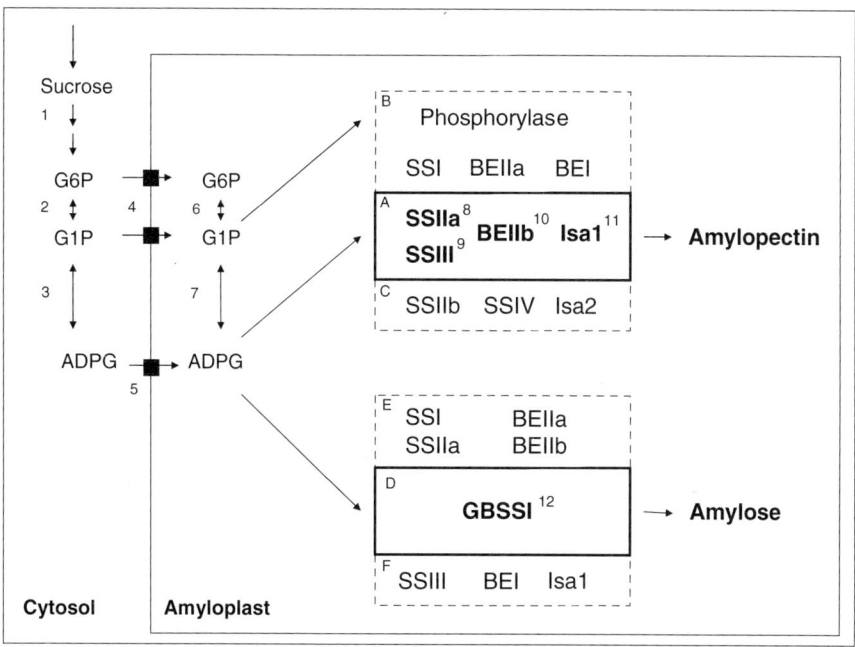

Figure 10.2 Starch biosynthesis pathway in the cereal endosperm. 1. Sucrose synthase (*sh1*). 2. Cytosolic phosphoglucomutase (interconverts glucose-1-phosphate and glucose-6-phosphate). 3. Cytosolic ADPglucose pyrophosphorylase (*bt2, sh2*) 4. Hexose-phosphate transporter. 5. Brittle-1 ADPglucose transporter (*bt1*). 6. Plastidic phosphoglucomutase. 7. Plastidic ADPglucose pyrophosphorylase. 8. Starch synthase IIa (*su2*). 9. Starch synthase III (*du1*). 10. Branching enzyme IIb (*ae1*). 11. Isoamylase1 (*su1*). 12. Granule-bound starch synthase I (*wx1*). Box A. Core set of amylopectin biosynthetic enzymes for which there is genetic evidence of a role that cannot be fully complemented by another gene. Box B. The set of enzymes present in the cereal endosperm amyloplast for which activity has been demonstrated but no mutant with a significant impact on starch synthesis has been identified. Box C. mRNA identified in the cereal endosperm but no enzyme characterized or mutant identified. Box D. The enzyme required for amylose synthesis. Box E. Enzymes present in the starch granule, but a role in amylose synthesis is unproven. Box F. Enzymes present in the cereal endosperm; however the enzyme is not located in the granule and no role in amylose synthesis has been demonstrated. Sucrose synthase catalyses a reaction between sucrose and UDP, forming UDPglucose and fructose. Phosphoglucomutase interconverts glucose-1-phosphate and glucose-6-phosphate. Starch synthase transfers the glucosyl moiety of ADPglucose to the nonreducing end of an α,1-4 glucan polymer, forming an additional α,1-4 linkage. Other activities are as defined in Figure 10.1.

of the enzyme demonstrated that ADPglucose pyrophosphorylase is subject to complex allosteric regulation by metabolites [10]. In vascular plants, the enzyme is activated by 3PGA and inhibited by inorganic phosphate. In other systems, such as bacteria and cyanobacteria, the allosteric activators differ in order to reflect the differing metabolic pools and pathways operating in those organisms, but typically include activators such as F6P and inhibitors such as AMP (for a review, see [5]). Bacterial enzymes tend to be homotetrameric [5], whereas higher plant enzymes are

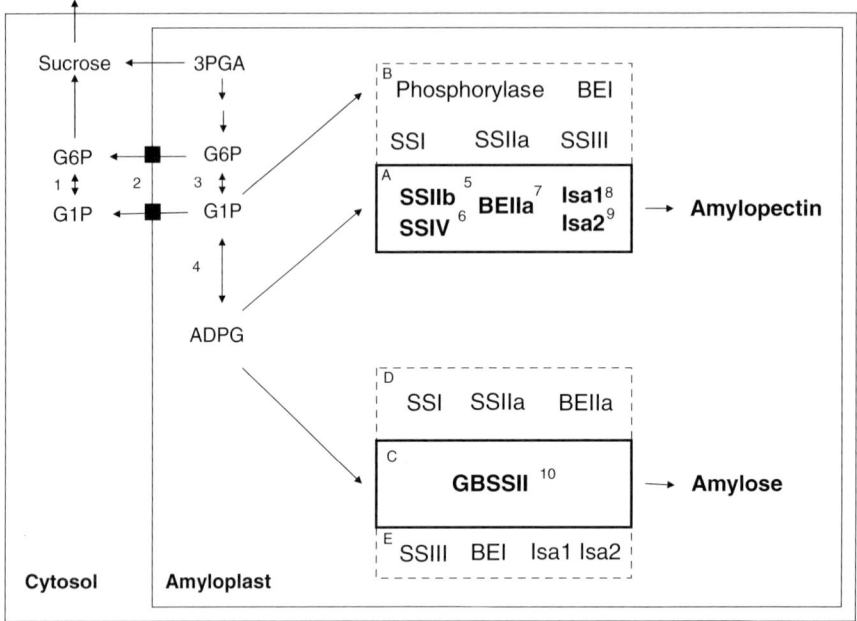

Figure 10.3 Putative starch biosynthesis pathway in the cereal leaf. 1. Cytosolic phosphoglucomutase 2, 3. Hexose-phosphate transporters. 4. Plastidic ADPglucose pyrophosphorylase. 5. Starch synthase IIb (based on mRNA expression data). 6. Starch synthase IV (based on mRNA expression data). 7. Branching enzyme IIa (based on mRNA expression data, Blauth et al. [50]). 8. Isoamylase1 (su1). 9. Isoamylase2. 10. Granule-bound starch synthase II. Box A. A proposed set of amylopectin biosynthetic enzymes in cereal leaf required for normal amylopectin synthesis. Box B. A set of genes expressed in the cereal leaf amyloplast at low levels which may have a role in leaf starch synthesis. Box C. Enzyme thought to be required for amylose synthesis in cereal leaf. Box D. Enzyme which may be present in the leaf starch granule but a role in amylose synthesis is unproven. Box E. Enzymes unlikely to be present in the starch granule and no role in amylose synthesis has been demonstrated. Isoamylases cleave α,1-6 linkages in glycogen or amylopectin substrates, but not pullulan. Other enzyme activities are as defined in Figures 10.1 and 10.2.

heterotetrameric, containing two 'large' and two 'small' subunits [8]. There are marked differences between the regulatory sensitivity of leaf chloroplast and endosperm-located ADPglucose pyrophosphorylases, with the enzymes in the chloroplast being highly dependent on 3PGA concentration when compared with the endosperm-located ADPglucose pyrophosphorylases [11]. This dependence is in keeping with a requirement for sensitive response to flux in the highly dynamic photosynthetic environment of the leaf, when compared with the endosperm where starch accumulation occurs over weeks.

Early reports demonstrated that in maize [12] and barley [13] endosperm, there were both plastidic and cytosolic forms of ADPglucose pyrophosphorylase. In wheat, Burton et al. [14] examined the nature of the small subunit genes encoding ADPglucose pyrophosphorylase and concluded that the cytosolic and plastidial small subunits were highly likely to be the products of separate genes. This conclusion was further reinforced by the analysis of a barley mutant, RisØ 16, which lacks the cytosolic small subunit of ADPglucose pyrophosphorylase, but the plastidial ADPglucose pyrophosphorylase activity was unaffected [15]. The low starch phenotype of the mutant suggests that a large part of the flux into the starch biosynthesis pathway proceeds via the cytosolic ADPglucose pyrophosphorylase. This result is consistent with observations of the low starch *bt2* and *sh2* mutants in maize endosperm where mutations abolish cytoplasmic ADPglucose pyrophosphorylase but a trace amount of plastidic enzyme remains. The distribution of genes for the large subunits of ADPglucose pyrophosphorylase is less well defined. The Arabidopsis genome contains six genes coding for ADPglucose pyrophosphorylases of which two are classified as small subunit types and four are classified as large subunits [16, 17]. Of the small, subunit genes, ApS1 is expressed in all tissues of the plant. Among the large subunit genes, ApL1 is present in source tissues, whereas ApL3 and ApL4 are present in sink leaves. The expression of ApL3 and ApL4 can be induced by sugars, and the gene products are less sensitive to allosteric regulators when compared with the forms expressed in source leaves [16, 17].

Studies on the maize *brittle-1* mutant [18] and the *lys5* mutation in barley [19] demonstrate the brittle-1/*lys*5 gene product is responsible for the transport of ADPglucose from the cytosol into the plastid in cereal endosperm. There are conflicting results concerning the role of extraplastidial ADPglucose pyrophosphorylases in tissues other than the cereal endosperm. Beckles *et al.* [20] concluded that a cytosolic ADPglucose pyrophosphorylase was a feature of the graminaceous endosperm, but not other storage organs. Evidence has been presented for a plastidial location of ADPglucose pyrophosphorylase in tomato fruit [21]. In contrast, Baroja-Fernandez [22] using transgenic potato plants expressing an adenosine diphosphate sugar pyrophosphatase (ASPP) from *E. coli* found results consistent with cytosolic production of the majority of the ADPglucose in the tissue. Further research is required to resolve this issue.

10.3.2 Amylose synthesis

The synthesis of amylose occurs within the starch granule and requires the action of granule-bound starch synthase (GBSS). While other enzymes contribute to amylose synthesis, extensive genetic evidence across a range of species demonstrates that GBSS activity is absolutely required for amylose synthesis. In both monocotyledonous and dicotyledonous species, two forms of GBSS are present, designated GBSSIa and GBSSIb [23, 24]. GBSSIa is expressed in the endosperm only, whereas GBSSIb is expressed in tissues other than the endosperm. There may be differences between these enzymes with respect to the structure and function of the polymers synthesized. However, there have been no cross-complementation studies that allow concrete conclusions to be drawn.

While amylose is frequently described as a 'linear' molecule, there is good evidence suggesting that amylose does contain infrequent branch points (<0.5%)

and that some of the branches may be small (of the order of 15–25 glucose units), but can range up to chains with a DP of ca 3000. This observation suggests that one or more of the branching enzymes are also involved in amylose synthesis. However, neither the specific isoform(s) involved or their mode of action within the granule has been defined.

Amylose synthesis is believed to occur within the starch granule, presumably involving the action of granule bound enzymes (granule-bound starch synthase, GBBSS, plus one or more branching enzymes) [25]. Also present with the granules of many species are starch synthases (SSI and SSIIa), but, there is no clear definition of the roles, if any, that these enzymes play in amylose synthesis [26–28]. Interestingly, SSIIa mutants in both wheat and barley lack essentially all the SSI, branching enzyme (BEIIa and BEIIb) protein normally present in the granule and have a reduced content of GBSS [29, 30]. The manner of synthesis of amylose in the granule is unclear. It is presumed that an essentially immobile GBSS entrapped within the granule matrix synthesizes amylose from ADPglucose that diffuses into the granule. Evidence that GBSS proceeds via a 'processive' rather than 'dissociative' mechanism supports this hypothesis [31]. The emerging polymer is thought to thread its way into the existing matrix, finding gaps between crystalline amylopectin, and within and between amorphous regions of amylopectin. There is evidence to suggest that malto-oligosaccharides can stimulate amylose synthesis, both directly (by providing additional primer for synthesis) and indirectly, through stimulation of the catalytic capability of GBSS [31, 32]. Evidence for GBSS-catalyzed amylose synthesis through extension of amylopectin chains in *Chlamydomonas* [33] is discussed in a later section.

10.3.3 Amylopectin synthesis

Amylopectin synthesis is thought to occur at the interface of the developing granule and the stroma [34]. While some enzymes of amylopectin synthesis are trapped within the starch granule matrix, others, such as SSIII, isoamylase and BEI, are not.

In vascular plants, at least seven starch synthase genes are present: two forms of GBSS (GBSSI and GBSSII), SSI, two forms of SSII (SSIIa and SSIIb), SSIII and SSIV. The role of SSII has been examined in a number of species, including pea [35], wheat [30], barley [29], rice [36] and maize [37]. A consensus picture has emerged that demonstrates that loss of SSII leads to a phenotype in which there is a reduction in the percentage of amylopectin chains of intermediate length (DP12-18) and a concomitant increase in chains of DP 6-11, and above DP19. In all species except rice, there is a reduction in starch content associated with loss of SSII activity, which can be as severe as a 50% reduction depending on the species. Clearly neither SSI or SSIII can fully compensate for the loss of SSII activity.

In monocotyledonous species there are two forms of SSII, designated SSIIa and SSIIb. Gene expression analysis suggests that SSIIa is predominantly expressed in the endosperm, whereas SSIIb is more highly expressed in the leaf [38]. Consistent with this observation are the phenotypes associated with loss of SSIIa in rice, maize, barley and wheat. Neither SSI nor SSIIb mutants have been reported, so the role of these genes in both leaf and endosperm synthesis remains to be defined. Mutations which eliminate SSIII activity have been identified in maize [39] and in *Chlamydomonas reinhardtii* [40]. In each case, there is a reduction in starch synthesis, involving predominantly a reduction in the long chains of amylopectin.

The role of branching enzymes is to introduce the a-1,6 branch points into starch. In dicotyledonous plants, two examples of high amylose starches are known: wrinkled pea and potato. In pea, a high amylose phenotype was obtained through a lesion that inactivates BEII (also known as branching enzyme A) [41]. In potato, the antisense technology [42, 43], and a unique antibody expression approach [44], was used to generate high amylose starches. Suppression of either BEI or BEII alone [42] generated either no increase in amylose (BEI) or only a moderate increase in amylose (BEII, to approximately 30%). However, suppression of both BEI and BEII resulted in an amylose content greater than 70% [45]. Interestingly, Arabidopsis appears to lack a homolog of BEI, yet accumulates apparently normal leaf starch. Arabidopsis contains two forms of the BEII genes but their respective roles have not been individually defined. The number of individual genes encoding BEII-type enzymes in other dicotyledonous plants such as pea and potato is not known.

In monocotyledonous species, three forms of BE are found, designated BEI, BEIIa and BEIIb. In a range of monocotyledonous species, suppression of BEI resulted in no significant phenotype in wheat [46] and maize endosperm, or in maize leaf [47]. In rice, subtle changes in amylopectin chain length distribution were reported [48]. The observation that high amylose starches are found in rice and maize (*amylose extender*) through mutations down-regulating BEIIb activity has focused attention on this isoform of the enzyme [49]. In maize, a mutator element induced mutation of BEIIa does not have a clear phenotype in the endosperm, but does yield a highly altered starch structure in the leaf [50]. BEIIa expression in the endosperm is low relative to BEIIb and high in the leaf (where BEIIIb is not expressed). In wheat, BEIIa is expressed at higher levels when compared with BEIIb (Ahmed Regina, personal communication) and so the control over amylose content in these cereals may differ from maize.

In both monocotyledonous and dicotyledonous species, the debranching enzyme genes present include three forms of isoamylase and one pullulanase gene [51, 52]. Debranching enzymes specifically hydrolyze α–1,6 linkages of glucan polymers and can be categorized as isoamylases or pullulanases depending on their substrate specificities across a range of substrates including amylopectin, glycogen and pullulan [53]. In some tissues of dicotyledonous species, isoamylase1 (Isa1) and isoamylase2 (Isa2) are present in a heteromultimeric complex and mutation in either enzyme abolishes the complex formation and results in decreased amylopectin synthesis, and increased production of a glycogen-like polysaccharide known as phytoglycogen [52, 54, 55]. The role of Isa2 in vascular plants has not been defined. There are two general hypotheses for the role of isoamylases in higher plants. First is the glucan trimming model in which isoamylase (or more accurately, the isoamylase multimeric assembly) interacts with amylopectin and 'edits' the structure of the polymer such that it is competent for crystallization [4, 34]. An alternative hypothesis suggests that the activity of isoamylase is required to remove soluble glucan that would otherwise compete with amylopectin and granular starch synthesis [52, 56]. There is no definitive resolution of the role of the enzyme at this point However, it is interesting to note that alterations in Isoamylase1 expression levels in cereal through transgenic approaches [57] provide evidence of a correlation between isoamylase expression level and several aspects of starch granule structure, including chain length distribution and gelatinization temperature. These observations tend to support some direct role of isoamylase in shaping amylopectin structure. Further research is required to finalize this point.

There have been suggestions for roles of other enzymes in starch biosynthesis. The identification of mutants in the D-enzyme of *C. reinhardtii* with reduced starch has led to speculation on the role of this enzyme in higher plants [57, 58]. However, down regulation of the enzyme in Arabidopsis suggests that the role of D-enzyme, at least in Arabidopsis leaves, is primarily in starch degradation [60]. The role of starch phosphorylase has also been revisited [61]. However, in Arabidopsis leaves, no evidence was obtained for a significant phenotype when the plastidic phosphorylase was eliminated through mutation [62]. The role of cytoplasmic phosphorylase in not known, and is somewhat problematic given that it is unlikely to encounter polymeric starch under normal physiological conditions.

10.4 Starch synthesis and breakdown in leaves and tubers

Transient starch synthesis occurs in leaves through the incorporation of atmospheric CO_2 by photosynthesis during the illuminated period. It is degraded at night to supply energy and a carbon source to the whole plant and/or to sustain the accumulation of storage starch in sink tissues. Although clearly distinct in terms of physiology, both Arabidopsis leaves and potato tubers (*Solanum tuberosum*) are very useful models to study transient and storage starch metabolism respectively. The rapid development of genomics tools, together with the ease in manipulating starch synthesis and degradation in Arabidopsis leaves and potato, has facilitated rapid progress in this area of plant biology during the last decade. In this section, particular attention will be paid to (1) the function of starch debranching enzymes, (2) the importance of starch phosphorylation as a signal for degradation, (3) the involvement of β-amylases during starch mobilization and the further metabolization of maltose and finally (4) the recent view about the control of starch metabolism in leaves and tubers.

10.4.1 Isoamylases are directly involved in the synthesis of amylopectin but also during starch mobilization

The function of debranching enzymes (DBEs) in starch metabolism has been a subject of debate during the last few years. Several reports over recent years have suggested that DBEs, initially suspected to only fulfill a catabolic function, might also represent a major determinant of amylopectin synthesis [63–67]. However, two contradictory models have been postulated based on divergent interpretations of the observed phenotypes [34, 67]. Three genetically distinct forms of isoamylases and one pullulanase have been annotated in the nuclear genome of *A. thaliana*: *AtISA1* (At2g39930), *AtISA2* (At1g03310), *AtISA3* (At4g09020) and *AtPU1* (At5g04360) (see http:// www.starchmetnet.org for a complete description). Mutation at either *AtISA1* or *AtISA2* loci leads to deficiency in one specific form of isoamylase. This correlates with a strong decrease in starch biosynthesis and the accumulation of glycogen-like water-soluble polymers [68]. Moreover, amylopectin structure is profoundly modified in both mutants [68]. These observations support a direct function of this isoamylase in the synthesis mechanism of amylopectin and validate the discontinuous model previously suggested by Ball *et al.* [34]. AtISA1 and AtISA2 polypeptides are likely to belong to the same heteromultimeric complex as described in potato [69]. AtISA2 might correspond to the inactive subunit of the complex as essential amino acids for debranching activity are missing in the sequence [69]. However, this remains to be formally demonstrated in Arabidopsis. The function of the pullulanase seems to be at least partially redundant compared to the function of the heteromultimeric isoamylase encoded by *AtISA1* and *AtISA2* [68]. This result suggests a role for pullulanase in the synthesis of amylopectin in Arabidopsis, in common with other species such as maize and rice [51, 70].

Although its catalytic function is to cleave α-1,6 linkages, the biological function of AtISA3 seems to be essentially limited to starch mobilization at night in leaves. Mutation at this locus leads to a starch-excess phenotype, observed both in the light and dark cycles [68]. This starch-excess phenotype is commonly observed when starch-degrading enzymes are knocked down in Arabidopsis (for example *sex1* and *sex4* [71, 72]. However, a function for AtISA3 during amylopectin synthesis might not be yet excluded since starch granules are still synthesized at very low level even when the functions of both the pullulanase and the heteromultimeric isoamylase are simultaneously abolished in Arabidopsis.

10.4.2 *Glucan phosphorylation is the key signal of starch degradation in both leaves and tubers*

Although described a long time ago and considered, originally, as an exotic feature of starch, phosphorylation is now assumed to play a central role in the metabolism of this polymer. First discovered and described as R1 protein in potato [73], the enzyme responsible for starch phosphorylation is a high mass glucan water dikinase (GWD) that preferentially phosphorylates C6 positions of glucose residues and is specifically associated with the granule [74, 75]. Antisense inhibition of this GWD in potato leads to a strong reduction of the phosphorylation level of starch coupled with an excess of starch in the leaves of the plant even after a prolonged period of darkness [73]. The same activity has been described in Arabidopsis leaves and elimination of this activity encoded by the *sex1* locus (At1g10760) leads to a starch-excess phenotype [71]. Therefore, GWD has been considered as the major determinant for starch mobilization in plants. However, the mechanism by which phosphorylation of starch can trigger its degradation is not yet understood. It has been shown recently that addition of ATP to isolated starch granules improves its degradation by starch associated proteins [76]. However, GWD does not represent the exclusive starch-phosphorylating enzyme involved in starch mobilization in Arabidopsis. Recently, a phosphoglucan water dikinase (PWD) encoded by gene At5g26570 has been described in Arabidopsis [77, 78]. This PWD preferentially introduces phosphate groups in C3 positions but only after prephosphorylation of starch with GWD. Interruption of the At5g26570 locus results in a starch excess phenotype together with the loss of C3 phosphorylation of starch. It is therefore proposed that both GWD and PWD proteins control starch turnover in Arabidopsis leaves in close collaboration with other starch-degrading enzymes that remain to be identified.

10.4.3 Starch degradation essentially occurs through β-amylolysis in leaves

The physiological function of β-amylases has remained elusive since the pioneering description of antisense potato plants where the activity of one chloroplastic β-amylase has been down regulated [79]. A starch excess phenotype in the leaves of these plants suggests a direct function of the chloroplastic β-amylase in starch mobilization. β-amylases are exo-amylases that release β-maltose from glucans through an inverting mechanism (β-amylases belong to glycoside hydrolase 14 (GH14) family of the carbohydrate active enzyme (CAZy) classification available at http://afmb.cnrs-mrs.fr/CAZY/). Therefore, if β-amylolysis denotes the major route for starch degradation in leaves, β-maltose metabolization should represent the next step toward the redistribution of energy and the exportation of carbon to other plant tissues. This issue has been recently addressed in Arabidopsis after the findings that β-maltose is the exported form of carbon from the chloroplast at night [80, 81] and that a specific chloroplastic maltose exporter named Mex1 (encoded by locus At5g17520) is responsible for this transfer [82]. After export, the cytosolic β-maltose seems to be subsequently converted to hexose and sucrose by a yet unknown mechanism. However, it has been suggested that a cytosolic glucanotransferase (named DPE2 and encoded by gene At2g40840), whose sequence is similar to that of the bacterial MalQ protein, is likely to be involved in this process [83, 84]. DPE2 transfers glucosyl residues from maltose to glycogen to produce longer glucans and glucose. A mutant of Arabidopsis that is defective in DPE2 activity overaccumulates both starch and maltose [83]. However, the acceptor molecule for maltose and DPE2 is still not identified in the cytosol and the mechanism by which the longer glucans produced by DPE2 are degraded is not known yet.

10.4.4 Starch metabolism is tightly controlled by several levels of regulation

Arabidopsis leaves and potato tubers have been extensively studied to uncover the regulatory processes of starch metabolism. It has been shown *in vivo* in potato tubers that the ADP-glucose pyrophosphorylase is tightly regulated not only by allosteric effectors produced during photosynthesis [8] but also by a redox mechanism [85, 86]. This redox mechanism might lead to a balanced state of enzyme conformation through, for instance, the formation or breakage of disulfide bonds between the different subunits of the holoenzyme. More recently, the formation of the starch granule growth rings has been investigated in potato tubers. Growth rings correspond to regularly alternating zones of amorphous and semicrystalline materials of the starch granule. The formation of these growth rings is neither under diurnal control nor influenced by a change in photoperiod length (24–40 h), suggesting a potential control by the circadian clock [87]. Such circadian oscillations have been described in Arabidopsis leaves at the level of GBSSI transcription and activity but not at the protein content level [88]. This may suggest that yet unidentified posttranslational mechanisms may be active in the cell to tightly regulate the activity of the enzyme, in common with ADP-glucose pyrophosphorylase.

10.5 Control of starch biosynthesis in monocotyledonous species

The majority of research on monocots has focused on the analysis of starch biosynthesis in the developing endosperm. This is largely because of the economic importance of starch in this tissue, but also because many monocots do not accumulate large amounts of starch in their leaves.

A range of systems have now been extensively studied, with much of the early work performed on maize. This is because of the utility of this species as a model system for plant genetics and the availability of induced and spontaneous mutants and the technology for the identification of mutator element induced mutations. However, through the sequencing of the rice genome, this species has emerged as a very powerful additional model for the analysis of starch synthesis in monocotyledonous species. Both wheat and barley have attracted considerable interest in recent years. Barley is used because it is an additional and useful system for forward genetics with large well characterized mutant collections that are available for reevaluation given the knowledge and tools now available. Wheat has also been a target because of its economic importance, but now, even wheat can be considered as a nonintractable system for reverse genetics through the development of tools such as TILLING [89].

The range of informative mutants in starch biosynthesis available in maize is shown in Table 10.1. The mutants are classified into two types: those identified through 'forward genetics' involving the identification of lesions in genes underpinning traits, and 'reverse genetics', in which mutations are induced in specific genes in order to identify their roles.

Mutations in nine genes have been identified by forward genetics. Four of the mutations affect substrate supply, namely *Shrunken-1* (a lesion in sucrose synthase), and mutations in two of the ADPglucose pyrophosphorylase subunits (Brittle-2 and *Shrunken-2* respectively). Mutations in the *waxy* gene have been known for many years and were amongst the first examples of mutations that were characterized such that the mutation, gene and protein encoded were linked [84]. The identification of the causal basis of *sugary-1* (isoamylase1) [64] and *dull1* (*du1*) [39] was a major achievement of its time and further cemented the

Table 10.1 Maize endosperm starch mutants

Mutation	Locus	Chromosome location	Causal gene	Reference
			Forward genetics	
Shrunken-1	Sh1	9S	Sucrose synthase	[9]
Brittle-2	Bt2	4S	Cytosolic ADPglucose pyrophosphorylase small subunit	[146]
Shrunken-2	Sh2		Cytosolic ADPglucose pyrophosphorylase large subunit	[147]
Brittle-1	Bt1	5L	Unknown (ADPglucose transporter?)	[148]
Waxy	Wx	9S	GBSSIa	[90]
Sugary-2	Su2	6L	SSIIa	[37]
Dull	Du1	10L	SSIII	[39]
Amylose extender	Ae1	5L	BEIIb	[149]
Sugary-1	Su1	4S	Isoamylase (Isa1)	[64]
			Reverse genetics	
ZPu1	Zpu1	2	Pullulanase	[51]
BEI	Sbe1		BEI	[47]
BEIIa	Sbe3	8L	BEIIa	[50]

importance of the maize system in defining the molecular basis of mutations for which there were no obvious candidate genes.

Given that multiple alleles of mutations identified through forward genetics have been identified, it would be reasonable to expect that mutations not yet identified either have no major phenotype or are lethal. The mutations identified by reverse genetics have in fact turned out to have quite subtle phenotypes, at least in the endosperm where the primary screening for starch mutants has been focused. For example, BEI has no obvious phenotype in either endosperm or leaf [47], elimination of BEIIa activity gave a high amylose phenotype in the leaf but not the endosperm, and a lesion in the pullulanase 7gene yields a significant phenotype only in the absence of isoamylase [51].

Mutations in a range of known genes are yet to be identified, again suggesting that their absence leads to either subtle or lethal phenotypes. These genes include SSI, SSIIb, SSIV, cytosolic phosphorylase, plastidic phosphorylase and the plastidic pyrophosphorylase genes.

Recently, Wilson *et al.* [91] have linked phenotypes in the maize endosperm to six starch biosynthetic genes: *amylose extender1* (*ae1*), *brittle endosperm2* (*bt2*), *shrunken1* (*sh1*), *sh2*, *sugary1* and *waxy1*. Starch composition and quality traits (pasting properties and amylose levels) were measured. Significant associations between *bt2*, *sh1*, and *sh2* and kernel composition traits were found. Starch pasting properties were associated with *ae1* and *sh2*, and *ae1* and *sh1* were associated with amylose levels.

10.5.1 Genes in the rice genome

The availability of rice genome sequences has facilitated an examination of the starch genes in the genomes of monocotyledonous species. Rice starch biosynthetic genes were identified using Blast and annotation based searches of the NCBI database, and are shown in Table 10.2. In rice, single isoforms of GBSSI, GBSSII, SSI, SBEI, SBEIIa, SBEIIb, ISA1, ISA2, ISA3, pullulanase, cytoplasmic phosphorylase

Table 10.2 Rice starch biosynthetic genes

Genes	Chromosome location	Accession number	Annotation	Ortholog	Tissue expression
ADPGPP SS	9	AP004756	Small subunit	Wheat ADPGPP small subunit	NA
	8	AP004459	Small subunit		NA
ADPGPP LS	1	AP004317	Large subunit	Wheat ADPGPP large subunit	NA
	3	AC096689	Large subunit		NA
	5	AC007858	Large subunit		
GBSSI	6	X53694	GBSSI	Wheat GBSSI, maize GBSSI	Endosperm, pollen
GBSSII	7	AP005325	GBSSII	Wheat GBSSII	Leaf
SSI	6	AB026295	SSI	Wheat SSI, maize SSI	Endosperm, leaf, root
SSII-1	10	AC0087547	NA		Endosperm, leaf, root
SSII-2	2	AP005297	SSIIb	Maize SSIIb	Mainly in leaf
SSII-3	6	AP003509	SSIIa	Maize SSIIa, Wheat SSIIa	Endosperm
SSIII-1	4	AL606645	SSIII	Wheat SSIII, maize SSIII	Leaf, root, endosperm
SSIII-2	8	AP004660	NA		Endosperm
SSIV-1	1	AY100470	NA		Endosperm, leaf
SSIV-2	5	AY100471	SSIV	Wheat SSIV	Leaf, endosperm
SBE1	6	AP004685	SBEI	Wheat SBEI, maize SBEI	Endosperm, leaf
SBE3	2	AP004879	SBEIIb	Wheat SBEIIb, maize SBEIIb	Endosperm
SBE4	4	AL731641	SBEIIa	Wheat SBEIIa, maize SBEIIa	Endosperm, leaf
ISA1	8	AP005509	Isoamylase1	Wheat isoamylase1	Endosperm, leaf
ISA2	5	AC132483			Endosperm, leaf
ISA3	9	AP005574			Endosperm, leaf
Pullulanase	4	Al662959	Pullulanase	Maize pullulanase	Endosperm
Starch phosphorylase	3	AC079887	Plastid starch phosphorylase	Wheat plastid starch phosphorylase	NA
Starch phosphorylase	1	AP004072	Cytosolic starch phosphorylase	Wheat cytosolic starch phosphorylase	NA

and plastidic phosphorylase are present. In contrast, multiple forms of SSIIa and SSIII and SSIV occur with varying copy number and varying relative expression levels in leaf, root and endosperm. These data are consistent with findings presented by Jiang *et al.* [92] and Hirose and Terao [93]. The biological drivers for the generation of multiple genes are unknown, although this multiplication of the gene families clearly provides a mechanism for independent control of expression of the gene in different tissues. Two ADPglucose pyrophosphorylase small subunit genes (as in Arabidopsis) and three ADPglucose pyrophosphorylase large subunit genes (compared with four in Arabidopsis) are found.

10.6 Starch synthesis in green algae

10.6.1 *Chlamydomonas reinhardtii defines the best microbial system to study plant starch metabolism*

Starch biosynthesis in most species is thought to be derived from a unique endosymbiotic event thought to have occurred over 1.5 billion years [94]. This endosymbiosis consisted of the capture of an ancestral cyanobacterium by a nonphotosynthetic flagellated eukaryote and gave birth to all present-day plastids [95]. Among the three lines that are directly derived from this unique event, the green algae are the only ones to accumulate polysaccharide within the plastid stroma. The terrestrial plants are known to be derived through evolution of the green algae and therefore, among all known eukaryote microorganisms, green algae were expected to be the simplest model systems that could account for starch biosynthesis in plants. Early work performed on this group of organisms clearly showed the presence of a polysaccharide structure closely related to that of vascular plant starch [96]. Some enzymes, such as ADP-glucose pyrophosphorylase [97, 98] or the phosphorylases [99], had been studied to some extent and set the stage for more detailed investigations. Among all green algae, *C. reinhardtii* defines, by far, the most intensively studied system (reviewed in Ball [100]). The so-called green yeast allows easy selection of mutants, crossing and genetic transformation. It is a facultative autotroph allowing for easy manipulation of not only photosynthesis but also respiration and organelle DNA transformation [101, 102]. It accumulates starch around the pyrenoid (a Rubisco-containing cellular structure involved in CO_2 concentration in aquatic plants) or close to the thylakoid membranes [103, 104]. Most importantly, starch seems to be completely dispensable for growth or germination of encysted zygotes, thus allowing the selection of the whole suite of viable starch defective mutants [105].

10.6.2 *The Chlamydomonas single cell can account for both transitory or storage starch synthesis*

Log phase cultures submitted to 12 h light and 12 h darkness growth regimes are subjected to recurrent phases of starch synthesis and degradation [104, 106]. In conditions where the cells are actively dividing, cell growth and division acts as a powerful sink for carbon, driving the carbon out of the plastid and into the cytoplasm for building novel cellular material. Under these conditions the synthesis of starch acts as a transitory overflow for photosynthate at particular times of the circadian cycle. This starch is recurrently degraded at other times of the cycle when the demand for carbon in the cytoplasm generally exceeds its supply through photosynthesis. Such a system mimics the recurrent starch synthesis and degradation occurring in plant leaf chloroplasts, where degradation is thought to be triggered at night to ensure a constant supply of carbon for the nonphotosynthetic roots. Indeed, the structure of starch accumulated under these conditions closely resembles that of plant leaf transitory starch [107]. Therefore these log phase cultures can, to some extent, be used to study transitory starch metabolism. Upon starvation of nitrogen, the cells engage in one or at most two successive rounds of division, become nonphotosynthetic and stop dividing. Under these conditions, cellular material is actively degraded and carbon redirected to the plastid where it is engaged in active starch biosynthesis [108]. This situation mimics the starch storing organs of vascular plants, such as the potato tuber or the cereal endosperm. The polysaccharide structure under these conditions was found to be identical to that of maize endosperm starch [107]. Both nitrogen starvation and active growth are natural physiological conditions faced by the *C. reinhardtii* single cell. These conditions are exceedingly easy to trigger and master in laboratory conditions where cultures can be produced and studied within 5 days with sufficient biomass to achieve standard biochemical characterization, a situation unparalleled by any model land plant system.

10.6.3 What have we learned from Chlamydomonas?

A genetic screen based on the spraying of iodine vapors on the nitrogen starved cell patches has been extensively used to isolate mutants defective for various aspects of starch metabolism. Eleven loci have been identified by forward genetic approaches. Mutants of the *Chlamydomonas* heterotetrameric ADPglucose pyrophosphorylase have been produced for both small and large subunits of the enzyme [105, 109]. Studies on those of the large subunit have, at the time, contributed to prove the requirement *in vivo* of 3-phosphoglycerate activation of ADPglucose pyrophosphorylase in order to obtain significant starch synthesis [103]. In addition, these mutants have helped to establish that controlling the flux of ADP-glucose into starch had a profound consequence on the polysaccharide structure which, under low glycoside-nucleotide supply, adopts the structure of transitory starch [104]. This conclusion was subsequently verified in pea and potato [111, 112]. The characterization of mutants of GBSSI, the enzyme responsible for amylose synthesis, have suggested that GBSSI is an important contributor to amylopectin synthesis [113]. The isolation of a soluble starch synthase defective mutant that was later proved to be analogous to the vascular plant SSIII enzymes contributed to establish that distinct soluble starch synthases had specific roles in building different size classes of amylopectin chains [40]. However, the most important finding came with the description of low starch mutants that completely substituted starch synthesis with a small amount of glycogen [66]. These mutants proved to lack a high mass isoamylase complex because of the absence of one of its subunits [114–117]. The phenotype of the *Chlamydomonas* mutant appears much more severe than those of the vascular plant counterparts, and led the authors to propose that debranching of an amylopectin precursor (preamylopectin) defined the step that distinguished glycogen from starch synthesis [34]. This step is deemed to be mandatory to obtain semicrystalline material aggregating into macrogranular insoluble structures. Again, the isolation of other mutants of *C. reinhardtii* strongly suggested the presence of an active heteromultimeric isoamylase whose quaternary structure defined an important component of the polysaccharide aggregation mechanism [115, 116]. This suggestion was subsequently proved by the isolation of the corresponding genes and products in potato [69]. More recently, mutants defective for the D-enzyme (an α-1,4 glucanotransferase) established the first link between this family of enzymes and starch metabolism [55, 58, 59]. Because of the low starch phenotype of these mutants, the D-enzyme was inferred to play an important role in starch biosynthesis by recycling chains produced through isoamylase. Finally, the selection of mutants defective for one plastidial form of starch phosphorylase displayed the first evidence *in vivo* for a function of this enzyme in starch catabolism. *C. reinhardtii* also harbors a GBSSI enzyme whose specific activity is between 10-fold and 100-fold superior to those reported for the plant enzymes. This property was used to establish synthesis *in vitro* of amylose with purified starch granules [33]. These typical biochemical approaches established that the *C. reinhardtii* enzyme synthesizes amylose chains by extension of an amylopectin chain followed by cleavage of mature amylose chain at the site of its synthesis. They also established that GBSSI does not contribute only negatively to the polysaccharide's crystallinity but is also responsible for the formation of a low but significant number of B-type crystals [118].

10.6.4 Similarity and differences between starch metabolism in plants and algae

The full genome sequence of two unicellular green algae *C. reinhardtii* and *Ostreococcus tauri* has recently been made available. *O. tauri*, the smallest of all eukaryotes, is a member of the prasinophytes, a subdivision of green algae that are thought to have diverged at the earliest stage during evolution of the green plants. Bioinformatic mining of these resources coupled with functional studies performed in both organisms clearly establish that green algae use a pathway very similar, if not identical, to those of vascular land plants [119]. In fact it is truly astonishing to realize that an organism such as *O. tauri*, which has streamlined the complexity of many pathways because of its tiny genome size, has conserved the whole 30 or so genes thought to be active for starch metabolism in higher plants. This finding has two important consequences. First it validates green algae as model systems for studies dealing with plant starch metabolism. Second it demonstrates that the complexity of starch metabolism, and in particular the number of isoforms concerned for each type of enzyme activity, is a very ancient feature that evolved shortly after endosymbiosis. The maintenance of this complexity in *O. tauri* suggests further that the whole suite of enzymes work as an exceptionally networked unit precluding the loss of any of its multiple components to accommodate the small genome size. This comes in stark contrast to the streamlining and simplification of other networks such as those governing the cell division cycle in this organism [120]. The *O. tauri* and *C. reinhardtii* genomes do have two distinctive features with respect to vascular plant starch metabolism. First they appear to contain a distinctive set of plastidial phosphorylases, and second they appear to lack glycogenin genes. Glycogenin is inferred to be active in the priming of plant starch synthesis [121]. It must be stressed that the absence of these genes in the algal genomes should be treated with caution. Indeed, the genome sequences are not entirely sequenced and assembled, and so a unique glycogenin sequence might have escaped detection. If the absence of glycogenin is confirmed, it would suggest that either this protein is not involved in polysaccharide priming in green plants, or there exist several distinct pathways of polysaccharide synthesis priming and of granule seeding. In *O. tauri,* the chloroplast contains, in most cases, a single granule at the center of the tiny plastid. Microscopic observations performed during plastid division establish that partition of this single granule is performed into the two dividing plastids through localized degradation [119]. Thus this study on *O. tauri* may suggest that targeted degradation of starch granules might be the way the seeds of novel granules are produced.

10.6.5 *The future of starch research in green algae*

At the beginning of the 1990s, *C. reinhardtii* defined the best system to tackle genetic dissection of starch metabolism. However, in the last 8 years Arabidopsis has attracted far more attention compared with *Chlamydomonas.* This mainly reflects the production of two unique resources that are available only in Arabidopsis. First, the Arabidopsis full genome sequence became available 5 years before *C. reinhardtii.* Second and above all, multiple mutant libraries have been generated in Arabidopsis consisting of T-DNA insertions whose flanked sequences have been determined (the FST resource). These libraries cover over 90% of all genes discovered in the genome sequence. It is possible nowadays to order seed heterozygote for any gene suspected to be involved in starch metabolism. This has diverted groups from the study of other organisms to focus attention on Arabidopsis. One limitation is that Arabidopsis is mainly useful for the study of mature leaf starch synthesis which defines only a subset of physiological constraints that have led to the selection of the starch metabolism network. *C. reinhardtii* will probably remain one of the best systems if not the best system to perform forward genetic mutant screens. Its usefulness will however very much depend on the pace at which the reverse genetic tools such as an FST resource are built. At the time of writing this chapter, there is no way of telling if these resources will be as extensive and as rapidly available as those generated for Arabidopsis by a far greater number of research groups.

10.7 Starch synthesis in other systems

10.7.1 Bacterial cells may have a primitive starch synthesizing machinery

All present evidence points to the presence of a pathway in green plants that was derived from bacterial glycogen synthesis. The photosynthetic eukaryotes are indeed thought to have acquired at least part of their polysaccharide synthesis machinery through endosymbiosis of the chloroplast. Most bacteria were initially thought to accumulate a storage polysaccharide structure that resembled structures reported for fungi and animal cells, with no evidence for the presence of semicrystalline structures. However, this picture is now slowly changing with a very recent report of the existence of amylopectin-like polysaccharide in a small number of cyanobacteria [122] which behaves as a high mass polymer with a chain-length distribution that is intermediate between amylopectin and glycogen. It is presently not known if these polymers aggregate into insoluble semicrystalline granules. In one of the very few cases where the structure of carbohydrate storage granules has been investigated, the authors have demonstrated that large size granules in *Cyanothece* correlated with the presence of polymers that were apparently more related to glycogen [123]. However, mathematical modeling suggests that glycogen particles could not theoretically exceed a diameter of 40 nm [124]. It remains possible, therefore, that these organisms store polymers that resemble amylopectin rather than glycogen. Nevertheless, the vast majority of bacteria, including the cyanobacterial model species, accumulate glycogen. Therefore, the comparison of glycogen metabolism in these organisms to that of the cyanobacteria accumulating amylopectin-like molecules will probably be enlightening.

In the meantime glycogen accumulating model bacteria such as *E. coli* and *Synechocystis* have yielded a number of important insights concerning storage polysaccharide metabolism (for reviews of glycogen metabolism in bacteria, see [5, 125, 126]). In *E. coli* in particular, mutants have been obtained for the unique genes coding for the ADP-glucose requiring glycogen synthase, ADP-glucose pyrophosphorylase and branching enzyme. These have been a great help in establishing the nature and function of the bacterial pathway (reviewed [125]), while in the glycogen accumulating *Synechocystis*, mutants of branching enzyme have been recently reported [127].

However these genetic studies are somewhat incomplete since the effect of many other relevant genes has not been investigated through mutant approaches. Recently, Dauvillée *et al.* have reported the effect of disruptions of bacterial glycogen debranching enzyme GlgX [6]. This result is relevant to the understanding of starch metabolism in plants since the GlgX protein belongs to the same family of enzymes as the isoamylases which are thought to be responsible for amylopectin crystallization [34]. The glycogen excess phenotypes of the mutants were consistent with the previously inferred function in glycogen catabolism of this enzyme [128]. Indeed, the debranching specificity of this enzyme was shown to be restricted to the branches from outer chains consisting of four glucose residues. These are the structures generated by the action of glycogen phosphorylase during polysaccharide breakdown. This work shows that glycogen accumulating species display a debranching enzyme with a marked specificity that distinguishes them from the preamylopectin trimming enzymes. The work performed on the bacterial pathway has been a past inspiration for the studies performed on algae and plants. Presently, the emphasis on bacteria has been toward establishing a number of 3D structures for the important enzymes of the glycogen-starch pathway such as the branching enzyme [129] or glycogen synthase [130]. Another important issue of relevance to our understanding of starch synthesis is the nature of the polysaccharide priming mechanism in bacteria. As in green algae, no evidence for the presence of glycogenin could be found. Interestingly, a recent report demonstrates the presence of a priming activity of the bacterial glycogen synthase [131].

10.7.2   UDPglucose-based systems that produce starch

Green algae and land plants are the only organisms that store starch within the plastid among the three lineages derived from the primary endosymbiosis event of a cyanobacterial-like cell. The other two lineages (the glaucophytes and the red algae) store the same type of polysaccharide in the cytoplasm. The name 'floridean' starch has been coined to describe the type of storage polysaccharide accumulated by red algae (reviewed by Viola *et al.* [132]). This material was thought to differ from the classic type of starch structure because of the absence of amylose. However, since the initial description of floridean starch, amylose has been found in the starch of several distinct red algal species [133], and a schematic view of starch biosynthesis in red algae is shown in Figure 10.4. The nature of the pathway involved has been somewhat controversial, as earlier studies favored an ADP-glucose based pathway, while more recent work pointed to the existence of an UDP-glucose pathway [134]. Red algae have themselves been the subject of either a single or multiple endosymbiotic events by other nonphotosynthetic eukaryotes. This offers a convenient explanation for the finding of floridean starch in the cytoplasm of dinoflagellates and apicomplexa parasites.

Coppin *et al.* [135], using bioinformatics and functional studies on the apicomplexa parasite *Toxoplasma gondii*, clearly established the mosaic nature of the starch pathway. *T. gondii* contained genes coding for several enzymes that are either only found in, or display phylogenies which relate them to the glycogen accumulating eukaryotes such as fungi or animals (for example, glycogenin, indirect debranching enzyme, branching enzyme [136], UDP-glucose utilizing glycogen (starch) synthase). Other genes encode enzymes that are typical of, or display

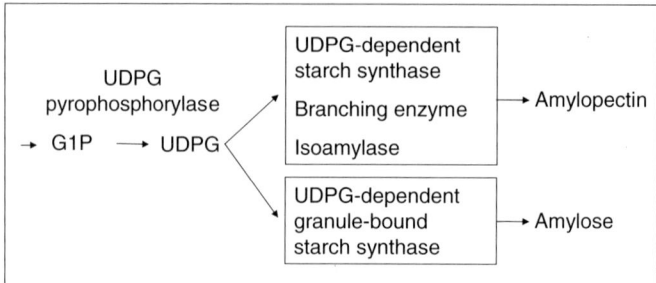

Figure 10.4 The putative starch synthesis pathway in the cytosol of red algae. The proposed role of isoamylase is speculative rather than based on loss of function studies; however, the enzyme is known to be present in red algae synthesizing starch (Coppin et al. [135]) [129]. Enzyme activities are as defined in Figures 10.1, 10.2 and 10.3.

phylogenies that relate them to, the bacterial pathway (α-1,4 glucanotransferase, isoamylase). Together with functional studies, these results establish floridean starch metabolism as UDPglucose based with a suite of enzymes that can clearly distinguish this pathway from those operating in the green algae and terrestrial plants. Of particular interest is the restricted number of isoforms found in the genomes of both *T. gondii* and the red algae *Cyanidioschyzon merolae* [135]. The finding of isoamylase in floridean starch accumulating organisms confirms the requirement for this enzyme for semicrystalline polysaccharide biogenesis. Further, Coppin et al. [135] hypothesize that both the UDPglucose and ADPglucose based pathways are mosaics of genes of prokaryotic and eukaryotic origin that resulted from the confrontation of the two pathways within the endosymbiont and in the cytoplasm of the host after endosymbiosis of the chloroplast. The astonishingly simple pathway revealed in *T. gondii* might make this organism an unexpected model to understand the specific features of starch granule biogenesis.

10.8 Control of starch biosynthesis

The majority of the discussion of starch synthesis in this chapter has focused on the elucidation of differences in the biosynthetic pathway, compartmentation and gene expression in differing species and tissues. Over the past 5 years, evidence has begun to accumulate concerning the fine transcriptional and posttranscriptional control of starch biosynthesis, processes that could be expected to influence both the amount of starch synthesized in a given time period, and the composition of that starch. The types of regulation that have been postulated to be important in starch biosynthesis are:

(1) *Circadian control of gene expression*: Observations from a range of species indicate that enzymes involved in starch synthesis and degradation are

under circadian control. In rice, Dian et al. [137] identified a GBSS gene (designated GBSSII), showed that the product of this gene was present in the starch granule and concluded that the enzyme was responsible for leaf starch synthesis. The expression of this gene was shown to be under circadian control [137]. In sweet potato, Wang et al. [138] concluded that GBSSI expression in leaves was also regulated by the circadian clock. In Arabidopsis leaves, Tenorio et al. [88] showed that GBSSI expression oscillated on a diurnal light/dark cycle. In a more comprehensive analysis of gene expression across the starch synthesis pathway, Smith et al. [139] used affymetrix arrays to investigate oscillations in starch gene expression in Arabidopsis leaves. They concluded that GBSSI and SSII oscillated on a diurnal light/dark cycle. Further studies are required to study whether similar processes are occurring in storage tissues such as tubers and the endosperm.

(2) *Redox regulation*: ADPglucose pyrophosphorylase from potato tuber has been shown by Fu et al. [140] to be partially inactivated by oxidation of disulfide bonds between the small ADPglucose pyrophosphorylase subunits. In subsequent work, Tiessen et al. [86] and Hendriks et al. [141] extended these observations to identify redox control of ADPglucose pyrophosphorylase in leaves of potato, pea and Arabidopsis. Tiessen et al. [86] also showed that redox control of ADPglucose pyrophosphorylase occurred in response to sucrose availability. Preliminary reports suggest that ADPglucose pyrophosphorylase from barley and wheat endosperms is also regulated by redox control [3].

(3) *Phosphorylation*: In wheat endosperm, Tetlow et al. [142] showed that several enzymes (BEI, BEIIa, BEIIb and plastidic phosphorylase) were phosphorylated (on serine residues) in amyloplasts labeled with ^{32}P-ATP, and dephosphorylation of these enzymes resulted in a reduction in enzyme activity. A range of other phosphoproteins have been identified in stromal extracts, including SSIIa and an unidentified form of starch synthase. The significance of protein phosphorylation as a general mechanism for regulation of starch synthesis in plants requires further research.

(4) *Complex formation*: Tetlow et al. [142] also demonstrated through immunoprecipitation experiments that complexes of starch biosynthetic enzymes exist in the wheat endosperm, including a complex containing SBEI, SBEIIb and starch phosphorylase. As noted previously, isoamylases in *C. reinhardtii* [116], potato [69] and Arabidopsis [54, 68] have been demonstrated to exist in multimeric aggregates. It appears that protein–protein interactions leading to the formation of these complexes may be a general feature of starch biosynthesis. However, further work is required to define the number, composition, function and formation/disintegration behavior of such complexes. The presence of multimeric enzyme complexes provides a further point of difference between glycogen synthesis in bacteria and higher plant starch synthesis. It is conceivable that such complexes are important in providing substrate channeling or chaperonin-like protection

of nascent amylopectin molecules that are important to the crystallization or amylopectin and the synthesis of starch granules.

10.9 Opportunities for the manipulation of starch synthesis and structure

The study of starch synthesis and its regulation has two general purposes. The first is to provide an understanding of the mechanisms underpinning the process. The second is to provide the information necessary to direct efforts to manipulate starch structure and function in ways that contribute to economic and societal goals. Some of the discussion in this review has alluded to the use of genetic manipulation to investigate the role of specific members of gene families involved in starch synthesis and such modifications of cereal starches for food and industrial use have recently been reviewed [2, 143]. The end use targets of starch modification include:

(1) *Food industry use*: Modification of starch for the food industry use involves the range of starch from waxy to high amylose and includes starches, such as sugary-2 and dull, with intermediate amylose contents but altered amylopectin structure and functionality.
(2) *Delivery of nutritional benefits*: The development of novel starches focuses on high amylose starches, which are believed to provide human health benefits through the provision of resistant starch and reduced glycemic index.
(3) *Industrial end uses*: Starches for industrial use range from readily digestible starches for ethanol and feedstock generation to high amylose starches for biodegradable plastics, films and adhesives.
(4) *Animal feeds*: The general focus of starch modification for animal feeds is to increase the rate of feed conversion as measured by metabolizable energy.
(5) *Pharmaceutical applications*: The focus of manipulation is on the development of starches that can provide specific functionalities such as slow release of encapsulated compounds. One additional method for modifying starch structure and functionality using genetic engineering technologies is the use of starch binding proteins. Several recent studies have shown that such proteins can be fused to other functionalities, providing a 'payload' system for delivering functionality to starches and conferring properties such as controlled release [144]. It should be noted that genetic modification is just one mechanism for enhancing starch functionality for specific applications. The current methods of choice for enhancing starch functionalities in most fields of use are thermophysical and chemical modification. However, with an enhanced understanding of starch synthesis and the ability to rapidly generate diversity, opportunities are likely to arise for developing more valuable applications of native starches. The opportunities will be driven not only by knowledge of the core genes of the starch biosynthesis pathway, but also by how they are regulated and how that regulation

itself can be modulated. A primary, and early, example of this approach is the use of unregulated ADPglucose pyrophosphorylase in engineering of potato starch synthesis [145].

10.10 Conclusions

The past decade has seen a rapid expansion of our knowledge of starch biosynthesis and its regulation. This rapid development has been fueled by the genome sequencing initiatives which have clarified the gene families found in the key species, Arabidopsis and rice. These initiatives have allowed extensive expression profiling and mutagenesis programs to define the roles of individual members of the families. Furthermore, improved tools such as proteomics have allowed the identification of proteins and their posttranslational modifications. Techniques such as TILLING and high throughput RNAi techniques hold out the promise that the next decade will see all the known members of the starch biosynthetic enzyme gene families assayed for function in the key target species.

Solving the core problems of identifying the genes and their primary functions was expected to lead to a comprehensive understanding of starch biosynthesis. However, three factors suggest that there is considerable way to go yet before we can claim to understand starch synthesis. The first factor is the presence of interactions between starch biosynthetic enzymes, for example, the complex formed by isoamylase1 and isoamylase2, and the complexes defined by Tetlow *et al.* [142]. These interactions suggest that the phenotypes of some of the mutants that have been characterized may result not only from effects on the target genes, but from pleiotrophic effects on other proteins with which the target protein interacts. Secondly, the mechanisms through which the starch biosynthetic enzymes, or assemblies of starch biosynthetic enzymes, interact at the granule surface and in the interior of the starch granule to form the granular architecture are unknown. Given the saturation mutagenesis programs that have been undertaken, and the repetitive isolation of alleles of the major mutants, it seems unlikely that other proteins with major impact on starch synthesis will be found that mediate granule assembly. However, this does not rule out roles for proteins in fine tuning granule assembly that have not been picked up in the screens used to date. Finally, the events involved in the initiation of the starch granule, as well as the initiation of amylose and amylopectin molecules, remain poorly defined and these processes need further research. It is therefore not unreasonable to state that the next decade of starch research promises to be as exciting and informative as the past decade.

References

1. S.G. Ball and M.K. Morell (2003) From bacterial glycogen to starch: understanding the biogenesis of the plant starch granule. *Annual Reviews of Plant Biology* **54**, 207–233.
2. S. Jobling (2004) Improving starch for food and industrial applications. *Current Opinion in Plant Biology* **7**, 210–218.

3. I.J. Tetlow, M.K. Morell and M.J. Emes (2004) Recent developments in understanding the regulation of starch metabolism in higher plants. *Journal of Experimental Botany* **55**, 2131–2145.
4. A.M. Myers, M.K. Morell, M.G. James and S.G. Ball (2000) Recent progress toward understanding the biosynthesis of the amylopectin crystal. *Plant Physiology* **122**, 989–998.
5. M.A. Ballicora, A.A. Iglesias and J. Preiss (2003) ADP-glucose pyrophosphorylase, a regulatory enzyme for bacterial glycogen synthesis. *Microbiology and Molecular Biology Reviews* **67**, 213–225.
6. D. Dauvillee, I.S. Kinderf, Z.Y. Li et al. (2005) Role of the *Escherichia coli* glgX gene in glycogen metabolism. *Journal of Bacteriology* **187**, 1465–1473.
7. A.M. Smith (2001) The biosynthesis of starch granules. *Biomacromolecules* **2**, 335–341.
8. J. Preiss and M. Sivak (1996) Starch synthesis in sinks and sources. In: *Photoassimilate Distribution in Plants and Crops*, Dekker, New York, pp. 63–69.
9. P.S. Chourey and O.E. Nelson Jr. (1976) The enzymatic deficiency conditioned by the shrunken-1 mutations in maize. *Biochemical Genetics* **14**, 1041–1055.
10. H.P. Ghosh and J. Preiss (1966) Adenosine diphosphate glucose pyrophosphorylase: a regulatory enzyme in the biosynthesis of starch in spinach leaf chloroplasts. *Journal of Biological Chemistry* **241**, 4491–4504.
11. L.A. Kleczkowski, P. Villand, E. Lüthi, O.A. Olsen and J. Preiss (1993) Insensitivity of barley endosperm ADP-glucose pyrophosphorylase to 3-phosphoglycerate and orthophosphate regulation. *Plant Physiology* **101**, 179–186.
12. K. Denyer, F. Dunlap, T. Thorbjørnsen, P. Keeling and A.M. Smith (1996) The major form of ADPglucose pyrophosphorylase in maize endosperm is extraplastidial. *Plant Physiology* **112**, 779–783.
13. T. Thorbjørnsen, P. Villand, K. Denyer, O.A. Olsen and A.M. Smith (1996) Distinct isoforms of ADPglucose pyrophosphorylase occur inside and outside the amyloplasts in barley endosperm. *Plant Journal* **10**, 243–250.
14. R.A. Burton, P.E. Johnson, D.M. Beckles et al. (2002) Characterization of the genes encoding the cytosolic and plastidial forms of ADP-glucose pyrophosphorylase in wheat endosperm. *Plant Physiology* **130**, 1464–1475.
15. P.E. Johnson, N.J. Patron, A.R. Bottrill et al. (2003) A low-starch barley mutant, riso 16, lacking the cytosolic small subunit of ADP-glucose pyrophosphorylase, reveals the importance of the cytosolic isoform and the identity of the plastidial small subunit. *Plant Physiology* **131**, 684–696.
16. P. Crevillen, M.A. Ballicora, A. Merida, J. Preiss and J.M. Romero (2003) The different large subunit isoforms of *Arabidopsis thaliana* ADP-glucose pyrophosphorylase confer distinct kinetic and regulatory properties to the heterotetrameric enzyme. *Journal of Biological Chemistry* **278**, 28508–28515.
17. P. Crevillen, T. Ventriglia, F. Pinto, A. Orea, A. Merida and J.M. Romero (2005) Differential pattern of expression and sugar regulation of *Arabidopsis thaliana* ADP-glucose pyrophosphorylase-encoding genes. *Journal of Biological Chemistry* **280**, 8143–8149.
18. J.C. Shannon, F.M. Pien, H. Cao and K.C. Liu (1998) Brittle-1, an adenylate translocator, facilitates transfer of extraplastidial synthesized ADP-glucose into amyloplasts of maize endosperms. *Plant Physiology* **117**, 1235–1252.
19. N.J. Patron, B. Greber, B.F. Fahy, D.A. Laurie, M.L. Parker and K. Denyer (2004) The lys5 mutations of barley reveal the nature and importance of plastidial ADP-Glc transporters for starch synthesis in cereal endosperm. *Plant Physiology* **135**, 2088–2097.
20. D.M. Beckles, A.M. Smith and T. ap Rees (2001) A cytosolic ADP-glucose pyrophosphorylase is a feature of graminaceous endosperms, but not of other starch storing organs. *Plant Physiology* **125**, 818–827.
21. D.M. Beckles, J. Craig and A.M. Smith (2001) ADP-glucose pyrophosphorylase is located in the plastid in developing tomato fruit. *Plant Physiology* **126**, 261–266.
22. E. Baroja-Fernandez, F.J. Munoz, A. Zandueta-Criado et al. (2004) Most of ADPglucose linked to starch biosynthesis occurs outside the chloroplast in source leaves. *Proceedings of the National Academy of Sciences of the United States of America* **101**, 13080–13085.

23. T. Nakamura, P. Vrinten, K. Hayakawa and J. Ikeda (1998) Characterization of a granule-bound starch synthase isoform found in the pericarp of wheat. *Plant Physiology* **118**, 451–459.
24. A, Edwards, J.P. Vincken, L.C. Suurs *et al.* (2002) Discrete forms of amylose are synthesized by isoforms of GBSSI in pea. *Plant Cell* **14**, 1767–1785.
25. H. Tatge, J. Marshall, C. Martin, E.A. Edwards and A.M. Smith (1999) Evidence that amylose synthesis occurs within the matrix of the starch granule in potato tubers. *Plant Cell and Environment* **22**, 543–550.
26. K. Denyer, C. Sidebottom, C.M. Hylton and A.M. Smith (1993) Soluble isoforms of starch synthase and starch-branching enzyme also occur within starch granules in developing pea embryos. *Plant Journal* **4**, 191–198.
27. K. Denyer, C.M. Hylton, C.F. Jenner and A.M. Smith (1995) Identification of multiple isoforms of soluble and granule-bound starch synthase in developing wheat endosperm. *Planta* **196**, 256–265.
28. S. Rahman, B. Kosar-Hashemi, M.S. Samuel *et al.* (1995) The major proteins of wheat endosperm starch granules. *Australian Journal of Plant Physiology* **22**, 793–803.
29. M.K. Morell, B. Kosar-Hashemi, M. Cmiel (2003) Barley *sex6* mutants lack starch synthase IIa activity and contain a starch with novel properties. *Plant Journal* **34**, 173–185.
30. M. Yamamori, S. Fujita, K. Hayakawa, J. Matsuki and T. Yasui (2000) Genetic elimination of a starch granule protein, SGP-1, of wheat generates an altered starch with apparent high amylose. *Theoretical and Applied Genetics* **101**, 21–29.
31. K. Denyer, D. Waite, S. Motawia, B.L. Moller and A.M. Smith (1999) Granule-bound starch synthase I in isolated starch granules elongates malto-oligosaccharides processively. *Biochemical Journal* **340**, 183–191.
32. K. Denyer, D. Waite, A. Edwards, C. Martin and A.M. Smith (1999) Interaction with amylopectin influences the ability of granule-bound starch synthase I to elongate malto-oligosaccharides. *Biochemical Journal* **342**, 647–653.
33. M. Van de Wal, C. D'Hulst, J.P. Vincken, A. Buléon, R. Visser and S. Ball (1998) Amylose is synthesized *in vitro* by extension of and cleavage from amylopectin. *Journal of Biological Chemistry* **273**, 22232–22240.
34. S. Ball, H.P. Guan, M. James *et al.* (1996) From glycogen to amylopectin: a model explaining the biogenesis of the plant starch granule. *Cell* **86**, 349–352.
35. J. Craig, J.R. Lloyd, K. Tomlinson *et al.* (1998) Mutations in the gene encoding starch synthase II profoundly alter amylopectin structure in pea embryos. *Plant Cell* **10**, 413–426.
36. T. Umemoto, M. Yano, H. Satoh, A. Shomura and Y. Nakamura (2002) Mapping of a gene responsible for the difference in amylopectin structure between japonica-type and indica-type rice varieties. *Theoretical and Applied Genetics* **104**, 1–8.
37. X. Zhang, C. Colleoni, V. Ratushna, M. Sirghie-Colleoni, M.G. James and A.M. Myers (2004) Molecular characterization demonstrates that the Zea mays gene *sugary2* codes for the starch synthase isoform SSIIa. *Plant Molecular Biology* **54**, 865–879.
38. C. Harn, M. Knight, A. Ramakrishnan, H. Guan, P.L. Keeling and B.P. Wasserman (1998) Isolation and characterization of the zSSIIa and zSSIIb starch synthase cDNA clones from maize endosperm. *Plant Molecular Biology* **37**, 639–649.
39. M. Gao, J. Wanat, P.S. Stinard, M.G. James and A.M. Myers (1998) Characterization of *dull1*, a maize gene coding for a novel starch synthase. *Plant Cell* **10**, 399–412.
40. T. Fontaine, C. D'Hulst, M.L. Maddelein *et al.* (1993) Toward an understanding of the biogenesis of the starch granule. Evidence that Chlamydomonas soluble starch synthase II controls the synthesis of intermediate size glucans of amylopectin. *Journal of Biological Chemistry* **268**, 16223–1623.
41. M.K. Bhattacharyya, A.M. Smith, T.H. Ellis, C. Hedley and C. Martin (1990) The wrinkled-seed character of pea described by Mendel is caused by a transposon-like insertion in a gene encoding starch-branching enzyme. *Cell* **60**, 115–122.
42. S.A. Jobling, G.P. Schwall, R.J. Westcott *et al.* (1999) A minor form of starch branching enzyme in potato (*Solanum tuberosum* L.) tubers has a major effect on starch structure: cloning and characterisation of multiple forms of SBE A. *Plant Journal* **18**, 163–171.

43. R. Safford, S.A. Jobling, C.M. Sidebottom *et al.* (1998) Consequences of antisense RNA inhibition of starch branching enzyme activity on properties of potato starch. *Carbohydrate Polymers* **35**, 155–168.
44. S.A. Jobling, C. Jarman, M.M. Teh, N. Holmberg, C. Blake and M.E. Verhoeyen (2003) Immunomodulation of enzyme function in plants by single-domain antibody fragments. *Nature Biotechnology* **21**, 77–80.
45. G.P. Schwall, R. Safford, R.J. Westcott *et al.* (2000) Production of very-high amylose potato starch by inhibition of SBE A and SBE B. *Nature Biotechnology* **18**, 551–554.
46. A. Regina, B. Kosar-Hashemi, Z. Li *et al.* (2004) Multiple isoforms of starch branching enzyme 1 in wheat: lack of the major SBE 1 isoforms does not alter starch phenotype. *Functional Plant Biology* **31**, 591–601.
47. S.L. Blauth, K.N. Kim, J. Klucinec, J.C. Shannon, D.B. Thompson and M. Guiltinan (2002) Identification of Mutator insertional mutants of starch-branching enzyme 1 (sbe1) in *Zea mays* L. *Plant Molecular Biology* **48**, 287–297.
48. H. Satoh, A. Nishi, K. Yamashita *et al.* (2003) Starch-branching enzyme I-deficient mutation specifically affects the structure and properties of starch in rice endosperm. *Plant Physiology* **133**, 1111–1121.
49. K.N. Kim, D.K. Fisher, M. Gao and M.J. Guiltinan (1998) Molecular cloning and characterization of the Amylose-Extender gene encoding starch branching enzyme IIB in maize. *Plant Molecular Biology* **38**, 945–956.
50. S.L. Blauth, Y. Yao, J.D. Klucinec, J.C. Shannon, D.B. Thompson and M. Guiltinan (2001) Identification of Mutator insertional mutants of starch-branching enzyme 2a in corn. *Plant Physiology* **125**, 1396–1405.
51. J.R. Dinges, C. Colleoni, M.G. James and A.M. Myers (2003) Mutational analysis of the pullulanase-type debranching enzyme of maize indicates multiple functions in starch metabolism. *Plant Cell* **15**, 666–680.
52. R. Bustos, B. Fahy, C.M. Hylton *et al.* (2004) Starch granule initiation is controlled by a heteromultimeric isoamylase in potato tubers. *Proceedings of the National Academy of Sciences of the United States of America* **101**, 2215–2220.
53. E.Y. Lee, J.J. Marshall and W.J. Whelan (1971) The substrate specificity of amylopectin-debranching enzymes from sweet corn. *Archives of Biochemistry and Biophysics* **143**, 365–374.
54. T. Delatte, M. Trevisan, M.L. Parker and S.C. Zeeman (2005) Arabidopsis mutants Atisa1 and Atisa2 have identical phenotypes and lack the same multimeric isoamylase, which influences the branch point distribution of amylopectin during starch synthesis. *Plant Journal* **41**, 815–830.
55. F. Wattebled, J.P. Ral, D. Dauvillée *et al.* (2003) *STA11* a Chlamydomonas locus required for normal starch granule biogenesis encodes disproportionating enzyme: further evidence for a function of α-1,4 glucanotransferases during starch granule biosynthesis in green algae. *Plant Physiology* **132**, 137–145.
56. R.A. Burton, H. Jenner, L. Carrangis *et al.* (2002) Starch granule initiation and growth are altered in barley mutants that lack isoamylase activity. *Plant Journal* **31**, 97–112.
57. A. Kubo, S. Rahman, Y. Utsumi *et al.* (2004) Complementation of sugary-1 phenotype in rice endosperm with the wheat isoamylase1 gene supports a direct role for isoamylase1 in amylopectin biosynthesis. *Plant Physiology* **137**, 43–56.
58. C. Colleoni, D. Dauvillée, G. Mouille *et al.* (1999) Genetic and biochemical evidence for the involvement of α-1,4 glucanotransferases in amylopectin synthesis. *Plant Physiology* **120**, 993–1003.
59. C. Colleoni, D. Dauvillée, G. Mouille *et al.* (1999) Biochemical characterization of the *Chlamydomonas reinhardtii* α-1,4 glucanotransferase supports a direct function in amylopectin biosynthesis. *Plant Physiology* **120**, 1005–1014.
60. J.H. Critchley, S.C. Zeeman, T. Takaha, A.M. Smith and S.M. Smith (2001) A critical role for disproportionating enzyme in starch breakdown is revealed by a knock-out mutation in Arabidopsis. *Plant Journal* **26**, 89–100.
61. N. Schupp and P. Ziegler (2004) The relation of starch phosphorylases to starch metabolism in wheat. *Plant and Cell Physiology* **45**, 1471–1484.

62. S.C. Zeeman, D. Thorneycroft, N. Schupp et al. (2004) Plastidial alpha-glucan phosphorylase is not required for starch degradation in Arabidopsis leaves but has a role in the tolerance of abiotic stress. *Plant Physiology* **135**, 849–858.
63. O. Pan and O.E. Nelson (1984) A debranching enzyme deficiency in endosperms of the s*ugary-1* mutants of maize. *Plant Physiology* **74**, 324–328.
64. M.G. James, D.S. Robertson and A.M, Myers (1995) Characterization of the maize gene *sugary1*, a determinant of starch composition in kernels. *Plant Cell* **7**, 417–429.
65. Y. Nakamura, T. Umemoto, Y. Takahata et al. (1996) Changes in structure of starch and enzyme activities affected by sugary mutations. Possible role of starch debranching enzyme (R-enzyme) in amylopectin biosynthesis. *Physiologia Plantarum* **97**, 491–948.
66. G. Mouille, M.L. Maddelein, N. Libessart et al. (1996) Phytoglycogen processing: a mandatory step for starch biosynthesis in plants. *Plant Cell* **8**, 1353–1366.
67. S.C. Zeeman, T. Umemoto, W.L. Lue et al. (1998) A mutant of Arabidopsis lacking a chloroplastic isoamylase accumulates both starch and phytoglycogen. *Plant Cell* **10**, 1699–1712.
68. F. Wattebled, Y. Dong, S. Dumez et al. (2005) Mutants of Arabidopsis lacking a chloroplastic isoamylase accumulate phytoglycogen and an abnormal form of amylopectin. *Plant Physiology* **138**, 184–195.
69. H. Hussain, A. Mant, R. Seale et al. (2003) Three isoforms of isoamylase contribute different catalytic properties for the debranching of potato glucans. *Plant Cell* **15**, 133–149.
70. A. Kubo, N. Fujita, K. Harada, T. Matsuda, H. Satoh and Y. Nakamura (1999) The starch-debranching enzymes isoamylase and pullulanase are both involved in amylopectin biosynthesis in rice endosperm. *Plant Physiology* **121**, 399–410.
71. T.S. Yu, H. Kofler, R.E. Hausler et al. (2001) The Arabidopsis *sex1* mutant is defective in the R1 protein, a general regulator of starch degradation in plants, and not in the chloroplast hexose transporter. *Plant Cell* **13**, 1907–1918.
72. S.C. Zeeman, A. Tiessen, E. Pilling, K.L. Kato, A.M. Donald and A.M. Smith (2002) Starch synthesis in Arabidopsis. Granule synthesis, composition, and structure. *Plant Physiology* **129**, 516–529.
73. R. Lorberth, G. Ritte, L. Willmitzer and J. Kossmann (1998) Inhibition of a starch-granule-bound protein leads to modified starch and repression of cold sweetening. *Nature Biotechnology* **16**, 473–477.
74. G. Ritte, R. Lorberth and M. Steup (2000) Reversible binding of the starch-related R1 protein to the surface of transitory starch granules. *Plant Journal* **21**, 387–391.
75. G. Ritte, J.R. Lloyd, N. Eckermann, A. Rotmann, J. Kossmann and M. Steup (2002) The starch related R1 protein is an α-glucan, water dikinase. *Proceedings of the National Academy of Sciences of the United States of America*, **99**, 1766–1771.
76. R. Reimann, M. Hippler, B. Machelett and K.J. Appenroth (2004) Light induces phosphorylation of glucan water dikinase, which precedes starch degradation in turions of the duckweed *Spirodela polyrhiza*. *Plant Physiology* **135**, 121–128.
77. O. Kotting, K. Pusch, A. Tiessen, P. Geigenberger, M. Steup and G. Ritte (2005) Identification of a novel enzyme required for starch metabolism in Arabidopsis leaves. The phosphoglucan, water dikinase. *Plant Physiology* **137**, 242–252.
78. L. Baunsgaard, H. Lutken, R. Mikkelsen, M.A. Glaring, T.T. Pham and A. Blennow (2005) A novel isoform of glucan, water dikinase phosphorylates pre-phosphorylated alpha-glucans and is involved in starch degradation in Arabidopsis. *Plant Journal* **41**, 595–605.
79. A. Scheidig, A. Frohlich, S. Schulze, J.R. Lloyd and J. Kossmann (2002) Downregulation of a chloroplast-targeted beta-amylase leads to a starch-excess phenotype in leaves. *Plant Journal* **30**, 581–591.
80. S.E. Weise, K.S. Kim, R.P. Stewart and T.D. Sharkey (2005) Beta-maltose is the metabolically active anomer of maltose during transitory starch degradation. *Plant Physiology* **137**, 756–761.
81. S.E. Weise, A.P. Weber and T.D. Sharkey (2004) Maltose is the major form of carbon exported from the chloroplast at night. *Planta* **218**, 474–482.
82. T. Niittyla, G. Messerli, M. Trevisan, J. Chen, A.M. Smith and S.C. Zeeman (2004) A previously unknown maltose transporter essential for starch degradation in leaves. *Science* **303**, 87–89.

83. T. Chia, D. Thorneycroft, A. Chapple et al. (2004) A cytosolic glucosyltransferase is required for conversion of starch to sucrose in Arabidopsis leaves at night. *Plant Journal* **37**, 853–863.
84. Y. Lu and T.D. Sharkey (2004) The role of amylomaltase in maltose metabolism in the cytosol of photosynthetic cells. *Planta* **218**, 466—473.
85. M.A. Ballicora, J.B. Frueauf, Y. Fu, P. Schurmann and J. Preiss (2000) Activation of the potato tuber ADP-glucose pyrophosphorylase by thioredoxin. *Journal of Biological Chemistry* **275**, 1315–1320.
86. A. Tiessen, J.H.M. Hendriks, M. Stitt et al. (2002) Starch synthesis in potato tuber is regulated by post-translational redox modification of ADP-glucose pyrophosphorylase. *Plant Cell* **14**, 2191–2213.
87. E. Pilling and A.M. Smith (2003) Growth ring formation in the starch granules of potato tubers. *Plant Physiology* **132**, 365–371.
88. G. Tenorio, A. Orea, J.M. Romero and A. Merida (2003) Oscillation of mRNA level and activity of granule-bound starch synthase I in Arabidopsis leaves during the day/night cycle. *Plant Molecular Biology* **51**, 949–958.
89. A.J. Slade, S.I. Fuerstenberg, D. Loeffler, M.N. Steine and D. Facciotti (2005) A reverse genetic, nontransgenic approach to wheat crop improvement by TILLING. *Nature Biotechnology* **23**, 75–81.
90. M. Shure, S. Wessler and N. Fedoroff (1983) Molecular identification and isolation of the Waxy locus in maize. *Cell* **35**, 225–233.
91. L.M. Wilson, S.R. Whitt, A.M. Ibanez, T.R. Rocheford, M.M. Goodman and E.S. Buckler 4th (2004) Dissection of maize kernel composition and starch production by candidate gene association. *Plant Cell* **16**, 2719–2733.
92. H. Jiang, W. Dian, F. Liu and P. Wu (2004) Molecular cloning and expression analysis of three genes encoding starch synthase II in rice. *Planta* **218**, 1062–1070.
93. T. Hirose and T. Terao (2004) A comprehensive expression analysis of the starch synthase gene family in rice (*Oryza sativa* L.). *Planta* **220**, 9–16.
94. H.S. Yoon, J.D. Hackett, C. Ciniglia, G. Pinto and D. Bhattacharya (2004) A molecular timeline for the origin of photosynthetic eukaryotes. *Molecular Biology and Evolution* **21**, 809–818.
95. G.I. McFadden and G.G. van Dooren (2004) Evolution: red algal genome affirms a common origin of all plastids. *Curent Biology* **14**, R514–R516.
96. E. Hirst, D. Manners and I.R. Pennie (1972) α-1,4 -D-glucans part XXI – the molecular structure of starch-type polysaccharide from *Haematococcus pluvialis* and *Tetraselmis carteriiformis*. *Carbohydrate Research* **22**, 5–11.
97. J. Preiss and E. Greenberg (1967) Biosynthesis of starch in *Chlorella pyrenoidosa*: I purification and properties of the adenosine diphosphoglucose: α-1, 4-glucan, α-4-glucosyl transferase from *Chlorella*. *Archives of Biochemistry and Biophysics* **118**, 702–708.
98. Y. Nakamura and M. Imamura (1985) Regulation of ADP-glucose pyrophosphorylase from *Chlorella vulgaris*. *Plant Physiology* **78**, 601–605.
99. Y. Nakamura and M. Imamura (1983) Characteristics of a glucan phosphorylase from *Chlorella vulgaris*. *Phytochemistry* **22**, 835–840.
100. S.G. Ball (2002) The intricate pathway of starch biosynthesis and degradation in the monocellular alga *Chlamydomonas reinhardtii*: starch biosynthesis and degradation in *Chlamydomonas reinhardtii*. *Australian Journal of Chemistry* **55**, 1–11.
101. P.A. Lefebvre and C.D. Silflow (1999) *Chlamydomonas*: the cell and its genomes. *Genetics* **151**, 9–14.
102. J.D.Rochaix (2004) Genetics of the biogenesis and dynamics of the photosynthetic machinery in eukaryotes. *Plant Cell* **16**, 1650–1660.
103. K-H. Süss, I. Prokhorenko and K. Adler (1995) *In situ* association of Calvin cycle enzymes, ribulose-1,5-bisphosphate carboxylase/oxygenase activase, ferredoxin-NADP$^+$ reductase, and nitrite reductase with thylakoid and pyrenoid membranes of *Chlamydomonas reinhardtii* chloroplasts as revealed by immunoelectron microscopy. *Plant Physiology* **107**, 1387–1397.
104. C. Thyssen, R. Schlichting and C. Giersch (2001) The CO_2-concentrating mechanism in the physiological context: lowering the CO_2 supply diminishes culture growth and economises starch utilisation in *Chlamydomonas reinhardtii*. *Planta* **213**, 629–639.

105. C. Zabawinski, N. Van den Koornhuyse, C. D'Hulst *et al.* (2001) Starchless mutants of *Chlamydomona reinhardtii* lack the small subunit of an heterotetrameric ADPglucose pyrophosphorylase. *Journal of Bacteriology* **183**, 1069–1077.
106. U. Klein (1987) Intracellular carbon partitioning in *Chlamydomonas reinhardtii*. *Plant Physiology* **85**, 892–897.
107. N. Libessart, M.L. Maddelein, N. Van Den Koornhuyse *et al.* (1995) Storage, photosynthesis and growth: the conditional nature of mutations affecting starch synthesis and structure in *Chlamydomonas reinhardtii*. *Plant Cell* **7**, 1117–1127.
108. S.G. Ball, L. Dirick, A. Decq, J.C. Martiat and R.F. Matagne (1990) Physiology of starch storage in the monocellular alga *Chlamydomonas reinhardtii*. *Plant Science* **66**, 1–9.
109. S. Ball, T. Marianne, L. Dirick, M. Fresnoy, B. Delrue and A. Decq (1991) *Chlamydomonas reinhardtii* low-starch mutant is defective for 3-phosphoglycerate activation and orthophosphate inhibition of ADP-glucose pyrophosphorylase.. *Planta* **85**, 17–26.
110. N. Van den Koornhuyse, N. Libessart, B. Delrue *et al.* (1996) Control of starch composition and structure through substrate supply in the monocellular alga *Chlamydomonas reinhardtii*. *Journal of Biological Chemistry* **271**, 16281–16288.
111. B.R. Clarke, K. Denyer, C.F. Jenner and A.M. Smith (1999) The relationship between the rate of starch synthesis, the adenosine 5'-diphosphoglucose concentration and the amylose content of starch in developing pea embryos. *Planta* **209**, 324–329.
112. J.R. Lloyd, F. Springer, A. Buleon B. Muller-Rober, L. Willmitzer and J. Kossmann (1999) The influence of alterations in ADP-glucose pyrophosphorylase activities on starch structure and composition in potato tubers.. *Planta* **209**, 230–238.
113. B. Delrue, T. Fontaine, F. Routier *et al.* (1992) Waxy *Chlamydomonas reinhardtii:* monocellular algal mutants defective in amylose biosynthesis and granule-bound starch synthase activity accumulate a structurally modified amylopectin. *Journal of Bacteriology* **174**, 3612–3620.
114. D. Dauvillée, V. Mestre, C. Colleoni *et al.* (2000) The debranching enzyme complex missing in glycogen accumulating mutants of *Chlamydomonas reinhardtii* displays an isoamylase-type specificity. *Plant Science* **157**, 145–156.
115. D. Dauvillee, C. Colleoni, G. Mouille *et al.* (2001) Two loci control phytoglycogen production in the monocellular green alga *Chlamydomonas reinhardtii*. *Plant Physiology* **125**, 1710–1722.
116. D. Dauvillee, C. Colleoni, G. Mouille *et al.* (2001) Biochemical characterization of wild-type and mutant isoamylases of *Chlamydomonas reinhardtii* supports a function of the multimeric enzyme organization in amylopectin maturation. *Plant Physiology* **125**, 1723–1731.
117. M.C. Posewitz, S.L. Smolinski, S. Kanakagiri, A. Melis, M. Seibert and M.L. Ghirardi (2004) Hydrogen photoproduction is attenuated by disruption of an isoamylase gene in *Chlamydomonas reinhardtii*. *Plant Cell* **16**, 2151–2163.
118. F. Wattebled, A. Buléon, B. Bouchet *et al.* (2002) Granule-bound starch synthase: a major enzyme involved in the biogenesis of B-crystallites in starch granules. *European Journal of Biochemistry* **269**, 3810–3820.
119. J.P. Ral, E. Derelle, C. Ferraz *et al.* (2004) Starch division and partitioning a mechanism for granule propagation and maintenance in the picophytoplanktonic green alga *Ostreococcus tauri*. *Plant Physiology* **136**, 3333–3340.
120. S. Robbens, B. Khadaroo, A. Camasses *et al.* (2005) Genome-wide analysis of core cell cycle genes in the unicellular green alga *Ostreococcus tauri*. *Molecular Biology and Evolution* **22**, 589–597.
121. M. Chatterjee, P. Berbezy, D. Vyas, S. Coates and T. Barsby (2004) Reduced expression of a protein homologous to glycogenin leads to reduction of starch content in Arabidopsis leaves. *Plant Science* **168**, 501–509.
122. Y. Nakamura, J. Takahashi, A. Sakurai *et al.* (2005) Some cyanobacteria synthesize semi-amylopectin type alpha-polyglucans instead of glycogen.. *Plant and Cell Physiology* **46**, 539–545.
123. M.A. Schneegurt, D.M. Sherman and L.A. Sherman (1997) Composition of the carbohydrate granules of the cyanobacterium, Cyanothece sp. strain ATCC 51142. *Archives of Microbiology* **167**, 89–98.

124. E. Melendez-Hevia, T.G. Waddell and D.E. Shelton (1993) Optimization of molecular design in the evolution of metabolism: the glycogen molecule. *Biochemical Journal* **295**, 477–483.
125. J. Preiss (1984) Bacterial glycogen synthesis and its regulation. *Annual Reviews of Microbiology* **38**, 419–458.
126. J. Preiss and T. Romeo (1994) Molecular biology and regulatory aspects of glycogen biosynthesis in bacteria. *Progress in Nucleic Acid Research and Molecular Biology* **47**, 299–329.
127. S.H. Yoo, M.H. Spalding and J.L. Jane (2002) Characterization of cyanobacterial glycogen isolated from the wild type and from a mutant lacking of branching enzyme. *Carbohydrate Research* **337**, 2195–2203.
128. T.N. Palmer, G. Wober and W.J. Whelan (1973) The pathway of exogenous and endogenous carbohydrate utilization in *Escherichia coli*: a dual function for the enzymes of the maltose operon. *European Journal of Biochemistry* **39**, 601–612.
129. M.C. Abad, K. Binderup, J. Rios-Steiner, R.K. Arni, J. Preiss and J.H. Geiger (2002) The X-ray crystallographic structure of *Escherichia coli* branching enzyme. *Journal of Biological Chemistry* **277**, 42164–42170.
130. A. Buschiazzo, J.E. Ugalde, M.E. Guerin, W. Shepard, R.A. Ugalde and P.M. Alzari (2004) Crystal structure of glycogen synthase: homologous enzymes catalyze glycogen synthesis and degradation. *EMBO Journal* **23**, 3196–3205.
131. J.E. Ugalde, A.J. Parodi and R.A. Ugalde (2003) *De novo* synthesis of bacterial glycogen: *agrobacterium tumefaciens* glycogen synthase is involved in glucan initiation and elongation. *Proceedings of the National Academy of Sciences of the United States of America* **100**, 10659–10663.
132. R. Viola, P. Nyvall and M. Pedersen (2001) The unique features of starch metabolism in red algae. *Proceedings of the Royal Society of London* B **268**, 1417–1422.
133. D.A. McCracken and J.R. Cain (1981) Amylose in floridean starch. *New Phytolytogist* **88**, 67–71.
134. P. Nyvall, J. Pelloux, H.V. Davies, M. Pedersen and R. Viola (1999) Purification and characterisation of a novel starch synthase selective for uridine 5′-diphosphate glucose from the red alga *Gracilaria tenuistipitata*. *Plant*, **209**, 143–152.
135. A. Coppin, J.S. Varre, L. Lienard *et al.* (2005) Evolution of plant-like crystalline storage polysaccharide in the protozoan parasite *Toxoplasma gondii* argues for a red alga ancestry. *Journal of Molecular Evolution* **60**, 257–267.
136. A.O. Lluisma and M.A. Ragan (1998) Cloning and characterization of a nuclear gene encoding a starch-branching enzyme from the marine red alga *Gracilaria gracilis*. *Current Genetics* **34**, 105–111.
137. W. Dian, H. Jiang, Q. Chen, F. Liu and P. Wu (2003) Cloning and characterization of the granule-bound starch synthase II gene in rice: gene expression is regulated by the nitrogen level, sugar and circadian rhythm. *Planta* **218**, 261–268.
138. S.-J. Wang, K.-W. Yeh and C.-Y. Tsai (2001) Regulation of starch granule-bound starch synthase I gene expression by circadian clock and sucrose in the source tissue of sweet potato. *Plant Science* **161**, 635–644.
139. S.M. Smith, D.C. Fulton, T. Chia *et al.* (2004) Diurnal changes in the transcriptome encoding enzymes of starch metabolism provide evidence for both transcriptional and posttranscriptional regulation of starch metabolism in Arabidopsis leaves. *Plant Physiology* **136**, 2687–2699.
140. Y. Fu, M.A. Ballicora, J.F. Leykam and J. Preiss (1998) ‚Mechanism of reductive activation of potato tuber ADP-glucose pyrophosphorylase. *Journal of Biological Chemistry* **273**, 25045–25052.
141. J.H.M. Hendriks, A. Kolbe, Y. Gibon, M. Stitt and P. Geigenberger (2003) ADP-glucose pyrophosphorylase is activated by posttranslational redox-modification in response to light and to sugars in leaves of Arabidopsis and other plant species. *Plant Physiology* **133**, 1–12.
142. I.J. Tetlow, R. Wait, Z. Lu *et al.* (2004) Protein phosphorylation in amyloplasts regulates starch branching enzyme activity and protein–protein interactions. *Plant Cell* **16**, 694–708.
143. M.K. Morell and A.M. Myers (2005) Towards the rational design of cereal starches. *Current Opinion in Plant Biology* **6**, 204–210.

144. Q. Ji, J.P. Vincken, L.C. Suurs and R.G. Visser (2003) Microbial starch-binding domains as a tool for targeting proteins to granules during starch biosynthesis. *Plant Molecular Biology* **51**, 789–801.
145. D.M. Stark, K.P. Timmermann, G.F. Barry, J. Preiss and G.M. Kishore (1992) Regulation of the amount of starch in plant tissues by ADP glucose pyrophosphorylase. *Science* **258**, 287–292.
146. J.R. Shaw, J.M. Bae and L.C. Hannah (1997) Elucidation of the alteration in the mutant brittle2-c (BT2-c). *Maize Genetics Conference Abstracts,* 39 (http://www.maizegdb.org/cgi-bin/displayrefrecord.cgi?id=133757).
147. M.R. Bhave, S. Lawrence, C. Barton and L.C. Hannah (1990) Identification and molecular characterization of shrunken-2 cDNA clones of maize. *Plant Cell* **2**, 581–588.
148. T.D. Sullivan, L.I. Strelow, C.A. Illingworth, R.L. Phillips and O.E. Nelson Jr. (1991) Analysis of maize brittle-1 alleles and a defective suppressor-mutator-induced mutable allele. *Plant Cell* **3**, 1337–1348.
149. P.L. Dang and C.D. Boyer (1989) Comparison of soluble starch synthases and branching enzymes from leaves and kernels of normal and amylose extender maize. *Biochemical Genetics* **27**, 521–532.

11 The organization and control of plant mitochondrial metabolism

Allison E. McDonald and Greg C. Vanlerberghe

11.1 Introduction

Plant mitochondria are well known for their central role in energy metabolism but it is also clear that other metabolic and cellular activities occur in these organelles. Our understanding of this organelle has advanced considerably in recent years, and an important development has been the use of proteomic approaches to investigate the protein complement [1] (also see Chapter 2). The mitochondrial proteomes of cells from tissues of several plant species have now been investigated, in many cases leading to novel discoveries. Approximately 400 different proteins have been directly identified in highly purified mitochondria, about 70 of which lack similarity to any protein of known function. Nonetheless, it is estimated that the organelle may contain 2000–3000 different gene products, indicating that, despite some progress, our understanding of plant mitochondria is still far from complete [1]. The purpose of this chapter is to outline the central components of mitochondrial metabolism and highlight recent advances in our understanding of metabolism and its control in this organelle.

11.2 Organization of the tricarboxylic acid cycle and mitochondrial electron transport chain

The following describes the major components of energy metabolism in plant mitochondria. Their organization is summarized in Figure 11.1.

11.2.1 Tricarboxylic acid cycle and associated enzymes

Pyruvate dehydrogenase (PDH) links glycolysis to the tricarboxylic acid (TCA) cycle by catalyzing: pyruvate + NAD^+ + CoA $\rightarrow$ acetyl-CoA + NADH + CO_2. This multienzyme complex consists of three primary components: pyruvate dehydrogenase (E1; consisting of E1α and E1β), dihydrolipoyl acetyl-transferase (E2) and dihydrolipoyl dehydrogenase (E3). It includes the bound co-factors thiamine pyrophosphate, lipoic acid and FAD. When PDH was reduced in tobacco anthers using an antisense gene and tapetum-specific promoter, plants were male sterile, indicating the importance of PDH during reproductive development [2].

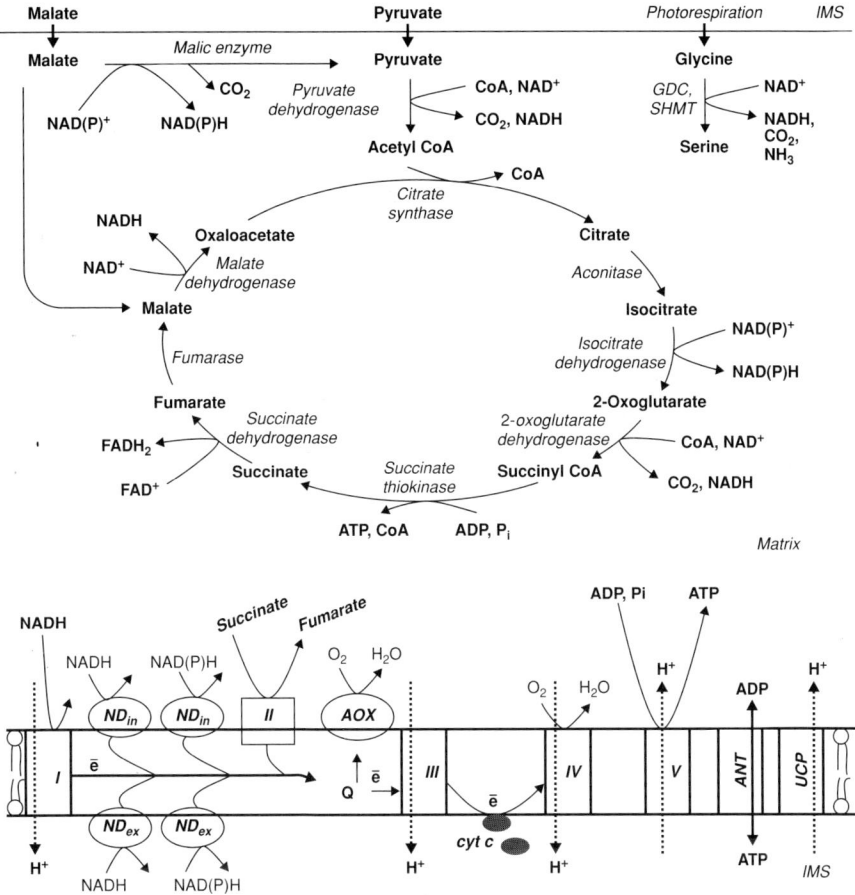

Figure 11.1 Organization of the TCA cycle and electron transport chain in plant mitochondria (see text for details).

Mitochondrial PDH is the only TCA cycle enzyme whose regulatory properties have been investigated in considerable detail [3]. Activity is subject to product inhibition by NADH and acetyl-CoA (both with K_i of ~20 μM). Importantly, PDH is controlled by reversible phosphorylation of two Ser residues of the E1α-subunit, with phosphorylation inhibiting activity. The phosphorylation state is controlled by PDH kinase and phosphopyruvate dehydrogenase phosphatase, both of which are associated with the PDH complex. Phosphopyruvate dehydrogenase phosphatase does not appear to be subject to control, indicating that PDH phosphorylation status is primarily dependent upon PDH kinase activity. PDH kinase is inhibited by pyruvate and ADP, while being activated by NH_3, through an increased affinity for ATP. An important consequence of this is an inactivation (phosphorylation) of PDH in the light due to increased matrix ATP, increased NH_3 (from photorespiration) and increased NADH levels. In contrast, high pyruvate levels may overcome some of this inhibition.

Citrate synthase catalyzes: oxaloacetate + acetyl-CoA → citrate + CoA. The plant enzyme exists as a homodimer composed of 50 kDa subunits [4, 5]. Little is known about regulatory properties of the enzyme. Surprisingly, reduction of citrate synthase activity in transgenic potato plants to 6% or less of normal levels had little impact on shoot growth or tuber yield [5]. However, flower development was aborted in plants with less than 30% of normal levels, presumably due to the high energetic demands associated with this development.

Aconitase (ACO) catalyzes: citrate → isocitrate. As in animals, plants contain both mitochondrial and cytosolic aconitases, but it is unresolved whether these represent different gene products or a dual-targeting of gene products. Interestingly, the animal cytosolic enzyme has an additional role in controlling Fe homeostasis. This 'Fe regulatory protein' role of ACO is due to its ability to bind mRNAs containing a Fe-responsive element consensus sequence, thereby regulating their stability or translation. Residues involved in mRNA binding are conserved in the plant cytosolic enzyme, indicating that it may play a similar role to its animal counterpart, but this is yet to be investigated [6]. Relatively few molecular or biochemical analyses of the plant mitochondrial ACO have been undertaken [6, 7–9]. Analyses that have been undertaken have shown that various molecules commonly associated with plant stress are capable of inactivating ACO *in vitro* by impairing the active site Fe–S cluster. These include H_2O_2 [7], nitric oxide [6] and salicylic acid [10]. However, few data are available to assess if and when such inactivation may occur *in vivo* [11].

Isocitrate dehydrogenase (ICDH) catalyzes: isocitrate + $NAD(P)^+$ → 2-oxoglutarate + NAD(P)H. Plant mitochondria contain both NAD^+- and $NADP^+$-specific isoforms of ICDH [12, 13], with the $NADP^+$-specific isoforms also present in several other compartments. While NAD^+-ICDH is generally considered the predominant isoform of the TCA cycle, relatively few biochemical studies of the enzyme have been reported, but it is known that the enzyme is subject to product inhibition by NADH [14].

2-oxoglutarate dehydrogenase (OGDH) catalyzes: 2-oxoglutarate + NAD^+ + CoA → succinyl-CoA + NADH + CO_2. Analogous to PDH, OGDH has three principal components: 2-oxoglutarate dehydrogenase (E1), dihydrolipoamide succinyl-transferase (E2) and dihydrolipoamide dehydrogenase (E3). Unlike PDH, OGDH E1 is not subject to control by phosphorylation, but in common with other TCA cycle dehydrogenases, OGDH activity is inhibited by NADH [14]. It is also activated by AMP.

Proteins within the OGDH complex have been characterized in potato [15]. Interestingly, this study found that a large proportion of OGDH is associated with the membrane fraction, perhaps in direct association with complex I, and similar to that observed in animals. It was also found that association of E3 within the OGDH complex is relatively weak, compared with the association of E3 within the PDH complex. Since these two complexes may share a common pool of E3, this finding may have implications for OGDH function.

Succinate thiokinase catalyzes: succinyl CoA + ADP + Pi → succinate + ATP + CoA. Relatively little molecular or biochemical characterization of the plant enzyme has been undertaken.

Succinate dehydrogenase (SDH) is both a TCA cycle enzyme (catalyzing: succinate + $FAD^+ \rightarrow$ fumarate + $FADH_2$) and a component of the electron transport chain (ETC), passing electrons from $FADH_2$ to ubiquinone (Q). It is the only TCA cycle enzyme that is an integral membrane protein of the inner mitochondrial membrane (IMM)(see complex II below).

Fumarase catalyzes: fumarate + $H_2O \rightarrow$ malate. Plant fumarase is of the class II type in which the enzyme exists as a homotetramer. There are only a few reports of molecular and biochemical characterization of the plant enzyme [16, 17].

Malate dehydrogenase (MDH) catalyzes: malate + $NAD^+ \rightarrow$ oxaloacetate + NADH. Since the reaction equilibrium lies strongly in favor of malate, operation of the TCA cycle depends upon the nonreversible citrate synthase reaction to remove oxaloacetate. MDH enzymes in other cellular compartments have been studied in some detail, but there are relatively few molecular or biochemical characterizations of the mitochondrial enzyme [18].

Malic enzyme (ME) catalyzes: malate + $NAD(P)^+ \rightarrow$ pyruvate + NAD(P)H + CO_2. While not considered a TCA cycle enzyme, the significance of ME is that it may represent an alternative mechanism to generate intramitochondrial pyruvate. This could support TCA cycle activity under conditions when substrate other than pyruvate (i.e., malate) is being imported into the mitochondrion. ME consists of two similar subunits, can exist in a range of oligomeric forms and is activated by fumarate and CoA [19, 20]. The different aggregation states of ME also display different kinetic properties. The effects of these regulatory properties on flux *in vivo* remain unknown.

Studies on maize root tips indicate that pyruvate generation via ME is a very minor flux relative to pyruvate generation via pyruvate kinase [21, 22]. In an attempt to assess the impact of ME activity on respiratory metabolism, Jenner *et al.* [23] generated transgenic potato tubers in which the maximum catalytic activity of ME was reduced by up to 60%. This level of reduction had no effect on TCA cycle flux, although a change in the level of some glycolytic metabolites was seen. Further studies are required to determine what metabolic conditions may enhance pyruvate generation via ME.

The combined action of the glycine decarboxylase complex (GDC) and serine hydroxymethyltransferase (SHMT) catalyzes: 2 glycine + NAD^+ + $H_2O \rightarrow$ serine + CO_2 + NH_3 + NADH. This is not part of the TCA cycle, but rather is a matrix-localized component of the photorespiratory pathway in photosynthetic cells. In the light, this activity represents the major intramitochondrial source of NADH, particularly given that PDH is inactivated in the light by phosphorylation (see above).

The GDC complex has been studied in great molecular and biochemical detail [24]. The dihydrolipoamide dehydrogenase component of this complex is also shared by PDH and OGDH (see above). Since mitochondrial glycine metabolism is dependent upon large amounts of the enzyme cofactors lipoic acid and tetrahydrofolate, all the enzymes necessary for the biosynthesis of these cofactors are present in plant mitochondria [25, 26]. GDC activity will be subject to product inhibition by serine and NADH, unless these products are rapidly removed.

As is apparent from the above discussion, there is a paucity of detailed molecular and biochemical characterization for many TCA cycle and associated enzymes. Hence, there is still much general speculation regarding control of carbon flux through the cycle. However, the regulation of PDH and other dehydrogenases by the NADH/NAD$^+$ ratio is likely to be more certain and an important aspect of control [3, 14].

11.2.2 Electron transport chain complexes I–V

As analytical techniques improve, information regarding the protein composition of the ETC complexes (I–V) of the IMM continues to improve. Such analyses have unexpectedly uncovered many plant-specific proteins of unknown function in these complexes. These proteins may have regulatory roles or may represent additional activities of these complexes not directly related to electron transport. A good example of such bifunctionality is complex III, which includes the mitochondrial processing peptidase that removes the N-terminal targeting signal of proteins targeted to the mitochondria [27].

Complex I (NADH dehydrogenase) catalyzes the oxidation of matrix NADH, coupled with reduction of Q and proton pumping from the matrix to intermembrane space (IMS). Electron transfer within the complex involves a flavin adenine mononucleotide and several Fe–S clusters. The complex is inhibited by rotenone and by diphenyleneiodonium.

Recently, 2D blue-native/SDS-PAGE coupled with mass spectrometry has further elucidated the protein composition of complex I from plants [28]. This study identified 30 different proteins in *Arabidopsis thaliana* and 24 different proteins in rice. Of these proteins, 14 are orthologs of the core components of the functional bacterial complex and 9 represent accessory proteins highly conserved amongst eukaryotes. Interestingly, both plant species also contained orthologs of GRIM-19, a mammalian protein identified as component of complex I and a regulator of apoptosis [29]. This suggests a potential link in plants and animals between the mitochondrial ETC and cell death pathways. Of further interest was the finding of several plant-specific complex I proteins, not found in other organisms. Their function is unknown, but some may represent proteins functioning in the storage or sensing of Fe [28].

Complex II (SDH) is the only TCA cycle enzyme tightly associated with the IMM. It catalyzes the oxidation of succinate, coupled with reduction of FAD. Electrons are then transferred to Q via several Fe–S clusters. A heme *b* group of unknown function is also present.

As with complex I, the subunit composition of complex II has been recently investigated in Arabidopsis using 2D blue-native/SDS-PAGE coupled with mass spectrometry [30]. The complex is presently thought to consist of eight proteins (SDH1-SDH8). SDH1-4 are the four classic subunits found in both bacteria and eukaryotes. The soluble SDH1 and SDH2 subunits carry FAD and Fe–S clusters, respectively, while the hydrophobic SDH3 and SDH4 subunits anchor the complex to the IMM. Unexpectedly, the complex also retains four additional plant-specific proteins (SDH5-8) not seen in other organisms. The function of these proteins is unknown.

Plant SDH is well known to be activated by ATP. This was recently confirmed by a novel approach in which SDH activation state was assessed in isolated mitochondria by using simultaneous measurements of oxygen uptake rate and Q reduction state [31]. This study found that SDH was also activated by succinate, ADP and complex V inhibitors while being inactivated by K^+ and uncouplers. The results are consistent with SDH activity being modulated in a novel way by membrane potential [31].

Complex III (cytochrome (cyt) c reductase) catalyzes the oxidation of Q, coupled with reduction of cyt c and proton translocation from the matrix to IMS. The electron transfer (Q-cycle) components include cyt b, Fe-S centers and cyt c_1. Antimycin A and myxothiazol are inhibitors, each blocking electron transport at different sites within the Q-cycle. Complex III consists of 10 subunits, some of which are involved in the processing of proteins being imported into the mitochondrion [32].

Complex IV (cyt oxidase) catalyzes the oxidation of cyt c, coupled with reduction of O_2 to H_2O and proton translocation from matrix to IMS. Electron transfer involves a copper sulfur cluster, cyt a, and a binuclear center consisting of cyt a_3 and a Cu atom. CN and CO inhibit the complex by binding to cyt a_3.

Interestingly, complex IV appears to exist in two forms: IVa (300 kDa) and IVb (220 kDa), in which IVa contains additional unique proteins [33]. The functional significance of these two forms is unknown. Millar *et al.* [30] found that complex IV consisted of eight proteins similar to those described in other organisms as well as six additional small proteins of unknown function. Because of difficulties in resolving this complex, the association of these six proteins with complex IV requires further confirmation.

Complex V (ATP synthase) accomplishes oxidative phosphorylation by facilitating the translocation of protons down their electrochemical gradient (from IMS to matrix) coupled with the synthesis of ATP from ADP + Pi. ATP synthase consists of the F_0 complex that forms a proton channel through the IMM and the F_1 catalytic complex that is anchored to F_0 and extends into the matrix. Oligomycin inhibits the complex by preventing proton translocation.

In several plant species, ATP synthase is shown to contain 13 subunits, 6 defining F_1 and 7 defining F_0 [33, 34]. The entire complex can be present in a dimeric form [33].

11.2.3 Additional electron transport chain and associated components

Besides complex I, the plant ETC contains alternate rotenone-resistant NAD(P)H dehydrogenases [35]. These likely include: (i) an NADH-specific enzyme on the internal side of the IMM ($NADH_{in}$); (ii) an NAD(P)H-utilizing enzyme on the internal side of the IMM ($NAD[P]H_{in}$); (iii) an NADH-specific enzyme on the external side of the IMM ($NADH_{ex}$) and (iv) an NAD(P)H-utilizing enzyme on the external side of the IMM ($NAD[P]H_{ex}$). The internal dehydrogenases have access to matrix NAD(P)H while the external dehydrogenases have access to extra-mitochondrial NAD(P)H. All of these dehydrogenases are rotenone-resistant (in comparison to the rotenone-sensitive Complex I) and all will act to reduce the energy yield of respiration since they are non-proton pumping and bypass the proton-pumping complex I.

The various dehydrogenase activities outlined above have been well documented in isolated mitochondria or submitochondrial particles, and proteins that could account for these activities have been purified. The identification of corresponding genes is beginning to be elucidated, but this process is still incomplete. Analysis of an Arabidopsis T-DNA knockout line has identified a gene encoding an NAD(P)H$_{in}$ [36]. This evidence is based upon the high homology of this protein to proteins characterized in microorganisms, a large decline in NAD(P)H$_{in}$ activity in mitochondria from the knockout line and demonstration that an *in vitro* translation product of the gene is imported into mitochondria and localized to the inside of the IMM. Recently, Michalecka *et al.* [37] identified the gene encoding an NAD(P)H$_{ex}$. This was based on showing that overexpression of the gene product in transgenic plants resulted in increased external NADPH oxidation in isolated mitochondria. Several of the genes encoding known or potential alternate NAD(P)H dehydrogenases include amino acid sequence similar to EF-hand motifs for Ca^{2+} binding [38]. This is in keeping with the observation that both the external activities and the NAD(P)H$_{in}$ activity are Ca^{2+}-dependent [35].

Besides complex IV, all plants contain an additional terminal oxidase called alternative oxidase (AOX) that catalyzes the oxidation of Q and reduction of O_2 to H_2O [39]. AOX is also non-proton pumping and since it bypasses proton-pumping complexes III and IV, electron flow to AOX will dramatically reduce the energy yield of respiration. Site-directed mutagenesis has been used to identify conserved Glu and His residues that coordinate the di-iron center of AOX, as well as to identify Tyr residues essential for activity [40, 41]. AOX is encoded by a multigene family but the functional significance of the different gene products is not yet clear [42]. AOX genes appear to be widely distributed in nature as they have been recently found in numerous marine eubacteria [43] as well as several animal invertebrate phyla [44].

The current molecular model of plant AOX indicates that it is an interfacial membrane protein on the matrix side of the IMM and that it exists as a homodimer (the state of which impacts the kinetic properties of the enzyme) (Figure 11.2) [45, 46]. Specifically, the dimer may be either noncovalently linked (reduced form) or covalently linked by a regulatory disulfide bond between the two monomers (oxidized form) [47]. A conserved Cys residue toward the N-terminus and within the matrix is responsible for this disulfide bond [48, 49]. Reduction of the disulfide bond is facilitated by the oxidation of specific TCA cycle substrates and, based upon the substrate specificity, it is hypothesized that specifically the generation of mitochondrial NADPH provides the reducing power for this regulatory reduction [50]. Consistent with this, overexpression of $NADP^+$-ICDH has recently been shown to favor conversion of AOX from its oxidized to reduced form [51]. Once reduced, AOX is sensitive to activation by specific α-keto acids, most notably pyruvate [50, 52]. It has been shown that pyruvate activation is due to its ability to interact with the exposed sulfhydryls of the regulatory Cys to produce a thiohemiacetal [48]. Activated AOX has been shown to effectively compete with the cyt pathway for electrons.

The above studies indicate that AOX activity is increased by a two-step process involving reduction of the regulatory disulfide bond and subsequent activation by

THE ORGANIZATION AND CONTROL OF PLANT MITOCHONDRIAL METABOLISM 297

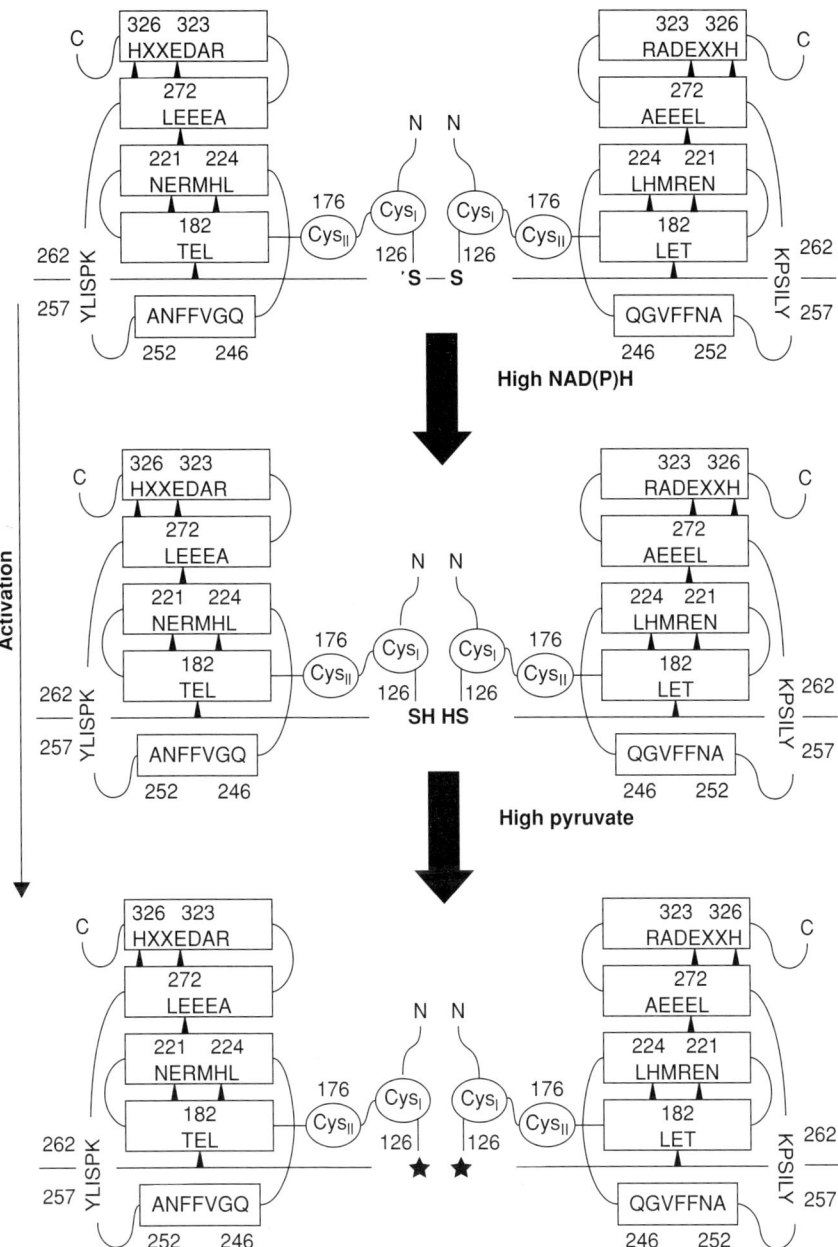

Figure 11.2 A working model for the structure and control of the plant mitochondrial AOX. AOX is an interfacial membrane protein on the matrix side of the IMM and is present as a homodimer. It is activated by a two-step process involving reduction of a regulatory disulfide bond (between Cys_I residues of the two subunits), followed by interaction of the component sulfhydryls with pyruvate to generate a thiohemiacetal (indicated by the solid stars). Putative iron-binding residues (Glu and His residues) are indicated by black triangles. Residue numbering corresponds to a tobacco AOX gene (accession number S71335) (see text for details).

pyruvate (Figure 11.2). In this way, the partitioning of electrons to AOX is controlled in a feed-forward fashion by upstream metabolism and is responsive to both the redox (NADPH) and carbon (α-keto acid) status of the matrix. While this regulatory model is widely supported by extensive studies in both isolated plant mitochondria and yeast or bacterial expression systems, the significance of this control *in vivo* is yet to be directly demonstrated. Also, some plant AOX proteins lack the regulatory Cys and appear to exhibit other regulatory properties [53] or possess the regulatory Cys but appear not to require pyruvate for activation [54]. In addition, while absence of the regulatory Cys (due to site-directed mutagenesis) produced an enzyme that lacked pyruvate activation and displayed little activity in isolated mitochondria (as expected), the enzyme nonetheless displayed high activity *in vivo* [49]. The above studies indicate that our understanding of AOX activation remains incomplete.

Because of its non-energy-conserving nature, considerable effort has been put toward understanding the physiological role of AOX respiration, in addition to its well established role to generate heat in highly specialized thermogenic tissues such as the voodoo lily inflorescence [55]. In general, AOX provides flexibility to a metabolic process that otherwise couples carbon metabolism to electron transport and ATP synthesis. This could prevent redox and carbon imbalances, hence promoting metabolic homeostasis. However, it is not clearly established under what metabolic, environmental or developmental conditions such a role for AOX is of importance. It is also unclear what the full consequences of AOX activity are: what is the 'metabolic cost' of such a non-energy-conserving activity? For example, to what degree does it compromise overall plant productivity? In contrast, what is the 'metabolic advantage' of such activity? For example, to what extent does the promotion of metabolic homeostasis by AOX impact tolerance to stress conditions, which typically act to disrupt homeostasis? Some of these questions have been investigated using transgenic tobacco cells lacking AOX. For example, it has been suggested that AOX respiration represents an important mechanism to modulate growth rate in response to nutrient availability [56] and that a lack of AOX increases susceptibility to programmed cell death [57, 58]. It should be kept in mind that AOX could also provide a means to maintain electron transport if cyt oxidase were directly inhibited. For example, the potent cyt oxidase inhibitors cyanide, nitric oxide and sulfide are all normal products of metabolism and may be produced directly within the mitochondrion. AOX also provides a 'heme-independent' respiratory pathway that may be important during anoxia to air transitions, when the cyt-containing respiratory complexes are not yet functional [59].

Plant uncoupling proteins (UCPs) are members of a large family of mitochondrial anion carriers that are integral proteins of the IMM [60]. Similar to animals, the plant UCPs can catalyze proton conductance from the IMS to matrix, hence uncoupling electron transport from ATP synthesis. Arabidopsis contains three putative UCPs: UCP1 (formerly called PUMP), UCP2 and UCP4, a naming based on orthology to animal UCPs [61]. An important question is how the proton conducting activity of the plant UCPs is controlled, and a major advance in our understanding of this control is discussed later.

11.3 Posttranslational control of mitochondrial metabolism and function

New high-throughput proteomic approaches have not only been used to document the protein complement of plant mitochondria but have also given an insight into posttranslational means by which metabolism and function in this organelle may be controlled. Some of these new findings are highlighted below.

11.3.1 Phosphorylation – dephosphorylation

Reversible phosphorylation is a posttranslational modification that regulates the activity of many enzymes in primary metabolism. Table 11.1 lists the plant mitochondrial proteins shown to be phosphorylated. Many of these proteins have only been identified recently by using 2D gel electrophoresis followed by mass spectrometry [62]. It includes proteins involved in the TCA cycle, electron transport, oxidative phosphorylation and responses to stress. However, the reversible inactivation

Table 11.1 Phosphoproteins of the plant mitochondrial proteome identified to date (see text for details).

Protein	Reference
TCA cycle and associated enzymes	
Pyruvate dehydrogenase (E1α subunit)	[3]
Aconitase	[62]
NAD-isocitrate dehydrogenase	[62]
Succinate thiokinase (α subunit)	[62]
Succinate thiokinase (β subunit)	[62]
Malate dehydrogenase	[62]
Malic enzyme (62 kD subunit)	[62]
Malic enzyme (59 kD subunit)	[62]
Formate dehydrogenase	[63]
Respiratory chain complexes	
Complex II (flavoprotein α subunit)	[62]
Complex III (β-MPP subunit)	[62]
ATP synthase (F_o b subunit)	[64]
ATP synthase (F_1 δ´ subunit)	[64]
ATP synthase (F_1 α subunit)	[62]
ATP synthase (F_1 β subunit)	[62]
Stress-related proteins	
HSP22	[66]
HSP60	[62]
HSP70	[67]
HSP90	[62]
Superoxide dismutase	[62]
Nucleoside diphosphate kinase	[65]

of PDH by phosphorylation (see above) represents the only case in which the regulatory significance of phosphorylation has been clearly elucidated. For the other proteins in Table 11.1, the potential regulatory significance is yet to be established [62–67]. Interestingly, while phosphorylation of the branched chain keto-acid dehydrogenase complex represents a prominent, well-characterized example of regulatory phosphorylation in animal mitochondria, to date the plant counterpart has not been identified as phosphorylated. As discussed in detail by Moorhead *et al.* (Chapter 5), the prerequisite for reversible regulatory phosphorylation is the existence of specific kinases and protein phosphatases capable of phosphorylating and dephosphorylating specific target proteins. Several studies have characterized such activities in different mitochondrial compartments but the molecular identity of these activities and their target proteins are yet to be elucidated [68–70]. The exception is PDH, in which specific kinase and phosphatase activities are tightly associated with the PDH complex (see above). Of the phosphoproteins listed in Table 11.1 there is at least one example (HSP70, [67]) in which the protein is capable of autophosphorylation. Alternatively, HSP22 was shown not to be capable of autophosphorylation, indicating involvement of a kinase [66].

14-3-3 proteins are phosphoserine/phosphothreonine – binding proteins involved in the control of primary plant metabolism (Chapter 5). Such proteins are present in the matrix or inner face of the IMM and were found to interact with ATP synthase in a phosphorylation-dependent manner [71]. Interaction was with the catalytic F_1 β-subunit and could only be seen if ATP synthase was isolated in the presence of a phosphatase inhibitor. The interaction was shown to reduce ATP synthase activity but neither the mechanism of this inhibition nor its physiological relevance is yet understood. It could represent a means to prevent ATP hydrolysis under conditions when the electrochemical gradient across the IMM collapses (such as anoxia) and the catalytic activity of ATP synthase switches from ATP synthesis to ATP hydrolysis. Interestingly, a number of ATP synthase subunits appear to be subject to phosphorylation (Table 11.1).

11.3.2 Dithiol-disulfide interconversion

Thioredoxins (Trxs) are small proteins with a redox-active site consisting of two Cys residues in close proximity (see Chapter 6 for additional details). Two Trx systems exist. In one, oxidized Trx which is reduced by a ferredoxin-Trx reductase using ferredoxin as a source of electrons. In another, the disulfide is reduced by an NADPH-dependent Trx reductase. Reduced Trx can then act as a source of reductant in select enzymatic reactions or it can reduce specific disulfide bonds in target proteins, thereby altering the activity or functional properties of these target proteins.

While the Trx systems (and many of their target proteins) present in chloroplast and cytosol of plants have been well described for some time, a mitochondria-localized system is only recently being elucidated. Using the Arabidopsis genome, Laloi *et al.* [72] identified and described a novel plant mitochondrial Trx (Trx-*o*) as well as a mitochondrial NADPH-dependent Trx reductase. Recently, characterization of an additional Arabidopsis mitochondrial Trx (similar to the Trx-*h* of the cytosol), as well as an additional Trx reductase, indicates that the system is complex [73, 74]. Such complexity is similar to that already known for the chloroplast and cytosol systems.

The complex nature of Trx systems is not surprising given the increasing range of processes for which these systems are being shown to take part. Recently, Balmer et al. [75] used mass spectrometry to identify mitochondrial Trx-binding proteins that eluted from a Trx-affinity column. This approach identified 50 soluble proteins (more than the number identified by similar approaches in the chloroplast), almost all of which could also be reduced by externally supplied Trx, as shown using a fluorescence gel electrophoresis technique. Further, most of these proteins were shown to contain conserved Cys residues, consistent with these residues being Trx targets. The proteins were divided into 12 functional categories and included many enzymes in energy metabolism (TCA cycle enzymes; proteins in ETC complexes I, III and IV; subunits of ATP synthase; adenylate kinase; nucleoside diphosphate kinase), amino acid metabolism, sulfur metabolism and stress-related reactions, among many others. Interestingly, the two matrix enzymes involved in photorespiration (GDC and SHMT) were identified, indicating that Trx might provide a regulatory link between chloroplast and mitochondrial metabolism [75]. The large number of stress-related proteins is in keeping with growing evidence that Trxs play an important role during stress. This may include the restoration of activities following reactive oxygen species (ROS)-induced inactivation, the provision of electrons for ROS scavenging systems and the control of ROS-scavenging enzymes [76, 77]. Interestingly, AOX (a protein implicated as important in dampening ROS generation by the ETC, see below) can be reduced and activated by externally supplied Trx [47, 73]. It is yet to be established whether this is a mode of AOX reduction *in vivo*. Clearly, a wealth of mitochondrial processes and target proteins whose activity or function may be linked to the mitochondrial redox state via NADPH and Trx are being uncovered. This may provide a rationale for the numerous modes of NADPH generation possible in plant mitochondria [78].

11.3.3 Other oxidative modifications

With the exception of Cys dithiol-disulfide interconversion (see above), little is known about the potential to regulate the activity or function of plant proteins by specific oxidative modifications. Proteins can display a wide range of oxidative modifications, both reversible and nonreversible [79]. While the regulatory value of nonreversible modifications seems questionable, it does not preclude the importance of such modifications to cellular metabolism, if they indeed impact the activity or functional properties of such proteins. Most oxidative modifications are viewed as being primarily induced by ROS and their by-products.

The potential impact of protein oxidation on mitochondrial metabolism is of particular interest since the ETC is a major source of ROS (see below). In fact, since ROS are being generated at ETC complexes; these complexes may themselves be important first targets of ROS-induced protein oxidation. The first steps toward understanding the significance of protein oxidation to mitochondrial function will be: (a) to identify those proteins most susceptible to oxidation and determine the oxidative modifications involved and (b) to determine the impact of such oxidations on the functional properties of the proteins.

Recent studies using proteomic analyses have begun to investigate the above questions. Sweetlove *et al.* [80] examined the impact of oxidative stress (induced by supplying Arabidopsis suspension culture cells with H_2O_2, menadione or antimycin A) on the mitochondrial proteome. A set of respiratory proteins were identified that consistently appeared as degradation products following any of the oxidative stress treatments. It was hypothesized that these represent proteins particularly susceptible to oxidative modification, which was then targeting these proteins for degradation. Another study examined the effect of oxidative stress on the mitochondrial proteome of rice [81]. Importantly, this study sought to identify proteins containing carbonyl groups, a specific form of oxidative modification. Identified proteins that were common to each of the above studies included the TCA cycle enzymes ACO and succinate thiokinase, F_1 subunits of ATP synthase, cyt *c* and the voltage-dependent anion channel [80, 81]. These and some others represent good candidate proteins for oxidative modification *in vivo*. While the ability of ACO to be inactivated by oxidative stress has been previously established [7], the effect of oxidative stress on the activity or function of these other candidate proteins has not yet been tested.

11.3.4 Supramolecular complexes

The use of mild solubilization protocols and blue-native polyacrylamide gel electrophoresis has established that the ETC complexes of yeast and mammalian mitochondria are not simply singular units randomly distributed within the IMM, but rather that different complexes could assemble together into supramolecular structures [82]. Kinetic analyses are also consistent with the existence of such superstructures [83]. Studies in many different systems (fungi, animal, plant) have now found similar results and such supercomplexes have become known as 'respirasomes'. These studies have begun to establish which complexes may enter into such associations with one another and what the stoichiometry of each complex is within these supercomplexes. In animals, complexes I, III and IV have all been shown to enter into such associations in various stoichiometries with one another. A potential advantage of 'respirasomes' is that they could facilitate the channeling of substrate between ETC components and possibly reduce the frequency of unwanted side reactions. Studies also suggest that association of complexes I and III in animals is necessary to maintain proper assembly or stability of complex I [84]. Evidence also indicates that the cardiolipin component of complexes may be essential for their ability to associate with one another [85]. Cardiolipin is a phospholipid specific to the IMM in both plants and animals. The first eukaryotic gene for cardiolipin synthase was recently identified in Arabidopsis and, as expected, encodes a mitochondrial-localized protein [86].

Initial studies with nongreen tissue of Arabidopsis, potato, bean and barley indicated associations between complexes I and III in stoichiometries of I_1III_2 and I_2III_4 [33]. The amount of complex I associated with complex III varied from 50% in bean and potato to 90% in barley. The use of gentler techniques demonstrated that complex

IV also associated into a supercomplex consisting of $I_1III_2IV_4$ in potato tuber mitochondria [87]. However, a supercomplex of I_1III_2 was still the most abundant form. Most recently, supercomplexes containing more abundant amounts of complex IV have been seen in mitochondria from green leaves [88]. These studies suggest that the presence of complexes I, III and IV in 'respirasomes' is a conserved feature amongst animals, fungi and plants. A notable feature of the plant ETC is the presence of the additional non-energy-conserving components, raising the question of whether these might also be present in supercomplexes. To date, there are conflicting reports as to whether plant AOX might be present in such supercomplexes [33, 89] and there is no evidence for the presence of an alternate NAD(P)H dehydrogenase. Much more work is necessary to determine if and how supercomplexes regulate in a dynamic way the rate or path of electron transport in plants. For example, it is unknown whether association of complexes I and III might limit the access of Q to AOX.

While there is as yet no evidence of associations with ATP synthase in plants, work in animals suggests that ATP synthase can exist in a complex with both a Pi transporter and the adenine nucleotide translocator (ANT) that exchanges ADP and ATP across the IMM. This 'ATP synthasome' could provide a mechanism whereby the substrates (ADP and P_i) for ATP synthesis are delivered directly to ATP synthase upon import across the IMM and the product of ATP synthase (ATP) is then delivered directly to ANT for export [90].

The concept of organizing a metabolic pathway into a supramolecular structure could also apply to the TCA cycle, although there is no direct evidence for this in plants. However, studies with animal mitochondria do support this concept of a TCA cycle 'metabolon'. For example, Haggie and Verkman [91] tagged four TCA cycle enzymes at their C-termini with green fluorescent protein so that their mobility characteristics could be directly observed *in vivo*. This study provided biophysical evidence that the four enzymes were indeed physically associated.

In animals, it is well established that some glycolytic enzymes can physically associate with the mitochondrion. Recently, proteomic analyses of highly purified Arabidopsis mitochondria have similarly demonstrated the presence of 7 of the 10 enzymes of glycolysis associated with the organelle [92]. Protease protection assays, enzyme activity analyses and metabolism of radiolabeled precursors supported association of the entire glycolytic pathway with the outer membrane. Yellow fluorescent protein fusion proteins of cytosolic aldolase and enolase also confirmed *in vivo* their close association with mitochondria [92]. It remains to be established whether these associations primarily act to channel pyruvate to the mitochondrion or whether they have other unrelated functions.

11.4 Integration of mitochondrial metabolism with other metabolic pathways

In this section, a few examples are provided as to how components of the core respiratory machinery (TCA cycle and ETC) are integrated with other metabolic activities, both within and outside the mitochondrion.

11.4.1 Mitochondrial metabolism during photosynthesis

It is generally accepted that optimal photosynthetic metabolism is dependent upon interactions with respiratory pathways, and the interdependence of chloroplast and mitochondrial metabolism has been the subject of several reviews [e.g. 93–95].

A key function of the mitochondrion during photosynthesis is the metabolism of photorespiratory glycine in the matrix by GDC and SHMT. Since the NADH generated by this metabolism is produced in equal quantities to that required for hydroxypyruvate reduction in the peroxisome, one possibility is the shuttling of this mitochondrial reductant to the peroxisome. However, different lines of evidence support the idea that a significant proportion of the NADH is instead oxidized by the mitochondrial ETC. In this case, chloroplast reductant would need to be shuttled to the peroxisome, a process that could help to buffer the chloroplast redox state and optimize photosynthesis. This idea is supported by analyses of tobacco mutants lacking complex I [96]. In these mutants, photosynthesis is compromised under photorespiratory conditions (and other transient conditions) due to the removal of complex I as an electron sink. Interestingly, while photosynthetic rate was reduced in these plants, respiration rate was not significantly impaired due to induction and presumably increased operation of the alternate NAD(P)H dehydrogenases. The inability of these to compensate for complex I during photosynthesis is likely due to their lower affinity for NAD(P)H, resulting in a more highly reduced cellular pool of pyridine nucleotides. However, other studies have shown that expression of NAD(P)H$_{in}$ is strongly light-dependent suggesting that, while it may not be able to compensate for complex I in the light, it nonetheless takes part in light-induced metabolism [97].

Recent genomic and biochemical analyses suggest that the mitochondrion may contain more enzymes involved in photorespiration than previously thought. For example, glutamine synthetase is dual-targeted to both leaf mitochondria and chloroplasts [98] and Arabidopsis mitochondria contain a putative glycolate dehydrogenase that, similarly to algae, could oxidize photorespiratory glycolate to glyoxylate within this organelle [99].

Recent analyses of mutant or transgenic plants with altered levels of specific TCA cycle enzymes have provided new insights into the role of mitochondrial metabolism in the light. Carrari *et al.* [100] characterized a tomato mutant with lowered levels of both mitochondrial and cytosolic ACO. Feeding experiments with $^{14}CO_2$ indicated a reduction in TCA cycle carbon flow. Surprisingly, these plants exhibited higher rates of photosynthesis and carbon assimilation, maintained a higher pool of sucrose and produced a higher fruit yield than wild-type (wt) plants. The results were consistent with the idea that the TCA cycle normally competes with the sucrose biosynthetic pathway for carbon. Importantly, the results also suggest that TCA cycle carbon flow coupled with oxidative phosphorylation is not a critical source of ATP for sucrose synthesis.

In another study, transgenic tomato plants with reduced levels of mitochondrial MDH were also found to have increased rates of photosynthesis and yield in comparison to wt plants [101]. Metabolite profiling indicated that the most striking change in the transgenics was a large increase in the pool of ascorbate (Asc).

Further, the increased photosynthetic rate of the mutant could be phenocopied by incubation of wt leaf discs with Asc. These results are interesting given that Asc biosynthesis may be linked to electron transport through complex I (see below). Taken together, the results suggest a novel link between TCA cycle carbon flow and photosynthesis and involving the important antioxidant Asc.

11.4.2 Ascorbate biosynthesis

While Asc is a critical component of the antioxidant system in plants (see below), its pathway(s) of biosynthesis is not yet entirely established. Nonetheless, one important biosynthetic pathway is known to terminate with the conversion of L-galactono-1,4-lactone to Asc by L-galactono-1,4-lactone dehydrogenase (GLDH). This enzyme resides in the IMM and has been shown to use oxidized cyt c (cyt c_{ox}) as its electron acceptor [102]. Hence, Asc synthesis can stimulate electron transport to oxygen, can be inhibited by CN (which acts to deplete the pool of cyt c_{ox}) and can be stimulated by antimycin A (which acts to increase the pool of cyt c_{ox}) [103]. Interestingly, it has recently been established that GLDH physically associates with complex I and that Asc synthesis can be partially inhibited by the complex I inhibitor rotenone, despite the fact that rotenone is expected to increase the pool of cyt c_{ox} [104]. This suggests that complex I function is a necessary prerequisite for optimal Asc synthesis by the associated GLDH. While details of the relationship between complex I and the synthesis of Asc remain unknown, the above findings have uncovered an intriguing relationship between electron transport and the biosynthesis of an important ROS-scavenger. This new model for Asc biosynthesis is summarized in Figure 11.3.

11.4.3 Mitochondrial fatty acid synthesis

Fatty acids are synthesized by the combined enzymatic activities of acetyl-CoA carboxylase (ACCase) and fatty acid synthase (FAS). While fatty acid synthesis is predominantly a plastid-localized process, the possibility of some mitochondrial-

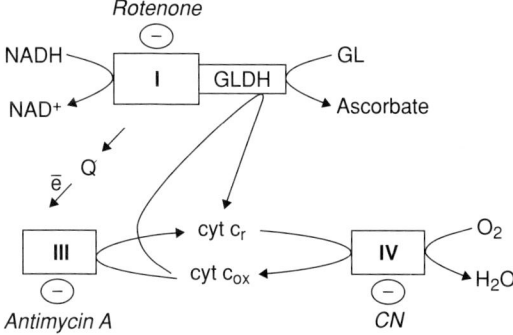

Figure 11.3 A working model for the biosynthesis of Asc in plant mitochondria (see text for details).

localized synthesis was supported by the discovery of acyl carrier protein (ACP) in mitochondria [105], the incorporation by mitochondria of 2-^{14}C malonate into fatty acids [106] and by the capacity of mitochondria to produce malonyl- and acetyl-ACP [26]. More recent results indicate that plant mitochondria do indeed possess FAS activity and that the pathway generates primarily C_8 and C_{14}–C_{16} fatty acyl chains [107, 108]. While the C_8 chains are precursors for lipoic acid (a cofactor in several matrix enzymes), the function of the longer acyl chains is unknown but likely relates to remodeling or repair of membrane phospholipids.

Previous studies with dicotyledonous species have failed to identify ACCase in mitochondria, suggesting that mitochondrial fatty acid synthesis will require the import of malonate. However, using blue-native/SDS-PAGE combined with mass spectrometry, abundant ACCase has been found in mitochondria of wheat and barley [109]. Further work is necessary to determine whether this represents a dual-targeted ACCase (to both plastid and mitochondria) and to determine whether its presence in mitochondria of monocotyledonous species and absence in mitochondria of dicotyledonous species is a general feature in plants.

11.4.4 *The glyoxylate cycle and lipid respiration*

Upon germination, oil-storing seeds convert stored triacylglycerols to sucrose, which then supports early growth of the seedling. This process involves the glyoxylate cycle which allows the acetyl-CoA generated by glyoxysomal β-oxidation of fatty acids to be converted into succinate. Succinate metabolism in the mitochondrion generates OAA, which then supports the synthesis of sucrose via cytosolic gluconeogenesis. During this germinated oilseed metabolism, the activities of decarboxylative steps in the TCA cycle are thought to be low, thus favoring gluconeogenesis over TCA cycle carbon flow [110].

The two unique and sequential glyoxysomal enzymes of the glyoxylate cycle are isocitrate lyase (ICL, generating glyoxylate and succinate from isocitrate) and malate synthase (MLS, generating malate from glyoxylate and acetyl-CoA). Recently, the analyses of Arabidopsis mutants lacking these enzymes have not only given an insight into the role of the glyoxylate cycle in lipid to sucrose conversion, but also have allowed an evaluation of the capacity of lipid breakdown to directly support mitochondrial respiration. In mutants lacking ICL, germinating seedlings lost the ability to convert stored lipid to sucrose and exhibited features characteristic of carbohydrate starvation, confirming the role of the glyoxylate cycle in this process [111, 112]. However, in the presence of exogenous sugar, these seedlings were nonetheless able to break down the lipid stores (in the absence of any gluconeogenic carbon flow), likely then using the carbon directly in respiration. This suggests a mechanism allowing carbon flow from lipid breakdown in the glyoxysome to the mitochondrion, other than a glyoxylate cycle. One possibility is that citrate and OAA shuttle between the glyoxysome and mitochondrion, generating a TCA cycle that effectively spans the two organelles [113]. Such a mechanism may be important in mature plant tissues that retain a functional fatty acid β-oxidation pathway but lack a glyoxylate cycle. In some cases, such fatty acid breakdown in mature tissues has been shown to support respiration [114].

In mutants lacking MLS, germination and the conversion of lipid to sucrose were largely unaffected [112]. In this case, succinate produced by the ICL could still feed into the TCA cycle and support OAA synthesis while glyoxylate (the substrate of MLS) was instead metabolized to sugars, likely by the photorespiratory enzymes known to coexist with glyoxylate cycle enzymes [112]. Both the ICL and MLS mutants emphasize the remarkable flexibility apparent in plant respiratory metabolism.

11.5 Mitochondrial metabolism of reactive oxygen species

11.5.1 Mitochondrial ROS generation

Aerobic metabolism is inevitably associated with the generation of ROS such as superoxide (O_2^-), hydrogen peroxide (H_2O_2) and hydroxyl radical. Since ROS can damage lipid, protein and DNA, their cellular levels are managed through both scavenging and avoidance mechanisms [115]. While the chloroplast ETC is thought to represent the major source of ROS from plant metabolism, there is a growing realization that the mitochondrial ETC is also an important source of ROS, perhaps the dominant source in nonphotosynthetic tissues, as well as photosynthetic tissue in the dark. Perhaps reflecting this, a recent study in wheat leaves suggests that mitochondria contain considerably more oxidatively modified proteins than do the chloroplast or peroxisome [116]. Many forms of stress promote ROS generation in plants, partly due to the ability of such stresses to disrupt metabolic homeostasis. The importance of mitochondrial ROS in this regard is suggested by studies showing that overexpression of a mitochondrial ROS-scavenging enzyme can improve stress tolerance [117, 118].

11.5.2 Mechanisms to scavenge mitochondrial ROS

Mittler *et al.* [119] did a comprehensive analysis of the Arabidopsis genome to identify and at least putatively localize the large network of proteins involved in managing cellular levels of ROS. Table 11.2 summarizes the findings of this study in terms of the potential ROS-scavenging complement of the mitochondrion. (*The table also defines many of the abbreviations to follow.*) The list includes enzymes that directly scavenge O_2^- (i.e., SOD), as well as those that scavenge H_2O_2 using either Asc (i.e., APX), GSH (i.e., GPX) or reduced sulfhydryls (i.e., PrxR) as a source of electrons. Also included are the enzymes capable of regenerating Asc (i.e., MDAR, DHAR, GLR), GSH (i.e., GR) or reduced sulfhydryls (i.e., Trx). Only a fraction of these putative mitochondrial localized proteins have been confirmed to reside in the mitochondrion. However, the basic scavenging systems of the mitochondrion are being elucidated, as outlined below.

The four enzymes APX, MDAR, DHAR and GR, along with the low molecular weight antioxidants Asc and GSH, constitute the Asc-glutathione cycle. Jiménez *et al.* [120] established the presence of all these components in pea leaf mitochondria.

Table 11.2 The major ROS-scavenging components proposed to be localized in mitochondria of Arabidopsis (see text for details)

Enzyme/reaction	Number of genes in Arabidopsis	Number with mitochondrial localization
Superoxide dismutase (SOD) $O_2^- + O_2^- + 2H^+ \rightarrow H_2O_2 + O_2$	8	1
Ascorbate peroxidase (APX) $2\,Asc + H_2O_2 \rightarrow 2\,MDA + 2H_2O$	9	3
Monodehydroascorbate reductase (MDAR) $MDA + NADPH + H^+ \rightarrow Asc + NADP^+$	5	2
Dehydroascorbate reductase (DHAR) $DHA + 2\,GSH \rightarrow Asc + GSSG$	5	1
Glutathione reductase (GR) $GSSG + NADPH \rightarrow 2GSH + NADP^+$	2	1
Glutathione peroxidase (GPX) $H_2O_2 + 2GSH \rightarrow 2H_2O + GSSG$	8	2
Peroxiredoxin (PrxR) $2P\text{-}SH + H_2O_2 \rightarrow P\text{-}S\text{-}S\text{-}P + 2H_2O$	11	2
Thioredoxin (Trx) $P\text{-}S\text{-}S\text{-}P + 2H^+ \rightarrow 2P\text{-}SH$	32	13
Glutaredoxin (GLR) $DHA + 2\,GSH \rightarrow Asc + GSSG$	27	15

More recently, the Arabidopsis Asc-glutathione cycle has been characterized at the molecular level [121]. Interestingly, this study showed dual-targeting of some gene products to both the chloroplast and mitochondrion. This indicates that the antioxidant defense capacity of these two organelles may be coordinately controlled. GPX is another potentially important ROS scavenging system in the mitochondrion and also represents a potentially important means to repair lipid damage, since these enzymes are also known to reduce lipid peroxides. Of the eight GPX genes in Arabidopsis, two appear to represent mitochondrial isoforms and, interestingly, one of these (GPx6) was the family member that showed the strongest increase in expression when induced by several different stress conditions [122]. Analyses of Arabidopsis mutants lacking a mitochondrial PrxR have indicated that this protein does contribute to the ROS-scavenging capacity of the matrix. Such plants appeared to compensate for this deficiency by increasing the levels of mitochondrial APX and GPX [123].

In addition to the principal components of the mitochondrial ROS-scavenging network described above, several other components also deserve mention. Metal ions present in the cell in oxidized forms (Fe^{3+}, Cu^{2+}) are reduced in the presence of O_2^- and can subsequently catalyze the conversion of H_2O_2 to hydroxyl radical (the Fenton reaction). Hence, cells contain metal-binding proteins such as ferritins (Fe-binding) and blue copper proteins (Cu-binding). The Arabidopsis genome includes up to four genes encoding ferritin proteins and up to nine genes encoding blue copper proteins. Of these, the mitochondrion may contain up to two ferritin and four blue copper proteins. Recently, the presence of ferritin in plant mitochondria has been confirmed [124]. Since both MDAR and GR utilize NADPH, a mitochondrial source of this electron donor is critical for operation of the Asc-glutathione cycle. In animals, the mitochondrial $NADP^+$-dependent ICDH represents a critical source of NADPH for the antioxidant defenses of the mitochondrion [125]. Such an enzyme is also present in plant mitochondria but whether it is critical for ROS-scavenging is not yet known.

11.5.3 Mechanisms to avoid mitochondrial ROS generation

Complexes I and III likely represent the primary sites of ROS generation by the mitochondrial ETC in both animals and plants [115, 126]. However, the relative importance of these two sites of generation, particularly in plants, is unknown. The factors that influence the rate of ROS generation are also largely unknown, although an important generalization can be made: ROS generation increases as ETC components become more highly reduced. ROS generation by isolated mitochondria is therefore increased under state 4 (ADP-limiting) conditions that increase membrane potential while being decreased by uncouplers that dissipate membrane potential [127]. ROS generation can also be increased by inhibition of specific sites in the ETC such as inhibition of complex III by antimycin A or inhibition of complex I by rotenone. Again, these inhibitors promote ROS generation by promoting overreduction of specific ETC components [115, 126].

As outlined earlier, the ETC of plants includes several unique components that impact the energetics of respiration. Another consequence of these components is that they may act to dampen ROS generation. By accepting electrons from Q, AOX could act to prevent overreduction of complexes I and III. Since AOX respiration is less tightly coupled to ATP production, this mechanism might be important under conditions in which cyt pathway activity is restricted by the rate of ATP turnover. Alternatively, it may be important if cyt pathway components are damaged by stress. The UCPs may also represent an important means to dampen mitochondrial ROS, owing to their ability to uncouple ETC activity. A role of AOX in dampening the generation of ROS is strongly supported by the finding that transgenic cells lacking AOX exhibit an increased rate of ROS release specifically from the mitochondrion [128]. Studies with isolated mitochondria also support the importance of AOX in dampening ROS generation [129]. Interestingly, AOX expression is induced by H_2O_2 [11] while activity of the UCPs is controlled by O_2^- (see below). These observations suggest positive feedback loops so that increases in cellular ROS enhance the activity of these ROS-dampening components of the ETC.

How the alternate NAD(P)H dehydrogenases may impact ROS generation is unknown. On the one hand, they may themselves represent sites of ROS generation, although no data are available regarding this. Alternatively, they may represent another means to dampen ROS generation since (1) their activity will bypass complex I, a known ROS producer and (2) unlike complex I, their activity does not contribute to the generation of membrane potential. It is also possible that one or more of these dehydrogenases could act in concert with AOX to provide a completely nonenergy conserving route of electron transport to O_2, but this hypothesis still needs to be tested.

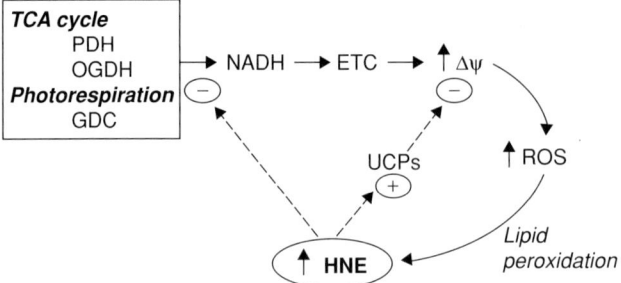

Figure 11.4 A working model for the control of mitochondrial energy metabolism and ROS production by the lipid peroxidation product HNE (see text for details). Δψ is membrane potential.

Recent studies of particular interest examined the ability of a product of ROS-induced membrane damage (the lipid peroxidation product 4-hydroxy-2-nonenal (HNE)) to inhibit mitochondrial enzyme activities [130, 131]. It was established that specifically the lipoic acid containing enzymes (PDH, OGDH, GDC) were inhibited by low levels of HNE due to the formation of lipoic acid-HNE Michael adducts. Of these enzymes, GDC was the most susceptible to inactivation, suggesting that photorespiratory glycine oxidation may be particularly prone to oxidative stress [131]. These studies suggest that photorespiration will not tolerate high levels of ROS formation and lipid peroxidation in the mitochondrion, and that mechanisms may therefore be engaged to limit such formation. In this regard, it is interesting that the expression of ETC components that might help limit ROS generation (the alternate NAD(P)H dehydrogenases) are induced by light [97].

Significantly, both UCPs and AOX appear to also be targets of HNE, although with contrasting effects on their activity. For some years, it has been known that UCP activity is stimulated by fatty acids and inhibited by nucleotides. Recently, it has been established that fatty acids do not directly activate UCPs but rather that matrix O_2^- (in the presence of fatty acids) activates UCPs [132]. Further, it appears that the effect of O_2^- is to generate HNE, which then acts as a potent activator of the UCPs [133]. Figure 11.4 describes a model for how HNE could act to control the mitochondrial generation of ROS. In this model, reductant generated in the matrix by the activity of the TCA cycle and photorespiratory pathways is oxidized by the ETC, thus increasing membrane potential. Increased membrane potential will stimulate ROS generation by the ETC, leading to increased levels of lipid peroxidation products such as HNE. However, the increase in HNE will subsequently reduce membrane potential (hence reducing ROS generation) since it acts to both inhibit the generation of reducing equivalents (by inhibiting PDH, OGDH and GDC) and stimulate dissipation of membrane potential (by activating UCPs). Interestingly, while AOX represents another important mechanism to reduce ROS generation by the ETC, it is also inactivated by HNE [134]. This may provide an explanation as to why plants maintain both these energy dissipating pathways.

11.5.4 Signalling functions of mitochondrial ROS

The consequence of mitochondrial-generated ROS for oxidative damage to mitochondrial components is an area of active research (see above). However, a second potentially important consequence of mitochondrial ROS is that they may act as signalling molecules that can be perceived (by either the mitochondrion or other parts of the cell), leading to specific cellular responses. For example, several different ETC mutants are described that alter the rate of ROS generation and which display specific constitutive changes in gene expression or aberrant gene expression in response to stress [128, 135, 136]. Such experiments implicate mitochondrial ROS as signalling molecules, but this will need to be confirmed by a description of the molecules involved in ROS perception and signal transduction. Such work is in its infancy in plants.

11.6 Additional stress-induced metabolic pathways associated with plant mitochondria

Some other aspects of mitochondrial metabolism are also often associated with stress. In the section below, we highlight recent findings on a select group of such pathways.

11.6.1 GABA shunt

The three unique enzymes of the γ-aminobutyrate (GABA) shunt (Figure 11.5) include Glu decarboxylase (GAD, converting Glu to GABA), GABA transaminase

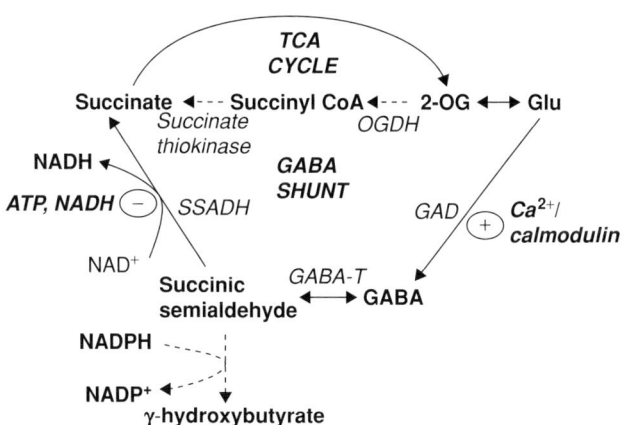

Figure 11.5 Organization and control of the GABA shunt in plant mitochondria. Note that each of the GABA shunt enzymes resides in the mitochondrial matrix, with the exception of GAD, which is cytosolic (see text for details).

(converting GABA to succinic semialdehyde) and succinic semialdehyde dehydrogenase (SSADH, converting succinic semialdehyde and NAD^+ to succinate and NADH). GAD is a cytosolic enzyme while GABA transaminase and SSADH are present in the mitochondrial matrix. The GABA shunt effectively converts 2-oxoglutarate (via Glu) to succinate via a pathway that bypasses the two usual TCA cycle steps (OGDH, succinate thiokinase) required for this conversion. Shunt activity may be controlled at the level of GAD and/or SSADH. GAD is controlled by a Ca^{2+}-calmodulin complex and, as expected, this appears to increase the GABA shunt flux in response to stress conditions [137]. SSADH activity is inhibited by ATP and NADH, linking pathway flux to the energy and redox status of the mitochondrial matrix [138]. Hence, shunt activity may be expected to increase under stress conditions that compromise energy metabolism.

Recently, it was shown that Arabidopsis mutants lacking SSADH accumulate ROS, particularly under stress conditions [139]. The source of these ROS is unknown, but their accumulation correlates with the accumulation of γ-hydroxybutyrate (GHB) in the mutant plants [140]. GHB is synthesized from the shunt intermediate succinic semialdehyde, which might be expected to accumulate in the mutant due to the lesion at SSADH. An enzyme capable of such GHB synthesis (γ-hydroxybutyrate dehydrogenase) has recently been characterized in Arabidopsis [141]. GHB synthesis is also favored under hypoxic or anoxic conditions when both its substrates (succinic semialdehyde and NAD[P]H) are expected to increase [141].

An untested possibility is that the GABA shunt provides a means to bypass a stress-induced lesion in the TCA cycle at either OGDH or succinate thiokinase. This could act to maintain normal energy metabolism and ROS levels.

11.6.2 Mitochondrial amino acid catabolism

Genomic, proteomic and biochemical analyses have now shown that most of the enzymes involved in the catabolism of the branched chain amino acids (Leu, Ile and Val) are present in plant mitochondria [142–145]. For example, the branched chain keto-acid dehydrogenase complex, which represents an early step common in the catabolism of each of these amino acids, is localized to the mitochondrion, as is a β-methylcrotonyl-CoA carboxylase required specifically for Leu catabolism [145]. These catabolic pathways appear to be induced by their substrates and particularly under conditions of sugar starvation [146, 147]. The oxidative metabolism of these amino acids may provide a source of respiratory substrates under stress or other conditions that compromise sugar availability.

11.6.3 Formate dehydrogenase

Formate dehydrogenase (FDH) catalyzes: formate + $NAD^+ \rightarrow CO_2$ + NADH. This matrix enzyme can be one of the most abundant proteins in mitochondria of nonphotosynthetic tissues but is much less abundant in photosynthetic tissues. However, leaf levels of FDH are greatly enhanced under stress conditions such as drought and cold, such that these mitochondria can now readily utilize formate as a respiratory substrate supporting oxidative phosphorylation [148]. Elucidation of the physiological role of FDH is complicated by an incomplete understanding of both the origins and metabolism of formate in different plant tissues [149]. Nonetheless, it has been shown that formate accumulates in both photosynthetic and nonphotosynthetic tissues of transgenic plants lacking FDH [150]. One untested possibility is that formate oxidation by FDH provides an important source of reducing power to the ETC under stress conditions in which other pathways such as the TCA cycle may have been compromised and in which the availability of formate may be increased. This could act to both support oxidative phosphorylation under stress and prevent the build-up of potentially toxic levels of formate.

Using sprouting potato tubers, Bykova *et al.* [63] recently showed an inverse relationship between the maximum catalytic activities of FDH and cyt oxidase in sprouting potato tubers. While cyt oxidase activity declined as a function of the steady-state oxygen concentration within the tuber, the FDH activity increased. These data suggest a role for FDH under hypoxic conditions, conditions that are known to induce FDH expression. One possibility is that some plant tissues maintain a formate-producing fermentation pathway, the formate from which is subsequently oxidized by FDH. In unicellular algae, the anaerobic enzyme pyruvate formate-lyase catalyzes the formation of acetyl-CoA and formate from pyruvate and CoA. While such an enzyme has not been identified in plants, the Arabidopsis genome does contain a conserved local sequence motif of pyruvate formate-lyase [63]. The combined action of pyruvate formate-lyase and FDH gives the same net result as the reaction catalyzed by PDH.

11.6.4 Mitochondrial aldehyde dehydrogenases

Aldehyde dehydrogenases (ALDHs) represent a large protein family capable of oxidizing a wide range of aldehydes to their corresponding acids, coupled with the reduction of $NAD(P)^+$. The plant species studied (including tobacco, maize, Arabidopsis and rice) each have two confirmed or predicted mitochondrial ALDHs that have distinct (but overlapping) substrate specificities and distinct expression patterns [151–154].

Besides having an important housekeeping function in preventing accumulation of various aldehydes to toxic levels, it is likely that the mitochondrial ALDHs also play more specific roles in metabolism and stress tolerance. One potentially harmful aldehyde is acetaldehyde, an intermediate in the conversion of pyruvate to ethanol during fermentation. During reaeration, accumulated ethanol can be converted back to acetaldehyde, causing injury. Recently, expression and activity analyses of ALDH during and following hypoxia suggested that one of the mitochondrial ALDHs may play a key role in preventing posthypoxic acetaldehyde-induced damage [154]. ALDH converts acetaldehyde to acetate, which can then be converted to acetyl-CoA by acetyl-CoA synthetase. Besides preventing toxic accumulation of acetaldehyde, this pathway could provide a ready supply of acetyl-CoA and NAD(P)H as aerobic respiration recovers. Other important sources of reactive aldehydes result from lipid peroxidation and defectively functioning TCA cycle enzymes, processes that may increase in the mitochondrion under various stress conditions [155]. Accordingly, several studies of mitochondrial proteomes have shown induction of ALDHs in response to stress [56].

Mitochondrial ALDH activity appears to be critical for normal pollen development [151] and was in fact identified as a 'nuclear restorer gene' against cytoplasmic male fertility [156]. In pollen, ALDH may be part of a necessary metabolic bypass of PDH for the conversion of pyruvate to acetyl-CoA [157].

11.6.5 Mitochondrial glycerol-3-phosphate dehydrogenase

Long ago, it was observed that isolated plant mitochondria oxidized glycerol-3-phosphate (G-3-P) coupled with ETC activity. It appears that this is due to the presence of an FAD-dependent G-3-P dehydrogenase (FAD-GPDH) recently shown to be associated with the membrane fraction of Arabidopsis mitochondria [158]. In animals, a G-3-P shuttle transfers reducing equivalents from the cytosol to the mitochondrial ETC. This shuttle consists of a cytosolic NAD-dependent GPDH (reducing dihydroxyacetone phosphate to G-3-P using NADH) and an FAD-GPDH on the outer face of the IMM that oxidizes G-3-P back to dihydroxyacetone phosphate coupled with electron flow from $FADH_2$ to the ETC. Since these shuttle components appear to also be present in plants, it represents another potential means (besides the alternate external NAD(P)H dehydrogenases) to deliver extra-mitochondrial reducing power to the ETC, although this still requires confirmation. Interestingly, activity of the plant FAD-GPDH is likely independent of Ca^{2+} since it lacks the EF-hand Ca^{2+}-binding domain seen in the animal enzyme. One possibility is that a G-3-P shuttle moves reducing equivalents under nonstressed conditions while the Ca^{2+}-activated alternate external NAD(P)H dehydrogenases supplement this activity under stress conditions.

11.6.6 Nucleoside diphosphate kinase

Nucleoside diphosphate kinases (NDPKs) catalyze the transfer of a phosphate group from nucleoside triphosphates to nucleoside diphosphates: $N_1TP + N_2DP \rightarrow N_1DP + N_2TP$. This activity is important, for example, in maintaining the balance between ATP and GTP. In plants, NDPKs are known to exist in the cytosol, plastid and mitochondrion [159]. The mitochondrial isoform has been localized to the intermembrane space as well as being tightly bound to the inner membrane, possibly via interaction with other proteins such as the ANT [159, 160]. Interaction with the ANT suggests that NDPK activity may in some way facilitate or regulate the exchange of ATP and ADP across the IMM [160]. Besides their primary metabolic role, some NDPKs have been suggested to play regulatory roles, due to their ability to interact with other proteins, possibly modifying their function. It is therefore interesting that the plant mitochondrial NDPK was recently shown to interact with an unidentified 86 kD protein synthesized in response to heat shock [161].

11.6.7 Root organic anion exudation

Organic acids such as citrate and malate are often excreted by plant roots in response to nutrient deficiency (particularly Pi) or in response to toxic soil levels of cations (particularly Al). These organic acids are almost certainly present in the cytosol in their dissociated (anion) form and can hence be transported passively from cell to soil down their electrochemical gradient. Plant growth is often limited by the *soluble* Pi content of the soil. The organic anions increase the concentration of soluble Pi in the rhizosphere by competing with Pi for binding sites in the soil and by forming stronger complexes with soil Al^{3+}, Fe^{3+} and Ca^{2+} than does Pi. By these mechanisms, organic anion exudation has been shown to both improve Pi nutrition and dampen Al toxicity [162].

Over a plant's lifetime, organic anion exudation can represent a significant loss of carbon [162]. Hence, exudation must be tightly controlled at the level of organic acid synthesis and/or passive transport to the rhizosphere. Increased exudation has often been associated with changes in the activity of cytosolic and mitochondrial enzymes involved in organic acid metabolism. This has included upregulation of phospho*enol*pyruvate carboxylase, MDH and citrate synthase (to increase malate/citrate synthesis), sometimes combined with decreases in ACO and ICDH (to decrease malate/citrate catabolism). Nonetheless, a clear picture of how these changes might regulate organic acid synthesis and increase exudation has not emerged [162].

Another approach has been to generate transgenic plants with manipulated levels of the enzymes involved in organic acid metabolism. In some cases, this has resulted in increased organic acid efflux, combined with improved P acquisition or Al^{3+} tolerance [163, 164]. Such results suggest that genetic engineering of organic acid metabolism might provide a means to increase the tolerance of crop plants toward nutrient deficiencies and inorganic ion toxicities. Importantly, the anion channels responsible for organic acid efflux are also being described, although their molecular nature is not yet known [165, 166].

11.7 Concluding remarks

Mitochondria are a central organelle for carbon and energy metabolism in plants. Nonetheless, there remain important gaps in our understanding of the components of carbon and energy metabolism in this organelle and how these components are controlled. There is also increasing evidence that plant mitochondria may be subject to stress-induced damage and that the functional state of mitochondria is an important factor in maintaining critical aspects of cellular function such as redox homeostasis. However, in comparison to our advanced understanding of how photosynthesis and chloroplast function are adjusted and maintained as environmental and metabolic conditions fluctuate, we have a relatively poor understanding of how respiration and mitochondrial function are similarly maintained. This remains an important challenge for the future.

Acknowledgements

Research grant support from the Natural Sciences and Engineering Research Council of Canada (to G.C.V.) and from an Ontario Graduate Scholarship (to A.E.M.) is gratefully acknowledged.

References

1. A.H. Millar, J.L. Heazlewood, B.K. Kristensen, H.-P. Braun and I.M. Møller (2005) The plant mitochondrial proteome. *Trends in Plant Science* **10**, 36–43.
2. R. Yui, S. Iketani, T. Mikami and T. Kubo (2003) Antisense inhibition of mitochondrial pyruvate dehydrogenase E1α subunit in anther tapetum causes male sterility. *The Plant Journal*, **34**, 57–66.
3. A. Tovar-Méndez, J.A. Miernyk and D.D. Randall (2003) Regulation of pyruvate dehydrogenase activity in plant cells. *European Journal of Biochemistry* **270**, 1043–1049.
4. E.A. Unger, J.M. Hand, A.R. Cashmore and A.C. Vasconcelos (1989) Isolation of a cDNA encoding mitochondrial citrate synthase from *Arabidopsis thaliana*. *Plant Molecular Biology* **13**, 411–418.
5. V. Landschutze, L. Willmitzer and B. Müller-Röber (1995) Inhibition of flower formation by antisense repression of mitochondrial citrate synthase in transgenic potato plants leads to a specific disintegration of the ovary tissues of flowers. *The EMBO Journal* **14**, 660–666.
6. D.A. Navarre, D. Wendehenne, J. Durner, R. Noad and D.F. Klessig (2000) Nitric oxide modulates the activity of tobacco aconitase. *Plant Physiology* **122**, 573–582.
7. F. Verniquet, J. Gaillard, M. Neuburger and R. Douce (1991) Rapid inactivation of plant aconitase by hydrogen peroxide. *Biochemical Journal* **276**, 643–648.
8. J. Jordanov, F. Courtois-Verniquet, M. Neuburger and R. Douce (1992) Structural investigations by extended X-ray absorption fine structure spectroscopy of the iron center of mitochondrial aconitase in higher plant cells. *Journal of Biological Chemistry* **267**, 16775–16778.
9. P. Peyret, P. Perez and M. Alric (1995) Structure, genomic organization, and expression of the *Arabidopsis thaliana* aconitase gene. Plant aconitase shows significant homology with mammalian iron-responsive element-binding protein. *Journal of Biological Chemistry* **270**, 8131–8137.
10. M. Ruffer, B. Steipe and M.H. Zenk (1995) Evidence against specific binding of salicylic acid to plant catalase. *FEBS Letters* **377**, 175–180.
11. G.C. Vanlerberghe and L. McIntosh (1996) Signals regulating the expression of the nuclear gene encoding alternative oxidase of plant mitochondria. *Plant Physiology* **111**, 589–595.
12. S. Galvez, O. Roche, E. Bismuth, S. Brown, P. Gadal and M. Hodges (1998) Mitochondrial localization of a NADP-dependent isocitrate dehydrogenase isoenzyme by using the green fluorescent protein as a marker. *Proceedings of the National Academy of Sciences the United States of America* **95**, 7813–7818.
13. M. Lancien, P. Gadal and M. Hodges (1998) Molecular characterization of higher plant NAD-dependent isocitrate dehydrogenase: evidence for a heteromeric structure by the complementation of yeast mutants. *The Plant Journal* **16**, 325–333.
14. N. Pascal, R. Dumas and R. Douce (1990) Comparison of the kinetic behavior toward pyridine nucleotides of NAD^+-linked dehydrogenases from plant mitochondria. *Plant Physiology* **94**, 189–193.
15. A.H. Millar, S.A. Hill and C.J. Leaver (1999) Plant mitochondrial 2-oxoglutarate dehydrogenase complex: purification and characterization in potato. *Biochemical Journal* **343**, 327–334.
16. G. Nast and B. Müller-Röber (1996) Molecular characterization of potato fumarate hydratase and functional expression in *Escherichia coli*. *Plant Physiology* **112**, 1219–1227.
17. R.H. Behal and D.J. Oliver (1997) Biochemical and molecular characterization of fumarase from plants: purification and characterization of the enzyme – cloning, sequencing, and expression of the gene. *Archives of Biochemistry and Biophysics* **348**, 65–74.

18. M.K. Hayes, M.H. Luethy and T.E. Elthon (1991) Mitochondrial malate dehydrogenase from corn: purification of multiple forms. *Plant Physiology* **97**, 1381–1387.
19. C.B. Grissom, P.F. Canellas and R.T. Wedding (1983) Allosteric regulation of the NAD malic enzyme from cauliflower: activation by fumarate and coenzyme A. *Archives of Biochemistry and Biophysics* **220**, 133–144.
20. S.D. Grover and R.T. Wedding (1984) Modulation of the activity of NAD-malic enzyme from *Solanum tuberosum* by changes in oligomeric state. *Archives of Biochemistry and Biophysics* **234**, 418–425.
21. M. Dieuaide-Noubhani, G. Raffard, P. Canioni, A. Pradet and P. Raymond (1995) Quantification of compartmented metabolic fluxes in maize root tips using isotope distribution from ^{13}C- or ^{14}C-labeled glucose. *Journal of Biological Chemistry* **270**, 13147–13159.
22. S. Edwards, B.-T. Nguyen, B. Do and J.K.M. Roberts (1998) Contribution of malic enzyme, pyruvate kinase, phospho*enol*pyruvate carboxylase, and the Krebs cycle to respiration and biosynthesis and to intracellular pH regulation during hypoxia in maize root tips observed by nuclear magnetic resonance imaging and gas chromatography-mass spectrometry. *Plant Physiology*, **116**, 1073–1081.
23. H.L. Jenner, B.M. Winning, A.H. Millar, K.L. Tomlinson, C.J. Leaver and S.A. Hill (2001) NAD malic enzyme and the control of carbohydrate metabolism in potato tubers. *Plant Physiology* **126**, 1139–1149.
24. R. Douce, J. Bourguignon, M. Neuburger and F. Rébeille (2001) The glycine decarboxylase system: a fascinating complex. *Trends in Plant Science* **6**, 167–176.
25. M. Neuburger, F. Rébeille, A. Jourdain, S. Nakamura and R. Douce (1996) Mitochondria are a major site for folate and thymidylate synthesis in plants. *Journal of Biological Chemistry* **271**, 9466–9472.
26. V. Gueguen, D. Macherel, M. Jaquinod, R. Douce and J. Bourguignon (2000) Fatty acid and lipoic acid biosynthesis in higher plant mitochondria. *Journal of Biological Chemistry* **275**, 5016–5025.
27. H.-P. Braun, M. Emmermann, V. Kruft and U.K. Schmitz (1992) The general mitochondrial processing peptidase from potato is an integral part of cytochrome c reductase of the respiratory chain. *The EMBO Journal* **11**, 3219–3227.
28. J.L. Heazlewood, K.A. Howell and A.H. Millar (2003) Mitochondrial complex I from Arabidopsis and rice: orthologs of mammalian and fungal components coupled with plant-specific subunits. *Biochimica et Biophysica Acta* **1604**, 159–169.
29. I.M. Fearnley, J. Carroll, R.J. Shannon, M.J. Runswick, J.E. Walker and J. Hirst (2001) GRIM-19, a cell death regulatory gene product, is a subunit of bovine mitochondrial NADH-ubiquinone oxidoreductase (Complex I). *Journal of Biological Chemistry* **276**, 38345–38348.
30. A.H. Millar, H. Eubel, L. Jänsch, V. Kruft, J.L. Heazlewood and H.-P. Braun (2004) Mitochondrial cytochrome c oxidase and succinate dehydrogenase complexes contain plant specific subunits. *Plant Molecular Biology* **56**, 77–90.
31. C. Affourtit, K. Krab, G.R. Leach, D.G. Whitehouse and A.L. Moore (2001) New insights into the regulation of plant succinate dehydrogenase: on the role of the proton motive force. *Journal of Biological Chemistry* **276**, 32567–32574.
32. H.-P. Braun, V. Kruft and U.K. Schmitz (1994) Molecular identification of the ten subunits of cytochrome-*c* reductase from potato mitochondria. *Planta* **193**, 99–106.
33. H. Eubel, L. Jänsch and H.-P. Braun (2003) New insights into the respiratory chain of plant mitochondria. Supercomplexes and a unique composition of Complex II. *Plant Physiology* **133**, 274–286.
34. L. Jänsch, V. Kruft, U.K. Schmitz and H.-P. Braun (1996) New insights into the composition, molecular mass and stoichiometry of the protein complexes of plant mitochondria. *The Plant Journal* **9**, 357–368.
35. A.G. Rasmusson, K.L. Soole and T.E. Elthon (2004) Alternative NAD(P)H dehydrogenases of plant mitochondria. *Annual Review of Plant Biology* **55**, 23–39.
36. C.S. Moore, R.J. Cook-Johnson, C. Rudhe *et al.* (2003) Identification of AtNDI1, an internal non-phosphorylating NAD(P)H dehydrogenase in Arabidopsis mitochondria. *Plant Physiology* **133**, 1968–1978.

37. A.M. Michalecka, S.C. Agius, I.M. Møller and A.G. Rasmusson (2004) Identification of a mitochondrial external NADPH dehydrogenase by overexpression in transgenic *Nicotiana sylvestris*. *The Plant Journal* **37**, 415–425.
38. A.G. Rasmusson, A.S. Svensson, V. Knoop, L. Grohmann and A. Brennicke (1999) Homologues of yeast and bacterial rotenone-insensitive NADH dehydrogenases in higher eukaryotes: two enzymes are present in potato mitochondria. *The Plant Journal* **20**, 79–87.
39. G.C. Vanlerberghe and S.H. Ordog (2002) Alternative oxidase: integrating carbon metabolism and electron transport in plant respiration. In: *Photosynthetic Nitrogen Assimilation and Associated Carbon and Respiratory Metabolism, Advances in Photosynthesis and Respiration*, Vol. 12. (eds C.H. Foyer and G. Noctor), Kluwer Academic, The Netherlands, pp. 173–191.
40. M.S. Albury, C. Affourtit, P.G. Crichton and A.L. Moore (2002) Structure of the plant alternative oxidase: site-directed mutagenesis provides new information on the active site and membrane topology. *Journal of Biological Chemistry* **277**, 1190–1194.
41. D.A. Berthold, N. Voevodskaya, P. Stenmark, A. Graslund and P. Nordlund (2002) EPR studies of the mitochondrial alternative oxidase: evidence for a diiron carboxylate center. *Journal of Biological Chemistry* **277**, 43608–43614.
42. M.J. Considine, R.C. Holtzapffel, D.A. Day, J. Whelan and A.H. Millar (2002) Molecular distinction between alternative oxidase from monocots and dicots. *Plant Physiology* **129**, 949–953.
43. A.E. McDonald and G.C. Vanlerberghe (2005) Alternative oxidase and plastoquinol terminal oxidase in marine prokaryotes of the Sargasso Sea. *Gene* **349**, 15–24.
44. A.E. McDonald and G.C. Vanlerberghe (2004) Branched mitochondrial electron transport in the Animalia: presence of alternative oxidase in several animal phyla. *IUBMB Life*, **56**, 333–341.
45. D.A. Berthold, M.E. Andersson and P. Nordlund (2000) New insight into the structure and function of the alternative oxidase. *Biochimica et Biophysica Acta* **1460**, 241–254.
46. J.N. Siedow and A.L. Umbach (2000) The mitochondrial cyanide-resistant oxidase: structural conservation amid regulatory diversity. *Biochimica et Biophysica Acta* **1459**, 432–439.
47. A.L. Umbach and J.N. Siedow (1993) Covalent and noncovalent dimers of the cyanide-resistant alternative oxidase protein in higher plant mitochondria and their relationship to enzyme activity. *Plant Physiology* **103**, 845–854.
48. D.M. Rhoads, A.L. Umbach, C.R. Sweet, A.M. Lennon, G.S. Rauch and J.N. Siedow (1998) Regulation of the cyanide-resistant alternative oxidase of plant mitochondria: identification of the cysteine residue involved in α-keto acid stimulation and intersubunit disulfide bond formation. *Journal of Biological Chemistry* **273**, 30750–30756.
49. G.C. Vanlerberghe, L. McIntosh and J.Y.H. Yip (1998) Molecular localization of a redox-modulated process regulating plant mitochondrial electron transport. *The Plant Cell* **10**, 1551–1560.
50. G.C. Vanlerberghe, D.A. Day, J.T. Wiskich, A.E. Vanlerberghe and L. McIntosh (1995) Alternative oxidase activity in tobacco leaf mitochondria: dependence on tricarboxylic acid cycle-mediated redox regulation and pyruvate activation. *Plant Physiology* **109**, 353–361.
51. G.R. Gray, A.R. Villarimo, C.L. Whitehead and L. McIntosh (2004) Transgenic tobacco (*Nicotiana tabacum* L.) plants with increased expression levels of mitochondrial $NADP^+$-dependent isocitrate dehydrogenase: evidence implicating this enzyme in the redox activation of the alternative oxidase. *Plant and Cell Physiology* **45**, 1413–1425.
52. A.H. Millar, J.T. Wiskich, J. Whelan and D.A. Day (1993) Organic acid activation of the alternative oxidase of plant mitochondria. *FEBS Letters* **329**, 259–262.
53. R.C. Holtzapffel, J. Castelli, P.M. Finnegan, A.H. Millar, J. Whelan and D.A. Day (2003) A tomato alternative oxidase protein with altered regulatory properties. *Biochimica et Biophysica Acta* **1606**, 153–162.
54. P.G. Crichton, C. Affourtit, M.S. Albury, J.E. Carré and A.L. Moore (2005) Constitutive activity of *Sauromattum guttatum* alternative oxidase in *Schizosaccharomyces pombe* implicates residues in addition to conserved cysteines in α-keto acid activation. *FEBS Letters* **579**, 331–336.

55. D.M. Rhoads and L. McIntosh (1991) Isolation and characterization of a cDNA clone encoding an alternative oxidase protein of *Sauromattum guttatum* (Schott). *Proceedings of the National Academy of Sciences of the United States of America* **88**, 2122–2126.
56. S.M. Sieger, B.K. Kristensen, C.A. Robson *et al.* (2005) The role of alternative oxidase in modulating carbon use efficiency and growth during macronutrient stress in tobacco cells. *Journal of Experimental Botany* **56**, 1499–1515.
57. C.A. Robson and G.C. Vanlerberghe (2002) Transgenic plant cells lacking mitochondrial alternative oxidase have increased susceptibility to mitochondria-dependent and -independent pathways of programmed cell death. *Plant Physiology* **129**, 1908–1920.
58. G.C. Vanlerberghe, C.A. Robson and J.Y.H. Yip (2002) Induction of mitochondrial alternative oxidase in response to a cell signal pathway down-regulating the cytochrome pathway prevents programmed cell death. *Plant Physiology* **129**, 1829–1842.
59. A.H. Millar, A.E. Trend and J.L. Heazlewood (2004) Changes in the mitochondrial proteome during the anoxia to air transition in rice focus around cytochrome-containing respiratory complexes. *Journal of Biological Chemistry* **38**, 39471–39478.
60. N. Picault, M. Hodges, L. Palmieri and F. Palmieri (2004) The growing family of mitochondrial carriers in Arabidopsis. *Trends in Plant Science* **9**, 138–146.
61. C. Hourton-Cabassa, A.R. Matos, A. Zachowski and F. Moreau (2004) The plant uncoupling protein homologues: a new family of energy-dissipating proteins in plant mitochondria. *Plant Physiology and Biochemistry* **42**, 283–290.
62. N.V. Bykova, H. Egsgaard and I.M. Møller (2003) Identification of 14 new phosphoproteins involved in important plant mitochondrial processes. *FEBS Letters* **540**, 141–146.
63. N. V. Bykova, A. Stensballe, H. Egsgaard, O.N. Jensen and I.M. Møller (2003) Phosphorylation of formate dehydrogenase in potato tuber mitochondria. *Journal of Biological Chemistry* **278**, 26021–26030.
64. A. Struglics, K.M. Fredlund, I.M. Møller and J.F. Allen (1998) Two subunits of the F_oF_1 – ATPase are phosphorylated in the inner mitochondrial membrane. *Biochemical and Biophysical Research Communications* **243**, 664–668.
65. A. Struglics and G. Håkansson (1999) Purification of a serine and histidine phosphorylated mitochondrial nucleoside diphosphate kinase from *Pisum sativum*. *European Journal of Biochemistry* **262**, 765–773.
66. A.A. Lund, D.M. Rhoads, A.L. Lund, R.L. Cerny and T.E. Elthon (2001) *In vivo* modifications of the maize mitochondrial small heat stress protein, HSP22. *Journal of Biological Chemistry* **276**, 29924–29929.
67. J.A. Miernyk, N.B. Duck, N.R. David and D.D. Randall (1992) Autophosphorylation of the pea mitochondrial heat-shock protein homolog. *Plant Physiology* **100**, 965–969.
68. C. Pical, K.M. Fredlund, P.X. Petit, M. Sommarin and I.M. Møller (1993) The outer membrane of plant mitochondria contains a calcium-dependent protein kinase and multiple phosphoproteins. *FEBS Letters* **336**, 347–351.
69. A. Struglics, K.M. Fredlund, I.M. Møller and J.F. Allen (1999) Phosphoproteins and protein kinase activities intrinsic to inner membranes of potato tuber mitochondria. *Plant and Cell Physiology* **40**, 1271–1279.
70. A. Struglics, K.M. Fredlund, Y.M. Konstantinov, J.F. Allen and I.M. Møller (2000) Protein phosphorylation/dephosphorylation in the inner membrane of potato tuber mitochondria. *FEBS Letters* **475**, 213–217.
71. T.D. Bunney, H.S. van Walraven and A.H. de Boer (2001) 14-3-3 protein is a regulator of the mitochondrial and chloroplast ATP synthase. *Proceedings of the National Academy of Sciences of the United States of America* **98**, 4249–4254.
72. C. Laloi, N. Rayapuram, Y. Chartier, J.-M. Grienenberger, G. Bonnard and Y. Meyer (2001) Identification and characterization of a mitochondrial thioredoxin system in plants. *Proceedings of the National Academy of Sciences of the United States of America* **98**, 14144–14149.
73. E. Gelhaye, N. Rouhier, J. Gérard *et al.* (2004) A specific form of thioredoxin *h* occurs in plant mitochondria and regulates the alternative oxidase. *Proceedings of the National Academy of Sciences of the United States of America* **101**, 14545–14550.

74. J.-P. Reichheld, E. Meyer, M. Khafif, G. Bonnard and Y. Meyer (2005) AtNTRB is the major mitochondrial thioredoxin reductase in *Arabidopsis thaliana*. *FEBS Letters* **579**, 337–342.
75. Y. Balmer, W.H. Vensel, C.K. Tanaka *et al.* (2004) Thioredoxin links redox to the regulation of fundamental processes of plant mitochondria. *Proceedings of the National Academy of Sciences of the United States of America* **101**, 2642–2647.
76. Y. Balmer, A. Koller, G. del Val, W. Manieri, P. Schürmann and B.B. Buchanan (2003) Proteomics gives insight into the regulatory function of chloroplast thioredoxins. *Proceedings of the National Academy of Sciences of the United States of America* **100**, 370–375.
77. J.H. Wong, Y. Balmer, N. Cai *et al.* (2003) Unraveling thioredoxin-linked metabolic processes using proteomics. *FEBS Letters* **547**, 151–156.
78. I.M. Møller and A.G. Rasmusson (1998) The role of NADP in the mitochondrial matrix. *Trends in Plant Science* **3**, 21–27.
79. I.M. Møller and B.K. Kristensen (2004) Protein oxidation in plant mitochondria as a stress indicator. *Photochemical and Photobiological Sciences* **3**, 1–7.
80. L.J. Sweetlove, J.L. Heazlewood, V. Herald *et al.* (2002) The impact of oxidative stress on Arabidopsis mitochondria. *The Plant Journal* **32**, 891–904.
81. B.K. Kristensen, P. Askerlund, N.V. Bykova, H. Egsgaard and I.M. Møller (2004) Identification of oxidized proteins in the matrix of rice leaf mitochondria by immunoprecipitation and two-dimensional liquid chromatography-tandem mass spectrometry. *Phytochemistry* **65**, 1839–1851.
82. H. Schägger and K. Pfeiffer (2000) Supercomplexes in the respiratory chains of yeast and mammalian mitochondria. *The EMBO Journal* **19**, 1777–1783.
83. C. Bianchi, M.L. Genova, G.P. Castelli and G. Lenaz (2004) The mitochondrial respiratory chain is partially organized in a supercomplex assembly: kinetic evidence using flux control analysis. *Journal of Biological Chemistry* **279**, 36562–36569.
84. H. Schägger, R. de Coo, M.F. Bauer, S. Hofmann, C. Godinot and U. Brandt (2004) Significance of respirasomes for the assembly/stability of human respiratory chain Complex I. *Journal of Biological Chemistry* **279**, 36349–36353.
85. M. Zhang, E. Mileykovskaya and W. Dowhan (2002) Gluing the respiratory chain together: cardiolipin is required for supercomplex formation in the inner mitochondrial membrane. *Journal of Biological Chemistry* **277**, 43553–43556.
86. K. Katayama, I. Sakurai and H. Wada (2004) Identification of an *Arabidopsis thaliana* gene for cardiolipin synthase located in mitochondria. *FEBS Letters* **577**, 193–198.
87. H. Eubel, J. Heinemeyer and H.-P. Braun (2004) Identification and characterization of respirasomes in potato mitochondria. *Plant Physiology* **134**, 1450–1459.
88. F. Krause, N.H. Reifschneider, D. Vocke, H. Seelert, S. Rexroth and N.A. Dencher (2004) "Respirasome"-like supercomplexes in green leaf mitochondria of spinach. *Journal of Biological Chemistry* **279**, 48369–48375.
89. R. Navet, W. Jarmuszkiewicz, P. Douette, C.M. Sluse-Goffart and F.E. Sluse (2004) Mitochondrial respiratory chain complex patterns from *Acanthamoeba castellanii* and *Lycopersicon esculentum*: comparative analysis by BN-PAGE and evidence of protein-protein interaction between alternative oxidase and Complex III. *Journal of Bioenergetics and Biomembranes* **36**, 471–479.
90. Y.H. Ko, M. Delannoy, J. Hullihen, W. Chiu and P.L. Pedersen (2003) Mitochondrial ATP synthasome: cristae-enriched membranes and a multiwell detergent screening assay yield dispersed single complexes containing the ATP synthase and carriers for Pi and ADP/ATP. *Journal of Biological Chemistry* **278**, 12305–12309.
91. P.M. Haggie and A.S. Verkman (2002) Diffusion of tricarboxylic acid cycle enzymes in the mitochondrial matrix *in vivo*: evidence for restricted mobility of a multienzyme complex. *Journal of Biological Chemistry* **277**, 40782–40788.
92. P. Giegé, J.L. Heazlewood, U. Roessner-Tunali *et al.* (2003) Enzymes of glycolysis are functionally associated with the mitochondrion in Arabidopsis cells. *The Plant Cell* **15**, 2140–2151.
93. M.H.N. Hoefnagel, O.K. Atkin and J.T. Wiskich (1998) Interdependence between chloroplasts and mitochondria in the light and the dark. *Biochimica et Biophysica Acta* **1366**, 235–255.
94. P. Gardeström, A.U. Igamberdiev and A.S. Raghavendra (2002) Mitochondrial function in the light and significance to carbon-nitrogen interactions. In: *Photosynthetic Nitrogen*

Assimilation and Associated Carbon and Respiratory Metabolism, Advances in Photosynthesis and Respiration, Vol. 12. (eds C.H. Foyer and G. Noctor), Kluwer Academic, The Netherlands, pp. 151–172.
95. R. van Lis R and A. Atteia (2004) Control of mitochondrial function via photosynthetic redox signals. Photosynthesis Research **79**, 133–148.
96. C. Dutilleul, S. Driscoll, G. Cornic, R. De Paepe, C.H. Foyer and G. Noctor (2003) Functional mitochondrial Complex I is required by tobacco leaves for optimal photosynthetic performance in photorespiratory conditions and during transients. Plant Physiology **131**, 264–275.
97. M.A. Escobar, K.A. Franklin, Å.S. Svensson, M.G. Salter, G.C. Whitelam and A.G. Rasmusson (2004) Light regulation of the Arabidopsis respiratory chain: multiple discrete photoreceptor responses contribute to induction of Type II NAD(P)H dehydrogenase genes. Plant Physiology **136**, 2710–2721.
98. M. Taira, U. Valtersson, B. Burkhardt and R.A. Ludwig (2004) *Arabidopsis thaliana* GLN2-encoded glutamine synthetase is dual targeted to leaf mitochondria and chloroplasts. The Plant Cell **16**, 2048.
99. R. Bari, R. Kebeish, R. Kalamajka, T. Rademmacher and C. Peterhänsel (2004) A glycolate dehydrogenase in the mitochondria of *Arabidopsis thaliana*. Journal of Experimental Botany **55**, 623–630.
100. F. Carrari, A. Nunes-Nesi, Y. Gibon, A. Lytovchenko, M.E. Loureiro and A.R. Fernie (2003) Reduced expression of aconitase results in an enhanced rate of photosynthesis and marked shifts in carbon partitioning in illuminated leaves of wild species tomato. Plant Physiology **133**, 1322–1335.
101. A. Nunes-Nesi, F. Carrari, A. Lytovchenko *et al.* (2005) Enhanced photosynthetic performance and growth as a consequence of decreasing mitochondrial malate dehydrogenase activity in transgenic tomato plants. Plant Physiology **137**, 611–622.
102. E. Siendones, J.A. González-Reyes, C. Santos-Ocaña, P. Navas and F. Córdoba (1999) Biosynthesis of ascorbic acid in kidney bean. L-galactono-γ-lactone dehydrogenase is an intrinsic protein located at the mitochondrial inner membrane. Plant Physiology **120**, 907–912.
103. C.G. Bartoli, G.M. Pastori and C.H. Foyer (2000) Ascorbate biosynthesis in mitochondria is linked to the electron transport chain between complexes III and IV. Plant Physiology **123**, 335–343.
104. A.H. Millar, V. Mittova, G. Kiddle *et al.* (2003) Control of ascorbate synthesis by respiration and its implications for stress responses. Plant Physiology **133**, 443–447.
105. D.K. Shintani and J.B. Ohlrogge (1994) The characterization of a mitochondrial acyl carrier protein isoform isolated from *Arabidopsis thaliana*. Plant Physiology **104**, 1221–1229.
106. H. Wada, D. Shintani and J. Ohlrogge (1997) Why do mitochondria synthesize fatty acids? Evidence for involvement in lipoic acid production. Proceedings of the National Academy of Sciences of the United States of America **94**, 1591–1596.
107. R. Yasuno, P. von Wettstein-Knowles and H. Wada (2004) Identification and molecular characterization of the β-ketoacyl-[acyl carrier protein] synthase component of the Arabidopsis mitochondrial fatty acid synthase. Journal of Biological Chemistry **279**, 8242–8251.
108. J.G. Olsen, A.V. Rasmussen, P. von Wettstein-Knowles and A. Henriksen (2004) Structure of the mitochondrial β-ketoacyl-[acyl carrier protein] synthase from Arabidopsis and its role in fatty acid synthesis. FEBS Letters **577**, 170–174.
109. M. Focke, E. Gieringer, S. Schwan, L. Jänsch, S. Binder and H.-P. Braun (2003) Fatty acid biosynthesis in mitochondria of grasses: malonyl-coenzyme A is generated by a mitochondrial-localized acetyl-coenzyme A carboxylase. Plant Physiology **133**, 875–884.
110. K.L. Falk, R.H. Behal, C. Xiang and D.J. Oliver (1998) Metabolic bypass of the tricarboxylic acid cycle during lipid mobilization in germinating oilseeds. Plant Physiology **117**, 473–481.
111. P.J. Eastmond, V. Germain, P.R. Lange, J.H. Bryce, S.M. Smith and I.A. Graham (2000) Postgerminative growth and lipid catabolism in oilseeds lacking the glyoxylate cycle. Proceedings of the National Academy of Sciences of the United States of America **97**, 5669–5674.
112. J.E. Cornah, V. Germain, J.L. Ward, M.H. Beale and S.M. Smith (2004) Lipid utilization, gluconeogenesis, and seedling growth in Arabidopsis mutants lacking the glyoxylate cycle enzyme malate synthase. Journal of Biological Chemistry **279**, 42916–42923.

113. C. Salon, P. Raymond and A. Pradet (1988) Quantification of carbon fluxes through the tricarboxylic acid cycle in early germinating lettuce embryos. *Journal of Biological Chemistry* **263**, 12278–12287.
114. M. Dieuaide, R. Brouquisse, A. Pradet and P. Raymond (1992) Increased fatty acid β-oxidation after glucose starvation in maize root tips. *Plant Physiology* **99**, 595–600.
115. I.M. Møller (2001) Plant mitochondria and oxidative stress: electron transport, NADPH turnover, and metabolism of reactive oxygen species. *Annual Review of Plant Physiology and Plant Molecular Biology* **52**, 561–591.
116. C.G. Bartoli, F. Gómez, D.E. Martínez and J.J. Guiamet (2004) Mitochondria are the main target for oxidative damage in leaves of wheat (*Triticum aestivum* L.). *Journal of Experimental Botany* **55**, 1663–1669.
117. C. Bowler, L. Slooten, S. Vandenbranden *et al.* (1991) Manganese superoxide dismutase can reduce cellular damage mediated by oxygen radicals in transgenic plants. *The EMBO Journal* **10**, 1723–1732.
118. U. Basu, A.G. Good and G.J. Taylor (2001) Transgenic *Brassica napus* plants overexpressing aluminum-induced mitochondrial manganese superoxide dismutase cDNA are resistant to aluminum. *Plant, Cell and Environment* **24**, 1269–1278.
119. R. Mittler, S. Vanderauwera, M. Gollery and F. Van Breusegem (2004) Reactive oxygen gene network of plants. *Trends in Plant Science* **9**, 490–498.
120. A. Jiménez, J.A. Hernández, L.A. del Río and F. Sevilla (1997) Evidence for the presence of the ascorbate-glutathione cycle in mitochondria and peroxisomes of pea leaves. *Plant Physiology* **114**, 275–284.
121. O. Chew, J. Whelan and A.H. Millar (2003) Molecular definition of the ascorbate-glutathione cycle in Arabidopsis mitochondria reveals dual targeting of antioxidant defenses in plants. *Journal of Biological Chemistry* **278**, 46869–46877.
122. M. Rodriguez Milla, A. Maurer, A. Rodriguez Huete and J.P. Gustafson (2003) Glutathione peroxidase genes in Arabidopsis are ubiquitous and regulated by abiotic stress through diverse signalling pathways. *The Plant Journal* **36**, 602–615.
123. I. Finkemeier, M. Goodman, P. Lamkemeyer, A. Kandlbinder, L.J. Sweetlove and K.-J. Dietz (2005) The mitochondrial type II peroxiredoxin F is essential for redox homeostasis and root growth of *Arabidopsis thaliana* under stress. *Journal of Biological Chemistry* **280**, 12168–12180.
124. M. Zancani, C. Peresson, A. Biroccio *et al.* (2004) Evidence for the presence of ferritin in plant mitochondria. *European Journal of Biochemistry* **271**, 3657–3664.
125. S.-H. Jo, M.-K. Son, H.-J. Koh *et al.* (2001) Control of mitochondrial redox balance and cellular defense against oxidative damage by mitochondrial $NADP^+$-dependent isocitrate dehydrogenase. *Journal of Biological Chemistry* **276**, 16168–16176.
126. A. Bernacchia, A. Biondi, M.L. Genova and G. Lenaz (2004) The various sources of mitochondrial oxygen radicals: a minireview. *Toxicology Mechanisms and Methods* **14**, 25–30.
127. S.S. Korshunov, V.P. Skulachev and A.A. Starkov (1997) High protonic potential actuates a mechanism of production of reactive oxygen species in mitochondria. *FEBS Letters* **416**, 15–18.
128. D.P. Maxwell, Y. Wang and L. McIntosh (1999) The alternative oxidase lowers mitochondrial reactive oxygen production in plant cells. *Proceedings of the National Academy of Sciences of the United States of America* **96**, 8271–8276.
129. A. Camacho, R. Moreno-Sanchez and I. Bernal-Lugo (2004) Control of superoxide production in mitochondria from maize mesocotyls. *FEBS Letters* **570**, 52–56.
130. A.H. Millar and C.J. Leaver (2000) The cytotoxic lipid peroxidation product, 4-hydroxy-2-nonenal, specifically inhibits decarboxylating dehydrogenases in the matrix of plant mitochondria. *FEBS Letters* **481**, 117–121.
131. N.L. Taylor, D.A. Day and A.H. Millar (2002) Environmental stress causes oxidative damage to plant mitochondria leading to inhibition of glycine decarboxylase. *Journal of Biological Chemistry* **277**, 42663–42668.
132. M.J. Considine, M. Goodman, K.S. Echtay *et al.* (2003) Superoxide stimulates a proton leak in potato mitochondria that is related to the activity of uncoupling protein. *Journal of Biological Chemistry* **278**, 22298–22302.

133. A.M.O. Smith, G. Ratcliffe and L.J. Sweetlove (2004) Activation and function of mitochondrial uncoupling protein in plants. *Journal of Biological Chemistry* **279**, 51944–51952.
134. A.M. Winger, A.H. Millar and D.A. Day (2005) Sensitivity of plant mitochondrial terminal oxidases to the lipid peroxidation product 4-hydroxy-2-nonenal (HNE). *Biochemical Journal* **387**, 865–870.
135. B. Lee, H. Lee, L. Xiong and J.-K. Zhu (2002) A mitochondrial complex I defect impairs cold-regulated nuclear gene expression. *The Plant Cell* **14**, 1235–1251.
136. G. Noctor, C. Dutilleul, R. De Paepe and C.H. Foyer (2004) Use of mitochondrial electron transport mutants to evaluate the effects of redox state on photosynthesis, stress tolerance and the integration of carbon/nitrogen metabolism. *Journal of Experimental Botany* **55**, 49–57.
137. W.A. Snedden, N. Koutsia, G. Baum and H. Fromm (1996) Activation of a recombinant petunia glutamate decarboxylase by calcium/calmodulin or by a monoclonal antibody which recognizes the calmodulin binding domain. *Journal of Biological Chemistry* **271**, 4148–4153.
138. K.B. Busch and H. Fromm (1999) Plant succinic semialdehyde dehydrogenase: cloning, purification, localization in mitochondria, and regulation by adenine nucleotides. *Plant Physiology* **121**, 589–597.
139. N. Bouché, A. Fait, D. Bouchez, S.G. Møller and H. Fromm (2003) Mitochondrial succinic-semialdehyde dehydrogenase of the γ-aminobutyrate shunt is required to restrict levels of reactive oxygen intermediates in plants. *Proceedings of the National Academy of Sciences of the United States of America* **100**, 6843–6848.
140. A. Fait, A. Yellin and H. Fromm (2005) GABA shunt deficiencies and accumulation of reactive oxygen intermediates: insight from Arabidopsis mutants. *FEBS Letters* **579**, 415–420.
141. K.E. Breitkreuz, W.L. Allan, O.R. Van Cauwenberghe *et al.* (2003) A novel γ-hydroxybutyrate dehydrogenase: identification and expression of an Arabidopsis cDNA and potential role under oxygen deficiency. *Journal of Biological Chemistry* **278**, 41552–41556.
142. P. Baldet, C. Alban, S. Axiotis and R. Douce, (1992) Characterization of biotin and 3-methyl-crotonyl-coenzyme A carboxylase in higher plant mitochondria. *Plant Physiology* **99**, 450–455.
143. M.D. Anderson, P. Che, J. Song, B.J. Nikolau and E.S. Wurtele (1998) 3-methylcrotonyl-coenzyme A carboxylase is a component of the mitochondrial leucine catabolic pathway in plants. *Plant Physiology* **118**, 1127–1138.
144. Y. Fujiki, T. Sato, M. Ito and A. Watanabe (2000) Isolation and characterization of cDNA clones for the E1β and E2 subunits of the branched-chain α-ketoacid dehydrogenase complex in Arabidopsis. *Journal of Biological Chemistry* **275**, 6007–6013.
145. N.L. Taylor, J.L. Heazlewood, D.A. Day and A.H. Millar (2004) Lipoic acid-dependent oxidative catabolism of α-keto acids in mitochondria provides evidence for branched-chain amino acid catabolism in Arabidopsis. *Plant Physiology* **134**, 838–848.
146. Y. Fujiki, M. Ito, I. Nishida and A. Watanabe (2001) Leucine and its keto acid enhance the coordinated expression of genes for branched-chain amino acid catabolism in Arabidopsis under sugar starvation. *FEBS Letters* **499**, 161–165.
147. Y. Fujiki, M. Ito, T. Itoh, I. Nishida and A. Watanabe (2002) Activation of the promoters of Arabidopsis genes for the branched-chain α-keto acid dehydrogenase complex in transgenic tobacco BY-2 cells under sugar starvation. *Plant and Cell Physiology* **43**, 275–280.
148. C. Hourton-Cabassa, F. Ambard-Bretteville, F. Moreau, J.D. de Virville, R. Rémy and C. Colas des Francs-Small (1998) Stress induction of mitochondrial formate dehydrogenase in potato leaves. *Plant Physiology* **116**, 627–635.
149. A.U. Igamberdiev, N.V. Bykova and L.A. Kleczkowski (1999) Origins and metabolism of formate in higher plants. *Plant Physiology and Biochemistry* **37**, 503–513.
150. F. Ambard-Bretteville, C. Sorin, F. Rébeille, C. Hourton-Cabassa and C. Colas des Francs-Small (2003) Repression of formate dehydrogenase in *Solanum tuberosum* increases steady-state levels of formate and accelerates the accumulation of proline in response to osmotic stress. *Plant Molecular Biology* **52**, 1153–1168.
151. R.G.L. op den Camp and C. Kuhlemeier (1997) Aldehyde dehydrogenase in tobacco pollen. *Plant Molecular Biology* **35**, 355–365.

152. F. Liu and P.S. Schnable (2002) Functional specialization of maize mitochondrial aldehyde dehydrogenases. *Plant Physiology* **130**, 1657–1674.
153. D.S. Skibbe, F. Liu, T.-J. Wen *et al.* (2002) Characterization of the aldehyde dehydrogenase gene families of *Zea mays* and Arabidopsis. *Plant Molecular Biology* **48**, 751–764.
154. H. Tsuji, N. Meguro, Y. Suzuki, N. Tsutsumi, A. Hirai and M. Nakazono (2003) Induction of mitochondrial aldehyde dehydrogenase by submergence facilitates oxidation of acetaldehyde during re-aeration in rice. *FEBS Letters* **546**, 369–373.
155. J. Bardel, M. Louwagie, M. Jaquinod *et al.* (2002) A survey of the plant mitochondrial proteome in relation to development. *Proteomics* **2**, 880–898.
156. F. Liu, X. Cui, H.T. Horner, H. Weiner and P.S. Schnable (2001) Mitochondrial aldehyde dehydrogenase activity is required for male fertility in maize. *The Plant Cell* **13**, 1063–1078.
157. S. Mellema, W. Eichenberger, A. Rawyler, M. Suter, M. Tadege and C. Kuhlemeier (2002) The ethanolic fermentation pathway supports respiration and lipid biosynthesis in tobacco pollen. *The Plant Journal* **30**, 329–336.
158. W. Shen, Y. Wei, M. Dauk, Z. Zheng and J. Zou (2003) Identification of a mitochondrial glycerol-3-phosphate dehydrogenase from *Arabidopsis thaliana*: evidence for a mitochondrial glycerol-3-phosphate shuttle in plants. *FEBS Letters* **536**, 92–96.
159. L.J. Sweetlove, B. Mowday, H.F. Hebestreit, C.J. Leaver and A.H. Millar (2001) Nucleoside diphosphate kinase III is localized to the inter-membrane space in plant mitochondria. *FEBS Letters* **508**, 272–276.
160. C. Knorpp, M. Johansson and A.-M. Baird (2003) Plant mitochondrial nucleoside diphosphate kinase is attached to the membrane through interaction with the adenine nucleotide translocator. *FEBS Letters* **555**, 363–366.
161. M.L. Escobar Galvis, S. Marttila, G. Håkansson, J. Forsberg and C. Knorpp (2001) Heat stress response in pea involves interaction of mitochondrial nucleoside diphosphate kinase with a novel 86-kilodalton protein. *Plant Physiology* **126**, 69–77.
162. P.R. Ryan, E. Delhaize and D.L. Jones (2001) Function and mechanism of organic anion exudation from plant roots. *Annual Review of Plant Physiology and Plant Molecular Biology* **52**, 527–560.
163. H. Koyama, A. Kawamura, T. Kihara, T. Hara, E. Takita and D. Shibata (2000) Overexpression of mitochondrial citrate synthase in *Arabidopsis thaliana* improved growth on a phosphorus-limited soil. *Plant and Cell Physiology* **41**, 1030–1037.
164. V.M. Anoop, U. Basu, M.T. McCammmon, L. McAlister-Henn and G.J. Taylor (2003) Modulation of citrate metabolism alters aluminum tolerance in yeast and transgenic canola overexpressing a mitochondrial citrate synthase. *Plant Physiology* **132**, 2205–2217.
165. M. Kollmeier, P. Dietrich, C.S. Dauer, W.J. Horst and R. Hedrich (2001) Aluminum activates a citrate-permeable anion channel in the aluminum-sensitive zone of the maize root apex: a comparison between an aluminum-sensitive and an aluminum-resistant cultivar. *Plant Physiology* **126**, 397–410.
166. E. Diatloff, M. Roberts, D. Sanders and S.K. Roberts (2004) Characterization of anion channels in the plasma membrane of Arabidopsis epidermal root cells and the identification of a citrate-permeable channel induced by phosphate starvation. *Plant Physiology* **136**, 4136–4149.

12 Photosynthetic carbon–nitrogen interactions: modelling inter-pathway control and signalling

Christine H. Foyer, Graham Noctor and Paul Verrier

12.1 Introduction

Plants have four basic requirements for growth and biomass production. These are: light, carbon dioxide, water and mineral elements. Among the last, the requirement for nitrogen (N) is the most important, as amino acids are the building blocks of protein, chlorophyll, nucleotides and numerous other metabolites and cell components. The fact that N is required by plants in greater quantities than any other mineral element largely reflects the abundant N invested in the photosynthetic apparatus, which uses carbon dioxide, water and inorganic N to produce sugars, organic acids and amino acids, the basic building blocks of biomass accumulation. The availability of N is thus a significant determinant of both photosynthetic capacity and crop yield. Plants can absorb and assimilate various forms of N, though in many forms of agriculture the high amounts of added nitrate and the presence of nitrifying bacteria mean that nitrate is the principal form available to the roots of crop plants.

Nitrogen is acquired from the soil through uptake of nitrate and ammonium catalysed by a range of low and high affinity transport systems. Mycorrhizal fungi associated with roots can aid in the uptake process, particularly in N-poor soils. Presumably the fungi benefit from reduced carbon present in the root exudates. A more extensive elaboration of the mutually beneficial exchange of carbon for N between organisms is the symbiotic association between the roots of leguminous plants and soil rhizobia that results in the development of specific organs, called nodules, whose primary function is N fixation. The products of symbiotic N fixation (amides in temperate legumes and ureides in tropical legumes) are exported from the nodules to the rest of the plant where they are incorporated into essential macromolecules such as proteins. The elaborate inter-organism exchange of C and N metabolites that facilitates this symbiosis is not covered in this short chapter, but it follows the basic principles of supply and demand that underpin inter-pathway flux controls and metabolite cross-talk in the orchestration of gene expression. This ensures an appropriate plant C/N balance relative to developmental and metabolic controls. Substrates, metabolites and products are important in the control of the expression and activity of key enzymes that allows coordination of C and N assimilation, as they act as 'signals' of C/N status. Moreover, these signalling metabolites have a crucial role in the control of whole plant growth and development. For example, nitrate is not only a powerful signal regulating root and shoot growth [1] but its addition to nodulated legumes causes rapid nodule senescence and prevents the formation of new nodules [2]. Therefore, N assimilation is integrated with photosynthetic and respiratory C metabolism between tissues and organs as well as at the intracellular and intercellular levels. A complex network of controls, involving signals emanating from nitrate, carbohydrates, and key metabolites such as Gln, Glu and Asp all integrated with hormone signalling, permits the plant to tailor the capacity for N assimilation and C metabolism to nutrient availability and requirements.

12.2 Integration of C and N metabolism in leaves

Nitrate and ammonium are assimilated in both roots and shoots, with the relative contribution of each tissue being highly species dependent. In roots, imported sucrose provides energy and C skeletons for N assimilation. Alternatively, nitrate and ammonium are transported to the leaves via the xylem and incorporated into amino acids largely through photosynthetic N assimilation. A key point concerning N assimilation is the requirement for reductant, in the same way as CO_2 fixation or the quantitatively less important assimilation of sulfate. This distinguishes N use from that of other potentially limiting elements that are either not covalently incorporated into organic forms (e.g., potassium) or incorporated in an oxidised form (e.g., phosphorus). Reductant is particularly important when N is available in the highly oxidised form, that is nitrate. Accordingly, leaf nitrate assimilation is highly dependent on light, flux rates in the dark being much lower than those in the light [3, 4]. Although leaf nitrate reduction rate is usually only a fraction of the rate of C fixation, the assimilation of nitrate into amino acids has a high requirement for reductant (10 mol electrons per mole: Figure 12.1a). Nitrate assimilation rates are variable and depend on many developmental and environmental factors, particularly leaf age, with young leaves typically displaying the highest N-assimilation rates. In these conditions, N assimilation can represent a significant sink for photosynthetic energy [5]. Indeed, increases in photosynthetic CO_2 fixation with increasing irradiance are accompanied by enhanced N uptake [6], while N assimilation is decreased at low CO_2 [7] or during depletion of carbohydrates in extended darkness [8]. Since the leaf is the predominant site of N assimilation in many crop species, the following discussion concerns the factors that coordinate N assimilation with C metabolism in leaves, with a view to modelling this interaction.

The high leaf capacity of N assimilation is associated with high activities of primary N-assimilation pathway enzymes such as nitrate reductase (NR [1]). Most of the reductant required to drive N assimilation comes directly from the photosynthetic electron transport chain via reduced ferredoxin (Figure 12.1a). NADH required for NR activity is supplied either by the photosynthetic electron transport chain, through the operation of redox shuttles, or via the respiratory oxidation of fixed C (Figure 12.1a). As in all other tissues, reductive N assimilation in leaves is integrated with respiratory carbon oxidation (Figure 12.1b). Organic acids produced from carbohydrates or sugar-phosphates are not only required as amino group acceptors in amino acid synthesis but their production also serves to balance

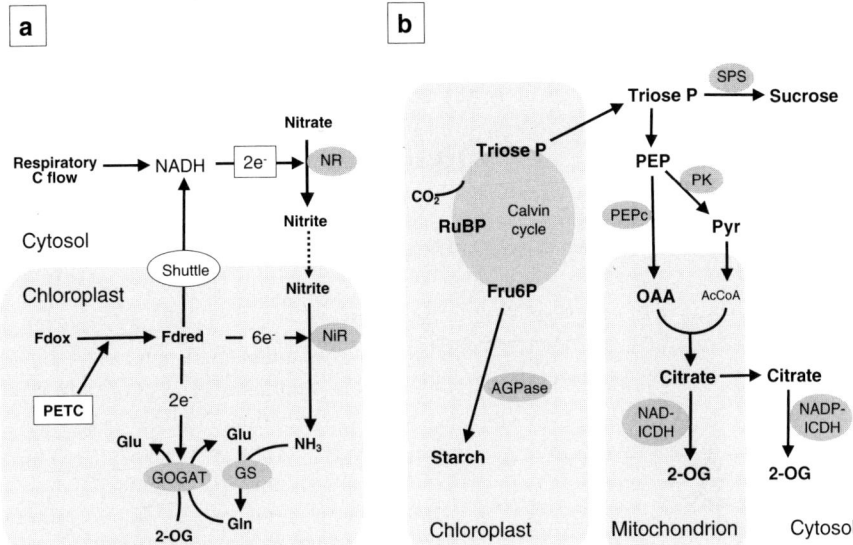

Figure 12.1 Reductant demands of leaf nitrate reduction (a) and simplified scheme showing interactions between carbohydrate synthesis and respiratory production of organic acids for amino acid synthesis (b). Key enzymes are shown in ellipses. Arrows depict transport and/or one or more enzymatic steps. The abbreviations are: AcCoA, acetyl CoA; AGPase, ADP-glucose pyrophosphorylase; Fdox, oxidised ferredoxin; Fdred, reduced ferredoxin; Fru6P, fructose 6-phosphate; Gln, glutamine; Glu, glutamate; GOGAT, glutamine:2-oxoglutarate aminotransferase; GS, glutamine synthetase; ICDH, isocitrate dehydrogenase; NiR, nitrite reductase; NR, nitrate reductase; 2-OG, 2-oxoglutarate; PEPc, phosphoenolpyruvate carboxylase; PETC, photosynthetic electron transport chain; PK, pyruvate kinase; RuBP, ribulose 1,5-bisphosphate; SPS, sucrose phosphate synthase.

cellular pH changes during the nitrate reduction process. Evidence highlighting the close functional relationship between respiratory C flow and primary N assimilation comes from studies where a N source is re-supplied to N-starved plant cells, causing a rapid marked stimulation of respiratory C flow [9–11]. The impact of N supply on overall cellular C metabolism is most marked in algae, where C skeletons are recruited from other sources to enable rapid uptake of this crucial resource that often limits growth in aquatic habits. In unicellular algae, switching from limiting to abundant N causes a marked inhibition of photosynthesis and concomitant stimulation of respiration [12]. In multi-cellular plants, changes are less marked, no doubt reflecting subtler, more complex control [13]. Nevertheless, the pathways of C metabolism are highly sensitive to N availability, acting to ensure sufficient supply of organic acids for amino acid synthesis. The partitioning of assimilated C between synthesis of organic acids, starch and sucrose is markedly affected by N availability. The extensive regulation that occurs in response to N involves both transcriptional and post-translational controls [14, 15]. Starch synthesis, for example, may be severely restricted by down-regulation of ADP-glucose pyrophosphorylase transcripts in the presence of nitrate.

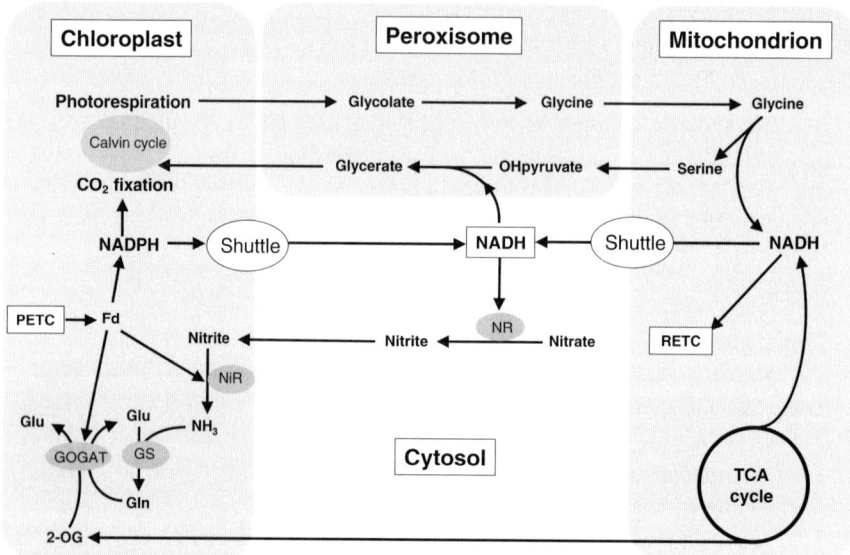

Figure 12.2 Intracellular reductant cycling during photorespiration and nitrate reduction. Key enzymes are shown in ellipses. Arrows depict transport and/or one or more enzymatic steps. Redox shuttles from the chloroplast and mitochondrion occur by metabolite exchange, notably triose phosphate for 3-phosphoglycerate or malate for oxaloacetate. For reasons of simplicity, intercompartmental transfer of ammonia, 2-OG and Glu in the photorespiratory nitrogen cycle is not shown. TCA, tricarboxylic acid; RETC, respiratory electron transport chain. Other abbreviations are as in Figure 12.1.

In addition to integration with respiratory C flow, N assimilation in the leaves of C_3 species is also intimately associated with photorespiratory C flow, an interaction that further complicates the photosynthetic C and N interaction (Figure 12.2). Under most conditions, the photorespiratory cycle is much more rapid than primary nitrate reduction and can influence N assimilation in at least two ways. First, ammonia liberated during the conversion of Gly to Ser in leaves is recycled through Gln and Glu [16], and this ammonia recycling probably occurs through the same isoforms of glutamine synthetase (GS) and glutamine:2-oxoglutarate aminotransferase (GOGAT) as in primary ammonia assimilation [17]. Second, photorespiration involves reductant cycling: NADH is produced by Gly oxidation in the mitochondria and required (in equal amounts) in peroxisomal glycerate synthesis. The extent to which NADH produced in Gly oxidation is reoxidised by the mitochondrial electron transport chain and/or in glycerate formation may be key in determining the cytosolic NADH concentration and thus influencing the rate of nitrate reduction (Figure 12.2). A further possible complication that remains largely uncharacterised is the role of γ-carbonic anhydrase in plant mitochondria. Mitochondrial γ-carbonic anhydrase is localised on the NADH dehydrogenase, complex I [18]. It is tempting to suggest that this enzyme might function to recover at least some of the CO_2 liberated as a result of the photorespiratory Gly to Ser conversion in the mitochondria. The interactions between N assimilation, photosynthesis and respiration discussed and modelled below are based on the following central concepts.

1. *Biomass production is fundamentally determined by the balance between photosynthesis and respiration.* A large portion of the carbon assimilated during photosynthesis is used in maintenance respiration for the upkeep of existing cellular components and in growth respiration to produce new components. A considerable proportion of growth respiration involves oxidation of assimilated reduced carbon in order to produce the organic acids that form the carbon skeletons required for N assimilation (Figure 12.1b).
2. *Nitrogen assimilation is activated and stimulated by light and is exquisitely sensitive to the availability of reductant.* Photosynthetic C and N assimilation are coordinately activated in the light and deactivated in the dark. Light causes post-translational modulation of the activity of NR and also provides the reducing power necessary for the reductive incorporation of nitrate into amino groups. The overall effect of light on organic acid synthesis is uncertain, since respiration may be partially inhibited in the light. Whether there is preferential inhibition of reductant-generating dissimilatory respiration over assimilatory (primarily precursor-generating) pathways remains unclear, though light activation of the C_3 isoform of phospho*enol*pyruvate (PEP) carboxylase (PEPc) [14] and down-regulation of pyruvate dehydrogenase [19] strongly suggest that illumination affects the balance between these two processes. Allosteric control by key metabolites is also important in regulating the relative activities of PEPc and cytosolic pyruvate kinase (PK), as anaplerosis demands that these two enzymes use PEP in a coordinated fashion [20–22].
3. *Amino acids are the central hub around which other processes such as nitrogen assimilation, associated carbon metabolism, photorespiration and export of organic nitrogen are modulated.* Specific major amino acids are modulated differentially by photorespiration and nitrogen assimilation [23]. As discussed further below, minor amino acids can show marked diurnal rhythms and their contents fluctuate in a coordinated manner.

12.3 Control of nitrate assimilation rates and the C/N interaction

Much attention has focused on signalled changes in the capacity of key enzymes in the coordination of C/N metabolism, and this must be a key factor in achieving appropriate C/N balance during development or as a function of nutritional status. Considerable attention has also been paid to the importance of post-translational modification of key enzymes, as well as allosteric control of their activities by metabolite activators and inhibitors. Less attention has focused on energetic factors. Here, we consider all three aspects.

Leaf amino acid contents increase as the N supply is increased [1, 13, 24]. Short-term effects probably mainly reflect increased substrate supply and/or enzyme activation while longer term changes are also due to modified expression of enzymes such as NR [1]. Short term changes notably include (1) post-translational modification of the activities of enzymes such as PEPc and NR through protein phosphorylation [14, 25] and (2) allosteric modulation by metabolite effectors [11, 21, 22]. Longer term changes in gene expression result from the monitoring of C and N metabolites that modulate the transcriptome according to nutritional and metabolic status [26, 27].

A key question to consider is the importance of NR capacity and/or NR phosphorylation state in setting the rate of nitrate reduction. Unlike PEPc and sucrose phosphate synthase (SPS), the phosphorylation state of the NR protein (and hence activation state) does not respond directly to nitrate availability [28, 29]. N-availability-dependent increases in amino acid contents are highly specific in the short term [23], the immediate enhancement of leaf amino acids upon feeding N being almost entirely due to Gln accumulation [30]. Similar effects on Gln have been found in tobacco mutants and transformants with varying NR activities, in which Gln showed an almost linear increase with increasing NR capacity [1]. Leaf Gln contents, or the Gln:Glu ratio, is therefore a good indicator of the balance between the capacity for C and N assimilation, which is low when C:N is high and vice versa [23]. Other metabolite ratios, such as Asp:Glu, may also be markers for C/N partitioning [22], though in leaves from C_3 species, Asp contents appear to be strongly influenced by photorespiratory flux [31]. Despite extensive study the factors that ultimately limit nitrate reduction in leaves are far from resolved [30, 32, 33].

The C/N interaction takes place within a context of energy use and production involving cooperation between different cellular compartments [13, 23, 34–36]. The notion that NADH availability may be important in limiting N assimilation is an old one and receives support from the likely very low cytosolic NADH concentration, which may be several-fold lower than the K_MNADH of NR. The ability of chloroplasts or mitochondria to deliver NAD(P)H to the cytosol could be crucial in determining the rate of N assimilation [37, 38]. Low photosynthesis rates limit NR activity through decreased reductant availability, as well as through post-translational inactivation [33, 39]. Nitrate reduction is significantly stimulated by anoxia and is favoured in conditions in which the complex redox cycling of photorespiration is active [40]. Moreover, because amino acid synthesis requires simultaneous C oxidation and nitrate reduction, any perturbation of redox cycling could radically alter C/N partitioning through concerted and opposing effects on the two processes. Organic acid synthesis may also be subject to redox modulation in the C/N interaction because the production of compounds such as 2-oxoglutarate (2-OG) requires oxidation through respiratory pathways involving the cytosol and mitochondria (Figure 12.1b; [41, 23]). Analysis of the C/N interaction in the *Nicotiana sylvestris* CMS mutant, which lacks mitochondrial complex I and respires through alternative respiratory chain dehydrogenases [42, 43], has shown that the modulation of respiratory electron transport chain impacts strongly on the C/N network and that the observed effects on metabolite profiles are largely independent of changes in enzyme capacities [44, 45]. Nitrate assimilation rates were enhanced in the mutant, especially in the dark, and N-rich amino acids accumulated in CMS leaves, suggesting that modulation of mitochondrial metabolism can drive the leaf towards a N-rich state through enhanced availability of NADH [45].

The GS-GOGAT cycle functions both in primary ammonia assimilation and in the reincorporation of NH_3 released by Gly decarboxylation during photorespiration [16, 46]. Amino groups are subsequently transferred beyond the GS/GOGAT cycle, predominantly via Glu, to other amino acids such as Asp and Ala. Leaf amino acid contents are determined by a complex interplay of factors. In addition to photorespiration, nitrate and/or ammonia availability, NR capacity and NADH availability, the flux of C into amino acid pools, either from photorespiration (via glycolate) or through glycolysis and the tricarboxylic acid cycle is also important. The effects of primary N assimilation and photorespiration on leaf amino acids are tightly linked [47, 48]. Gln, Gly and Ser all accumulated throughout the photoperiod in tobacco plants growing at constant irradiance [49]. Exchange between amino acid pools via transamination reactions and utilisation of both C and N metabolites for synthesis of end-products is also important in control. Other impacts could include amino acid catabolism, protein degradation and non-photorespiratory ammonia cycling via enzymes such as phenylalanine ammonia lyase. Finally, as end-products, amino acids are exported and imported from leaves at different stages in development.

12.4 Pathway coordination and the C/N signal transduction network

The above discussion has emphasised the crucial importance of N assimilation in plant metabolism. This macronutrient interacts in an integrated manner with other resources, particularly C and other nutrients in respect to light, circadian rhythms, etc. The coordination of C and N metabolism is controlled by a repertoire of signals, including nitrate, ammonia, sugars, amino acids, particularly Gln, Glu and Asp, and organic acids such as 2-OG and malate [50–52]. As with the orchestration of other complex physiological processes, the transduction of these signals from perception to response involves cross-talk within a complex network incorporating both intra- and extra-cellular communication [53, 54]. Moreover, signalling includes effects occurring both locally (e.g., leaf nitrate contents control C partitioning [15]) and systemically (e.g., leaf nitrate contents modulate shoot–root allocation [1, 52]). Long distance signalling of information concerning light and CO_2 occurs between mature and developing leaves [55] and it is therefore logical to assume that similar information on nitrate and N status passes between leaves of different stages to modulate their physiology and development. Sucrose, starch, nitrate and amino acids are highest in the youngest leaves and decline with leaf age, while glucose, fructose and ammonium increase with leaf age and are predominant in senescing leaves, suggesting that gradients in these metabolites might provide metabolic information that modulates gene expression relative to local and systemic leaf and organ development.

In recent years, the potential role of nitrate reduction, particularly with regard to the different forms of NR, in the production of another systemic signal, nitric oxide (NO), has received much attention because of the potential roles of NO as a signalling molecule [56]. Nitrate reduction in conditions of low oxygen has been considered to form an alternative respiratory pathway aiding cell survival under anoxia [57]. It is interesting that in this regard nitrate is viewed as an intermediate electron acceptor, assisting in the maintenance of cellular redox and energy status [57].

Sugars elicit transcriptional and post-translational controls that limit the rate of nitrate assimilation, amino acid metabolism and photosynthesis. Because coordinated rates of nitrate reduction, CO_2 fixation and carbon partitioning are essential to the appropriate partitioning of assimilate for starch, sucrose and amino acid production, such control has a crucial function in adjustment of leaf activities of enzymes such as NR, SPS, ADP-glucose pyrophosphorylase, cytosolic PK and PEPc [25, 58]. Hexose and sucrose-specific mechanisms control gene expression [59]. Moreover, complex signalling networks link sugar (particularly glucose) signalling networks to plant stress perception hormones, particularly ethylene and abscisic acid (ABA; [60, 61]. While there remains some resistance to the concept of direct molecular overlap and connection between sugar and ABA-signalling pathways [59], there is strong genetic evidence in support of an integration of sugar and ethylene-signalling pathways [60]. Considerable evidence now shows extensive cross-talk not only between the sugar and N-signalling pathways but also those of ethylene and ABA [61, 62]. Indeed, genes involved in ABA synthesis and signalling [63, 64] and also in ethylene perception [65] are critical for sugar and nitrate signalling. In recent years, micro-array technologies have revealed that the expression levels of a large number of genes are changed in response to alterations in nitrate and sugar availability [26, 66, 67]. Genes associated with N assimilation are generally up-regulated when C is abundant and down-regulated when C skeletons are limiting or organic N is abundant [51].

The nitrate assimilation pathway has been intensively investigated and the expression of most of the genes have been shown to be coordinately controlled with respect to N source and intracellular amounts of amino acids, light, hormones and carbon status [48, 53]. However, there is still relatively little information available on regulatory genes and only the *Nit2* gene from the unicellular green alga *Chlamydomonas reinhardtii*, which is repressed by ammonium, has thus far been shown to have a major positive regulatory effect on nitrate assimilation, allowing functional genomics studies to be undertaken [68].

In addition to tight control over the capacity for primary N assimilation at the gene and protein regulation levels, there may also be coordination of the different pathways of minor amino acid synthesis particularly with regard to C status [69, 70]. Unequivocal evidence for coordinated control of genes encoding enzymes of amino acid biosynthesis in plants is scarce [71, 72]. However, a protein kinase, AtGCN2, that is structurally and functionally related to GCN2, a component of the regulatory system responsible for general amino acid control in budding yeast, has been identified in *Arabidopsis thaliana* [73]. Such mechanisms could act to coordinate minor amino acid synthesis in response to C/N supply and demand. Metabolite data provide evidence for some form of general control of minor amino acid synthesis in plants. Leaf carbohydrate accumulation tends to correlate with increases in minor amino acids, and feeding sucrose also leads to a general increase in these compounds [58]. Despite considerable variability, minor amino acid, contents vary in a concerted fashion in wheat, potato and barley leaves, again suggesting that the different synthetic pathways are under some form of common control [69]. Leaf contents of several minor amino acids were increased in transformed poplar leaves that accumulated glutathione as a result of enhanced glutathione synthesis capacity in the chloroplast, where most minor amino acids are synthesised [74]. Recently, enzymes involved in the synthesis of several amino acid synthesis pathways in *C. reinhardtii* have been identified as thioredoxin targets *in vitro* [75]. It is therefore possible that post-translational control through enzyme thiol status contributes to general control of minor amino acid synthesis pathways, perhaps in response to light and/or carbohydrate status.

Bacteria coordinate the C/N metabolic interface through a protein designated PII, which is one of the most widespread and conserved regulatory proteins known. It sits at the interface of signal sensing and control of primary N assimilation through the GS/GOGAT cycle. PII-like proteins are present in plants [76, 77] but their precise roles remain unclear, particularly in relation to the roles of other regulatory components such as 14-3-3-proteins [78] (also see Chapter 5), Dof transcription factors, 'two-component regulatory systems' and 'multi-step His-Asp phosphorelay' systems, which all have putative roles in the sensing of N status (see [70]).

12.5 Modelling the C/N interaction

Given the complexity of the C/N interaction and the multi-layered nature of its regulation, modelling is a promising approach to understanding how the system is integrated and influenced by the environment. We have thus developed simple models with the aim of assessing and defining the limits of the importance of known mechanisms of C/N regulation in different conditions. First, we have used stoichiometric modelling to investigate energetic interactions between C and N assimilation. Reactions integral to C and N assimilation require and produce ATP and reductant at very different stoichiometries. Whereas C assimilation into sugar phosphate requires 4 electrons and 3 ATP, assimilation of nitrate into amino acids demands 10 electrons but only 1 ATP. N assimilation therefore requires 7.5 times more reductant per ATP than C assimilation. In addition, the respiratory flow associated with carbohydrate and sugar phosphate oxidation may produce reductant and ATP at widely different ratios. We therefore considered the ATP:NADPH requirement of C_3 photosynthesis with regard to contributions from nitrate assimilation and its associated respiratory activity [34]. Even though a relatively small part of overall redox exchange during active photosynthesis, NAD(P)H generated via carbon oxidation has the potential to generate abundant amounts of ATP outside the chloroplast, particularly in the mitochondria, enhancing cellular ATP:ADP ratios and favouring large oscillations in energy availability [79]. Potentially detrimental oscillations in cellular ATP:ADP ratios will result from even slight imbalances in the rate of generation and utilisation of ATP. In the absence of compensatory changes, an imbalance of only 1% in the generation and utilisation of ATP is predicted to cause marked perturbations in cellular ATP:ADP ratios [79]. Hence, regulation involving metabolic valves and damping mechanisms is crucial to prevent large oscillations in energy availability that will cause the photosynthetic C and N pathways to stall. Important processes are the ability of the chloroplast to produce ATP and NADPH at different ratios, the capacity of the mitochondria to convert NADH to ATP with different yields and the presence of transporters that facilitate rapid efficient inter-compartmental transfer of pyridine nucleotides and adenylates.

In a second approach, we explored how the energetic interaction between C and N metabolism will vary in importance as a function of leaf development and the nutritional status of the plant. Such variability demands plasticity of assimilate partitioning in leaves to accommodate large changes in supply and demand [5]. When N is available a significant proportion of photosynthetic energy must be devoted to N assimilation. However, developmental changes in nutritional demand and changing N availability make this proportion extremely variable [5]. Amino acid synthesis is often considered to a minor sink for carbon. Nevertheless, assimilation of one molecule of ammonia requires one molecule of 2-OG to form the product of the GS-GOGAT pathway, Glu (C/N = 5). Other C skeletons are then required as oxo-acid scaffolds that are converted into amino acids, chiefly by Glu-dependent transaminations. The main physiological fates of assimilated N are protein synthesis (in young expanding leaves) and amino acid export (mature or senescing leaves). The average C/N of the 20 protein amino acids is 4.35. Other nitrogenous products synthesised primarily in young leaves typically have higher C/N ratios (e.g., chlorophyll, C/N = 13.75). The C demand linked to N assimilation will be particularly high, therefore, in young tissues, because the rates of N assimilation and the C/N of the ultimate product are relatively high. Even in older tissues the demand for C could be appreciable. Taking N assimilation as 5% of the rate of net CO_2 fixation, about 20% of fixed C needs to be allocated to amino acid synthesis. These values suggest that considerable respiratory fluxes must operate with N assimilation, though associated gas exchange may not be very high. For 2-OG production through the simplified route shown in Figure 12.1b, the rate of respiratory CO_2 release would only be one-sixth of that occurring during full TCA cycle activity. Synthesis of other oxo-acids could involve even less CO_2 evolution (e.g., pyruvate formation for Ala synthesis). If net production of all protein amino acids occurs at N assimilation equal to 5% of net CO_2 fixation, the minimum required respiratory CO_2 release would probably be around 4% of net C fixed, i.e. negligible. The respiratory fluxes that occur during anaplerosis may therefore be difficult to estimate by gas exchange, and this problem is greatly exacerbated when estimates are to be made during photosynthesis.

A third aspect of the C/N interaction that we have explored using modelling is the relationship between photorespiratory flux and leaf contents of major amino acids [31]. As metabolomic and metabolite profiling methods become ever more powerful, a major limitation will involve the interpretation of changes in metabolite contents. Establishing how tissue contents of key C and N compounds change as a function of leaf status and environment is a key challenge to identifying reliable metabolite markers that provide meaningful information on physiological status. Amino acid profiling was combined with estimates of photorespiratory flux from measured CO_2 exchange by using established models of C_3 photosynthesis described in detail in Von Caemmerer [80]. The contents of certain amino acids (notably Gly, Ser, but also Ala and Asp) are highly influenced by photorespiration, whereas the initial products of ammonia incorporation (Glu and Gln) are much more weakly correlated with this process [31]. These data suggested that the key factor affecting metabolite ratios around the GS/GOGAT pathway is the relative supply of ammonia and 2-OG, and that minor imbalances in nitrate reduction (producing ammonia) and anaplerotic respiration (producing 2-OG) can markedly affect Gln/Glu ratios, even against the background of higher absolute flux of ammonia and 2-OG cycling in photorespiration. Our current modelling project, which we now discuss, has taken the GS/GOGAT pathway as its point of departure.

12.5.1 Construction of a tentative model to explore the sensitivities in the GS and GOGAT reactions

The GS-GOGAT pathway is at the crossroads of the C/N interaction and the response of this pathway to changes in incoming or outgoing fluxes will likely impact strongly on overall leaf metabolism and function. As part of a project aiming to simulate the complexity of the C/N during the C_3 photosynthetic process, we have produced a preliminary model of the key reactions of nitrate and ammonia assimilation. Various methodologies are available to construct such models and to explore the dynamics of the associated pathways. Hybrid Petri-Nets as used by the Cell Illustrator package provides a good visualisation of the system, but has a tendency to hide the details of the underlying processes. The biochemical systems theory (BST) approach ably described by Torres and Voit [81] utilises power-law representations of the reactions and basic mass action systems to formulate a set of differential equations representing the system of reactions under investigation. This is a powerful approach that enables, after some consideration, the parameters to be selected based on measured events and known kinetics of the reactions. In addition the BST approach is very amenable to being re-worked to add into the system further compounds, reactions and events, including inhibitors and feedback loops.

The system of reactions considered in our preliminary model of nitrate and ammonia assimilation is shown in Figure 12.3. In this simplistic model, the fate of products not directly utilised in the next step of the pathway is ignored as is the existence of substrates that, although essential to the operation of the pathway, are not of specific interest in the model. Although there are known activators and inhibitors of the enzymatic reactions (as discussed above), these are ignored in the present model and the system will be assumed to be in a near steady state. In addition the reactions can be assumed to be irreversible which is very close to reality. For purposes of simplicity, the present model assumes that the mitochondrial NAD-dependent isocitrate dehydrogenase is the sole producer of 2-OG from isocitrate (see discussion above).

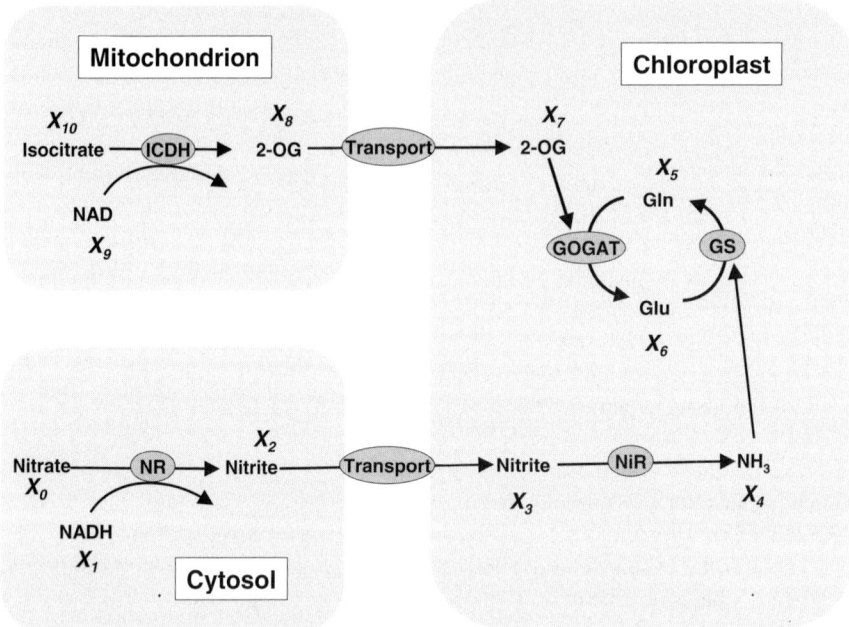

Figure 12.3 Compartmentalised simple schematic of nitrate and ammonia assimilation pathway as used for modelling. Each distinct compound is given a variable name (e.g., X_2 for nitrite in the cytosol). Many substrates and products are omitted in the modelling for simplicity. Although their dynamics could have an effect on the operation of the system, their presence would only add confusion to this simplified model. Mostly, missing products would simply be consumed. Changing reaction rates and enzyme kinetics will alter the overall dynamics of the system, changes in concentrations of the substrate pools at start-up give indications of the way the system behaves under differing availabilities.

Using the S-system approach to create a mathematical description of this pathway model, we are able to construct the following set of differential equations to describe the rate of change of any given compound over time:

$$X_0 = \text{constant}$$
$$X_1 = \text{constant}$$
$$X_2' = \alpha_2 X_0^{g2,3} X_1^{g2,1} - \beta_2 X_2^{h2,2}$$
$$X_3' = \alpha_3 X_2^{g3,2} - \beta_3 X_3^{h3,3}$$
$$X_4' = \alpha_4 X_3^{g4,3} - \beta_4 X_4^{h4,4}$$
$$X_5' = \alpha_5 X_4^{g5,4} X_6^{g5,6} - \beta_5 X_5^{h5,5}$$
$$X_6' = \alpha_6 X_7^{g6,7} X_5^{g6,5} - \beta_6 X_6^{h6,6}$$
$$X_7' = \alpha_7 X_8^{g7,8} - \beta_7 X_7^{h7,7}$$
$$X_8' = \alpha_8 X_{10}^{g8,10} X_9^{g8,9} - \beta_8 X_8^{h8,8}$$

$$X_9 = \text{constant}$$
$$X_{10} = \text{constant}$$

where α_i and β_i for $i = 0, \ldots, n$ are rate constants and gi,j and hi,j are kinetic constants which need to be determined in order to fully construct our simulation of the system.

The parameters $h2,2$ and $g3,2$ can be set to unity since the transport of nitrite is simply a percentage of the production in the cytosol. The transport efficiency is taken as 0.98 ($=\beta_2$) which allows for some small degradation of nitrite. Several transporters may be involved in transferring 2-OG from the mitochondrion into the chloroplast: for the purposes of the present model, 2-OG transport has been simplified by taking an overall transfer efficiency of 30%. Thus, assuming that transport is not inhibited, we can take the kinetic rate as unity ($=h8,8$) and also set $\beta_8 = 0.3$. Each of the above equations can also be re-written as a flux balance of the form $X' = V^+ - V^-$ where V^+ is the flux producing X and V^- represents the loss of X. In order that the overall mass is conserved in the system, we can see that $V_7^+ = V_8^-$ which implies that $g7,8 = h8,8$ and $\alpha_7 = \beta_8$. Similarly, $\alpha_3 = \beta_2$ and $g3,2 = h2,2$. Assuming that we have a steady state system, we can assume that $h3,3 \approx K_M/(K_M + S)$ where K_M is the Michaelis–Menton constant and S is the substrate concentration (mM). Flux balance enables us to evaluate similarly other h and g parameters in a crude manner from the steady state, but making allowance for the fact that the true evaluation of g requires a more detailed approach. However, for our purposes, the steady state approach will enable a simulation to start to iterate to a steady state. Table 12.1 shows the starting values used in the model. These are considered to be realistic approximations of likely steady state substrate concentrations *in vivo*. For many of the enzymes K_M values can be obtained from literature results or from the BRENDA database of enzymatic reactions (http://www.brenda.uni-koeln.de/), and

Table 12.1 Estimated *in vivo* substrate concentrations used as starting values in the model

Compound	Compartment	Concentration (mM)[a]
Nitrate	Cytosol	2–5
Nitrite	Chloroplast, cytosol	0.05–1
Ammonia	Chloroplast, mitochondria	1–2
Glutamine	Chloroplast	10–30
Glutamate	Choroplast, peroxisome	10–30
2-OG	Mitochondria, chloroplast, peroxisome	0.1–0.2
Gly	Mitochondria, peroxisome	5–10
Ser	Mitochondria, peroxisome	5–10
Glyoxylate	Peroxisome	1
Isocitrate	Mitochondria	0.1–0.2
NAD^+	Cytosol	1
NADH	Cytosol	0.001
NAD^+	Mitochondria	0.6
NADH	Mitochondria	0.2

[a] References: [82–84] or estimates based on whole tissue concentrations.

those used to calculate the h and g parameters are shown in Table 12.2. The remaining rate constants were calculated using the enzyme activities, but also taking some account of the effect of missing substrates and products in the determination of the rate. Noting that $V_j^+ = \alpha X_i^{gj,i} X_k^{gj,k}$ we can use the likely relative enzyme activities (Table 12.3) to determine α for the steady state concentrations. Using these values and adjusting (by an informed guess), for the 'missing' products and substrates, we can arrive at a first approximation to the remaining α constants. The values for the remaining β constants can be obtained through flux balance calculations.

By this approach, the metabolite pools shown in Figure 12.3 can be modelled by completing our system of ordinary differential equations as follows:

$$X_0 = 3$$
$$X_1 = 0.001$$
$$X_2' = 10.6 X_0^{0.003} X_1^{0.75} - 0.98 X_2$$
$$X_3' = 0.98 X_2 - 0.68 X_3^{0.8}$$
$$X_4' = 0.68 X_3^{0.8} - 0.11 X_4^{0.23}$$
$$X_5' = 0.15 X_4^{0.23} X_6^{0.25} - 0.23 X_5^{0.024}$$
$$X_6' = 0.45 X_7^{0.73} X_5^{0.024} - 0.15 X_6^{0.25}$$
$$X_7' = 0.3 X_8 - 0.087 X_7^{0.73}$$
$$X_8' = 0.22 X_{10}^{0.57} X_9^{0.35} - 0.3 X_8$$
$$X_9 = 0.6$$
$$X_{10} = 0.15$$

Table 12.2 Enzyme K_M values used in the models

K_M (mM)	Substrate	Enzyme[a]	Species	Reference
0.003	NADH	NR	–	[85]
0.01	Nitrate	NR	Arabidopsis	[86]
0.28	Nitrite	NiR	Spinach	[87]
0.22	Ammonia	GS2	Lupin	[88]
6.7	Glu	GS2	Spinach	[89]
0.5	Gln	GOGAT	Tomato	[90]
0.33	2-OG	GOGAT	Rice	[91]
0.33	NAD	ICDH	Pea	[92]
0.2	Isocitrate	ICDH	Pea	[84]
2	Ser	SGAT	Spinach	[93]
0.15	Glyoxylate	SGAT	Spinach	[93]
2	Glu	GGAT	Spinach	[93]
0.15	Glyoxylate	GGAT	Spinach	[93]

[a] GGAT, glutamate:glyoxylate aminotransferase; GOGAT, glutamate synthase; GS, glutamine synthetase; ICDH, isocitrate dehydrogenase; NiR, nitrite reductase; NR, nitrate reductase; SGAT, serine:glyoxylate aminotransferase.

Table 12.3 Enzyme capacities used in the models

Enzyme	Activity (μmol g^{-1}FW h^{-1})[a]
ICDH	10–50
NR	10–50
NiR	200
GS2	400
GOGAT	400
Serine:glyoxylate aminotransferase	400
Glutamate:glyoxylate aminotransferase	400
Glycine decarboxylase	400
Serine hydroxymethyl transferase	400

[a]Values are approximate, based on likely relative maximum fluxes of photorespiration, respiration and nitrogen assimilation in C_3 leaves during rapid photosynthesis, the latter estimated at about 500–600 μmol g^{-1}FW h^{-1}.

This set of equations can be solved by any ODE solver system to produce data on how pool sizes are predicted to change over time as a function of flux from nitrate and NADH (in the cytosol) and isocitrate and NAD (in the mitochondria). For the steady state substrate levels used above, Figure 12.4 (curves A) shows the iterated solution over time for the 2-OG, ammonia, Glu and Gln concentrations. Other substrates and products remain close to their initial values. While Gln and 2-OG rise, ammonia decreases substantially despite ongoing formation from nitrate. This likely reflects the high capacity of the GS-GOGAT pathway compared to NR in leaves of C_3 plants. It is therefore necessary to consider a more physiological condition in which photorespiratory N cycling is producing ammonia via Gly decarboxylation and 2-OG is being regenerated from Glu by peroxisomal aminotransferases. If we include these additional pathways in our modelling system (Figure 12.5), and repeat the modelling process as described above, we can derive a further set of equations that model this more complex system. To parameterise this model, a number of crude estimates have been made to determine the rate and kinetic parameters. As for the scheme shown in Figure 12.3, transport processes have been simplified as far as possible. This second model can be described by the following equations:

$$X_0 = 3$$
$$X_1 = 0.001$$
$$X_2' = 10.6 X_0^{0.003} X_1^{0.75} - 0.98 X_2$$
$$X_3' = 0.98 X_2 - 0.68 X_3^{0.97}$$
$$X_4' = 0.2 X_3^{0.18} X_{11} - 0.11 X_4^{0.23}$$
$$X_5' = 0.15 X_4^{0.23} X_6^{0.25} - 0.23 X_5^{0.024}$$
$$X_6' = 0.6 X_7^{0.73} X_5^{0.024} - 0.02 X_6^{0.25} X_{12}$$

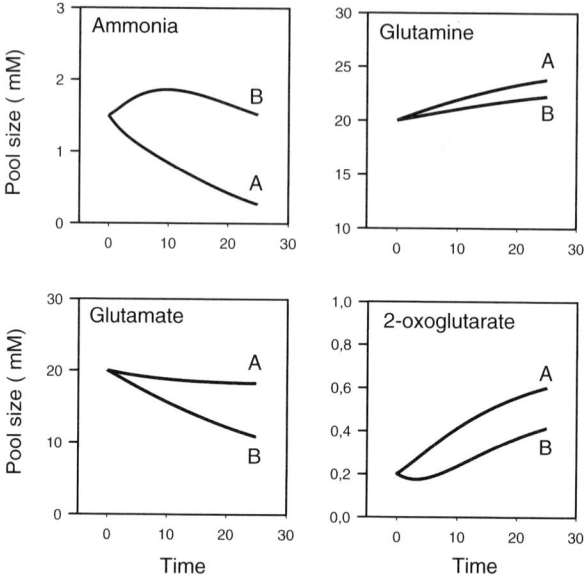

Figure 12.4 Modelled changes in chloroplastic pools of GS-GOGAT pathway metabolites during nitrate reduction in the absence (A) or presence (B) of the photorespiratory nitrogen cycle. For curves labelled A, data are generated by modelling according to the simple scheme shown in Figure 12.3, where constant inputs are nitrate and NADH in the cytosol and isocitrate and NAD in the mitochondria. Curves labelled B were produced using the same parameters but by adding glyoxylate as a constant substrate input in the peroxisomes (Figure 12.5). Units of time are arbitrary.

$$X_7' = 0.4 X_8 X_{13} - 0.087 X_7^{0.73}$$

$$X_8' = 0.22 X_{10}^{0.57} X_9^{0.35} - 0.3 X_8$$

$$X_9 = 0.6$$

$$X_{10} = 0.15$$

$$X_{11}' = 0.08 X_{16}^{0.45} - 0.2 X_{11}$$

$$X_{12}' = 0.02 X_6 - 0.2 X_{12}^{0.16}$$

$$X_{13}' = 0.04 X_{14}^{0.8} X_{12}^{0.4} - 0.2 X_{13}$$

$$X_{14} = 1.0$$

$$X_{15}' = 0.8 X_{14}^{0.99} X_{12}^{0.35} X_{18}^{0.3} - 0.3 X_{15}$$

$$X_{16}' = 0.6 X_{15} - 0.8 X_{16}^{0.5}$$

$$X_{17}' = 0.2 X_{16}^{0.4} - 0.4 X_{17}$$

$$X_{18}' = 0.4 X_{17} - 0.4 X_{18}^{0.3}$$

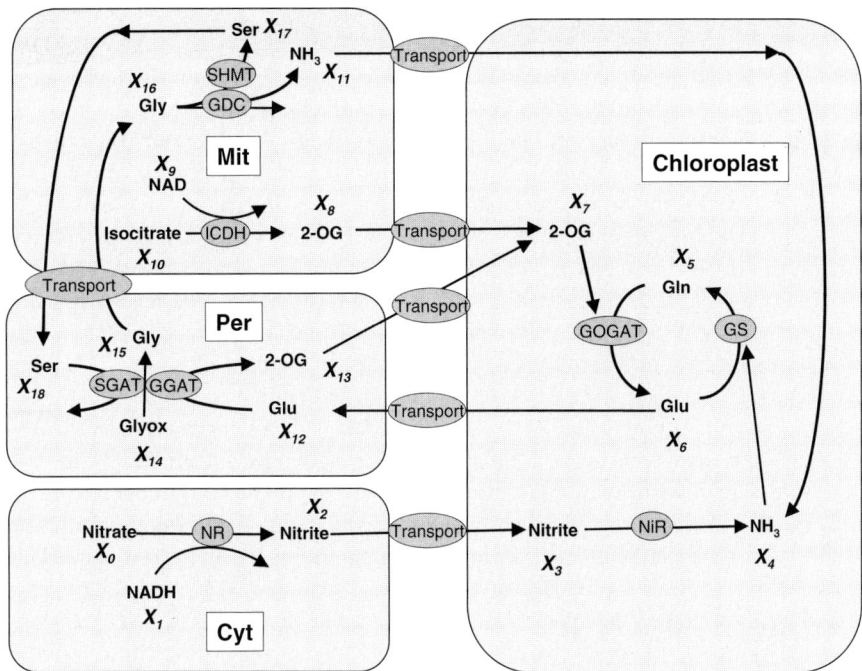

Figure 12.5 More complex model of nitrate and ammonia assimilation that adds the principal features of the photorespiratory N cycle to the model shown in Figure 12.3. In particular, this means that glyoxylate figures as an additional constant substrate input. Mit, mitochondrion; Per, peroxisomes; Cyt, cytosol.

This system was solved numerically with the additional starting points of $X_2 = X_3 = 0.007$, $X_4 = 1.5$, $X_5 = X_6 = 20$, $X_7 = 0.2$, $X_8 = X_{17} = 0.15$, $X_{11} = 2$, $X_{12} = 15$, $X_{13} = 0.15$, $X_{15} = X_{16} = 7.5$ (all values in mM). Changes of GS-GOGAT metabolite pools are shown in Figure 12.4, curves B. In this case we find that, as for the simpler model, Gln accumulates and the Glu pool decreases. When photorespiration is active, however, the ammonia pool remains stable and changes in 2-OG are less dramatic (Figure 12.4, compare curves A and B). Clearly, the data are very preliminary, and more rigorous parameterisation is required. Nevertheless, these first results perhaps suggest a potential effect of photorespiratory N cycling in stabilising metabolite pools involved in ammonia assimilation. Some aspects of the results generated are qualitatively in agreement with measured literature data, e.g., the decrease in Glu that occurs in some systems on resumption of nitrate assimilation ([11] and references cited therein). Work towards a more realistic parameterisation of this model is in progress with a view to incorporating the major pathways of C and N metabolism in the leaves of C_3 species.

12.6 Conclusions and perspectives

Under field conditions parameters associated with C and N metabolism have a close relationship to yield and are thus potentially important current targets for analyses using quantitative trait loci and other genetic approaches. To date numerous studies using forward and reverse genetics in addition to classical biochemistry and physiology have revealed the enormous complexity of the C/N interaction and its intimate association and impact on plant hormone signalling. While many aspects of this control remain to be characterised, the emerging importance of systems biology in agriculture as in other fields necessitates that current information is used in strategies for prescriptive modelling that might be applied to crop plants. In the above analysis of the C/N interaction we have highlighted the importance of amino acids as the focal point around which primary N assimilation and associated carbon metabolism and photorespiration and export occur. We have modelled the C/N interaction around the GS-GOGAT pathway as this is the point at which C metabolites and ammonia converge and from which downstream biosyntheses fan out. The preliminary model concerns only the key reactions of nitrate and ammonia assimilation with visualisation using BST to formulate a set of differential equations representing the reaction systems. This initial model, while as yet modest and imprecise, is nevertheless functional. Data produced by such approaches could be useful, for example, in defining relationships between processes and pathways and in providing frameworks within which to interpret complex data sets such as those produced by metabolomic studies. A second type of application could concern examination of fluxes to minor amino acids of crucial importance to certain crops. For instance, the production of theanine, the major free amino acid in tea, is of major commercial importance to tea growers. The factors that control theanine synthesis from glutamic acid (or pyroglutamic acid) and ethylamine in a reaction catalysed by theanine synthetase (L-glutamic acid ethylamine ligase) are crucial to the tea producer as they are important to the flavour of the final tea product. To date, growers can only speculate on how various environmental and metabolic cues influence theanine content and hence the value of the tea crop. Providing that accurate values for required parameters can be added from experimental measurements, the model described here could be elaborated to address these and other key questions that are essential to agro-industries as well as have intrinsic scientific interest.

Acknowledgement

Rothamsted Research is funded by the Biotechnology and Biological Sciences Research Council (UK).

References

1. W.R. Scheible, M. Lauerer, E.D. Schulze, M. Caboche and M. Stitt (1997) Accumulation of nitrate in the shoot acts as a signal to regulate shoot–root allocation in tobacco. *Plant Journal* **11**, 671–691.
2. A. Puppo, K. Groten, F. Bastian *et al.* (2005) Legume nodule senescence: roles for redox and hormone signalling in the orchestration of the natural aging process. *New Phytologist* **165**, 683–701.

3. M. Aslam, R.C. Huffaker, D.W. Rains and K.P. Rao (1979) Influence of light and ambient carbon dioxide concentration on nitrate assimilation by intact barley seedlings. *Plant Physiology* **63**, 1205–1209.
4. A.J. Reed, D.T. Canvin, J.H. Sherrard and R.H. Hageman (1983) Assimilation of [^{15}N]nitrate and of [^{15}N]nitrite in leaves of five plant species under light and dark conditions. *Plant Physiology* **71**, 291–294.
5. C.E. Lewis, G. Noctor, D. Causton and C.H. Foyer (2000) Regulation of assimilate partitioning in leaves. *Australian Journal of Plant Physiology* **27**, 507–517.
6. F. Gastal and B. Saugier (1989) Relationships between nitrogen uptake and carbon assimilation in whole plants of tall fescue. *Plant Cell and Environment* **12**, 407–418.
7. G.H. Pace, R.J. Volk and W.A. Jackson (1990) Nitrate reduction in response to CO_2-limited photosynthesis. Relationship to carbohydrate supply and nitrate reductase activity in maize seedlings. *Plant Physiology* **92**, 286–292.
8. T.W. Rufty, C.T. MacKown and R.J. Volk (1989) Effects of altered carbohydrate availability on whole plant assimilation of $^{15}NO_3^-$. *Plant Physiology* **89**, 457–463.
9. P.O. Larsen, K.L. Cornwell, S.L. Gee and J.A. Bassham (1981) Amino acid synthesis in photosynthesizing spinach cells. Effects of ammonia on pool sizes and rates of labeling from $^{14}CO_2$. *Plant Physiology* **68**, 292–299.
10. J.S. Paul, K.L. Cornwell and J.A. Bassham (1981) Effects of ammonia on carbon metabolism in photosynthesizing isolated mesophyll cells from *Papaver somniferum* L. *Planta* **142**, 49–54.
11. H.C. Huppe and D.H. Turpin (1994) Integration of carbon and nitrogen metabolism in plant and algal cells. *Annual Review of Plant Physiology and Plant Molecular Biology* **45**, 577–607.
12. I.R. Elrifi and D.H. Turpin (1986) Nitrate and ammonium induced photosynthetic suppression in N-limited *Selenastrum minutum*. *Plant Physiology* **81**, 273–279.
13. C.H. Foyer, G. Noctor, M. Lelandais *et al.* (1994) Short-term effects of nitrate, nitrite and ammonium assimilation on photosynthesis, carbon partitioning and protein phosphorylation in maize. *Planta* **192**, 211–220.
14. M.L Champigny and C.H Foyer (1992) Nitrate activation of cytosolic protein kinases diverts photosynthetic carbon from sucrose to amino acid biosynthesis. Basis for a new concept. *Plant Physiology* **100**, 7–12.
15. W.R. Scheible, A. Gonzáles-Fontes, M. Lauerer, B. Müller-Röber, M. Caboche and M Stitt (1997) Nitrate acts as a signal to induce organic acid metabolism and repress starch metabolism in tobacco. *Plant Cell* **9**, 783–798.
16. A.J. Keys, I.F. Bird, M.J. Cornelius, P.J. Lea, B.J. Miflin and R.M. Wallsgrove (1978) Photorespiratory nitrogen cycle. *Nature* **275**, 741–743.
17. B. Hirel and P.J Lea (2002) The biochemistry, molecular biology and genetic manipulation of primary ammonia assimilation. In: *Photosynthetic Nitrogen Assimilation and Associated Carbon and Respiratory Metabolism* (eds C.H. Foyer and G. Noctor), Kluwer Academic, Dordrecht, The Netherlands, pp. 151–172.
18. G. Parisi, M. Perales, M. Fornasari *et al.* (2004) Gamma carbonic anhydrases in plant mitochondria. *Plant Molecular Biology* **55**, 193–207.
19. J. Gemel and D.D. Randall (1992) Light regulation of leaf mitochondrial pyruvate dehydrogenase complex. Role of photorespiratory carbon metabolism. *Plant Physiology* **100**, 908–914.
20. G.C. Vanlerberghe, K.A. Schuller, R.G. Smith, R. Feil, W.C. Plaxton and D.H. Turpin (1990) Relationship between NH_4^+ assimilation rate and *in vivo* phospho*enol*pyruvate carboxylase activity. Regulation of anaplerotic carbon flow in the green alga *Selenastrum minutum*. *Plant Physiology* **94**, 284–290.
21. T.F. Moraes and W.C Plaxton (2000) Purification and characterization of phosphoenolpyruvate carboxylase from *Brassica napus* (rapeseed) suspension cell cultures. Implications for phosphoenolpyruvate carboxylase regulation during phosphate starvation, and the integration of glycolysis with nitrogen assimilation. *European Journal of Biochemistry*, **267**, 4465–4476.
22. C.R. Smith, V.L. Knowles and W.C. Plaxton (2000) Purification and characterization of cytosolic pyruvate kinase from *Brassica napus* (rapeseed) suspension cell cultures. Implications for the

integration of glycolysis with nitrogen assimilation. *European Journal of Biochemistry* **267**, 4477–4485.
23. C.H. Foyer, M. Parry and G. Noctor (2003) Markers and signals associated with nitrogen assimilation in higher plants. *Journal of Experimental Botany* **54**, 585–593.
24. S. Khamis, T. Lamaze, Y. Lemoine and C.H. Foyer (1990) Adaptation of the photosynthetic apparatus in maize leaves as a result of nitrogen limitation. Relationships between electron transport and carbon assimilation. *Plant Physiology* **94**, 1436–1443.
25. W.M. Kaiser and S.C. Huber (1994) Post-translational regulation of nitrate reductase in higher plants. *Plant Physiology* **106**, 817–821.
26. R. Wang, K. Guegler, S.T. LaBrie and N.M. Crawford (2000) Genomic analysis of a nutrient response in Arabidopsis reveals diverse expression patterns and novel metabolic and potential regulatory genes induced by nitrate. *Plant Cell* **12**, 1491–1509.
27. P.M. Palenchar, A. Kouranov, L.V. Lejay and G.M. Coruzzi (2004) Genome-wide patterns of carbon and nitrogen regulation of gene expression validate the combined carbon and nitrogen (CN)-signalling hypothesis in plants. *Genome Biology* **5**, R91.
28. J.L. Huber, S.C. Huber, W.H. Campbell and M.G. Redinbaugh (1992) Reversible light/dark modulation of spinach leaf nitrate reductase activity involves protein phosphorylation. *Archives of Biochemistry and Biophysics* **296**, 58–65.
29. S. Ferrario, M.H. Valadier and C.H. Foyer (1996) Short-term modulation of nitrate reductase activity by exogenous nitrate in *Nicotiana plumbaginofolia* and *Zea mays* leaves. *Planta* **199**, 366–371.
30. C.H. Foyer, J.C. Lescure, C. Lefebvre, J.F. Morot-Gaudry, M. Vincentz and H. Vaucheret (1994) Adaptations of photosynthetic electron transport, carbon assimilation, and carbon partitioning in transgenic *Nicotiana plumbaginofolia* plants to changes in nitrate reductase activity. *Plant Physiology* **104**, 171–178.
31. L. Novitskaya, S. Trevanion, S.D. Driscoll, C.H. Foyer and G. Noctor (2002) How does photorespiration modulate leaf amino acid contents. A dual approach through modelling and metabolite analysis. *Plant, Cell and Environment* **25**, 821–836.
32. L. Lejay, I. Quilleré, Y. Roux *et al.* (1997) Abolition of post-transcriptional regulation of nitrate reductase partially prevents the decrease in leaf nitrate reduction when photosynthesis is inhibited by CO_2 deprivation, but not in darkness. *Plant Physiology* **115**, 623–630.
33. W.M. Kaiser, M. Stoimenova and H.M. Man (2002) What limits nitrate reduction in leaves? In: *Photosynthetic Nitrogen Assimilation and Associated Carbon and Respiratory Metabolism* (eds C.H. Foyer and G. Noctor), Kluwer Academic, Dordrecht, The Netherlands, pp. 63–70.
34. G. Noctor and C.H. Foyer (1998) A re-evaluation of the ATP:NADPH budget during C_3 photosynthesis. A contribution from nitrate assimilation and its associated respiratory activity? *Journal of Experimental Botany* **49**, 1895–1908.
35. P. Gardeström, A.U. Igamberdiev and A.S. Raghavendra (2002) Mitochondrial functions in the light and significance to carbon–nitrogen interactions. In: *Photosynthetic Nitrogen Assimilation and Associated Carbon and Respiratory Metabolism* (eds C.H. Foyer and G. Noctor), Kluwer Academic, Dordrecht, The Netherlands, pp. 151–172.
36. A.S. Raghavendra and K. Padmasree (2003) Beneficial interactions of mitochondrial metabolism with photosynthetic carbon assimilation. *Trends in Plant Science* **8**, 546–553.
37. S. Krömer and H.W. Heldt (1991) Respiration of pea leaf mitochondria and redox transfer between the mitochondrial and extramitochondrial compartment. *Biochimica et Biophysica Acta* **1057**, 42–50.
38. I. Hanning and H.W. Heldt (1993) On the function of mitochondrial metabolism during photosynthesis in spinach (*Spinacia oleracea* L.) leaves. *Plant Physiology* **103**, 1147–1154.
39. W.M. Kaiser, A. Kandlbinder, M. Stoimenova and J. Glaab (2000) Discrepancy between nitrate reduction in intact leaves and nitrate reductase activity in leaf extracts: what limits nitrate reduction in situ? *Planta* **210**, 801–807.
40. S. Rachmilevitch, A.B. Cousins and A.J. Bloom (2004) Nitrate assimilation in plant shoots depends on photorespiration. *Proceedings of the National Academy of Sciences of the United States of America* **101**, 11506–11510.

41. M. Hodges (2002) Enzyme redundancy and the importance of 2-oxoglutarate in plant ammonium assimilation. *Journal of Experimental Botany* **53**, 905–916.
42. S. Gutierres, M. Sabar, C. Lelandais *et al.* (1997) Lack of mitochondrial and nuclear-encoded subunits of complex I and alteration of the respiratory chain in *Nicotiana sylvestris* mitochondrial deletion mutants. *Proceedings of the National Academy of Sciences of the United States of America* **94**, 3436–3441.
43. M. Sabar, R. De Paepe and Y. De Kouchkovsky (2000) Complex I impairment, respiratory compensations, and photosynthetic decrease in nuclear and mitochondrial male sterile mutants of *Nicotiana sylvestris*. *Plant Physiology* **124**, 1239–1249.
44. G. Noctor, C. Dutilleul, R. De Paepe and C.H. Foyer (2004) Use of mitochondrial electron transport mutants to evaluate the effects of redox state on photosynthesis, stress tolerance and the integration of carbon/nitrogen metabolism. *Journal of Experimental Botany* **55**, 49–57.
45. C. Dutilleul, C. Lelarge, J.L. Prioul, R. De Paepe, C.H. Foyer and G. Noctor (2005) Mitochondria-driven changes in leaf NAD status exert a crucial influence on the control of nitrate assimilation and the integration of carbon and nitrogen metabolism. *Plant Physiology* **139**, 64–78.
46. B.J. Miflin and P.J. Lea (1982) Ammonia assimilation. In: *The Biochemistry of Plants*, Vol. 5 (ed. B.J. Miflin), Academic, New York, pp. 169–202.
47. M. Stitt and A. Krapp (1999) The interaction between elevated carbon dioxide and nitrogen nutrition: the physiological and molecular background. *Plant, Cell and Environment* **22**, 583–621.
48. M. Stitt, C. Müller, P. Matt *et al.* (2002) Steps towards an integrated view of nitrogen metabolism. *Journal of Experimental Botany* **53**, 959–970.
49. W.R. Scheible, A. Krapp and M. Stitt (2000) Reciprocal diurnal changes of phosphoenolpyruvate carboxylase expression and NADP-isocitrate dehydrogenase expression regulate organic acid metabolism during nitrate assimilation in tobacco leaves. *Plant, Cell and Environment* **23**, 1155–1167.
50. P. Brechlin, A. Unterhalt, R. Tischner and G. Mack (2000) Cytosolic and chloroplastic glutamine synthetase of sugarbeet (*Beta vulgaris*) respond differently to organ ontogeny and nitrogen source. *Physiologia Plantarum* **108**, 263–269.
51. G.M. Coruzzi and L. Zhou (2001) Carbon and nitrogen sensing and signalling in plants: emerging 'matrix effects'. *Current Opinion in Plant Biology* **4**, 247–253.
52. B.G. Forde (2002) Local and long-range signalling pathways regulating plant responses to nitrate. *Annual Review of Plant Biology* **53**, 203–224.
53. G. Coruzzi and D.R Bush (2001) Nitrogen and carbon nutrient and metabolite signalling in plants. *Plant Physiology* **125**, 61–64.
54. J.E. Taylor and M.R. McAinsh (2004) Signalling crosstalk in plants: emerging issues. *Journal of Experimental Botany* **55**, 147–149.
55. J.A. Lake, W.P. Quick, D.J. Beerling and F.I. Woodward (2001) Plant development – signals from mature to new leaves. *Nature* **411**, 154–159
56. S.J. Neill, R. Desikan and J.T. Hancock (2003) Nitric oxide signalling in plants. *New Phytologist* **159**, 11–35.
57. A.U. Igamberdiev and R.D. Hill (2004) Nitrate, NO and haemoglobin in plant adaptation to hypoxia: an alternative to classic fermentation pathways. *Journal of Experimental Botany* **55**, 2473–2482.
58. R. Morcuende, A. Krapp, V. Hurry and M Stitt (1998) Sucrose feeding leads to increased rates of nitrate assimilation, increased rates of α-oxoglutarate synthesis, and increased synthesis of a wide spectrum of amino acids in tobacco leaves. *Planta* **206**, 394–409.
59. R.R. Finkelstein and W. Gibson (2002) ABA and sugar interactions regulating development: cross-talk or faces in a crowd? *Current Opinion in Biology* **5**, 26–32.
60. P. Leon and J. Sheen (2003) Sugar and hormone connections. *Trends in Plant Science* **8**, 110–116.
61. De Smet, L. Signora, T. Beeckman, D. Inze, C.H. Foyer and H. Zhang (2003) An ABA-sensitive lateral root developmental checkpoint in Arabidopsis. *The Plant Journal* **33**, 543–555.
62. R.R. Finkelstein and T.J Lynch (2000) Abscisic acid inhibition of radicle emergence but not seedling growth is suppressed by sugars. *Plant Physiology* **122**, 1179–1186.

63. F. Arenas-Huertero, A. Arroyo, L. Zhou, J. Sheen and P. Leon (2000) Analysis of Arabidopsis glucose insensitive mutants, gin 5 and gin 6, reveals a central role of the plant hormone ABA in the regulation of plant vegetative development by sugar. *Genes and Development* **14**, 2085–2096.
64. L. Signora, I. De Smet, C.H. Foyer and H. Zhang (2001) ABA plays a central role in mediating the regulatory effects of nitrate on root branching in Arabidopsis. *Plant Journal* **28**, 655–662.
65. J. Price, T. Li, S. Kang, J. Na and J.C. Jang (2003) Mechanisms of glucose signalling during germination of *Arabidopsis thaliana*. *Plant Physiology* **132**, 1424–1438.
66. R. Wang, M. Okomoto, X. Xing and N.M. Crawford (2003) Microarray analysis of the nitrate response in Arabidopsis roots and shoots reveals over 1,000 rapidly responding genes and new linkages to glucose, trehalose-6-phosphate, iron, and sulfate metabolism. *Plant Physiology* **132**, 556–567.
67. C.D. Todd, P. Zeng, A.M. Huete, M.E. Hoyos and J.C. Polacco (2004) Transcripts of MYB-like genes respond to phosphorus and nitrogen deprivation in Arabidopsis. *Planta* **219**, 1003–1009.
68. D. Gonzalez-Ballester, A. de Montaigu, J.J. Higuera, A. Galvan and E. Fernández (2005) Functional genomics of the regulation of the nitrate assimilation pathway in *Chlamydomonas*. *Plant Physiology* **137**, 522–533.
69. G. Noctor, L. Novitskaya, P.J. Lea and C.H. Foyer (2002) Co-ordination of leaf minor amino acid contents in crop species: significance and interpretation. *Journal of Experimental Botany* **53**, 939–945.
70. C.H. Foyer and G. Noctor (2002) Photosynthetic nitrogen assimilation: inter-pathway control and signalling. In: *Photosynthetic Nitrogen Assimilation and Associated Carbon and Respiratory Metabolism*, Vol. 12 (eds C.H. Foyer and G. Noctor), Kluwer Academic, Dordrecht, The Netherlands, pp. 1–22.
71. D. Guyer, D. Patton and E. Ward (1995) Evidence for cross-pathway regulation of metabolic gene expression in plants. *Proceedings of the National Academy of Sciences of the United States of America*, **92**, 4997–5000.
72. J. Zhao, C.C. Williams and R.L. Last (1998) Induction of Arabidopsis tryptophan pathway enzymes and camalexin by amino acid starvation, oxidative stress, and an abiotic elicitor. *Plant Cell*, **10**, 359–370.
73. Y. Zhang, R.R. Dickinson, M. J. Paul and N.G. Halford (2003) Molecular cloning of an Arabidopsis homologue of GCN2, a protein kinase involved in co-ordinated response to amino acid starvation. *Planta* **217**, 668–675.
74. G. Noctor, A.C.M. Arisi, L. Jouanin and C.H. Foyer (1998) Manipulation of glutathione and amino acid biosynthesis in the chloroplast. *Plant Physiology* **118**, 471–482.
75. S.D. Lemaire, B. Guillon, P. Le Maréchal, E. Keryer, M. Miginiac-Maslow and P. Decottignies (2004) New thioredoxin targets in the unicellular photosynthetic eukaryote *Chlamydomonas reinhardtii*. *Proceedings of the National Academy of Sciences of the United States of America* **101**, 7475–7480.
76. G.B.G. Moorhead and C.S. Smith (2003) Interpreting the plastid carbon, nitrogen and energy status. A role for PII? *Plant Physiology* **133**, 492–498.
77. C.S. Smith, N. Morrice and G.B.G. Moorhead (2004) Lack of evidence for phosphorylation of *Arabidopsis thaliana* PII: implications for plastid carbon and nitrogen signalling. *Biochimica et Biophysica Acta* **1699**, 145–154.
78. G. Moorhead, P. Douglas, V. Cotelle *et al.* (1999) Phosphorylation-dependent interactions between enzymes of plant metabolism and 14-3-3 proteins. *Plant Journal* **18**, 1–12.
79. G. Noctor and C.H. Foyer (2000) Homeostasis of adenylate status during photosynthesis in a fluctuating environment. *Journal of Experimental Botany* **51**, 347–356.
80. S. Von Caemmerer (2000) *Biochemical Models of Leaf Photosynthesis*. CSIRO Collingwood. Australia
81. N.V. Torres and E.O. Voit (2002) Pathway analysis and optimization. In: *Metabolic Engineering*. Cambridge University Press, Cambridge, UK.
82. D. Heineke, B. Riens, H. Grosse *et al.* (1991) Redox transfer across the inner chloroplast envelope membrane. *Plant Physiology* **95**, 1131–1137.
83. H. Winter, G. Robinson and H.W. Heldt (1994) Subcellular volumes and metabolite concentrations in spinach leaves. *Planta* **193**, 530–535.

84. A.U. Igamberdiev and P. Gardeström (2003) Regulation of NAD- and NADP-dependent isocitrate dehydrogenases by reduction levels of pyridine nucleotides in mitochondria and cytosol of pea leaves. *Biochimica et Biophysica Acta* **1606**, 117–125.
85. W.H. Campbell (1999) Nitrate reductase structure, function and regulation: bridging the gap between biochemistry and physiology. *Annual Review of Plant Physiology and Plant Molecular Biology* **50**, 277–303.
86. W. Su, J.A. Mertens, K. Kanamaru, W.H. Campbell and N.M. Crawford (1997) Analysis of wild-type and mutant plant nitrate reductase expressed in the methylotrophic yeast *Pichia pastoris*. *Plant Physiology* **115**, 1135–1143.
87. E. Back, W. Burkhart, M. Moyer, L. Privalle and S. Rothstein (1988) Isolation of cDNA clones coding for spinach nitrite reductase: complete sequence and nitrate induction. *Molecular Genetics and Genomics* **212**, 20–26.
88. J. Chen and I.R Kennedy (1985) Purification and properties of lupin nodule glutamine synthetase. *Phytochemistry* **24**, 2167–2172.
89. M.C. Ericson (1985) Purification and properties of glutamine synthetase from spinach leaves. *Plant Physiology* **79**, 923–927.
90. C. Avila, F. Canovas, I. Nunez de Castro and V. Valpuesta (1984) Separation of two forms of glutamate synthase in leaves of tomato (*Lycopersicon esculentum*). *Biochemical and Biophysical Research Communications* **122**, 1125–1130.
91. A. Suzuki and P. Gadal (1982) Glutamate synthase from rice leaves. *Plant Physiology* **69**, 848–852.
92. C.A. McIntosh (1997) Partial purification and characteristics of membrane-associated NAD^+-dependent isocitrate dehydrogenase activity from etiolated pea mitochondria. *Plant Science* **129**, 9–20.
93. Y. Nakamura and N.E. Tolbert (1983) Serine:glyoxylate, alanine:glyoxylate, and glutamate:glyoxylate aminotransferase reactions in peroxisomes from spinach leaves. *Journal of Biological Chemistry* **258**, 7631–7638.

13 Control of sulfur uptake, assimilation and metabolism

Malcolm J. Hawkesford, Jonathan R. Howarth and Peter Buchner

13.1 Introduction

Sulfur (S) is taken up by the roots as sulfate, and for the most part distributed within the plant as sulfate, before being incorporated into essential sulfur-containing compounds for a range of functions in plants. Uptake and assimilation of sulfate, as well as partitioning into the various sinks, are controlled processes and in this chapter the major mechanisms contributing to this control are outlined and discussed.

The main sink for sulfur is for synthesis of the amino acids cysteine and methionine, required for protein synthesis. This predominantly determines the insoluble content of sulfur in plant tissues which occurs with nitrogen in a ratio of approximately 20:1. In certain instances there is some plasticity of this ratio, for example in the synthesis of either S-rich or S-poor seed storage proteins in response to variable S availability [1]. In addition, other sinks for sulfur include sulfolipids, glucosinolates and alliins (for reviews see [2, 3] and references therein). Sulfur can remain as sulfate, which accumulates in the vacuoles when sulfur is available in excess (for example [4]). This sulfate represents an important storage pool, and control of its utilisation is an important aspect of whole plant sulfur metabolism.

From the most simplistic viewpoint, sulfate uptake and reductive sulfate assimilation may be considered as a linear pathway from sulfate to cysteine, controlled by supply and demand. There are, however, multiple branch points, metabolic sinks, regulatory influences and developmental interactions involved and the assimilatory pathway in fact forms part of a metabolic web, emphasised by recent transcriptomic and metabolomic profiling studies [5–8]. In addition to the up-regulation of genes directly involved in S-metabolism, these studies revealed further far reaching effects on the expression of flavonoid, auxin and jasmonate biosynthetic pathway genes, particularly under conditions of S-deficiency [6].

As sulfate is the major sulfur pool in plants, sulfate uptake, storage and re-mobilisation are major considerations. Central to these processes are the membrane-located sulfate transporters, encoded by a multi-gene family. Many members of this family (14 in *Arabidopsis thaliana*) are regulated by transcriptional and post-translational mechanisms [9–12]. Sulfate transport into the cell has been estimated to have the greatest impact on overall sulfur assimilation [13]; however, flux is also controlled through the assimilatory pathways by means of a complex sensing

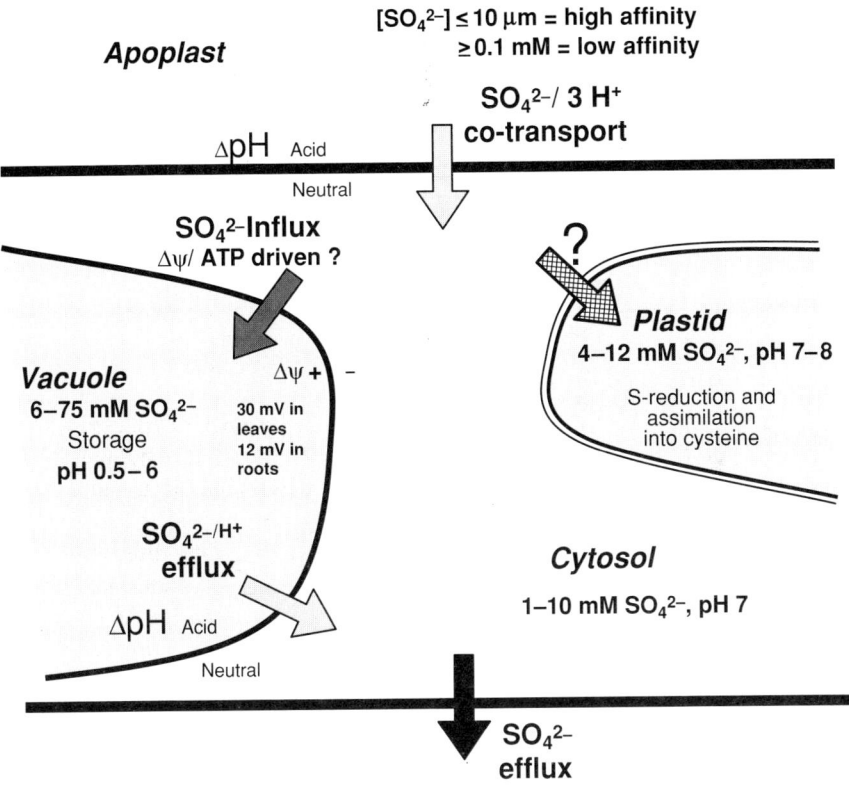

Figure 13.1 Trans-membrane sulfate movements, driving forces and suggested mechanisms (adapted and modified from Hawkesford and Wray [14]): cytoplasmic, vacuole, plastid sulfate concentrations, pH and tonoplast driving forces are indicated [15, 16]. Plasma membrane uptake is a proton-sulfate co-transport [17, 18]. Arrows indicate transmembrane fluxes of sulfate.

mechanism and by feedback loops. In addition, specific branches in the pathway are feedback regulated to control fluxes into the various sinks.

The major regulatory mechanisms controlling S-metabolism are:

- At the level of sulfate uptake and distribution, including loading and re-mobilisation from vacuoles (see Figure 13.1)
- The flux of sulfate through the reductive assimilatory pathway (see Figure 13.2)
- By coordination with demand, ultimately determined by N availability and photosynthetic capacity, which is achieved at the point of cysteine synthesis
- Control of fluxes into the various branch points and sinks

Transcriptional control plays a major part in the control of these processes in response to supply of, and demand for, sulfur. In the root, when demand outstrips

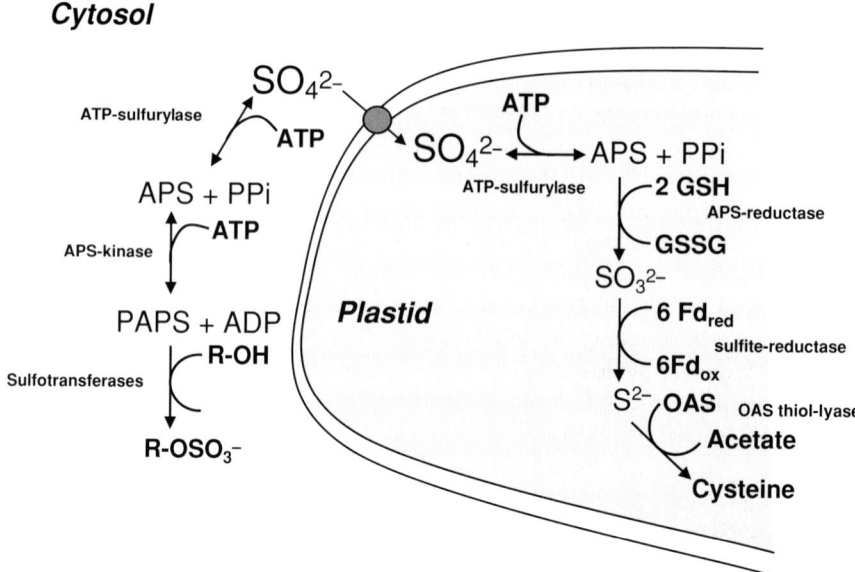

Figure 13.2 Compartmentation of major plant sulfur assimilation pathways.

supply, expression of transporters is increased to maximise acquisition. Conversely, when S-supply is in excess, expression of these genes is reduced and, as the transporter proteins turnover rapidly [14, 19, 20], transport capacity is decreased. Further sulfate which is taken up mostly enters the vacuoles, from where it may be subsequently re-mobilised. Therefore, the balance of influx and efflux from the vacuole as well as uptake into the cell and into the chloroplast, the site of reductive assimilation, will be the major determinants in controlling availability for the assimilatory pathway. In the chloroplast, control switches to the entry and flux through the initial stages of the assimilatory pathway, at the level of sulfate activation by ATP sulfurylase (E.C. 2.7.7.4) or reduction by APS reductase (APR) (E.C. 1.8.4.9). Control of these steps may be transcriptional and/or via enzyme regulation from downstream products of the pathway. The chief candidates in all mechanisms proposed to date are related to cysteine and glutathione synthesis, namely sulfide, O-acetyl L-serine (O-acetylserine; OAS), cysteine or glutathione (GSH). Substantial interest has focussed on cysteine synthesis, particularly as it is at the point of coordination with nitrogen assimilation, and the activity of enzymes directly responsible for cysteine synthesis.

Additional control points exist at the various branches of the pathway, controlling flux to alternative sinks, including glutathione, phytochelatins, methionine or sulfolipids. These are often combinations of transcriptional and post-translational mechanisms. Metabolic control, therefore, is achieved by the coordination of these control mechanisms to adapt to supply of and demand for sulfur.

13.2 Sulfate uptake and distribution

The initial transport step into the plant is against a concentration gradient from a usually dilute soil solution and is therefore an active process thought to be coupled to and dependent upon the proton gradient [21]. After this uptake sulfate must be transported across several membrane systems as it is distributed within the cell and throughout the plant. Mechanisms for the energisation of these transport steps remain speculative [21]. Apart from the primary plasma membrane influx via the root epidermis and cortex cells, distinct transport steps are required due to the sub-cellular compartmentation of sulfate reduction in the plastids, for storage and release from the vacuoles and for long distance transport (Figure 13.1). Since the first reported identification of a plant sulfate transporter in *Stylosanthes hamata* [22] many genes encoding sulfate transporters have been isolated and characterised [22–37]. Sequence phylogeny of the known putative sulfate transporter genes subdivides the plant sulfate transporter family into four closely related groups and a fifth more diverse, but clearly related group [38]. The emerging picture of a large gene family emphasises the critical roles the sulfate transporters have in controlling whole plant sulfur metabolism [38, 39]. Not all members of the family have been demonstrated to transport sulfate; however, as yet no other transport function has been shown. The sulfate transporter group belongs to the larger SLC26 group of anion transporters [40], although similarities with other groups of plant transporters is minimal.

Group 1, 2 and 3 sulfate transporters are located in the plasma membrane [9, 22, 29, 33, 36, 37]. Affinities for sulfate as well as spatial expression patterns and differential expression in response to nutritional status indicate different functions of the individual groups and also of individual transporters. Group 1 and 2 transporters differ in their transport kinetics. The group 1 transporters have a high affinity for sulfate with K_m in the range of 1.5–10 μM, whereas the group 2 transporters have low affinity with K_m between 99.2 μM and 1.2 mM [22, 27, 28, 33, 36, 37]. Within group 3, functional data are only available for Sultr3;5 of Arabidopsis. Expression of the Arabidopsis Sultr3;5 alone was not able to complement the yeast mutant strain CP154-7B, but when expressed in combination with the Arabidopsis Sultr2;1 as a double transformant, an increase in the V_{max} was detected, indicating an influence of Sultr3;5 on the Sultr2;1 transport activity as part of the low affinity sulfate transport system [9]. The nature of this interaction remains to be elucidated.

Spatial expression patterns and knock-out mutant analysis indicate different functions of the individual groups and isoforms. The expression of the high affinity transporters Sultr1;1 and Sultr1;2 in the root tip and root epidermis, root hairs and cortical cells of the mature Arabidopsis root [33, 36] suggest a role for these transporters in primary uptake of sulfate from the soil. Antisense and mutant analysis of Sultr1;1 and Sultr1;2 in Arabidopsis and localisation studies of homologues in tomato and barley support this function [26, 28, 30, 36, 41]. A specialised role is suggested for the third group 1 high affinity sulfate transporter Sultr1;3 of Arabidopsis which is localised solely in the phloem companion cells and which may mediate the redistribution of sulfate between sink and source organs [37]. The two low affinity sulfate transporters of group 2 are also exclusively expressed in the vascular tissue [33], indicating an involvement of both transporters in the vascular movement of sulfate.

Group 4 sulfate transporters are localised in the tonoplast [42] and the analysis of double knock mutants of the two Arabidopsis group 4 sulfate transporter genes (Sultr4;1 and Sultr4;2) strongly suggests a role in the efflux of sulfate from the vacuoles, releasing stored inorganic sulfate to the cytosol [42]. Genes encoding vacuole sulfate influx transporters have not been identified; however, a tonoplast localisation of one of the group 5 sulfate transporters is suggested (Buchner and Hawkesford; unpublished) and is speculated to have a role in influx. This vacuolar stored sulfate will have an important role in cellular sulfate homeostasis and sulfate availability for the assimilatory pathway.

13.2.1 Transcriptional of transport

When demand exceeds availability, a decreased intracellular content of sulfate, cysteine and glutathione is observed which is concomitant with increasing transporter activity [27]. Gene and protein expression studies have confirmed that increases in transporter activity occur predominantly at the transcriptional level [14, 18, 22, 27, 33, 36, 42]. A breakdown of the expression patterns of the genes of the group 1, 2 and 4 sulfate transporters reflects a dual pattern of regulation: (i) cell specific expression of two group 1 (Sultr1;2 and Sultr1;3), one group 2 (Sultr2;2) and one group 4 (Sultr4;1) transporters increased in response to S-limiting conditions and (ii) expression of the other group 1, 2 and 4 transporters only under S-limiting conditions [43]. At least in roots the sulfate transporters of group 1, 2 and 4 are co-regulated under sulfate deprivation to provide maximum uptake of sulfate, release of stored sulfate from vacuoles and vascular transport capacity. The expression of the group 3 sulfate transporters is not modulated by the sulfur-nutritional status of the plant [9, 24, 33].

Given the importance of transcriptional control, analysis of sulfur responsive promoters and the identification of regulatory factors involved in transcriptional activation are required. Studies on *NIT3* nitrilase [44] and the β-subunit of β-conglycinin [45] provided initial data, indicating the presence of regulatory regions in the 5′-promoter sequences of these sulfur-responsive genes. Localisation of the sulfur-deficiency response elements was narrowed to 317 bp and 235 bp regions upstream of the transcriptional start site, respectively. Within the −2777/−2761 promoter region of the Arabidopsis Sultr1;1, a sulfur responsive *cis*-acting SURE element (<u>s</u>ulfur <u>r</u>esponsive <u>e</u>lement) was identified which was essential for the regulation of Sultr1;1 expression under sulfur-deficient conditions [46]. The SURE core sequence (GGAGACA) was also found in the promoter regions of the sulfate transporters Sultr2;1 and Sultr4;2, and APR3 of Arabidopsis. In contrast, SURE elements were not present in the promoter region of all sulfur-deficiency regulated genes, including genes involved in sulfate transport and assimilation, indicating additional regulatory mechanisms may be involved [46].

The SURE dependent expression was down-regulated by supply of cysteine and glutathione, suggesting that the internal sulfur status represented by the thiol content controls expression of the corresponding genes in Arabidopsis [46]. Alternatively, addition of thiols may disturb turnover of sulfate, affecting conversion of downstream assimilatory pathways. In contrast to cysteine and glutathione, *O*-acetylserine, which has been reported as a positive effector for the regulation of sulfur responsive genes [5, 27, 47, 48], did not show any effect on modulating SURE dependent expression. From these results, it is suggested that more than one mechanism is involved in the sulphur related regulatory processes of gene expression.

In the green alga, *Chlamydomonas reinhardtii*, a Snf1-like Ser/Thr kinase, Sac3, is involved in the regulation of sulfate uptake and arylsulfatase activities [49]. Recently it has been shown that the sulfur-deficiency inducible gene expression of Sultr1;1 in Arabidopsis required a protein phosphatase and an upstream regulatory factor [10], indicating that phosphorylation and de-phosphorylation are important in the regulation of sulfur-deficiency responsive genes in plants and algae.

13.2.2 Post-translation controls

Sulfate transporters in plants and animals are structurally conserved and the C-terminal region shares a significant similarity with the *Bacillus* sp. anti-anti-sigma protein SpoIIAA, and is referred to as the STAS domain (sulfate transporter and anti-sigma antagonist). Although the exact function of the STAS has not been elucidated, mutations in the STAS domain of human sulfate transporters result in serious disease. Studies of STAS truncated Arabidopsis high affinity transporters and chimeric transporter constructs suggest that the function of the STAS domain is more concerned with plasma membrane localisation and stability than with regulation of transporter activity. The Arabidopsis sulfate transporter, Sultr1;2, with a deleted STAS domain was unable to complement the yeast sulfate transporter mutant strain CP154-7B. Fusing the STAS domain from other sulfate transporters to the STAS-deleted Sultr1;2 restored function and plasma membrane localisation; however, the kinetics of sulfate uptake in the transformants were dependent on the origin of the STAS domain probably due to significant differences among the STAS domains of the different Arabidopsis sulfate transporters. These results suggest that the STAS domain is essential, either directly or indirectly, for facilitating localisation of the transporters to the plasma membrane, but it also appears to influence the kinetic properties of the catalytic domain of transporters [50]. Substitutions made at the putative phosphorylation site, Thr-587, led to a complete loss of the sulfate transport function of Arabidopsis Sultr1;2. The reduction or suppression of sulfate transport of the Sultr1.2 mutants in yeast was not due to an incorrect targeting to the plasma membrane. Three-dimensional modelling and mutational analyses strengthen the hypothesis that the STAS domain is involved in protein–protein interactions that could control sulfate transport [51]. A similar protein–protein interaction was suggested for the Arabidopsis sulfate transporter isoforms Sultr2;1 and Sultr3;5 [9]. Further, evidence for post-translational regulation of sulfate transport is the observation that transcript abundance of the high affinity transporters increases far more drastically under sulfate deprivation when compared with the transport activity measured [12, 24, 27]. In some cases, the observed increase in activity in response to sulfur starvation reaches a maximum, or only shows a modest transient rise, whilst mRNA continues to increase in abundance in response to the stress [12, 27].

13.3 The assimilatory pathway – activation and reduction

Sulfate is assimilated both in the oxidised and the reduced form, occurring principally in the cytosol and plastid respectively (Figure 13.2). To facilitate these two assimilatory pathways, some but not all sulfur assimilation enzymes exist as isozymes localised in multiple subcellular compartments. Chemically inert inorganic sulfate requires activation via formation of 5´-adenylylsulfate (APS), the sole entry point of sulfate into cellular metabolic pathways. The activation reaction is catalysed by the enzyme ATP sulfurylase.

13.3.1 Cytosolic pathways

When incorporated in the oxidised form, sulfate is used for esterification of a variety of compounds including polysaccharides [52], sulfated flavonoids [53], brassinosteroids [54], glucosinolates and many others [55]. All sulfotransferases (E.C. 2.8.2) that carry out sulfate esterification are cytosolic enzymes [53], indicating the need for activated sulfate in the cytosol. This is achieved in a two-step activation process comprising a cytosolic isoform of ATP sulfurylase, [56–59], and the phosphorylation of the APS produced by APS kinase (E.C. 2.7.1.25) to 3´-phosphoadenosine 5´-phosphosulfate (PAPS), the substrate for the sulfonation reactions catalysed by sulfotransferases. A total of 18 sulfotransferase genes have been identified in Arabidopsis, localising to different sub-cellular compartments [60] but the functions of these many isoforms are not known.

There are four APS kinase genes in Arabidopsis, two of which are predicted to be located in the plastid. Characterisation of one of these plastidic isoforms (*Akn1*) indicated a high affinity for its substrate, APS and substrate inhibition at low ATP concentrations [61]. The enzyme is also modulated by redox state *in vitro*. The function of the plastid APS kinases is unclear as most sulfated esters are synthesised in the cytosol, and sulfolipid (sulfoquinovosyldiacylglyceride) synthesis requires APS, not PAPS [61].

13.3.2 Reductive assimilation in the plastid

13.3.2.1 Sulfate activation by ATP sulfurylase

Most sulfate is reduced to sulfide and incorporated as the thiol group of cysteine. Eight electrons are required to reduce SO_4^{2-} to S^{2-}. In vascular plants, APS is reduced by APS reductase to sulfite, which is subsequently reduced to sulfide by sulfite reductase (E.C. 1.8.7.1) for incorporation into *O*-acetylserine by *O*AS thiol-lyase (*O*ASTL; E.C. 2.5.1.47) forming cysteine. The plastids contain all the enzymes necessary for sulfate reduction and cysteine synthesis and are the exclusive location of the enzyme APS reductase [59, 62]. As a result, the plastids are the obligate site for reduction and sulfate must be transferred into the plastid. However, the mechanism and transporter responsible for sulfate influx into plastids remain to be identified.

In Arabidopsis, three plastid-localised ATP sulfurylase isoforms have been identified in addition to a fourth, putatively cytosolic isoform. Subcellular fractionation experiments indicated that the cytosolic and plastidic isoforms of ATP sulfurylase are differentially regulated during development. The activity of the plastidic isoform declines as plant age. In contrast, the cytosolic isoform increases during development [59]. Sulfate activation is considered to be the limiting step in the assimilation pathway [63], and overexpression of a plastidic ATP sulfurylase, encoded by the APS1 gene [64], in Indian mustard resulted in an approximately 2–2.5-fold increase in enzyme activity. As well as increased selenate reduction, the transgenic plants showed higher amounts of total sulfur as well as glutathione [65]. In contrast, an eight-fold greater ATP sulfurylase activity achieved by expression of the Arabidopsis plastidic APS2 gene [66] in Bright Yellow 2 tobacco suspension culture cells or tobacco plants did not effect sulfate flux and rate of sulfur metabolism. This suggests that the enzyme is under strict control by some products of its activity or downstream steps of the sulfate assimilation pathway [67].

13.3.2.2 Sulfate reduction by APS reductase
Biochemical studies *in vitro* revealed that the APS reductase enzyme APR1 of Arabidopsis is activated by oxidation, probably through the formation of a disulfide bond in its thioredoxin-like domain. The APR1 enzyme is 45-fold more active when expressed in a strain of *Escherichia coli* lacking thioredoxin reductase (trxB) than in a trxB(+) wild type. The enzyme is inactivated *in vitro* by treatment with disulfide reductants and is reactivated with thiol oxidants. Treatment of Arabidopsis seedlings with oxidised glutathione or paraquat induces APS reductase activity even when transcription or translation is blocked with inhibitors. The results suggest that a post-translational mechanism controls APS reductase. A model is proposed whereby redox control of APS reductase provides a rapidly responding, self-regulating mechanism to ultimately control glutathione synthesis [68]. Flux analysis through the pathway supports the idea that APS reductase has the greatest contribution to overall control of the intracellular pathway [13].

In general, ATP sulfurylase and APS reductase are expressed in all photosynthetic cells of C3 plants. However, in monocotyledonous C4 species, ATP sulfurylase and APS reductase are exclusively located in bundle sheath cells [69–71]. The bundle sheath specificity seems to be restricted to monocotyledonous C4 species, since in the dicotyledonous C4 species, *Flaveria trinervia*, APS reductase mRNA was detected in both mesophyll and bundle sheath cells [72]. In addition to photosynthetic tissues, ATP sulfurylase and APS reductase expression and activity is also found in roots [66, 73–75], but the cellular localisation in this tissue is unknown.

The sulfate reduction pathway established to operate in vascular plants is in contrast to that found in fungi and enteric bacteria such as *E. coli* and *Salmonella typhimurium*. In these bacterial species, APS is phosphorylated by APS kinase to

PAPS in order to be reduced in a thioredoxin-dependent reaction by PAPS reductase (EC 1.8.4.8) and then in a second reduction step, sulfite is reduced by NADPH-dependent sulfite reductase to sulfide [76, 77]. The existence of the PAPS-dependent pathway has been a long standing controversy in plants. While the purification of PAPS reductase from spinach has been reported [78], the Arabidopsis and rice genomes do not contain any genes homologous to the bacterial PAPS reductase, other than those encoding APS reductase. The existence of a PAPS reduction pathway was identified in the moss *Physcomitrella patens*, as an APS reductase knockout mutant of the moss was still able to grow on sulfate as the sole sulfur source. Although PAPS reductase activity was not measured in the mutant, the gene coding for this enzyme has been isolated [79]. The moss enzyme differed from the bacterial PAPS reductase in that it lacked the thioredoxin-like domain and a plastid targeting peptide, which would suggest the coexistence of both APS- and PAPS-dependent sulfate assimilation in the chloroplasts of *Physcomitrella*.

13.3.3. Transcriptional regulation and coordination with C and N pathways

A sulfate-deprivation-related increase of gene expression has been shown for at least two plastidic ATP sulfurylase isoforms and three APS reductase isoforms in Arabidopsis [30, 65, 80, 81]. In addition, the putative cytosolic Arabidopsis ATP sulfurylase isoform 2 and the plastidic isoform 4 of Arabidopsis were not regulated by the sulfur status [30, 82]. Data from plants other than Arabidopsis suggest similar patterns of expression [23, 26, 75, 83].

The sulfate-deprivation response of sulfate transporters as well as ATP sulfurylase and APS reductase gene expression and activity was shown to be down-regulated by nitrogen starvation [11, 48, 80, 84] or stimulated by nitrate application [10, 34]. In addition to the coordination of sulfate assimilation with nitrate metabolism, there is also a link with carbon assimilation. Sulfate transporter and APS reductase genes are induced by light [85, 86], an induction that can be mimicked by the addition of *O*-acetylserine (*O*AS) as well as sucrose or glucose to the nutrition medium [74, 86, 87]. As sucrose/glucose addition also induced nitrate and ammonium transport as well as nitrate reduction [74, 86, 88], the assimilation of sulfate and nitrate are clearly interconnected. The induction by *O*AS, formed from acetate and L-serine by the enzyme serine acetyltransferase (SAT; E.C.2.3.1.30) and subsequently a substrate for cysteine synthesis, links assimilatory sulfate reduction with carbohydrate and nitrogen metabolism and *O*AS has been proposed as a signalling molecule coordinating these three pathways [89]. Feeding experiments indicated that the regulation of sulfate transporter and APS reductase expression by sugars is independent of *O*AS regulation [74, 86] which may indicate that signals from nitrogen and carbon metabolism act synergistically and that positive signalling from sugars can override negative signalling from nitrate assimilation [74]. In contrast to the positive regulatory effect of *O*AS, cysteine and glutathione act as negative regulators [11, 13]. Glutathione is thought to be a phloem-translocated signal molecule that represses genes of sulfur assimilation [73].

On the other hand, observations of the sulfate uptake capacity in *Brassica* species indicate that the shoot to root signalling involved in the control of the pathway is unlikely to be mediated by the size of the thiol pool. Instead, it is more likely that the concentration of sulfate itself has a determining influence as a direct or an indirect signal [24].

Sulfate transport and assimilation in plants may also be regulated by plant hormones. Arabidopsis transcriptome analysis suggests that auxin and methyl jasmonate are involved in the sulfur-deficiency stress response [5–7]. Cytokinin has been shown to down-regulate the high-affinity sulfate transporter expression in Arabidopsis roots [90], but cellular levels of cytokinin do not significantly change in response to the sulfur status [91].

13.4 Control of flux through the assimilatory pathway – cysteine synthesis

Whereas transcriptional regulation affects overall control of sulfur assimilation, a finer metabolic control operates at the level of cysteine biosynthesis [92–95]. This control mechanism is important in cellular regulation under most conditions other than severe depletion. The transcriptional and catalytic control systems are intrinsically linked via the same metabolite pools.

In the final step of the reductive sulfate assimilatory pathway, sulfur in the form of sulfide is incorporated into the amino acid L-cysteine. By combining *O*AS and sulfide, this step represents the point of entry of S into organic combination and also the point of convergence between the nitrate and sulfate assimilation pathways. From L-cysteine, various pathways then operate for the production of methionine, proteins, glutathione, phytochelatins, vitamins and a wide variety of organic S-containing compounds.

The biosynthesis of L-cysteine in plants is carried out by two enzymes, SAT and *O*ASTL, which associate in a bi-enzyme complex and it is suggested that they have a key role in 'sensing' the sulfur-nutritional status in the cell. As shown in Figure 13.3, the first enzyme, SAT, acetylates the amino acid L-serine using acetyl coenzyme A to form *O*AS before the *O*-acetylserine is combined, by the action of *O*ASTL, with sulfide from the reductive sulfate assimilation pathway to form cysteine.

The biosynthesis of cysteine was first characterised in *E. coli* and *S. typhimurium* where the whole process of S-uptake and reduction is controlled by a Cys regulon of at least 16 genes [96, 97]. A single SAT enzyme (CysE) and two *O*ASTL enzymes (CysK/M) are responsible for cysteine synthesis. In plants, the situation is more complicated. SAT and *O*ASTL activities (unlike the preceding enzymes in the reductive sulfate assimilation pathway) are in the cytoplasm, mitochondria and chloroplasts [98–102]. The Arabidopsis genome contains five SAT genes and six *O*ASTL genes. Three of the *O*ASTL genes are thought to be involved in cysteine synthesis and three related genes involved in cyanoalanine synthesis [103–105]. The products of these genes are directed to the cytosol or the chloroplast/mitochondrial matrix by N-terminal targeting peptides [103, 106–110].

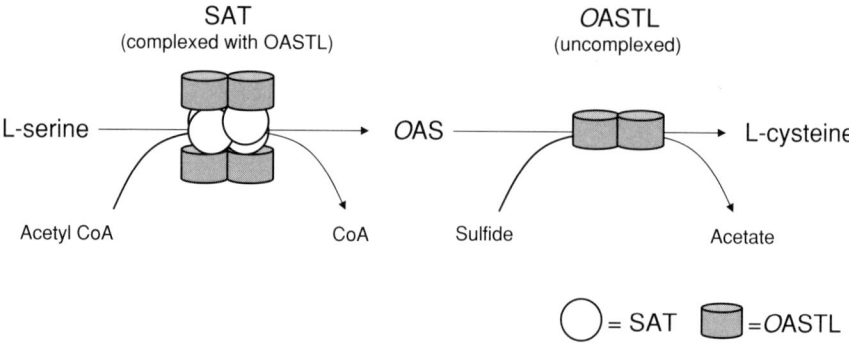

Figure 13.3 The reactions of L-cysteine synthesis from L-serine and sulfide in plants.

13.4.1. The 'cysteine synthase' complex

Early attempts to purify SAT from plants revealed that activity was associated with a protein of 300–350 kDa, which could be dissociated into smaller sub-units in the presence of *O*-acetylserine [111, 112]. In fact, a homo-tetrameric SAT molecule was associating with two molecules of a homo-dimeric *O*ASTL in a complex originally designated 'cysteine synthase' (CS) [96, 113, 114]. The structural relationship between SAT and *O*ASTL in the CS complex was elucidated using the yeast two-hybrid system. These experiments showed that the C-terminal β-sheet region of the SAT protein, which also possesses the active transferase domain, was responsible for the interaction of SAT with *O*ASTL, and SAT/SAT interactions were associated with a central α-helical region [92].

13.4.2 Metabolic control of the 'cysteine synthase' complex

The ratio of *O*ASTL to SAT in plant tissues is approximately 300:1 and a similar ratio of activity was required for the optimal production of cysteine *in vitro* [112, 115]. In cell-free extracts and with recombinant proteins overexpressed in bacteria, SAT was only active when complexed with *O*ASTL and was inactive in its uncomplexed form. Contrary to this, *O*ASTL was found to be highly active in its uncomplexed form yet virtually inactive in complex [93, 112]. From this it was clear that the majority of *O*ASTL is found as an active free-dimer and that only a small proportion would be in complex with SAT. These findings confirmed that *O*-acetylserine is released into solution from the CS complex (as previously found in bacteria [116]) rather than being channelled from SAT to *O*ASTL. Thus *O*ASTL appears to act as a regulatory factor for SAT activity in the bi-enzyme complex rather than having a catalytic role. The regulatory role the CS complex plays in controlling cysteine synthesis and S-assimilation in general was elucidated by combining knowledge of the activity of the free and bound sub-units with studies on the stability of the complex in the presence of its substrates. The complex was found to be stabilised in the presence of sulfide and disrupted into its component molecules in the presence of *O*-acetylserine [93, 113]. In effect the complex acts as a sensor for the S-status of the cellular compartment with SAT activity being controlled by the balance between S and N metabolism.

13.4.3   Control of SAT activity by cysteine feedback

In addition to the control of activity by association/dissociation of the complex with OASTL, plant SATs have also been shown to be sensitive to feedback inhibition by physiological concentrations of cysteine. SAT activity was reduced by 65% in *Phaseolus vulgaris* by 1 mM cysteine [117] and almost completely inhibited at this concentration in SAT from spinach chloroplasts [99]. Inhibition was also demonstrated in a heterologous system by expressing the cytosolic SAT-A gene from watermelon in *E. coli*. Inhibition of SAT activity was observed at approximately 3 µM in this case [118]. A subsequent assessment of the inhibition of the SAT gene family from Arabidopsis however revealed differing sensitivities for the cytosolic, chloroplastic and mitochondrial isoforms [106]. As in watermelon, the cytosolic (Sat-c) isoform was inhibited by micromolar concentrations of cysteine, but the chloroplastic (Sat-p) and mitochondrial (Sat-m) forms were found to be insensitive. As the range of plant SATs tested for feedback inhibition increased, it became clear that rather than cysteine sensitivity being limited to cytosolic isoforms, a high degree of interspecies variation occurred. Chloroplastic SAT activity showed cysteine sensitivity in pea and spinach at physiological concentrations, but the cytosolic and mitochondrial isoforms were insensitive [95, 119]. A glycine residue at position 277 in the watermelon SAT was shown, using site-directed mutagenesis, to be responsible for the feedback inhibition by cysteine [107]. Alignment of all plant cysteine sensitive and insensitive SAT sequences (where known) confirms the role of this amino acid in the feedback response, being present only in sensitive isoforms [120]. Recent research has shown that cysteine inhibits SAT by causing the dissociation of the CS complex, in the same way as *O*-acetylserine does, leading to the formation of inactive high molecular mass aggregates of SAT [95].

As in bacteria, cysteine sensitive isoforms of SAT are clearly involved in the control of cysteine synthesis in plants. However, the possibility that SAT insensitive isoforms may have evolved in cell compartments with high cysteine sink demand in a species-dependent manner should be considered.

13.4.4.   The role of O-acetylserine as an 'inducer' of gene expression

In addition to the ability of *O*-acetylserine to control the activity of SAT post-translationally by disrupting the CS complex, in plants, in common with bacteria, it may also be involved in the transcriptional regulation of gene expression. The position of the cysteine synthase complex at the junction between sulfur and nitrogen metabolism, as well as between inorganic and organic metabolites, makes it an important site for regulation of S metabolism. Such control was first demonstrated in prokaryotic systems, where a simple feedback mechanism ensures optimal cysteine production according to S availability and cysteine requirement. In bacteria, the SAT gene is constitutively expressed producing O-acetylserine for conversion to cysteine by OASTL. O-acetylserine acts as an inducer molecule for the Cys regulon promoting sulfate uptake and reduction. However, SAT enzyme activity is feedback inhibited by low levels of accumulated cysteine causing the Cys regulon to be 'switched off' by reduced O-acetylserine production under conditions of low cysteine requirement. This metabolic control is mediated by the CysB protein [121], which interacts at the promoter site, facilitating or blocking binding of RNA polymerase, depending on binding the protein with OAS or sulfide, respectively.

Whilst no plant homologue of the CysB regulatory protein has yet been detected by nucleotide or amino acid sequence database searching, a number of studies have sought evidence to support a similar model of transcriptional control mediated by O-acetylserine in plants. O-acetylserine is well documented as an inducer of reductive sulfate assimilation, and as already noted above, feeding experiments lead to enhanced activity of uptake, transporter expression and expression of other components of the assimilatory pathway. However, sulfate reduction was stimulated in *Lemna minor* by application of O-acetylserine under dark conditions. This was the first evidence that the sulfate reduction pathway could be decoupled from the light-dependent thioredoxin system by the effect of O-acetylserine as a signalling molecule [87]. O-acetylserine applications also increase the transcription of genes involved in sulfate uptake [27, 122], activation by ATP sulfurylase [122], reduction by APS reductase [48] and have also been shown to control general mRNA profiles in sulfur-starved Arabidopsis [6]. Thus a clear correlation exists between genes induced under sulfur starvation and those induced by application of O-acetylserine, which is indicative of the central importance of this metabolite pool [5].

Induction of sulfate assimilation occurred in the presence of sulfate and with increasing internal cysteine and glutathione levels [27]. These data supported the idea that O-acetylserine is a dominant 'signal' molecule. Additionally, an EMS mutagenised Arabidopsis line (mapped to a thiol reductase) that overaccumulated O-acetylserine showed increased transporter and APS reductase expression [123]. However, transgenic expression of cysE (serine acetyltransferase) in potato resulted in increased sulfate transporter and APS reductase expression, with concomitant cysteine and glutathione accumulation, but not with O-acetylserine accumulation [12]. Time course analysis of plant material subjected to limiting sulfur supply shows a poor correlation between induction of sulfate transporter and APS reductase gene expression and the increase in internal OAS concentration [12, 24]. Taken together these data suggest that artificially increasing O-acetylserine supply acts to increase flux to cysteine and other downstream products. This occurs by relieving any limitation in O-acetylserine precursor supply rather than acting as a direct inducer of gene expression, as is observed in bacteria. Supporting the idea that O-acetylserine is commonly limiting, the metabolite was also the major limiting factor for cysteine biosynthesis in transgenic tobacco plants overexpressing OASTL [124]. During sulfur limitation, the rises in O-acetylserine are a consequence of limited sulfide, so this compound may not be a good candidate for the 'signalling' metabolite. This leaves the role of this molecule as a major open question, with other potential signalling candidates as any molecule from sulfate to cysteine.

In contrast, in yeast, transcriptional control of the SUL and MET genes (which encode most of the S-uptake and assimilation pathway) is controlled by S-adenosyl methionine (SAM; AdoMet) concentration mediated by a multi-protein complex which binds to the respective promoters [125]. Homologous mechanisms have not been investigated in plants.

13.4.5 A model for control of cysteine synthesis

Evidently the plant CS complex is at the centre of control for sulfate uptake and reduction as well as cysteine synthesis. These regulatory properties are interlinked to ensure sulfate reduction and O-acetylserine production is optimised according to the cysteine requirement of the cellular compartment in which they operate. The result is a unique and elegant metabolic control system where cysteine supply is optimised whilst preventing the toxic accumulation of sulfide or cysteine under the condition of fluctuating S availability, changes in N metabolism or requirements for S-containing organic compounds. As shown in Figure 13.4, O-acetylserine, sulfide and cysteine modulate the activity of SAT by a combination of CS complex association/dissociation accompanied by corresponding effects on gene expression. The CS complex is therefore controlled by the substrates sulfide and O-acetylserine to ensure optimal cysteine synthesis to meet sink demands.

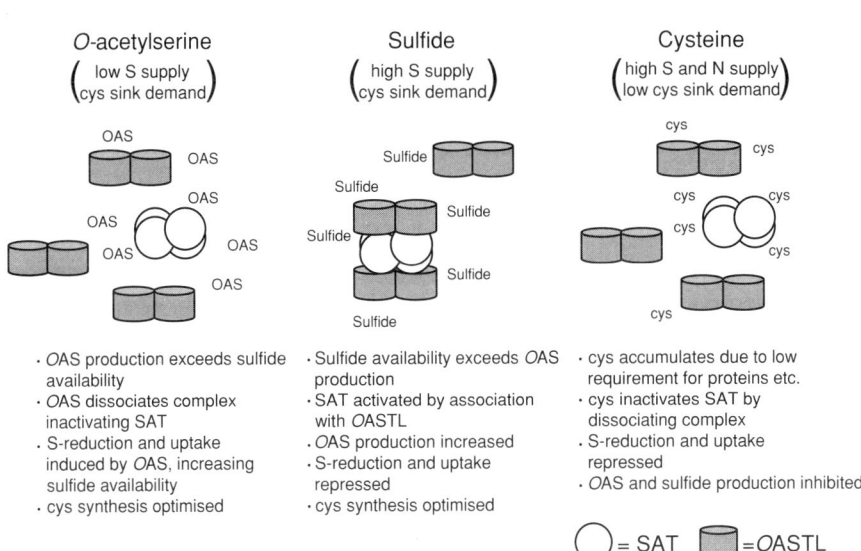

Figure 13.4 Control of cysteine biosynthesis: regulation of the SAT/OASTL complex in response to accumulation of the metabolites OAS, sulfide and cysteine. OAS and/or cysteine or some related metabolite act on expression of the sulfate transporters and components of the assimilatory pathway, principally APR.

In summary the regulatory model works as follows:

1. *O-acetylserine accumulation.* When sulfide availability is low, *O*AS represses its own synthesis by causing the CS complex to dissociate. Sulfide production is then stimulated by the accumulated *O*-acetylserine which may induce the transcription of genes involved in sulfate uptake and reduction.
2. *Sulfide accumulation.* When sulfide availability is in excess, CS complex formation is promoted, activating SAT which in turn increases the production of *O*-acetylserine for optimal cysteine synthesis.
3. *Cysteine accumulation.* When sink demand for cysteine (e.g. methionine, proteins, glutathione vitamins) is low, cysteine accumulation causes the CS complex to dissociate, thus decreasing the production of *O*-acetylserine. Consequently, sulfate uptake and reduction are repressed and cysteine production is prevented until the cysteine pool is exhausted.

13.5 Control of flux to the various sinks after cysteine biosynthesis

In the previous sections, mechanisms resulting in optimisation of cysteine biosynthesis were described. At this point there are various alternative sinks for the reduced S and the flux through the cysteine pools itself will be dictated by the strength of these various sinks. Flux to the sinks is controlled by the activities of specific steps in the pathway, often influenced by feedback loops mediated by pathway products acting at the transcriptional or post-translation level (see Figure 13.5). An important pathway utilising cysteine is the biosynthesis of methionine. The major sink for both cysteine and methionine is protein and the strength of this sink is determined developmentally, as well as by photosynthesis and by nitrogen availability. Glutathione is another important sink for cysteine. The cellular content of glutathione is estimated to be in the region of 0.1–0.2 mM in the cytosol and 2–4 mM in the chloroplast [126], but is subject to the sulfur nutritional status of the plant, and ultimately limited by the availability of sulfur. As a major sink for cysteine, factors affecting glutathione requirements have a major impact on fluxes through the pathway. Demand for glutathione will depend on development and environmental stress. It is also the major form in which cysteine can be transported around the plant, and may act as a store of reduced sulfur. Glutathione itself has an important role in regulating the redox state of the cell and acts as an anti-oxidant. Synthesis is increased in response to a number of stresses, including pathogen attack or any oxidative stress. Finally, glutathione is the precursor for phytochelatins. In response to metal exposure, the activity of phytochelatin synthase increases, increasing the flux through the glutathione and cysteine pools [127].

Glutathione is synthesised in a two-step ATP-requiring pathway, catalysed by γ-glutamyl cysteine synthetase (E.C. 6.3.2.2) and glutathione synthetase (E.C. 6.3.2.3). Transgenic studies [128–131] have clearly indicated that the first step has

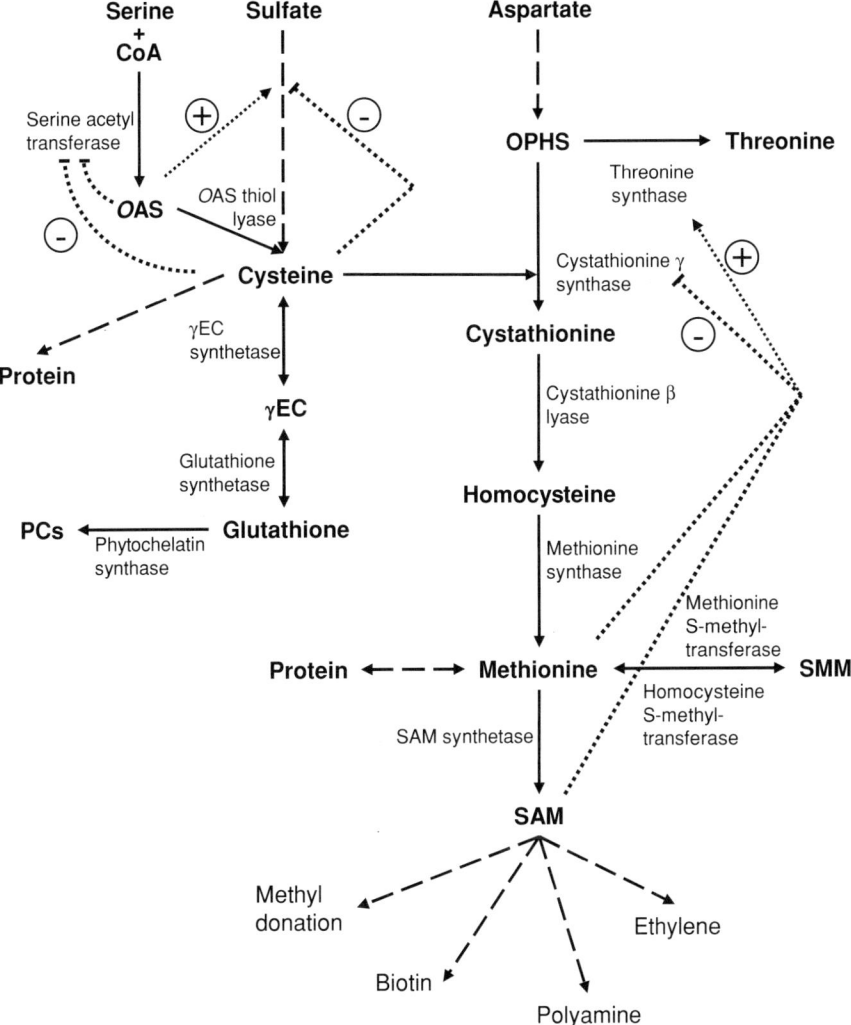

Figure 13.5 Pathways, sinks and feedback loops operating in the sulfate assimilatory pathway. Solid lines represent metabolite fluxes with the dashed line indicating multiple steps. The dotted lines represent feedback loops on gene expression or enzyme activity (see text). Only some SAT isoforms are cysteine sensitive (see text).

the greatest influence on limiting the flux through this pathway. Expression of genes for both steps is influenced by stress and demands for glutathione, although the mechanisms are not clear. There is also some evidence for post-transcriptional control of γ-glutamyl cysteine synthetase [132] occurring as part of these stress responses, acting to rapidly increase glutathione production.

13.5.1 Methionine biosynthesis

Methionine belongs to the aspartate family of amino acids which also includes lysine, threonine and isoleucine. The enzymology and molecular biology of this pathway has recently been thoroughly reviewed [133, 134]. The precursor for methionine is O-phosphohomoserine (OPH), which represents a branch in metabolism between the threonine and methionine biosynthetic pathways. For methionine biosynthesis (Figure 13.5), OPH combines with cysteine in a condensation reaction to form cystathionine, catalysed by cystathionine γ-synthase (E.C. 4.2.99.2). Cystathionine is then converted to homocysteine by cystathionine β-lyase (E.C. 4.4.1.8), before final methionine synthesis catalysed by methionine synthase (E.C. 2.1.1.14). As well as being a component of proteins, methionine is a precursor for SAM, an important methyl group donor for transmethylation reactions, including DNA, and is also a precursor for ethylene and for polyamine production. The sink demands for most of these reactions are not great as methionine is not consumed, but re-cycled. However the SAM pool may be controlled by its rate of synthesis, as the polyphosphate (pyrophosphate?) product of the reaction is inhibitory to the enzyme. Importantly SAM is an activator of threonine synthase (E.C. 4.2.99.2) [135] inducing an 8-fold increase in the rate of catalysis and a 25-fold decrease in the K_m for OPH. SAM may also have a role, along with methionine, in regulating expression of the genes of both the methionine and threonine branches by both repressing expression of cystathionine γ-synthase (at least in Arabidopsis, if not in potato [134]), and by inducing expression of threonine synthase. This leads to threonine synthesis, the alternate branch for OPH consumption. The net effect of this allosteric regulation of threonine biosynthesis and antagonistic transcriptional regulation of the two branches is to favour threonine synthesis over methionine and SAM production in the metabolic branch when sufficient sulfur supply leads to excess methionine synthesis. Both branches of the pathway are limited by homoserine availability [136].

13.6 Summary

Repression or de-repression of expression of genes encoding components of the uptake and assimilation pathway, principally the sulfate transporters and APS reductase, are apparently controlled by the balance of sulfate supply and sink demand (Figure 13.6). One or more signals for sink demand, whose identity is unclear, act at the cellular level. In a situation where cysteine or some associated metabolite pool size is perturbed, either through insufficient sulfur supply or increased demand, the cell responds accordingly. The cysteine synthase complex appears to have a central role in sensing nutritional status and catalytic activity is modulated by the relative concentrations of substrates and products. These same metabolites are likely candidates in the signal transduction pathways regulating gene expression. The sink tissues will be actively growing tissue and includes developing seeds. Source tissues will be the green tissues in which active assimilation is taking place or the roots where acquisition occurs. Apparent sink-to-source

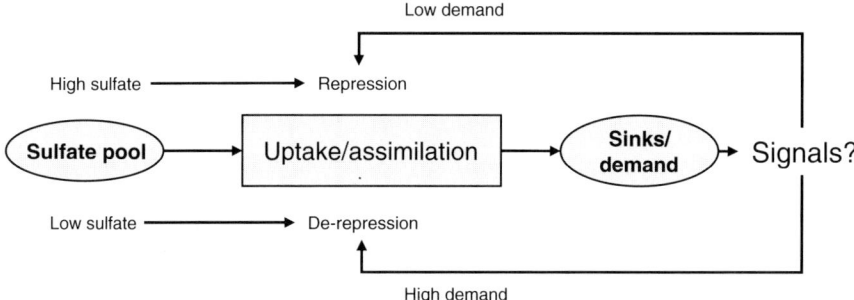

Figure 13.6 Control of sulfur metabolism is the results of a balance between supply of and demand for sulfur. Unknown metabolites are involved in the signalling of the feedback regulation.

communication is likely to be the result of a sequential perturbation of these essential metabolites, firstly in cells of sink tissues and ultimately in the root cells involved in acquisition. The actual identity of the signal metabolites still remains elusive. Some progress is being made on the missing elements of this model: the *cis* and *trans* acting factors linking metabolite pools to gene expression. The roles for post-transcriptional control are in fine tuning the network of sulfur distribution and utilisation in the context of specific metabolic pathways and the whole plant.

Acknowledgements

The authors are sponsored by grants from BBSRC and DEFRA (AR0911). Rothamsted Research receives grant-aided support from the Biotechnology and Biological Sciences Research Council of the UK.

References

1. F.J. Zhao, M.J. Hawkesford and S.P. McGrath (1999) Sulphur assimilation and effects on yield and quality of wheat. *Journal of Cereal Science* **30**, 1–17.
2. U. Wittstock and B.A. Halkier (2002) Glucosinolate research in the Arabidopsis era. *Trends in Plant Science* **7**, 263–270.
3. M.G. Jones, J. Hughes, A. Tregova, J. Milne, A.B. Tomsett and H.A. Collin (2004) Biosynthesis of the flavour precursors of onion and garlic. *Journal of Experimental Botany* **55**, 1903–1919.
4. M.M.A. Blake-Kalff, K.R. Harrison, M.J. Hawkesford, F.J. Zhao, and S.P. McGrath (1998) Distribution of sulfur within oilseed rape leaves in response to sulfur deficiency during vegetative growth. *Plant Physiology* **118**, 1337–1344.
5. M.Y. Hirai, T. Fujiwara, M. Awazuhara, T. Kimura, M. Noji and K. Saito (2003) Global expression profiling of sulfur-starved Arabidopsis by DNA macroarray reveals the role of *O*-acetyl-L-serine as a general regulator of gene expression in response to sulfur nutrition. *The Plant Journal* **33**, 651–663.

6. V. Nikiforova, J. Freitag, S. Kempa, M. Adamik, H. Hesse and R. Hoefgen (2003) Transcriptome analysis of sulfur depletion in *Arabidopsis thaliana*: interlacing of biosynthetic pathways provides response specificity. *The Plant Journal* **33**, 633–650.
7. A. Maruyama-Nakashita, E. Inoue, A. Watanabe-Takahashi, T. Yarnaya and H. Takahashi (2003) Transcriptome profiling of sulfur-responsive genes in Arabidopsis reveals global effects of sulfur nutrition on multiple metabolic pathways. *Plant Physiology* **132**, 597–605.
8. M.Y. Hirai, M. Yano, D.B. Goodenowe *et al.* (2004) Integration of transcriptomics and metabolomics for understanding of global responses to nutritional stresses in *Arabidopsis thaliana*. *Proceedings of the National Academy of Sciences of the United States of America* **101**, 10205–10210.
9. T. Kataoka, N. Hayashi, T. Yamaya and H. Takahashi (2004) Root-to-shoot transport of sulfate in Arabidopsis. Evidence for the role of SULTR3;5 as a component of low-affinity sulfate transport system in the root vasculature. *Plant Physiology* **136**, 4198–4204.
10. A. Maruyama-Nakashita, Y. Nakamura, A. Watanabe-Takahashi, T. Yamaya and H. Takahashi (2004) Induction of SULTR1;1 sulfate transporter in Arabidopsis roots involves protein phosphorylation/dephosphorylation circuit for transcriptional regulation. *Plant Cell Physiology* **45**, 340–345.
11. A. Maruyama-Nakashita, Y. Nakamura, T. Yamaya and H. Takahashi (2004) Regulation of high-affinity sulphate transporters in plants: towards systematic analysis of sulphur signalling and regulation. *Journal of Experimental Botany* **55**, 1843–1849.
12. L. Hopkins, S. Parmar, A. Błaszczyk, H. Hesse, R. Hoefgen, and M.J. Hawkesford (2005) *O*-acetylserine and the regulation of expression of genes encoding components for sulfate uptake and assimilation in potato. *Plant Physiology* **138**, 433–440.
13. P. Vauclare, S. Kopriva, D. Fell *et al.* (2002) Flux control of sulphate assimilation in *Arabidopsis thaliana*: adenosine 5'-phosphosulphate reductase is more susceptible than ATP sulphurylase to negative control by thiols. *The Plant Journal* **31**, 729–740.
14. M.J. Hawkesford and J.L. Wray (2000) Molecular genetics of sulphate assimilation. *Advances in Botanical Research* **33**, 159–223.
15. G. Kaiser, E. Martinoia, G. Schroppel-Maier and U. Heber (1989) Active-transport of sulfate into the vacuole of plant-cells provides halotolerance and can detoxify SO_2. *Journal of Plant Physiology* **133**, 756–763.
16. A.J. Miller, S.J. Cookson, S.J. Smith and D.M. Wells (2001) The use of microelectrodes to investigate compartmentation and the transport of metabolized inorganic ions in plants. *Journal of Experimental Botany* **52**, 541–549.
17. B. Lass and C.I. Ullrich-Eberius (1984) Evidence for proton/sulfate cotransport and its kinetics in *Lemna gibba* G1. *Planta* **161**, 53–60.
18. M.J. Hawkesford, J-C. Davidian and C. Grignon (1993) Sulfate proton cotransport in plasmamembrane vesicles isolated from roots of *Brassica napus* L-increased transport in membranes isolated from sulfur-starved plants. *Planta* **190**, 297–304.
19. H. Rennenberg, O. Kemper and B. Thoene (1989) Recovery of sulfate transport into heterotrophic tobacco cells from inhibition by reduced glutathione. *Physiologia Plantarum* **76**, 271–276.
20. D.T. Clarkson, M.J. Hawkesford, J-C. Davidian and C. Grignon (1992) Contrasting responses of sulfate and phosphate-transport in barley (*Hordeum vulgare* L) roots to protein-modifying reagents and inhibition of protein synthesis. *Planta* **187**, 306–314.
21. M.J. Hawkesford and A.J. Miller (2004) Ion-coupled transport of inorganic solutes. In: *Membrane Transport (Annual Plant Reviews* Vol. 15*)*, (ed. M. Blatt), Blackwell, Oxford, pp. 105–134.
22. F.W. Smith, P.M. Ealing, M.J. Hawkesford and D.T. Clarkson (1995) Plant members of a family of sulfate transporters reveal functional subtypes. *Proceedings of the National Academy of Sciences of the United States of America* **92**, 9373–9377.
23. A. Bolchi, S. Petrucco, P.L. Tenca, C. Foroni and S. Ottonello (1999) Coordinate modulation of maize sulfate permease and ATP sulfurylase mRNAs in response to variations in sulfur nutritional status: stereospecific down-regulation by L-cysteine. *Plant Molecular Biology* **39**, 527–537.

24. P. Buchner, C.E.E. Stuiver, S. Westermann et al. (2004) Regulation of sulfate uptake and expression of sulfate transporter genes in *Brassica oleracea* as affected by atmospheric H$_2$S and pedospheric sulfate nutrition. *Plant Physiology* **136**, 3396–3408.
25. P. Buchner, I. Prosser and M.J. Hawkesford (2004) Phylogeny and expression of paralogous and orthologous sulphate transporter genes in diploid and hexaploid wheats. *Genome* **47**, 526–534.
26. J. Howarth, P. Fourcroy, J-C. Davidian, F.W. Smith and M.J. Hawkesford (2003) Cloning of two contrasting high-affinity sulfate transporters from tomato induced by low sulfate and infection by the vascular pathogen *Verticillium dahliae*. *Planta* **218**, 58–64.
27. F.W. Smith, M.J. Hawkesford, P.M. Ealing et al. (1997) Regulation of expression of a cDNA from barley roots encoding a high affinity sulphate transporter. *The Plant Journal* **12**, 875–884.
28. N. Shibagaki, A. Rose, J.P. McDermott et al. (2002) Selenate-resistant mutants of *Arabidopsis thaliana* identify Sultr1;2, a sulfate transporter required for efficient transport of sulfate into roots. *The Plant Journal* **29**, 475–486.
29. H. Takahashi, N. Sasakura, M. Noji and K. Saito (1996) Isolation and characterization of a cDNA encoding a sulfate transporter from *Arabidopsis thaliana*. *FEBS Letters* **392**, 95–99.
30. H. Takahashi, M. Yamazaki, N. Sasakura et al. (1997) Regulation of sulfur assimilation in higher plants: a sulfate transporter induced in sulfate-starved roots plays a central role in *Arabidopsis thaliana*. *Proceedings of the National Academy of Sciences of the United States of America* **94**, 11102–11107.
31. H. Takahashi, W. Asanuma and K. Saito (1999) Cloning of an Arabidopsis cDNA encoding a chloroplast localizing sulphate transporter isoform. *Journal of Experimental Botany* **50**, 1713–1714.
32. H. Takahashi, N. Sasakura, A. Kimura, A. Watanabe and K. Saito (1999) Identification of two leaf-specific sulfate transporter in *Arabidopsis thaliana* (accession no. AB012048 and AB004060) (PGR99-154). *Plant Physiology* **121**, 686.
33. H. Takahashi, A. Watanabe-Takahashi, F.W. Smith, M. Blake-Kalff, M.J. Hawkesford and K. Saito (2000) The roles of three functional sulphate transporters involved in uptake and translocation of sulphate in *Arabidopsis thaliana*. *The Plant Journal* **23**, 171–182.
34. J.J. Vidmar, J.K. Schjoerring, B. Touraine and A.D.M. Glass (1999) Regulation of the hvst1 gene encoding a high-affinity sulfate transporter from *Hordeum vulgare*. *Plant Molecular Biology* **40**, 883–892.
35. J.J. Vidmar, A. Tagmount, N. Cathala, B. Touraine and J-C. Davidian (2000) Regulation of the *hvst1* gene encoding a high-affinity sulfate transporter from *Hordeum vulgare*. *FEBS Letters* **475**, 65–69.
36. N. Yoshimoto, H. Takahashi, F.W. Smith, T. Yamaya and K. Saito (2002) Two distinct high-affinity sulfate transporters with different inducibilities mediate uptake of sulfate in Arabidopsis roots. *The Plant Journal* **29**, 465–473.
37. N. Yoshimoto, E. Inoue, K. Saito, T. Yamaya and H. Takahashi (2003) Phloem-localizing sulfate transporter, Sultr1;3, mediates re-distribution of sulfur from source to sink organs in Arabidopsis. *Plant Physiology* **131**, 1511–1517.
38. M.J. Hawkesford (2000) Plant responses to sulphur deficiency and the genetic manipulation of sulphate transporters to improve S-utilization efficiency. *Journal of Experimental Botany* **51**, 131–138.
39. M.J. Hawkesford (2003) Transporter gene families in plants: the sulphate transporter gene family – redundancy or specialization? *Physiologia Plantarum* **117**, 155–165.
40. D.B. Mount and M.F. Romero (2004) The SLC26 gene family of multifunctional anion exchangers. *Pflügers Archiv European Journal of Physiology* **447**, 710–721.
41. A.L. Rae and F.W. Smith (2002) Localisation of expression of a high-affinity sulfate transporter in barley roots. *Planta* **215**, 565–568.
42. T. Kataoka, A. Watanabe-Takahashi, N. Hayashi et al. (2004) Vacuolar sulfate transporters are essential determinants controlling internal distribution of sulfate in Arabidopsis. *The Plant Cell* **16**, 2693–2704.
43. P. Buchner, H. Takahashi and M.J. Hawkesford (2004) Plant sulphate transporters: co-ordination of uptake, intracellular and long-distance transport. *Journal of Experimental Botany* **55**, 1785–1798.

44. A. Kutz, A. Müller, P. Hennig, W.M. Kaiser, M. Piotrowski and E.W. Weiler (2002) A role for nitrilase 3 in the regulation of root morphology in sulphur-starving *Arabidopsis thaliana*. *The Plant Journal* **30**, 95–106.
45. M. Awazuhara, H. Kim, D.B. Goto *et al.* (2002) A 235-bp region from a nutritionally regulated soybean seed-specific gene promoter can confer its sulfur and nitrogen response to a constitutive promoter in aerial tissues of *Arabidopsis thaliana*. *Plant Science* **163**, 75–82.
46. A. Maruyama-Nakashita, Y. Nakamura, A. Watanabe-Takahashi, E. Inoue, T. Yamaya and H. Takahashi (2005) Identification of a novel *cis*-acting element conferring sulfur deficiency response in Arabidopsis roots. *The Plant Journal* **42**, 305–314.
47. H. Kim, M.Y. Hirai, H. Hayashi, M. Chino, S. Naito and T. Fujiwara (1999) Role of O-acetyl-L-serine in the coordinated regulation of the expression of a soybean seed storage-protein gene by sulfur and nitrogen nutrition. *Planta* **209**, 282–289.
48. A. Koprivova, M. Suter, R. Op den Camp, C. Brunold and S. Kopriva (2000) Regulation of sulfate assimilation by nitrogen in Arabidopsis. *Plant Physiology* **122**, 737–746.
49. J.P. Davies, F.H. Yildiz and A.R. Grossman (1999) Sac3, an Snf1-like serine threonine kinase that positively and negatively regulates the responses of Chlamydomonas to sulfur limitation. *The Plant Cell* **11**, 1179–1190.
50. N. Shibagaki and A.R. Grossman (2004) Probing the function of STAS domains of the Arabidopsis sulfate transporters. *The Journal of Biological Chemistry* **279**, 30791–30799.
51. H. Rouached, P. Berthomieu, E. El Kassis *et al.* (2005) Structural and functional analysis of the C-terminal STAS (sulfate transporter and anti-sigma antagonist) domain of the *Arabidopsis thaliana* sulfate transporter SULTR1.2. The *Journal of Biological Chemistry* **280**, 15976–15983.
52. E.L. McCandless and J.S. Craigie (1979) Sulfated polysaccharides in red and brown algae. *Annual Reviews of Plant Physiology* **30**, 41–53.
53. L. Varin, F. Marsolais, M. Richard and M. Rouleau (1997) Sulfation and sulfotransferases 6: biochemistry and molecular biology of plant sulfotransferases, *The FASEB Journal* **11**, 517–525.
54. M. Rouleau, F. Marsolais, M. Richard *et al.* (1999) Inactivation of brassinosteroid biological activity by a salicylate-inducible steroid sulfotransferase from *Brassica napus*. *The Journal of Biological Chemistry* **274**, 20925–20930.
55. T. Leustek, M.N. Martin, J-A. Bick and J.P Davies (2000) Pathways and regulation of sulfur metabolism revealed through molecular and genetic studies. *Annual Review of Plant Physiology and Plant Molecular Biology* **51**, 141–166.
56. J.E. Lunn, M. Droux, J. Martin and R. Douce (1990) Localization of ATP sulfurylase and O-acetylserine(thiol)lyase in spinach leaves. *Plant Physiology* **94**, 1345–1352.
57. F. Renosto, H.C. Patel, R.L. Martin, C. Thomassian, G. Zimmerman and I.H. Segel (1993) ATP sulfurylase from higher plants: kinetic and structural characterization of the chloroplast and cytosol enzymes from spinach leaf. *Archives Biochemistry Biophysics* **307**, 272–285.
58. D. Klonus, R. Höfgen, L. Willmitzer and J.W. Riesmeier (1994) Isolation and characterization of 2 cDNA clones encoding ATP-sulfurylases from potato by complementation of a yeast mutant. *The Plant Journal* **6**, 105–112.
59. C. Rotte and T. Leustek (2000) Differential subcellular localization and expression of ATP sulfurylase and 5'-adenylylsulfate reductase during ontogenesis of Arabidopsis leaves indicates that cytosolic and plastid forms of ATP sulfurylase may have specialized functions. *Plant Physiology* **124**, 715–724.
60. M. Klein and J. Papenbrock (2004) The multi-protein family of Arabidopsis sulphotransferases and their relatives in other plant species. *Journal of Experimental Botany* **55**, 1809–1820.
61. C.H. Lillig, S. Schiffmann, C. Berndt, A. Berken, R. Tischka and J.D. Schwenn (2001) Molecular and catalytic properties of *Arabidopsis thaliana* adenylyl sulfate (APS)-kinase. *Archives of Biochemistry and Biophysics* **392**, 303–310.
62. C. Brunold and M. Suter (1989) Localization of enzymes of assimilatory sulfate reduction in pea roots. *Planta* **179**, 228–234.
63. T. Leustek (1996) Molecular genetics of sulfate assimilation in plants. *Physiologia Plantarum* **97**, 411–419.

64. T. Leustek, M. Murillo and M. Cervantes (1994) Cloning of a cDNA-encoding ATP sulfurylase from *Arabidopsis thaliana* by functional expression in *Saccharomyces cerevisiae*. *Plant Physiology* **105**, 897–902.
65. E.A.H. Pilon-Smith, S. Hwang, C.M. Lytle *et al.* (1999) Overexpression of ATP sulfurylase in Indian mustard leads to increased selenate uptake, reduction, and tolerance. *Plant Physiology* **119**, 123–132.
66. H.M. Logan, N. Cathala, C. Grignon and J-C. Davidian (1996) Cloning of a cDNA encoded by a member of the *Arabidopsis thaliana* ATP sulfurylase multigene family – expression studies in yeast and in relation to plant sulfur nutrition. *The Journal of Biological Chemistry* **271**, 12227–12233.
67. Y. Hatzfeld, N. Cathala, C. Grignon and J-C. Davidian (1998) Effect of ATP sulfurylase overexpression in bright yellow 2 tobacco cells. Regulation of ATP sulfurylase and SO_4^{2-} transport activities. *Plant Physiology* **116**, 1307–1313.
68. J.A. Bick, A.T. Setterdahl, D.B. Knaff *et al.* (2001) Regulation of the plant-type 5'-adenylyl sulfate reductase by oxidative stress. *Biochemistry* **40**, 9040–9048.
69. B.C. Gerwick, S.B. Ku and C.C. Black (1980) Initiation of sulfate activation: a variation in C4 photosynthesis in plants. *Science* **209**, 512–515.
70. D. Schmutz and C. Brunold (1984) Intercellular localization of assimilatory sulfate reduction in leaves of *Zea mays* and *Triticum aestivum*. *Plant Physiology* **74**, 866–870.
71. M. Burgener, M. Suter, S. Jones and C. Brunold (1998) Cyst(e)ine is the transport metabolite of assimilated sulfur from bundle-sheath to mesophyll cells in maize leaves. *Plant Physiology* **116**, 1315–1322.
72. A. Koprivova, M. Melzer, P. von Ballmoos, T. Mandel, C. Brunold and S. Kopriva (2001) Assimilatory sulfate reduction in C_3, C_3–C_4, and C_4 species of *Flaveria*. *Plant Physiology* **127**, 543–550.
73. A.G. Lappartient, J.J. Vidmar, T. Leustek, A.D.M. Glass and B. Touraine (1999) Inter-organ signalling in plants: regulation of ATP sulfurylase and sulfate transporter genes expression in roots mediated by phloem-translocated compound. *The Plant Journal* **18**, 89–95.
74. H. Hesse, N. Trachsel, M. Suter *et al.* (2003) Effect of glucose on assimilatory sulphate reduction in *Arabidopsis thaliana* roots. *Journal of Experimental Botany* **54**, 1701–1709.
75. L. Hopkins, S. Parmar, D.L. Bouranis, J.R. Howarth and M.J. Hawkesford (2004) Coordinated expression of sulfate uptake and components of the sulfate assimilatory pathway in maize. *Plant Biology* **6**, 408–414.
76. M.M. Jones-Mortimer (1968) Positive control of sulphate reduction in *Escherichia coli*. Isolation, characterization and mapping of cysteineless mutants of *E. coli* K12. *Biochemical Journal* **110**, 589–595.
77. N.M. Kredich (1971) Regulation of L-cysteine biosynthesis in *Salmonella typhimurium*. I. Effects of growth on varying sulfur sources and *O*-acetyl-L-serine on gene expression. *The Journal of Biological Chemistry* **246**, 3474–3484.
78. J.D. Schwenn (1989) Sulfate assimilation in higher-plants – a thioredoxin-dependent PAPS-reductase from spinach leaves. *Zeitschrift für Naturforschung* **44c**, 504–508.
79. A. Koprivova, A.J. Meyer, G. Schween, C. Herschbach, R. Reski and S. Kopriva (2002) Functional knockout of the adenosine 5'-phosphosulfate reductase gene in *Physcomitrella patens* revives an old route of sulfate assimilation. *The Journal of Biological Chemistry* **277**, 32195–32201.
80. Y. Yamaguchi, T. Nakamura, E. Harada, N. Koizumi and H. Sano (1999) Differential accumulation of transcripts encoding sulfur assimilation enzymes upon sulfur and or nitrogen deprivation in *Arabidopsis thaliana*. *Bioscience Biotechnology and Biochemistry* **63**, 762–766.
81. J.F. Gutierrez-Marcos, M.A. Roberts, E.I. Campbell and J.L. Wray (1996) Three members of a novel small gene-family from *Arabidopsis thaliana* able to complement functionally an *Escherichia coli* mutant defective in PAPS reductase activity encode proteins with a thioredoxin-like domain and "APS reductase" activity. *Proceedings of the National Academy of Sciences of the United States of America* **93**, 13377–13382.
82. Y. Hatzfeld, S. Lee, M. Lee, T. Leustek and K. Saito (2000) Functional characterization of a gene encoding a fourth ATP sulfurylase isoform from *Arabidopsis thaliana*. *Gene* **248**, 51–58.

83. S. Heiss, H.J. Schafer, A. Haag-Kerwer and T. Rausch (1999) Cloning sulfur assimilation genes of *Brassica juncea* L: cadmium differentially affects the expression of a putative low-affinity sulfate transporter and isoforms of ATP sulfurylase and APS reductase. *Plant Molecular Biology* **39**, 847–857.
84. C. Brunold and M. Suter (1984) Regulation of sulfate assimilation by nitrogen nutrition in the duckweed *Lemna minor* L. *Plant Physiology* **76**, 579–583.
85. S. Kopriva, R. Muheim, A. Koprivova et al. (1999) Light regulation of assimilatory sulphate reduction in *Arabidopsis thaliana*. *The Plant Journal* **20**, 37–44.
86. L. Lejay, X. Gansel, M. Cerezo *et al.* (2003) Regulation of root ion transporters by photosynthesis: functional importance and relation with hexokinase. *The Plant Cell* **15**, 2218–2232.
87. U. Neuenschwander, M. Suter and C. Brunold (1991) Regulation of sulfate assimilation by light and O-acetyl-L-serine in *Lemna minor* L. *Plant Physiology* **97**, 253–258.
88. C-L. Cheng, G.N. Acedo, M. Cristinsin and M. Conkling (1992) Sucrose mimics the light induction of Arabidopsis nitrate reductase gene transcription. *Proceedings of the National Academy of Sciences of the United States of America* **89**, 1861–1864.
89. C. Brunold (1993) Regulatory interactions between sulfate and nitrate assimilation. In: *Sulfur Nutrition and Sulphur Assimilation in Higher Plants*, (eds L.J. de Kok, I. Stulen, H. Rennenberg, C. Brunold and W.E. Rauser), SPB Academic, The Hague, The Netherlands, pp. 61–75.
90. A. Maruyama-Nakashita, Y. Nakamura, T. Yamaya and H. Takahashi (2004) A novel regulatory pathway of sulfate uptake in Arabidopsis roots: implication of CRE1/WOL/AHK4-mediated cytokinin-dependent regulation. *The Plant Journal* **38**, 779–789.
91. N. Ohkama, K. Takei, H. Sakakibara, H. Hayashi, T. Yoneyama and T. Fujiwara (2002) Regulation of sulfur-responsive gene expression by exogenously applied cytokinins in *Arabidopsis thaliana*. *Plant Cell Physiology* **43**, 1493–1501.
92. N. Bogdanova and R. Hell (1997) Cysteine synthesis in plants: protein–protein interactions of serine acetyltransferase from *Arabidopsis thaliana*. *The Plant Journal* **11**, 251–262.
93. M. Droux, M-L. Ruffet, R. Douce and D. Job (1998) Interactions between serine acetyltransferase and O-acetylserine (thiol) lyase in higher plants – structural and kinetic properties of the free and bound enzymes. *European Journal of Biochemistry* **255**, 235–245.
94. M. Wirtz, O. Berkowitz, M. Droux and R. Hell (2001) The cysteine synthase complex from plants. Mitochondrial serine acetyltransferase from *Arabidopsis thaliana* carries a bifunctional domain for catalysis and protein–protein interaction. *European Journal of Biochemistry* **268**, 686–693.
95. M. Droux (2003) Plant serine acetyltransferase: new insights for regulation of sulphur metabolism in plant cells. *Plant Physiology and Biochemistry* **41**, 619–627.
96. N.M. Kredich and G.M. Tomkins (1966) The enzymic synthesis of L-cysteine in *Escherichia coli* and *Salmonella typhimurium*. *The Journal of Biological Chemistry* **241**, 4955–4965.
97. N.M. Kredich (1996) Biosynthesis of cysteine. In: *Escherichia coli and Salmonella typhimurium. Cellular and molecular biology*, (eds F.C. Neidhardt, R. Urtiss, J.L. Ingraham, E.C.C. Lin, K.B. Low, B. Magasanik, W.S. Reznikoff, M. Riley, M. Schaechter and E. Umberger), ASM, Washington, DC, pp. 514–527.
98. I.K. Smith and J.F. Thompson (1969) The synthesis of O-acetylserine by extracts prepared from higher plants. *Biochemical and Biophysical Research Communications* **35**, 939–945.
99. C. Brunold and M. Suter (1982) Intracellular-localization of serine acetyltransferase in spinach leaves. *Planta*, **155**, 321–327.
100. M. Droux, J. Martin, P. Sajus and R. Douce (1992) Purification and characterization of O-acetylserine (thiol) lyase from spinach chloroplasts. *Archives of Biochemistry and Biophysics* **295**, 379–390.
101. J.E. Lunn, M. Droux, J. Martin and R. Douce (1990) Localization of ATP sulfurylase and O-acetylserine(thiol)lyase in spinach leaves. *Plant Physiology* **94**, 1345–1352.
102. N. Rolland, M. Droux and R. Douce (1992) Subcellular distribution of O-acetylserine(thiol)lyase in cauliflower (*Brassica oleracea* L.) inflorescence. *Plant Physiology* **98**, 927–935.
103. M. Droux (2004) Sulfur assimilation and the role of sulfur in plant metabolism: a survey. *Photosynthesis Research* **79**, 331–348.

104. A.G.S. Warrilow and M.J. Hawkesford (2000) Cysteine synthase (O-acetylserine (thiol) lyase) substrate specificities classify the mitochondrial isoform as a cyanoalanine synthase. *Journal of Experimental Botany* **51**, 985–993.
105. J. Jost, O. Berkowitz, M. Wirtz, L. Hopkins, M.J. Hawkesford and R. Hell (2000) Genomic and functional characterization of the oas gene family encoding O-acetylserine (thiol) lyases, enzymes catalyzing the final step in cysteine biosynthesis in *Arabidopsis thaliana*. *Gene* **253**, 237–247.
106. M. Noji, K. Inoue, N. Kimura, A. Gouda and K. Saito (1998) Isoform-dependent differences in feedback regulation and subcellular localization of serine acetyltransferase involved in cysteine biosynthesis from *Arabidopsis thaliana*. *The Journal of Biological Chemistry* **273**, 32739–32745.
107. M. Noji and K. Saito (2002) Molecular and biochemical analysis of serine acetyltransferase and cysteine synthase towards sulfur metabolic engineering in plants. *Amino Acids* **22**, 231–243.
108. J.R. Howarth, J.R. Dominguez-Solis, G. Gutierrez-Alcala, J.L. Wray, L.C. Romero and C. Gotor (2003) The serine acetyltransferase gene family in *Arabidopsis thaliana* and the regulation of its expression by cadmium. *Plant Molecular Biology* **51**, 589–598.
109. M. Wirtz, M. Droux and R. Hell (2004) O-acetylserine (thiol) lyase: an enigmatic enzyme of plant cysteine biosynthesis revisited in *Arabidopsis thaliana*. *Journal of Experimental Botany* **55**, 1785–1798.
110. C.G. Kawashima, O. Berkowitz, R. Hell, M. Noji and K. Saito (2005) Characterization and expression analysis of a serine acetyltransferase gene family involved in a key step of the sulfur assimilation pathway in Arabidopsis. *Plant Physiology* **137**, 220–230.
111. K. Nakamura and G. Tamura (1990) Isolation of serine acetyltransferase complexed with cysteine synthase from *Allium tuberosum*. *Agricultural and Biological Chemistry* **54**, 649–656.
112. M.L. Ruffet, M. Droux and R. Douce (1994) Purification and kinetic-properties of serine acetyltransferase free of O-acetylserine(thiol)lyase from spinach-chloroplasts. *Plant Physiology* **104**, 597–604.
113. N.M. Kredich, M.A. Becker and G.M. Tomkins (1969) Purification and characterisation of cysteine synthase, a bifunctional protein complex from *Salmonella typhimurium*. *The Journal of Biological Chemistry* **244**, 2428–2439.
114. K. Nakamura, A. Hayama, M. Masada, K. Fukushima and G. Tamura (1988) Purification and some properties of plant serine acetyltransferase. *Plant Cell Physiology* **29**, 689–693.
115. M-L. Ruffet, M. Lebrun, M. Droux and R. Douce (1995) Subcellular-distribution of serine acetyltransferase from *Pisum sativum* and characterization of an *Arabidopsis thaliana* putative cytosolic isoform. *European Journal of Biochemistry* **227**, 500–509.
116. P.F. Cook and R.T. Wedding (1977) Initial kinetic characterization of multienzyme complex, cysteine synthetase. *Archives of Biochemistry and Biophysics* **178**, 293–302.
117. I.K. Smith and J.F. Thompson (1971) Purification and characterization of L-serine transacetylase and O-acetyl-L-serine sulfhydrylase from kidney bean seedlings (*Phaseolus vulgaris*). *Biochimica et Biophysica Acta* **227**, 288–295.
118. K. Saito, H. Yokoyama, M. Noji and I. Murakoshi (1995) Molecular-cloning and characterization of a plant serine acetyltransferase playing a regulatory role in cysteine biosynthesis from watermelon. *The Journal of Biological Chemistry* **270**, 16321–16326.
119. M. Noji, Y. Takagi, N. Kimura et al. (2001) Serine acetyltransferase involved in cysteine biosynthesis from spinach: molecular cloning, characterization and expression analysis of cDNA encoding a plastidic isoform. *Plant Cell Physiology* **42**, 627–634.
120. J.R. Howarth, M.A. Roberts and J.L. Wray (1997) Cysteine biosynthesis in higher plants: a new member of the *Arabidopsis thaliana* serine acetyltransferase small gene-family obtained by functional complementation of an *Escherichia coli* cysteine auxotroph. *Biochimica et Biophysica Acta* **1350**, 123–127.
121. R.S. Monroe, J. Ostrowski, M.M. Hryniewicz and N.M. Kredich (1990) *In vitro* interactions of CysB protein with the CysK and CysJIH promoter regions of *Salmonella typhimurium*. *Journal of Bacteriology* **172**, 6919–6929.
122. D.T. Clarkson, E. Diogo and S. Amancio (1999) Uptake and assimilation of sulphate by sulphur deficient Zea mays cells: the role of O-acetyl-L-serine in the interaction between nitrogen and sulphur assimilatory pathways. *Plant Physiology and Biochemistry* **37**, 283–290.

123. N. Ohkama-Ohtsu, I. Kasajima, T. Fujiwara and S. Naito (2004) Isolation and characterization of an Arabidopsis mutant that overaccumulates *O*-acetyl-L-Ser. *Plant Physiology* **136**, 3209–3222.
124. K. Saito, M. Kurosawa, K. Tatsuguchi, Y. Takagi and I. Murakoshi (1994) Modulation of cysteine biosynthesis in chloroplasts of transgenic tobacco overexpressing cysteine synthase [O-acetylserine(thiol)-lyase]. *Plant Physiology* **106**, 887–895.
125. D. Thomas and Y. Surdin-Kerjan (1997) Metabolism of sulfur amino acids in *Saccharomyces cerevisiae*. *Microbiology and Molecular Biology Reviews* **61**, 503–532.
126. H. Rennenberg (1982) Glutathione metabolism and possible biological roles in higher plants. *Phytochemistry* **21**, 2771–2781.
127. C. Cobbett (2003) Metallothioneins and phytochelatins: molecular aspects. In: *Sulphur in Plants*, (eds Y.P. Abrol and A. Ahmad), pp.177–188. Kluwer Academic Publishers, Dordrecht.
128. G. Noctor, A.C.M. Arisi, L. Jouanin and C.H. Foyer (1998) Manipulation of glutathione and amino acid biosynthesis in the chloroplast. *Plant Physiology* **118**, 471–482.
129. G. Noctor, M. Strohm, L. Jouanin, K.J. Kunert, C.H. Foyer and H. Rennenberg (1996) Synthesis of glutathione in leaves of transgenic poplar overexpressing gamma-glutamylcysteine synthetase. *Plant Physiology* **112**, 1071–1078.
130. A.C. Arisi, G. Noctor, C.H. Foyer and L. Jouanin (1997) Modification of thiol contents in poplars (*Populus tremula* x P. alba) overexpressing enzymes involved in glutathione synthesis. *Planta* **203**, 362–372.
131. G. Creissen, J. Firmin, M. Fryer *et al.* (1999) Elevated glutathione biosynthetic capacity in the chloroplasts of transgenic tobacco plants paradoxically causes increased oxidative stress. *The Plant Cell* **11**, 1277–1291.
132. M.J. May, T. Vernoux, R. Sánchez-Fernández, M. van Montagu, and D. Inzé (1998) Evidence for posttranscriptional activation of gamma-glutamylcysteine synthetase during plant stress responses. *Proceedings of the National Academy of Sciences of the United States of America* **95**, 12049–12054.
133. S. Ravanel, B. Gakiere, D. Job and R. Douce (1998) The specific features of methionine biosynthesis and metabolism in plants. *Proceedings of the National Academy of Sciences of the United States of America* **95**, 7805–7812.
134. H. Hesse and R. Hoefgen (2003) Molecular aspects of methionine biosynthesis. *Trends in Plant Science* **8**, 259–262.
135. G. Curien, D. Job, R. Douce and R. Dumas (1998) Allosteric activation of Arabidopsis threonine synthase by S-adenosylmethionine. *Biochemistry* **37**, 13212–13221.
136. M. Lee, M.N. Martin, A.O. Hudson, J. Lee, M.J. Muhitch and T. Leustek (2005) Methionine and threonine synthesis are limited by homoserine availability and not the activity of homoserine kinase in *Arabidopsis thaliana*. *The Plant Journal* **41**, 685–696.

Index

14-3-3 proteins, 131, 132, 133, 134, 135, 136, 137, 139, 142, 143, 239, 245, 246, 300, 333
2,4-dioenoyl CoA reductase, 39

α-amylase, 237, 243
β-amylase, 227, 237, 267, 268
β-conglycinin, 352
β-glucuronidase (GUS), 100, 104
β-methylcrotonyl-CoA carboxylase, 312
γ-carbonic anhydrase (see also METC complex I), 328
γ-glutamyl cysteine synthase, 362

ABRE (abscisic acid response element), 12
abscisic acid, 151, 332
acetyl–CoA carboxylase (ACCase), 163, 305, 306
acetyl-CoA synthetase, 313
aconitase, 26, 32, 36, 42, 291, 292, 299, 302, 304, 315
    Fe regulatory protein, 292
acyl carrier protein (ACP), 306
adenosine 5′-phosphosulfate reductase (APSR), 174, 350, 354, 355, 356, 360, 364
    AtAPR1, 355
adenosine 5′-phosphosulfate kinase (APSK), 350, 354, 355
    Akn1 plastidic isoform, 354
adenylate kinase, 301
AdoHyc hydrolase, 94
ADP-glucose pyrophosphorylase (ADGP), 26, 33, 107, 263, 269, 271, 272, 273, 274, 280, 281
    ADPG mutants, 263, 270, 271
    bacterial, 260, 277
    cytosolic, 261
    plastidic, 88, 89, 90, 162, 163, 247, 258, 261, 262, 271, 327, 332

AGX-1 protein, 241
alcohol dehydrogenase (ADH), 13, 26, 169
aldehyde dehydrogenase (ALDH), 37, 142, 313, 314
    'nuclear restorer gene', 314
aldolase, 33, 42, 169
    aldolase-like protein, 238
    cytosolic, 26, 88, 202, 235, 236, 238, 248, 250, 303
    plastidic, 32, 39, 153, 168, 188, 189, 197, 202, 203, 205
algae, 190, 196, 197, 258, 304, 313, 327, 353
    green, 197, 200, 202, 204, 219, 229, 272, 273, 278, 279, 353
    red, 156, 191, 202, 278, 279
alkaline phosphatase, 26
alkylhydroperoxide reductase, 158
alliin, 348
allosteric regulation, 33, 45, 86, 88, 90, 140, 143, 199, 219, 229, 243, 244, 247, 260, 261, 269, 329, 330, 364
alternative oxidase (AOX), 167, 291, 296, 297, 298, 301, 303, 309, 310
alternative splicing (RNA), 223, 224
aminoacyl tRNA synthetase, 159
AMP-activated protein kinase (AMPK), 138, 139, 141
amyloplast, 33, 38, 142
Anacystis nidulans, 204
anaplerotic reactions, 229, 329, 335
ANN (artificial neural networks), 10, 11
antiporter,
    glutamine/glutamate, 93
    calcium/H$^+$ (CAX1), 172, 173
    sucrose/H$^+$, 102
apoplast, 103, 349
    cell walls, 34, 40, 86
    loading, 102
    transport, 98
apple (*Malus domesticus*), 78

Arabidopsis thaliana, 2, 3, 4, 5, 6, 11, 12, 15, 31, 33, 34, 36, 39, 40, 42, 43, 64, 65, 67, 68, 72, 74, 75, 76, 77, 78, 90, 92, 93, 94, 95, 99, 100, 104, 105, 106, 121, 122, 123, 124, 128, 129, 130, 131, 132, 133, 134, 138, 140, 141, 153, 156, 157, 158, 159, 162, 163, 166, 167, 169, 171, 172, 173, 175, 193, 194, 196, 199, 205, 220, 221, 223, 226, 227, 228, 229, 237, 238, 239, 240, 241, 242, 244, 245, 248, 249, 250, 259, 263, 265, 266, 267, 268, 269, 271, 276, 279, 280, 282, 294, 298, 300, 302, 304, 307, 308, 311, 313, 332, 338, 348, 351, 352, 353, 354, 355, 356, 357, 359, 360, 364
  Cvi-1 ecotype, 76
  T-DNA insertional mutants, 89, 101, 105, 276, 296
  Ws-2 ecotype, 76
  Sex1, 267, 268
  Sex4, 267
archaebacterium, 191
ascorbate-glutathione cycle, 37, 308, 309
asparagine synthetase, 15, 93
ATPase, 98
  AAA^+ protein family, 192, 193
ATP-citrate lyase, 132
ATP sulfurylase, 354, 360
  AtATPS1, 355
  AtATPS2, 355
  cytosolic, 350
  plastidic, 350, 354, 356
ATP synthase complex
  mitochondrial (ETC complex V), 32, 136, 291, 295, 299, 300, 301, 303
  chloroplastidic, 163, 164, 168
ATP synthase inhibitors (*see also* Plant UCPs),
  oligomycin, 295
aubergine (*Solanum melongena*), 223
auxin, 39, 151, 357

Bacillus sp., 353
barley (*Hordeum vulgare*), 33, 68, 99, 158, 169, 221, 241, 242, 243, 262, 263, 264, 265, 270, 280, 302, 333, 351
  Lys5 mutant, 263
  RisØ 16 mutant, 263

Bcr protein (14-3-3 target), 137
bean (*Phaseolus vulgaris*), 102, 302, 359
  transgenic, 74
Benson-Calvin cycle (*see* Calvin cycle)
biotin, 30
blue-copper proteins, 308
Botrytis allii, 77
branched chain keto-acid dehydrogenase complex, 300, 312
Brassica, 170, 357
  Brassica oleracea, 170
  cauliflower (*Brassica oleraceae* (Botrytis group)), 94
  Indian mustard (*Brassica juncea*), 355
  oilseed rape (*Brassica napus*), 100
  seeds, 107
brassinosteroids, 354
  brassinolide, 129, 130
brassinolide signalling,
  BES1, 130
  BIN2, 130
  BSU1 (nuclear protein phosphatase), 129, 130
  BSU1-like protein, 129
  BSL1, 129
  BSL2, 129
  BSL3, 129
BRCT (BRCA1 carboxy-terminal) domain, 131, 132, 133
bulbs, 97

C_4 photosynthesis, 96, 143, 187, 219, 220, 222, 224, 228
CAB gene, 15, 226
Caenorhabditis elegans, 121, 134
calcium-dependent protein kinases (CDPKs), 122, 140, 224, 225, 244
  CDPK3, 141, 239
  CDPK-related kinase, 122, 123
calmodulin, 124, 227, 311, 312
Calvin cycle, 38, 39, 47, 86, 87, 90, 95, 96, 105, 107, 150, 156, 163, 164, 187, 188, 189, 190, 194, 195, 197, 198, 199, 201, 202, 203, 204, 205, 206, 328, 333
CaMV 35*S* promoter, 101, 105
carbohydrate active enzyme (CAZy) classification, 268
carboxyarabinitol-1-phosphate (CA1P), 192

carboxypeptidase A, 191
cardiolipin, 302
castor oil plant (*Ricinus communis*), 32, 106, 172, 229
catalase, 167
CcdA, 165
CDSP32 (drought-induced stress protein), 168, 169
cell wall, *see* apoplast
cellulose synthase complex, 140
Chlamydomonas reinhardtii, 142, 156, 161, 165, 166, 172, 190, 191, 193, 199, 200, 203, 204, 205, 228, 264, 265, 266, 272, 273, 274, 275, 276, 280, 332, 333, 353
chloroplast
   envelope, 86, 87, 90, 92, 95, 96
   thylakoid membrane, 203
   transit peptide, 190
chromoplast, 38
CHS gene, 226
circadian rhythm, 200, 223, 225, 226, 227, 228, 273, 279, 331
   diurnal cycle, 4, 16, 17, 44, 235, 239, 243, 244, 269, 273, 280
   citrate synthase, 26, 40, 42, 291, 292, 293, 315
   citric acid cycle, 14, 26, 33, 36, 37, 42, 43, 46, 75, 92, 143, 290, 292, 293, 294, 296, 299, 301, 302, 303, 304, 305, 306, 307, 310, 312, 313, 314, 328, 331, 334
Cladosporium fulvum, 170
   Avr9, 170
Clusia minor, 221
Clusia spp., 226
control theory, 60, 61
Corynebacterium glutamicum, 74
cotton (*Gossypium hirsutum*), 203
CP12 protein, 195, 196, 197, 200, 204, 205, 206
CP12, 205
Crassulacean acid metabolism, 143, 187, 219, 220, 222, 224, 225, 226, 227, 228
cucumber (*Cucumis sativus*), 32
cyanobacteria, 159, 160, 161, 190, 191, 197, 198, 204, 245, 261, 272, 277
Cyanothece, 277
Cyanidioschyzon merolae, 279

cyclin-dependent kinases, 133
   CDK2, 137
cyclohexamide, 220
cyclophilin (*see also* immunophilins), 164
   AtCYP20-3, 164
Cys regulon, 360
   CysB, 360
   CysE, 357, 360
   CysK, 357
   CysM, 357
cystathionine β-lyase, 94, 363, 364
cystathionine γ-synthetase, 94, 363, 364
cysteine proteases, 98
cysteine synthase complex, 358, 359, 361, 362, 364
cytochrome P450, 13
cytokinin, 357
cytoplasmic male sterility/fertility, 32, 314, 330

D-enzyme, 266, 275
dehydroascorbate reductase, 307, 308
deoxy-arabino-heptulosonate 7-phosphate (DHAP) synthase, 162
deoxy-xylulose 5-phosphate (DOXP) pathway, 96
Digiteria sanguinalis, 224, 225
dihydrolipoamide reductase, 158
DNA microarrays, 2, 3, 5, 6, 8, 11, 12, 15, 16, 17, 128, 194, 242, 243, 245, 249, 279
DRE element, 12
Drosophila melanogaster, 129, 131, 134, 138

Edman degradation sequencing, 32
EF-hand motifs for Ca^{2+}-binding, 296
electron transport chain, chloroplastic (PETC), 47, 156, 194, 200, 326, 327, 328
   cytochrome b_6/f complex, 165, 166, 194
   DCMU, 200
   LHCII kinases, 166
   light-harvesting complex II (LHCII), 166
   photosystem I, 165
   photosystem II, 165, 166
   psaA-psaB (PSI), 165
   psbA (PSII), 165
   psbD (PSII), 165

electron transport chain, mitochondrial
(METC), 36, 43, 46, 47, 291, 294,
298, 299, 301, 304, 305, 307, 309,
311, 313, 314, 328
   alternate NAD(P)H dehydrogenase, 296,
303, 304, 309, 314, 330
   complex I, 32, 36, 37, 291, 292, 294,
301, 302, 303, 304, 305, 309, 328,
330
   complex II (see also SDH), 291, 293,
294, 295, 299, 302
   complex III, 37, 291, 295, 296, 299,
301, 302, 309
   complex IV, 37, 291, 295, 296, 301,
302, 303
   rotenone-resistant NAD(P)H
dehydrogenase, 295
electron transport chain inhibitors,
   antimycin A, 295, 302, 305, 309
   menadione, 302
   myxothiazol, 295
   rotenone, 294, 305, 309
electrophoresis,
   blue-native (BN-PAGE), 27, 28, 36, 294,
302, 306
   capillary, 67, 77
   fluorescence, 301
   two-dimensional (2DGE), 27, 28, 30,
32, 33, 35, 42, 45, 46, 299
elongation factor 2, 169
EMS mutants, 73
endoplasmic reticulum (ER), 35
   ER proteins, 173
   ER retention signals, 174
enolase, 26, 32, 40, 92, 94, 303
   sieve-element-localised, 32
epidermal growth factor (EGF), 47
Erwinia carotovora spp carotovora, 77
Escherichia coli, 152, 159, 164, 192,
219, 240, 243, 263, 277, 355,
357, 359
ethylene, 332, 363, 364
Euglena gracilis, 202
eukaryote translation factors,
   initiation factor 4A, 169
   initiation factor 4B, 138
   initiation factor 4E, 138
exchangers,
   oxaloacetic/malate, 92

fatty acid synthase (FAS), 305, 306
FC analysis, 8
FDR (false discovery rate), 9
ferredoxin-thioredoxin control, 188, 189,
198, 200
ferritin proteins, 308
FGC (functional gene categories), 12
FK506 (FKBP), 164
   AtFKP13 (peptidyl-prolyl *cis-trans*
isomerase), 164, 165
Flaveria trinervia, 221, 223, 224, 225,
228, 355
floral tissues, 78, 205, 222, 292
   anthers, 32, 290
   peduncle, 205
   pistil, 170
   pollen, 32, 170, 250, 272, 314
   stigma, 170
   tapetum, 290
floridean starch, 278, 279
fluorescent proteins,
   blue, 50
   green, 34, 50, 100, 156, 303
   yellow, 50, 303
formate dehydrogenase, 46, 299, 312, 313
forkhead associated (FHA) domain, 128, 132
FRET (fluorescence energy transfer), 49,
50, 78
fructan exohydrolase (FEH), 237
fructokinase, 17, 26, 235, 237
fructose-1,6-bisphosphatase, 169
   cytosolic, 17, 88, 90, 141, 197, 235,
236, 237, 238, 239, 240, 247, 248
   plastidic, 39, 156, 162, 163, 168, 187,
188, 189, 197, 198, 199, 200, 201,
203, 249
fructose-1,6-/sedoheptulose-1,7-
bisphosphatase (FBP/SBPase),
197, 198
fructose-2,6-bisphosphate, 141, 239, 240,
247, 248
fructosyltransferase, 235
fruit, 188, 237, 238, 250
fumarase, 26, 291, 293
fungi, 78, 197, 279, 355
   mycorrhizal association, 326
   Neurospora crassa, 197
Fusarium oxysporum, 77
fusicoccin, 136

GABA (γ-aminobutyric acid) shunt, 37, 311, 312
   γ-hydroxybutyrate dehydrogenase (GHBD), 312
   GABA transaminase (GABA-T), 311, 312
   glutamate decarboxylase (GAD), 311, 312
   succinic semialdehyde dehydrogenase (SSADH), 311, 312
galactinol synthase/α-galactosyltransferase, 235
L-galactono-1,4-lactone dehydrogenase (GLDH), 305
gas chromatography, 64, 67
GCN (general control non-derepressible) 2 protein, 137, 139, 332
GCRMA analysis, 7, 8
germin-like protein, 169
gibberellins, 243
GiGA pathway visualisation tool, 13, 14
GFP (green fluorescent protein), see fluorescent proteins
gliadin, 38
glucan water dikinase (GWD), 162, 163, 237, 268
glucanotransferase (DPE2), 269
glucokinase, 235, 237
gluconeogenesis, 33, 74, 197, 198, 238, 306
glucose-6-phosphate dehydrogenase, 26, 163, 188
glucosinolate, 348, 354
glucosinolate-myrosinase system, 62
glucosyltransferase, 235, 279
glutamate:glyoxylate aminotransferase (GGAT) (peroxisomal), 338, 339, 340
glutamate dehydrogenase, 16
glutamate synthase (glutamine:oxogluturate aminotransferase, GOGAT),91, 93, 327, 328, 336, 338, 339, 340
glutamine synthetase (GS), 91, 169
   GS1 (cytosolic), 142
   GS2 (plastidic), 92, 93, 142, 164, 304, 327, 328, 336, 338, 339, 340
   mitochondrial, 304
glutathione, 66, 74, 152, 153, 154, 167, 172, 307, 333, 350, 352, 355, 356, 357, 362, 363

glutathione reductase (GR), 158, 307, 308, 309
glutathione synthetase, 362, 363
glutaredoxin (GRX) (see also PDORs), 155, 159, 170, 171, 172, 173, 307, 308
glutelin, 174
glutenin, 38
glyceraldehyde-3-phosphate dehydrogenase (GAPDH), 33, 152
   cytosolic, 26, 40, 169, 194, 196
   GapA (GAPDH), 196, 205
   GapB (GAPDH), 196, 205
   non-phosphorylating (GAPN), 142, 143, 194
   plastidic, 39, 86, 163, 164, 187, 188, 189, 194, 195, 196, 197, 198, 200, 201, 203, 204, 205
glycerol-3-phosphate dehydrogenase (GPDH), 314
   FAD-dependent, 314
glycine decarboxylase complex (GDC), 48, 92, 93, 291, 293, 301, 304, 310, 339, 340
   dihydrolipoamide dehydrogenase (see also OGDH), 293
glycogen branching enzyme, 259, 260, 277, 278
   glgX, 260, 277
glycogen phosphorylase, 122, 124
   phosphorylase a, 122
   phosphorylase b, 122
glycogen synthase, 258, 259, 260, 277, 278, 279
glycogenin, 276, 279
glycolate dehydrogenase, 304
glycolysis, 14, 26, 32, 33, 40, 41, 42, 43, 90, 92, 95, 107, 164, 188, 201, 236, 290, 303, 331
glycoside hydrolase 14 (GH14) protein family, 268
glyoxylate cycle, 39, 75, 306, 307
glyoxysome, 39, 306
GO (gene ontology), 12
GOGAT, see glutamate synthase
golgi apparatus, 12, 35
grape (Vitis vinifera), 74
GRIM-19 protein, 294
GS/GOGAT reaction cycle, 91, 92, 331, 333, 334, 335, 336, 338, 340, 341, 342

HCF164 protein, 165
HEAT repeats, 123, 126
heat shock proteins (*see also* heat stress), 4, 314
 HSP22, 299, 300
 HSP60, 299
 HSP70, 158, 299, 300
 HSP90, 299
hexokinase, 17, 26, 106, 162, 235, 237, 250
hexose phosphate isomerase, 26
Hip (HS70-interacting protein), 158
histidine phosphatase, 132
histone deacetylase, 137
homocysteine *S*-methyltransferase, 363
HPLC (high performance liquid chromatography), 67
hypocotyl, 172
 elongation, 226

ICAT (isotope-coded affinity tag), 30, 35
IMAC (immobilized metal-ion affinity chromatography), 46
immunophilins, 164
IMS (information management systems), 5
invertase, 26, 104, 235, 250
 yeast, 74
ion channels
 calcium-permeable cation channels, 151, 224
isocitrate dehydrogenase,
 cytosolic, 91, 92, 315, 327
 glyoxosomal (NADP-specific), 39, 40
 plastidic, 92
 mitochondrial, 26, 291, 292, 296, 299, 309, 315, 327, 335, 336, 338, 339
isocitrate lyase, 39, 306, 307

jasmonic acid, 39, 77
 methyl jasmonate, 68, 75, 357

Kalanchoë daigremontiana, 226
Kalanchoë fedtschenkoi, 220, 221, 227
KEGG pathway maps, 15
Kelch-repeat domain, 129
kinetoplastid, 197
Kir6 (potassium channel protein), 137
kiwifruit (*Actinidia spp.*), 242
Krebs' cycle, *see* citric acid cycle

lactate dehydrogenase, 40
leaves, 37, 42, 64, 65, 66, 78, 97, 141, 143, 172, 174, 197, 198, 201, 205, 222, 223, 224, 225, 227, 228, 235, 236, 237, 239, 240, 243, 244, 245, 247, 250, 258, 262, 265, 266, 268, 269, 271, 272, 273, 279, 280, 304, 307, 326, 330, 331, 333, 335
 bundle sheath cells, 68, 95, 98, 219, 222, 225, 355
 development, 326, 331
 leaf-specific proteins, 33
 mesophyll cells, 95, 98, 99, 102, 105, 219, 222, 224, 225, 355
 movements, 226
 petioles, 100
 senescence, 62, 97, 98, 102, 331
 source, 102, 331
leucoplasts, 38
Lemna minor, 360
light quality,
 blue, 196
 red, 196
 white, 196
lignification, 78
lipoxygenase, 68, 69
loblolly pine (*Pinus taeda*), 221
Lotus japonicus, 78, 221
lupin (*Lupinus spp.*), 338

Magnaporthe grisea, 197
maize (*Zea mays*), 43, 94, 140, 219, 220, 221, 222, 224, 225, 227, 241, 242, 243, 245, 246, 248, 249, 260, 262, 263, 264, 265, 270, 272, 274, 293, 313
 amylose extender (*ae1*) mutant, 265, 270, 171
 BEI (*Sbe1*), 270
 BEIIa (*Sbe3*), 270
 Brittle-1 (*Bt1*) mutant, 263, 270, 271
 Brittle-2 (*Bt2*) mutant, 270, 271
 Dull (*Du1*), 270, 271, 281
 Shrunken-1 (*sh1*) mutant, 260, 270, 271
 Shrunken-2 (*sh2*) mutant, 270, 271
 Sugary-1 (*Su1*), 270, 271
 Sugary-2 (*Su2*), 270, 271, 281
 Waxy (*Wx*), 270, 271

malate dehydrogenase,
  cytosolic, 32, 169, 315
  mitochondrial, 26, 291, 293, 299, 304, 315
  plastidic, 156, 162, 163, 167, 249
  sieve-element-localised, 32
malate synthase, 39, 74, 306, 307
malic enzyme, 26, 291, 293, 299
  NAD-specific, 36, 219
  NADP-specific, 219
MalQ protein, 269
MAPK phosphatase (*see also* tyrosine phosphatases), 130
  cdc14, 130
  myotubularins, 130
  PRLs, 130
  PTEN, 130
  slingshots, 130
MAPMAN pathway visualisation tool, 14, 16
MAS (microarray suite), 6, 8, 9
mass spectrometry, 31, 32, 35, 42, 45, 46, 63, 64, 66, 67, 69, 76, 161, 204, 229, 294, 301, 306
  chromatography-coupled MS, 64, 65, 67, 68, 73, 76, 77, 78, 79
  Fourier-transform ion cyclotron MS, 68
  MALDI-TOF, 28, 29
  tandem MS/MS, 28, 29, 32, 33, 38, 47
mercuric reductase, 158
Mesembryanthemum crystallinum, 221, 227
methylerythritol 4-phosphate pathway, 96
methionine synthase, 94, 169, 363, 364
methionine *S*-methyltransferase, 363
methyl jasmonate, *see* jasmonic acid
MetNet pathway visualisation tool, 14
mevalonic acid pathway, 96
MH2 protein domain, 132
MIAME (minimum information about a microarray experiment), 3, 5, 11
Michaelis-Menten constant, 191, 330, 337, 338, 351
mistletoe (*Viscum album* L.), 244
monodehydroascorbate reductase (MDAR), 307, 308, 309
MuDPIT (multidimensional protein identification technology), 30, 32
Münch hypothesis (osmotic pressure-flow), 98

Mycobacterium leprae, 160
Mycosphaerella graminicola, 174
myrosinase, 62

nitrate reductase, 15, 16, 91, 139, 153, 326, 327, 328, 330, 332, 336, 338, 339, 340
  Nit2, 332
  nitrate reductase inhibitory protein, 139
nitric oxide, 153, 292, 298, 332
nitric oxidase synthase, 153
nitrilase (*NIT3*), 352
nitrite reductase, 91, 327, 328, 336, 338, 339, 340
N-terminal signal sequence, 33, 357
NMR (nuclear magnetic resonance), 63, 64, 66, 69, 74, 76, 77, 78, 79
  ^{1}H-NMR, 74
  ^{13}C-NMR, 75
  two-dimensional, 68
nucleus, 33, 40
  nuclear export signal, 137
  nuclear localisation signal, 137
nucleoside diphosphate kinase, 299, 301, 314

O-acetylserine-serine-glycine pathway, 76
O-acetylserine(thiol) lyase (*O*ASTL), 350, 354, 357, 358, 359, 360, 363
  *O*ASTL-SAT complex (*see* cysteine synthase complex)
Ostreococcus tauri, 275, 276
onion (*Allium cepa*), 77
oxidative pentose phosphate pathway, 26, 33, 41, 163, 164, 188, 189, 201, 236
oxoglutarate dehydrogenase complex (OGDH), 26, 291, 292, 310, 311, 312
  dihydrolipoamide dehydrogenase (*see also* GDC complex), 292, 293
  dihydrolipoamide succinyl-transferase, 292

PII (bacterial) protein, 333
  PII-like (plant) proteins, 333
paraquat, 355
PathMAPA pathway visualisation tool, 13, 14

Pathway Processor pathway visualisation tool, 13, 14
PDNN analysis, 7
pea (*Pisum sativum*), 37, 42, 102, 103, 104, 162, 163, 188, 195, 197, 198, 203, 204, 205, 239, 243, 244, 248, 264, 274, 280, 307, 338
  wrinkled, 265
peach, 174
peanut (*Arachis hypogea*), 166
peptidyl-prolyl *cis-trans* isomerase (PPIase), 164, 165
permease,
  amino acid, *see* transporters
peroxidase, 167, 169
  ascorbate peroxidase (APX), 169, 307, 308
  gluthathione peroxidase (GPX), 167, 307, 308
  peroxiredoxins (PrxR), 167, 168, 169, 307, 308
peroxisome, 33, 39, 93, 304, 307, 328, 337
  membrane, 40
phenylpropanoid biosynthesis, 75, 78, 95, 201
phenylalanine ammonia lyase (PAL),
  PAL1, 75
  PAL2, 75
phloem, 32, 98, 99
  loading, 100, 102, 235, 351
  transport, 97, 98, 158, 235, 356
  sieve element/companion cell complex (SE/CC), 99
  sieve element membrane, 100
phloem-xylem exchange, 97
phosphoadenosine 5′-phosphosulfate (PAPS) reductase, 356
phospho*enol*pyruvate (PEP) carboxykinase, 40, 219
phospho*enol*pyruvate (PEP) carboxylase, 26, 42, 74, 92, 107, 143, 219, 220, 222, 224, 225, 227, 228, 229, 230, 315, 327, 329, 330, 332
  bacterial-type PEPC, 228, 229, 230
  non-phosphorylatable PEPC, 228
  Ppc1 (eukaryotic-type PEPC), 228
  Ppc2 (bacterial-type PEPC), 228
phospho*enol*pyruvate (PEP) carboxylase kinase, 122, 123, 143, 219, 220, 222, 224, 225, 226, 227, 228, 229, 230
PEPC kinase-related kinase, 122, 123
PPCK gene family, 219, 220, 222, 223, 224, 225, 226, 227, 228, 229
phosphofructokinase (PFK1), 26, 32, 141, 235
  PPi-dependent (PFP), 90, 141, 235, 236, 240
phosphofructo-2-kinase/fructose-2,6-bisphosphatase (PFK2), 141, 239, 240, 247, 248
phosphoglucan:water dikinase (PWD), 237, 268
phosphogluconate dehydrogenase, 26
phosphoglucoisomerase,
  cytosolic, 235, 236, 240
  plastidic, 105
phosphoglucomutase, 33, 161
  cytosolic, 26, 88, 235, 236, 240, 261, 262
  pgm mutant, 15, 16
  plastidic, 15, 44, 105, 261, 262
phosphoglycerate kinase, 33, 42
  cytosolic, 26, 40, 92
  plastidic, 39, 86, 106, 164, 187, 188, 202, 203, 205
  sieve-element-localised, 32
phosphoglycerate mutase, 26, 32, 88, 92, 94
  sieve-element-localised, 32
phosphoinositide-3-kinase-like kinase (KIPP), 121, 123
  ATM (ataxia-telangiectasia mutated) protein, 123, 133
  ATR (ataxia-telangiectasia and RAD3 related), 123
  FAT domain, 123
  FATC domain, 123
  FRAP protein, 123
  TOR (target of rapamycin), 123, 137, 138, 139
  TRRAP protein, 123
phospholipase C,
  phosphoinositide-dependent, 224, 227
phosphopentose epimerase, *see* ribose isomerase
phosphopentose isomerase, *see* phosphoribose isomerase

phosphoribose isomerase (PRI), 39, 188, 203, 205
phosphoribulokinase (PRK), 39, 107, 163, 187, 188, 189, 195, 197, 199, 200, 201, 203, 204, 205, 206, 249
  PRK, 205
phospho-tyrosine binding domain Sre homolgy two (SH2), 123, 132
photorespiration, 39, 48, 76, 92, 93, 150, 291, 301, 304, 307, 310, 328, 329, 330, 331, 334, 335, 339, 341, 342
photosynthetic bacteria, 191
Physcomitrella patens, 106, 174, 356
phytochelatin, 350, 357, 362
phytochelatin synthase, 362, 363
PICOT, (protein kinase C-interacting cousin of trx), 172
Plantago major (plantain), 100
plant uncoupling proteins, 298, 309, 310
plasma membrane, 37, 97, 99, 103, 104
plasma membrane H$^+$-ATPase, 134, 136
plasmodesmata, 98
PM values, 7
Poaceae, 242
polyamines, 363, 364
poplar (*Populus spp.*), 153, 158, 159, 167, 168
  Populus trichocarpa, 242, 245
post-translational modifications (PTMs), 45, 46, 48, 51
  acetylation, 121, 190
  adenylylation, 121
  carbamylation, 192
  carbonylation, 48, 302
  farnesylation, 121
  formylation, 190
  glutathionylation, 152, 152, 238
  glycosylation, 237
  hydroxylation, 121, 152, 174
  methylation, 121, 152, 190
  nitrosylation, 153
  oxidation, 48, 152
  phosphorylation, 24, 25, 27, 43, 45, 46, 47, 50, 106, 121, 122, 123, 124, 126, 130, 131, 132, 134, 136, 137, 138, 139, 140, 141, 142, 143, 152, 165, 166, 170, 190, 219, 220, 223, 224, 225, 227, 228, 229, 238, 239, 241, 244, 245, 247, 248, 250, 259, 280, 291, 292, 293, 299, 300, 330, 353
  sumolyation, 121
  uridylylation, 121
potato (*Solanum tuberosum*), 37, 66, 74, 75, 99, 163, 168, 221, 223, 224, 239, 241, 248, 265, 267, 268, 274, 275, 280, 281, 292, 302, 303 333, 360
  transgenics, 74, 79, 89, 101, 106, 107, 198, 237, 239, 240, 263, 268, 292
  tubers, 74, 79, 101, 107, 162, 266, 269, 274, 280, 292, 293, 313
PRI-PRK-Rubisco complex, 205
PRK-GAPDH complex, 204
PRK-GAPDH-CP12 complex, 195, 197, 204, 205, 206
PRI-PRK-Rubisco-PGK-GAPDH complex, 203
PRI-PRK-Rubisco-PGK-GAPDH-FBPase complex, 203
programmed cell death (PCD), 35, 151, 298
prolyl hydroxylase, 174
protease inhibitors,
  (α-amylase/subtilisin inhibitor family, 169
  Bowman-Birk, 174
  horse gram inhibitor, 174
protein disulfide oxido-reducatases (PDORs), 154, 155, 158, 161, 172, 173, 175
protein disulfide isomerase (PDI) (*see also* PDORs), 155, 161, 166, 173, 174
protein kinase A (PKA), 133
protein kinase C (PKC), 133, 225
  PKC-like activity, 225
protein phosphatase 1 (PP1), 124, 125, 129
  PP1-MYPT1 complex, 125, 126
protein phosphatase 2A (PP2A), 124, 126, 127, 129, 138, 220, 244
  PR65, 126
  PR72 protein, 127
  TONNEAU2 (TON2), 127
  TON2-PP2A complex, 127
protein phosphatase 2B (PP2B), 124
protein phosphatase 2C (PP2C), 124, 127, 128
  AB1 (abscisic acid insensitive 1), 128
  AB2 (abscisic acid insensitive 2), 128

protein phosphatase (*Continued*)
  MP2C (*Medicago sativa* phosphatase 2C), 128
  POLTERGIEST (POL)-type phosphatase, 128
  KAPP (kinase-associated protein phosphatase), 128
protein phosphatase 4 (PP4), 124, 128, 129
protein phosphatase 5 (PP5), 124, 125, 128, 129
protein phosphatase 6 (PP6), 124, 128, 129
protein phosphatase 7 (PP7), 124, 128, 129
protein phosphatase Y, 129
protein phosphatase Z, 129
protein phosphatase inhibitors,
  calculin A, 129
  microcystin, 129
protein staining,
  Deep Purple, 27
  Sypro Ruby, 27
PTIP domain, 133
PttMYB76, 78
pumpkin (*Cucurbita spp.*), 32
pyrophosphatase, 240
pyruvate decarboxylase, 26
pyruvate dehydrogenase, 26, 32, 46, 92, 107, 290, 291, 292, 293, 294, 299, 310, 314, 329
  dihydrolipoyl acetyl-transferase, 290
  dihydrolipoyl dehydrogenase, 290
pyruvate dehydrogenase kinase (PDH kinase), 291
pyruvate dehydrogenase phosphatase, 291
pyruvate formate-lyase, 313
pyruvate kinase,
  cytosolic, 26, 36, 92, 293, 329, 332
  mitochondrial-association, 327
  plastidic, 96, 106
pyruvate:orthophosphate dikinase (PPDK), 95, 96

quantitative trait loci (QTL), 79, 249
quiescin-sulfhydryl oxidase, 174

R1 protein (*see also* GWD), 268
raf-1 protein kinase, 133, 137, 139

RB47 protein, 165
RB60 protein, 165, 166
reactive oxygen species (ROS), 150, 151, 153, 167, 301, 305, 308, 309, 310, 311, 312
real time RT-PCR, 3
reductive pentose phosphate pathway, *see* Calvin cycle
respirasome, 37, 302, 303
resurrection plant (*Craterostigma plantagineum*), 241
rhizomes, 97
Rhodobacter sphaeroides, 199
ribose epimerase, 38
ribosomal proteins, 164
ribosomal protein S6 kinase (S6K), 138
ribulose-1,5-bisphosphate carboxylase/oxygenase (rubisco), 39, 42, 86, 98, 102, 107, 168, 187, 188, 189, 190, 191, 192, 193, 194, 199, 202, 203, 205, 206, 207, 219, 273
  rbcL, 165
  rubisco activase, 163, 164, 190, 192, 193, 194
  rubisco large subunit, 164, 165
  rubisco small subunit, 164
ribulose-5-phosphate epimerase, 38, 39, 188, 200
rice (*Oryza sativa*), 3, 31, 32, 33, 34, 36, 68, 75, 131, 160, 172, 174, 220, 221, 222, 228, 229, 242, 245, 246, 259, 264, 265, 270, 271, 279, 282, 313, 338, 356
RMA (log-scale robust multi-array analysis), 7, 9
RNA binding proteins, 164
RNAi repression, 105, 106, 282
RNA polymerase, 360
RNA polymerase II phosphatases, 131
  FCP1 (TFIIF associated CTD phosphatase), 131, 132, 133
  FCP1-like proteins, 132, 133
  SCP1, 132
  SCP-like enzymes, 132
roots, 43, 77, 78, 172, 205, 222, 271, 272, 273, 293, 315, 326, 348, 349, 351, 352, 355, 357
  hairs, 151, 351
  nodules, 39, 78, 220, 326

rhizosphere, 315
root-specific proteins, 33
RP (simple rank product), 9

S-adenosylmethionine synthetase, 94, 363
SAGE (serial analysis of gene expression), 2
SAG12, 98
Salmonella typhimurium, 355, 357
salicylic acid, 77, 292
SAM (significance analysis of microarrays), 9, 10, 14
satsuma orange (Citrus unshiu), 241, 242
Scenedesmus obliquus, 204
sedoheptulose-1,7-bisphosphatase, 39, 163, 164, 188, 189, 197, 198, 199, 202, 206, 207
   cytosolic, 197
seeds, 74, 97, 102, 103, 106, 158, 169, 244, 258
   apoplast, 103, 104
   aleurone layer, 33, 158, 172
   coat/testa, 102
   cotyledons, 39, 103, 104, 172
   development, 104, 172
   embryo, 33, 102, 106, 158, 161
   endosperm, 33, 102, 143, 158, 161, 169, 174, 229, 262, 263, 265, 269, 271, 272, 274, 280
   germination, 39, 42, 158, 237, 238, 243, 250, 306
   oilseeds, 39, 107, 229, 306
   priming, 42
   seed-specific proteins, 33
   siliques, 205
   storage proteins, 348
Selenastrum minutum, 228, 229
serine acetyltransferase (SAT), 356, 357, 358, 359, 360, 361, 363
   SAT-A gene, 359
   Sat-c, 359
   Sat-m, 359
   Sat-p, 359
   SAT-OASTL complex (see cysteine synthase complex)
   SAT-SAT complex, 358
serine:glyoxylate transferase (SGAT) (peroxisomal), 338, 339, 340
serinehydroxymethyl transferase (SHMT), 291, 301, 302, 339, 340

serine/threonine kinases, 121, 122, 123, 124
serine/threonine receptor kinases, 122
   RLK5, 128
serine/threonine phosphatase, 127
serotonin-N-acetyltransferase, 136, 137
shikimic acid pathway, 94, 95
SILAC (stable isotope labelling with amino acids in cell culture), 30
Sinapis alba, 204
single nucleotide polymorphisms (SNPs), 249
sink organs, 102, 236, 331, 351
S-locus, 170
   S-locus cysteine-rich (SCR) protein, 170
   S-locus receptor kinase (SRK), 170
snf1 kinases, 139, 239, 244
   snf1-like kinases, 139, 162, 244, 353
sorghum (Sorghum bicolor), 163, 221, 222, 225, 242
soybean (Glycine max), 102, 103, 220, 221, 222, 223, 227
SOPS (standard operating procedures – in metabolomics), 70, 71
spider mite (Tetranychus urticae), 77
spinach (Spinacia oleracea), 94, 99, 141, 163, 172, 188, 191, 193, 196, 197, 200, 201, 203, 204, 205, 239, 241, 243, 244, 245, 248, 338, 356, 359
SpoIIAA sigma protein, 353
SPS-SPP metabolon, 246
starch branching enzyme (BE), 26, 141, 258, 278
   BEI, 261, 262, 264, 265, 271, 280
   BEIIa, 261, 262, 264, 265, 271, 280
   BEIIb, 261, 264, 265, 280
   BE mutants, 270
starch debranching enzymes (DBE) (isoamylase/pullulanase), 258, 265, 266, 267, 271, 272, 278, 279
   DBE mutants, 270, 271, 274
   isoamylase1, 261, 262, 266, 267, 271
   isoamylase1-isoamylase2 complex, 282
   isoamylase2, 261, 262, 266, 267, 271
   isoamylase3, 267, 271
   pullulanase, 267, 268, 271
starch phosphorylase, 227, 237, 261, 262, 266, 271, 272, 273, 275, 280

starch synthase, 26, 141, 258
starch synthase (soluble form) (SS), 261, 271, 272
　SS mutants, 270, 271, 274
　SSI, 261, 262, 264, 265, 271, 279
　SSIIa, 261, 262, 264, 265, 271, 279, 280
　SSIIb, 261, 262, 264, 265, 271
　SSIII, 261, 262, 264, 265, 271, 274
　SSIV, 261, 262, 271
　UDPG-dependent, 278
starch synthase (granule-bound), 263, 264, 279
　GBSSI, 261, 269, 271, 274, 275, 279
　GBSSII, 262, 271
　GBSS III, 142
　GBSS mutants, 270, 274
　UDPG-dependent, 278
STAS domain, 353
stem, 100, 172
stilbene synthase, 74
stress response, 298, 301, 308
　aluminium toxicity, 315
　anoxia, 43, 75, 140, 141
　cold, 12, 40, 62, 241, 243, 312
　drought/osmotic stress, 76, 77, 243, 312
　heat stress (*see also* HSPs), 62, 76, 77
　herbivore attack, 62
　hypoxia, 313
　metal exposure, 362
　nitrosative stress, 153
　nitrogen-deprivation, 274, 327, 356
　nutrient deprivation, 139, 315
　oxidative, 43, 48, 169, 302, 363
　pathogen attack, 77, 78, 362
　phosphate starvation, 141, 241, 315
　salt-stress, 225, 227
　sugar-starvation, 44
　sulfur-deprivation, 76, 352, 356, 360
　wounding, 62
stt7 (ser/thr kinase), 166
STN7 (ser/thr kinase), 166
Stylosanthes hamata, 351
succinate dehydrogenase (SDH) (*see also* ETC complex II), 26, 291, 293, 294, 295
succinyl CoA ligase (*see* succinate thiokinase)

succinate thiokinase, 26, 291, 292, 299, 302, 311, 312
sucrose synthase, 26, 140, 235, 236, 244, 250
　sucrose synthase mutants, 260, 261
sucrose-nonfermenting related kinase, 122
sucrose-phosphate synthase (SPS), 31, 88, 90, 140, 141, 234, 235, 241, 242, 243, 244, 245, 246, 248, 249, 250, 327, 330, 332
　A-type SPS, 242, 243, 244, 245
　B-type SPS, 242, 243, 244, 245
　C-Type SPS, 242, 243, 244, 245
　D-Type SPS, 242, 244, 245, 247
　SPP-like domain, 247
sucrose-phosphate synthase kinase, 244
sucrose-phosphate phosphatase (SPP), 88, 234, 235, 244, 245, 247, 248, 250
　SPP genes, 245, 246
　SPP glycosyltransferase domain, 246
　SPP HAD domain, 246
sugar beet (*Beta vulgaris*), 221
sugar cane (*Saccharum officinarum*), 3, 241, 242, 249
sugar phosphate phosphatase, 235
sulfate transporter, 348, 350, 351, 356, 357, 360, 361, 364
　Arabidopsis group 1, 351, 352, 353
　Arabidopsis group 2, 351, 352, 353
　Arabidopsis group 3, 351, 353
　Arabidopsis group 4, 352
　Arabidopsis group 5
　high affinity transporters (HAST), 349
　low affinity transporters (LAST), 349
sulfite reductase, 350, 356
sulfotransferases, 350, 354
superoxide dismutase, 46, 167, 299, 307
SURE (sulfur response element), 352
sweet pepper (*Capsicum annum*), 188
sweet potato (*Ipomoea batatas*), 279
symplastic transport, 98, 102, 103
symporters,
　sucrose/proton, 98
　proton-coupled amino acid symporter, 99, 103
Synechocystis, 156, 161, 172, 192, 277
　strain PCC6803, 204, 244, 245, 246

TAP (tandem affinity purification) tagging, 49
TCA cycle, *see* citric acid cycle
tetratricopeptide repeats, 129
theanine synthetase ($_L$-glutamic acid ethylamine ligase), 342
thiocalsin, 169
thioredoxin (Trx) (*see also* PDORs), 30, 47, 48, 88, 152, 155, 156, 159, 161, 161, 165, 167, 168, 169, 170, 172, 173, 174, 195, 198, 225, 300, 307, 360
  AtTDX, 158
  Trx-c (*see* Trx-h),
  Trx-f, 156, 157, 161, 163, 164, 166, 198
  Trx-h, 153, 156, 157, 158, 159, 167, 169, 170, 239, 300
  Trx-m, 156, 157, 161, 163, 164, 166, 168, 198
  Trx-o, 156, 157
  Trx-x, 156, 157, 161, 168
  Trx-y, 156, 157, 161
thioredoxin-binding proteins, 301
thioredoxin-like domain, 355
thioredoxin-like proteins, 170
  CITRX, 170
thioredoxin reductase, 47, 301
  Fd-Thioredoxin reductase (FTR), 160, 166, 300
  NADP-Thioredoxin reductase (NTR), 158, 159, 160, 161, 300
  NTRA, 167
  NTRC, 160
  trxB (*E. coli*), 355
threonine synthase, 363, 364
TILLING, 270, 282
tobacco (*Nicotiana spp.*), 73, 78, 99, 106, 188, 191, 193, 194, 196, 204, 223, 245, 290, 313, 330, 331, 355
  BY-2 cells, 355
  mutants, 304, 330
  transgenic, 89, 100, 194, 198, 200, 240, 360
tomato (*Lycopersicon esculentum*), 73, 74, 77, 78, 79, 170, 221, 223, 224, 242, 338, 351
  cf-2, 170
  cf-9, 170
  Lycopersicon pennelli, 79
  mutants, 304
  transgenic, 304
TOR-interacting proteins, 138
  GβL protein, 138
  KOG1 protein, 138
  LST8 protein
  regulatory associated protein of TOR (RAPTOR), 138
Toxoplasma gondii, 278, 279
transaldolase, 26
transcription factors,
  AP-1, 172
  CCA1 (circadian clock associated 1 gene), 226, 227
  Dof, 333
  LHY (late elongated hypocotyl gene), 226, 227
  NF-$_K$B, 172
  TFIIA, 131
  TFIIB, 131
  TFIID, 131
  TFIIE, 131
  TFIIF, 131
  TFIIH, 131
  TOC1 (timing of CAB expression 1 gene), 226, 227
transfer cells, 103, 104
transketolase,
  cytosolic, 26, 201
  plastidic, 39, 95, 188, 200, 201, 202, 206, 207
translocators,
  2-oxoglutarate/malate translocator (DiT1), 91, 92
  adenine nucleotide (ANT), 106, 107, 291, 303, 314
  cue1 mutant, 94, 95, 96
  glutamate/malate translocator (DiT2), 91, 92, 93
  phospho*enol*pyruvate/phosphate (PPT) translocator, 94
  PPT1, 94
  PPT2, 94
  triose-phosphate/phosphate translocator, 38, 87, 88, 89, 189, 235, 236, 238, 247
  tpt-1, 89
  xylulose 5-phosphate/phosphate translocator, 96

transporters (*see also* sulfate transporters),
   amino acid/amino acid permeases, 93, 100, 104
   Brittle-1 ADPglucose (*bt1*), 270
   Glucose-6-phosphate transporter (GPT), 105
   GPT1, 105, 106
   gpt1, 106
   GPT2, 105
   hexose-phosphate transporter, 90, 235, 261, 262
   isopentenyldiphosphate (IPP) transporter, 96
   maltose, 235, 237, 269
   monosaccharide, 104, 235
   phosphate (Pi)/proton transporter, 37
   pyruvate transporters, 96
   SUT/SUC transporters, 99, 100, 101, 103, 104, 250
   SLC26 (anion) group (*see also* sulfur transporters), 351
trehalose phosphate synthase, 140, 141, 245
trichome, 68
triose phosphate isomerase, 33, 42, 153
   cytosolic, 26, 32, 235, 238, 248, 250
   plastidic, 39, 86, 187, 188, 202, 235
tropic responses, 151
Trypanosoma brucei, 197
trypsin, 28, 29
tryptophan hydroxylase, 133
tubers (*see also* potato), 97, 237, 250, 258, 267
two-component regulatory system, 333
   His-Asp phosphorelay, 333
tyrosine hydroxylase, 133
tyrosine kinases, 121, 123, 124
tyrosine phosphatases (*see also* MAPK phosphatases), 123, 124, 130, 131, 132
   CDC25 phosphatases, 130, 131, 137
   human eyes absent protein, 131

ubiquitin-mediated protein degradation, 228, 229
   proteasomes, 140, 169, 228
   MG132 proteasomal inhibitor, 228
   UDP-galactose/glucose epimerase, 235
   UDP-glucose pyrophosphorylase, 26, 88, 235, 236, 240, 241, 248, 250, 278
   Ugp genes, 240, 241
UMP kinase,
   sieve-element-localised, 32

vacuole, 33, 98, 226, 235, 236, 237, 352
   tonoplast, 224, 226, 352
Vicia faba, 102, 104

watermelon (*Citrullus lanatus*), 359
WD40 protein domain, 132
wheat (*Triticum spp.*), 38, 142, 143, 158, 163, 169, 174, 197, 221, 222, 242, 243, 245, 248, 262, 264, 265, 270, 272, 280, 333
   Triticum durum, 174
WW protein domain, 132

xanthine oxidase, 153
Xanthobacter flavus, 202
Xenopus, 108
xylem,
   transport, 97, 326, 349
xylulose-5-phosphate epimerase, *see* ribulose-5-phosphate epimerase

yeast (*Saccharomyces spp.*), 36, 74, 94, 108, 121, 126, 127, 129, 130, 131, 134, 138, 141, 156, 168, 172, 302, 303, 353, 361
   CP154-7B mutant, 351, 353
   MET gene, 361
   two-hybrid screening, 49, 170
   SUL gene, 361